COHOMOLOGY OF DRINFELD MODULAR VARIETIES, PART I

Already published

1   W.M.L. Holcombe *Algebraic automata theory*
2   K. Petersen *Ergodic theory*
3   P.T. Johnstone *Stone spaces*
4   W.H. Schikhof *Ultrametric calculus*
5   J.-P. Kahane *Some random series of functions, 2nd edition*
6   H. Cohn *Introduction to the construction of class fields*
7   J. Lambek & P.J. Scott *Introduction to higher-order categorical logic*
8   H. Matsumura *Commutative ring theory*
9   C.B. Thomas *Characteristic classes and the cohomology of finite groups*
10 M. Aschbacher *Finite group theory*
11 J.L. Alperin *Local representation theory*
12 P. Koosis *The logarithmic integral I*
13 A. Pietsch *Eigenvalues and s-numbers*
14 S.J. Patterson *An introduction to the theory of the Riemann zeta-function*
15 H.J. Baues *Algebraic homotopy*
16 V.S. Varadarajan *Introduction to harmonic analysis on semisimple Lie groups*
17 W. Dicks & M. Dunwoody *Groups acting on graphs*
18 L.J. Corwin & F.P. Greenleaf *Representations of nilpotent Lie groups and their applications*
19 R. Fritsch & R. Piccinini *Cellular structures in topology*
20 H Klingen *Introductory lectures on Siegel modular forms*
21 P. Koosis *The logarithmic integral II*
22 M.J. Collins *Representations and characters of finite groups*
24 H. Kunita *Stochastic flows and stochastic differential equations*
25 P. Wojtaszczyk *Banach spaces for analysts*
26 J.E. Gilbert & M.A.M. Murray *Clifford algebras and Dirac operators in harmonic analysis*
27 A. Frohlich & M.J. Taylor *Algebraic number theory*
28 K. Goebel & W.A. Kirk *Topics in metric fixed point theory*
29 J.F. Humphreys *Reflection groups and Coxeter groups*
30 D.J. Benson *Representations and cohomology I*
31 D.J. Benson *Representations and cohomology II*
32 C. Allday & V. Puppe *Cohomological methods in transformation groups*
33 C. Soulé et al *Lectures on Arakelov geometry*
34 A. Ambrosetti & G. Prodi *A primer of nonlinear analysis*
35 J. Palis & F. Takens *Hyperbolicity and sensitive chaotic dynamics at homoclinic bifurcations*
36 M. Auslander, I. Reiten & S. Smalo *Representation theory of Artin algebras*
37 Y. Meyer *Wavelets and operators*
38 C. Weibel *An introduction to homological algebra*
39 W. Bruns & J. Herzog *Cohen-Macaulay rings*
40 V. Snaith *Explicit Brauer induction*
41 G. Laumon *Cohomology of Drinfeld modular varieties I*
42 E.B. Davies *Spectral theory and differential operators*
43 J. Diestel, H. Jarchow & A. Tonge *Absolutely summing operators*
44 P. Mattila *Geometry of sets and measures in Euclidean spaces*
45 R. Pinsky *Positive harmonic functions and diffusion*
46 G. Tenenbaum *Introduction to analytic and probabilistic number theory*
50 I. Porteous *Clifford algebras and the classical groups*

Cohomology of Drinfeld Modular Varieties, Part I

Geometry, counting of points and local harmonic analysis

Gérard Laumon
Université Paris–Sud

CAMBRIDGE
UNIVERSITY PRESS

Published by the Press Syndicate of the University of Cambridge
The Pitt Building, Trumpington Street, Cambridge CB2 1RP
40 West 20th Street, New York, NY 10011-4211, USA
10 Stamford Road, Oakleigh, Melbourne 3166, Australia

© Cambridge University Press 1996

First published 1996

Printed in Great Britain at the University Press, Cambridge

A catalogue record of this book is available from the British Library

Library of Congress cataloguing in publication data

Laumon, Gérard.
Cohomology of Drinfeld modular varieties/Gérard Laumon.
p. cm. – (Cambridge studies in advanced mathematics: 41)
Includes bibliographical references and index.
Contents: pt. 1. Geometry, counting of points, and local harmonic analysis
ISBN 0 521 47060 9 (pt. 1)
1. Drinfeld modular varieties. 2. Homology theory. I. Title.
II. Series.
QA251.L287 1996
512'.24 – dc20 94-27643 CIP

ISBN 0 521 47060 9 hardback

Contents

Preface ix

1. Construction of Drinfeld modular varieties 1

(1.0) Notations 1

(1.1) Endomorphisms of the additive group 1

(1.2) Drinfeld modules 3

(1.3) Level structures 5

(1.4) Modular varieties 7

(1.5) Deformation theory 9

(1.6) Hecke algebras, correspondences 12

(1.7) Hecke operators 15

(1.8) Comments and references 18

2. Drinfeld A-modules with finite characteristic 19

(2.0) Notations 19

(2.1) Isogenies 19

(2.2) Isogeny classes of Drinfeld modules 22

(2.3) Tate modules of a Drinfeld module 27

(2.4) Dieudonné modules 31

(2.5) Dieudonné module of a Drinfeld module 35

(2.6) First description of an isogeny class 43

(2.7) Isogeny classes as double coset spaces 47

(2.8) Comments and references 50

3. The Lefschetz numbers of Hecke operators 51

(3.0) Introduction 51

(3.1) The Lefschetz numbers of correspondences 51

(3.2) Counting of fixed points 52

(3.3) Where the orbital integrals come in 56

(3.4) Transfer of conjugacy classes 60

(3.5) Transfer of Haar measures 66

(3.6) The Lefschetz numbers as sums of twisted orbital integrals 72

(3.7) Comments and references 74

4. The fundamental lemma 75

(4.0) Introduction 75

(4.1) Satake isomorphism 76

(4.2) Base change homomorphism 83

(4.3) Orbital integrals 87

(4.4) Twisted orbital integrals 93

(4.5) Main theorem 98

(4.6) The elliptic case 102

(4.7) The general case 114

(4.8) Non-closed orbital integrals 117

(4.9) Comments and references 128

5. Very cuspidal Euler–Poincaré functions 129

(5.0) Introduction 129

(5.1) The function f 130

(5.2) Kottwitz's functions 133

(5.3) Elliptic orbital integrals of f 135

(5.4) K-invariant constant terms of f 142

(5.5) The function f is very cuspidal 152

(5.6) Non-elliptic orbital integrals of f 156

(5.7) Comments and references 156

6. The Lefschetz numbers as sums of global elliptic orbital integrals   158

7. Unramified principal series representations   160

(7.0) Introduction   160

(7.1) Parabolic induction and restriction   160

(7.2) Cuspidal representations   165

(7.3) Principal series representations   168

(7.4) Unramified principal series representations   178

(7.5) Spherical representations   187

(7.6) Comments and references   191

8. Euler–Poincaré functions as pseudocoefficients of the Steinberg representation   192

(8.0) Introduction   192

(8.1) The Steinberg representation   192

(8.2) Main theorem   207

(8.3) Some easy vanishing results   208

(8.4) Cohomological interpretation of $\text{tr}\pi(f)$ 213

(8.5) Unitarizable representations   220

(8.6) Proof of Howe and Moore's criterion of non-unitarizability   226

(8.7) Comments and references   248

Appendices

A. Central simple algebras   249

(A.0) Central simple algebras   249

(A.1) Bicommutant theorem   250

(A.2) Central simple algebras over local fields   251

(A.3) Central simple algebras over function fields   253

(A.4) Comments and references   255

B. Dieudonné's theory : some proofs   256

(B.1) Proof of (2.4.5)   256

(B.2) Proof of (2.4.6)   268

(B.3) Proof of (2.4.11)   272

(B.4) Comments and references   280

C. Combinatorial formulas   281

(C.0) Introduction   281

(C.1) q-binomial coefficients   281

D. Representations of unimodular, locally compact, totally discontinuous, separated, topological groups   284

(D.0) Introduction   284

(D.1) Smooth representations of H 284

(D.2) Admissible representations of H 290

(D.3) Induction and restriction   292

(D.4) Cuspidal representations of H 311

(D.5) Injective and projective objects in $\mathrm{Rep}_s(H)$; cohomology   322

(D.6) Unitarizable representations   327

(D.7) Decomposition of representations into tensor products   334

(D.8) Comments and references   336

References   337

Index   341

Preface

Shimura varieties are quasi-projective varieties over number fields that are associated with reductive groups and some other data. Similarly Drinfeld modular varieties are quasi-projective varieties over function fields that are associated with reductive groups and some other data. The cohomology of Shimura varieties is related to automorphic forms over number fields. In the same way, the cohomology of Drinfeld modular varieties is related to automorphic forms over function fields. The study of the relations between the cohomology of Drinfeld modular varieties and automorphic forms over function fields is the main theme of this book.

The Drinfeld modular varieties that I will consider were discovered by Drinfeld in 1973. They depend on the following data: a function field F with a finite field of constants, a place ∞ of F, a positive integer d and a non-zero ideal I of the ring A of functions in F which are regular outside ∞. The Drinfeld modular variety corresponding to (F, ∞, d, I) is denoted by M_I^d. The underlying reductive group is GL_d over F.

As the classical modular curve $X(N)$ (N a positive integer) parametrizes the elliptic curves with a level-N-structure, M_I^d parametrizes the so-called elliptic modules of rank d (I will use the terminology Drinfeld modules of rank d) with a level-I-structure. As $X(N)$ is affine and smooth of pure relative dimension 1 over $\operatorname{Spec}(\mathbb{Z}[1/N])$ as long as $N \geq 3$, M_I^d is affine and smooth of pure relative dimension $d-1$ over $\operatorname{Spec}(A) - V(I)$ as long as $I \subsetneqq A$. As one can define the Hecke correspondence T_p on $X(N)$ for each prime p which doesn't divide N, Drinfeld has defined the Hecke correspondences T_x^i ($i = 1, \ldots, d-1$) for each place $x \neq \infty$ of F which doesn't divide I.

When I varies the affine schemes M_I^d can be organized into a projective system. The limit M^d of this projective system is an affine scheme over F. On M^d, Drinfeld has defined an action of the adelic group $GL_d(\mathbb{A}^\infty)$ where $\mathbb{A}^\infty = F \otimes_A \widehat{A}$ is the ring of adeles of F outside ∞. For each I as before we have

$$F \otimes_A M_I^d = M^d / K_I^\infty$$

where K_I^∞ is the compact open subgroup

$$\operatorname{Ker}(GL_d(\widehat{A}) \to GL_d(\widehat{A}/I\widehat{A}))$$

of $GL_d(\mathbb{A}^\infty)$ and for each place $x \neq \infty$ of F which doesn't divide I and for each $i = 1, \ldots, d$, the Hecke correspondence T_x^i on $F \otimes_A M_I^d$ is induced by the action of

$$
\begin{pmatrix}
\varpi_x & & & & & \\
& \ddots & & & 0 & \\
& & \varpi_x & & & \\
& & & 1 & & \\
& 0 & & & \ddots & \\
& & & & & 1
\end{pmatrix}
\in GL_d(F_x) \subset GL_d(\mathbb{A}^\infty)
$$

on M^d (ϖ_x is a uniformizer of F at x and is repeated i times on the diagonal).

The cohomology of Drinfeld modular varieties that I will study in these notes is the ℓ-adic étale cohomology with compact supports

$$
H_c^*(\overline{F} \otimes_F M^d, \overline{\mathbb{Q}}_\ell) = \varinjlim_I H_c^*(\overline{F} \otimes_F M_I^d, \overline{\mathbb{Q}}_\ell)
$$

for some fixed prime ℓ and for some fixed algebraic closures $\overline{\mathbb{Q}}_\ell$ and $\overline{F}$ of $\mathbb{Q}_\ell$ and F respectively. On this cohomology, $\mathrm{Gal}(\overline{F}/F)$ and $GL_d(\mathbb{A}^\infty)$ act and these two actions commute.

MAIN PROBLEM. — *Describe the virtual* $\overline{\mathbb{Q}}_\ell[\mathrm{Gal}(\overline{F}/F) \times GL_d(\mathbb{A}^\infty)]$-*modules*

$$
\sum_n (-1)^n H_c^n(\overline{F} \otimes_F M^d, \mathbb{Q}_\ell)
$$

in terms of automorphic representations of GL_d *over* F.

The method that I will follow to attack this problem was initiated by Ihara and Langlands in the case of classical modular curves and greatly extended by Langlands, Kottwitz and many other mathematicians in the case of Shimura varieties. It was applied by Drinfeld himself to solve the above main problem for $d = 2$. Roughly speaking, it can be described in the following way.

The first step is to use the Grothendieck–Lefschetz trace formula to express the trace of the above virtual module in terms of a number of fixed points.

The second step is to compare this number of fixed points with the geometric side of a suitable Arthur–Selberg trace formula.

The third step is to compute explicitly the spectral side of this Arthur–Selberg trace formula.

Finally, the identification of the cohomological side of the Grothendieck–Lefschetz trace formula with the spectral side of the Arthur–Selberg trace formula gives the desired description of our virtual module.

In pursuing this program for Shimura varieties, one is faced with two main difficulties. The first one is related to the unstable conjugacy classes in a general reductive group. The second one is related to the non-compactness of the Shimura varieties.

Fortunately the first difficulty does not occur for Drinfeld modular varieties: in GL_d all the conjugacy classes are stable. From this point of view, Drinfeld modular varieties are much easier to study than Siegel modular varieties of principally polarized abelian varieties of dimension g (except if $g = 1$). But the second difficulty does occur: Drinfeld modular varieties are non-compact. This has two consequences. First of all we cannot apply directly the Grothendieck–Lefschetz trace formula. Secondly, the Arthur–Selberg trace formula is very complicated in this case. Following ideas of Deligne and Kazhdan, I will show how to overcome this difficulty in the second volume of this book.

Now let me briefly describe the content of the first volume of this book.

In the first chapter I review Drinfeld's definitions of the Drinfeld modules of rank d over a base scheme, of the characteristic of such a module and of the level-I-structures on such a module. Then I recall the construction and the basic properties of Drinfeld modular varieties. Finally I recall Drinfeld's construction of the Hecke operators on these varieties.

In the second chapter I review Drinfeld's description of the set of the Drinfeld modules of rank d for a given finite characteristic. This is completely analogous to the Honda–Tate theory for elliptic curves in finite characteristic.

Using this description of the set of points of the Drinfeld modular varieties in finite characteristic one obtains a formula for the number of fixed points of the product of a Frobenius power and a Hecke correspondence acting on this set. This formula is given in terms of twisted orbital integrals. This is the aim of chapter 3.

One cannot directly compare the formula for the number of fixed points with the geometric side of the Arthur–Selberg trace formula. An important step in this comparison is to replace the twisted orbital integrals by ordinary orbital integrals. This is the purpose of chapters 4 and 5.

In chapter 4 I recall the statement and Drinfeld's proof of a particular case of the so-called fundamental lemma. I also review some important properties of the orbital integrals for GL_d over a non-archimedean local field.

In chapter 5 I recall Kottwitz's construction of Euler–Poincaré functions and the computation of their orbital integrals. I also introduce the new notion of very cuspidal function and I show (with the help of Waldspurger) that a suitable linear combination f of Kottwitz's Euler–Poincaré functions is very cuspidal. The notion of very cuspidal function does not play an important role in the first volume of this book but will be crucial in the second volume.

Putting together the results of the previous chapters, one gets that the above number of fixed points is equal to the elliptic part of the Arthur–Selberg trace formula. This is done in chapter 6.

The last two chapters are devoted to a review of some fundamental results in local harmonic analysis which will be needed in the second volume of this book.

In chapter 7 I recall the construction and the main properties of the unramified principal series representations over a non-archimedean local field. In particular, I give the classification of the spherical representations and compute their traces.

In chapter 8, I prove that the very cuspidal function f which has been introduced in chapter 5 is a pseudo-coefficient of the Steinberg representation. In fact, let π_I $(I \subset \{1, \ldots, d-1\})$ be the irreducible constituents of the induced representation from the standard Borel subgroup B to GL_d of $\delta_B^{-1/2}$ (δ_B is the modulus character of B). Then I prove that

$$\text{tr } \pi_I(f) = (-1)^{|\Delta - I|}$$

for all $I \subset \Delta$, that

$$\text{tr } \pi(f) = 0$$

for each smooth irreducible representation π of GL_d which is not isomorphic to one of the π_I's and that among the π_I's only the trivial representation π_Δ and the Steinberg representation $\pi_\emptyset$ are unitarizable.

There are four appendices.

In appendix A I review the basic facts that are needed on central division algebras.

In appendix B I have included the proofs of the theorems of Dieudonné theory in equal characteristic that I have used in chapter 2.

Appendix C contains the proofs of combinatorial formulas that I use in chapter 4 (proof of (4.6.1)).

In the long appendix D I recall the basic results about smooth representations of locally compact, totally discontinuous, separated, topological groups.

This book originates from two graduate courses, one given during the fall of 1989 at the University of Minnesota and the other one during the spring of 1991 at Caltech.

I am grateful to the Department of Mathematics of the University of Minnesota and especially to W. Messing and S. Sperber for their support and their kind hospitality during the fall of 1989. I thank the colleagues and the students who attended my lectures.

I am grateful to the Department of Mathematics of Caltech and especially to D. Ramakrishnan for their kind hospitality and their support during the spring of 1991. The graduate course that I gave at Caltech was part of a special program organized jointly by D. Ramakrishnan and D. Blasius, H. Hida, J. Rogawski from UCLA. I thank them as well as the other mathematicians who attended my course.

I am indebted to D. Kazhdan and R. Kottwitz. In 1986 they lectured on similar subjects. Kazhdan's course at Harvard University dealt with the cohomology of Drinfeld modular varieties with values in certain local systems. The main topic of Kottwitz's course at the Université de Paris-Sud was the computation of the zeta function of Shimura varieties following Langlands' method. I have been strongly influenced by both courses and by the numerous discussions that I had with D. Kazhdan and R. Kottwitz. In particular, I have organized my course in the same way as R. Kottwitz.

I owe a special acknowledgment to J.-L. Waldspurger who helped me to complete the proof of the main theorem of chapter 5. I also had a lot of very useful discussions with him.

Thanks are also due to several other colleagues, in particular H. Carayol, L. Clozel, G. Henniart, M. Rapoport and J.-P. Wintenberger for many stimulating discussions and their help in solving some difficulties.

Special thanks go to M. Bonnardel and M. Le Bronnec who typed the manuscript with care while being introduced to TeX and to the editors who did a beautiful job.

1

Construction of Drinfeld modular varieties

(1.0) Notations

Let p be a prime number and let X be a smooth, projective and connected curve over the finite field $\mathbb{F}_p = \mathbb{Z}/p\mathbb{Z}$. We will denote by η the generic point of X and by $F = \kappa(\eta)$ its function field; we will identify the set of closed points $|X|$ of X with the set of places of F. For each $x \in |X|$, we will denote by F_x the completion of F at the place x, by $x : F_x \to \mathbb{Z} \cup \{\infty\}$ the discrete valuation of F_x, by $\mathcal{O}_x \subset F_x$ the corresponding valuation ring, by ϖ_x a prime element of $\mathcal{O}_x$ ($x(\varpi_x) = 1$), by $\kappa(x) = \mathcal{O}_x/(\varpi_x)$ the residue field of $\mathcal{O}_x$ and by $\deg(x)$ the degree of $\kappa(x)$ over $\mathbb{F}_p$ ($\kappa(x)$ has $p^{\deg(x)}$ elements).

Once for all we fix a place ∞ of F. The open subset $X - \{\infty\}$ of X is affine; let A be its ring of regular functions, i.e.

$$A = \{a \in F \mid x(a) \geq 0, \forall x \in X - \{\infty\}\}.$$

For each $a \in A - \{0\}$, we have

$$\deg(\infty)\infty(a) = -\dim_{\mathbb{F}_p}(A/(a)).$$

(1.1) Endomorphisms of the additive group

Let k be a ring of characteristic p. The ring $\mathrm{End}(\mathbb{G}_{a,k})$ of endomorphisms of the k-group scheme $\mathbb{G}_{a,k} = \mathrm{Spec}(k[t])$ is canonically isomorphic to the non-commutative polynomial ring $k[\tau]$ with the commutation rule

$$\tau \cdot \alpha = \alpha^p \cdot \tau \quad (\forall \alpha \in k)$$

(see [Or]); each $\alpha \in k$ is identified with the endomorphism

$$\mathbb{G}_{a,k} \xrightarrow{\ \alpha\ } \mathbb{G}_{a,k} \ , \quad t \mapsto \alpha t$$

and

$$\mathbb{G}_{a,k} \xrightarrow{\ \tau\ } \mathbb{G}_{a,k} \ , \quad t \mapsto t^p$$

is the Frobenius endomorphism. In particular, the group of automorphisms of the k-group scheme $\mathbb{G}_{a,k}$ is canonically identified with the group $k[\tau]^\times$ of invertible elements in $k[\tau]$.

LEMMA (1.1.1). — *The element* $\sum\limits_{n=0}^{N} \alpha_n \tau^n \in k[\tau]$ *is invertible if and only if* α_0 *is invertible in* k *and* $\alpha_1, \cdots, \alpha_N$ *are nilpotent.* $\qquad\square$

LEMMA (1.1.2). — *Let* d *be an integer* > 0 *and let*

$$\varphi = \sum_n \alpha_n \tau^n \ , \quad \psi = \sum_n \beta_n \tau^n$$

be elements in $k[\tau]$ *such that* α_d *and* β_0 *are invertible in* k *and the* α_n*'s* $(n > d)$ *and the* β_n*'s* $(n > 0)$ *are nilpotent. We set*

$$\psi \circ \varphi \circ \psi^{-1} = \sum_n \delta_n \tau^n.$$

Then we have the following properties:

(i) δ_d *is invertible in* k *and the* δ_n*'s* $(n > d)$ *are nilpotent,*

(ii) *there exists one and only one* ψ *such that* $\beta_0 = 1$ *and* $\delta_n = 0$ *for all* $(n > d)$.

Proof: If $\psi^{-1} = \sum\limits_n \gamma_n \tau^n$, then we have

$$\delta_n = \sum_{i+j+k=n} \beta_j \alpha_i^{p^j} \gamma_k^{p^{i+j}}$$

and $\beta_j \alpha_i^{p^j} \gamma_k^{p^{i+j}}$ is nilpotent (resp. invertible) as long as $j > 0$ or $i > d$ or $k > 0$ (resp. $i = d$ and $j = k = 0$). Part (i) follows immediately from this remark.

Let I be the ideal of k generated by the α_n's $(n > d)$ and the β_n's $(n > 0)$. Then there exists an integer $s \geq 1$ such that $I^s = 0$. By induction we can (and we will) assume that $s = 2$ (replace successively (k, I) by $(k/I^2; I/I^2)$, $(k/I^4, I^2/I^4), \cdots)$.

Now to prove the uniqueness of ψ it is sufficient to prove that $\psi \equiv 1$ when $\alpha_n = 0$ for all $n > d$ and $\delta_n = 0$ for all $n > d$. By induction we can assume that $\beta_n = 0$ for all $n > N$ and we need to prove that $\beta_N = 0$ (N is any integer > 0). But we have

$$\beta_N \alpha_d^{p^N} = \delta_d \beta_N^{p^d}$$

by equating the coefficients of τ^{N+d}, hence $\beta_N = 0$ ($\delta_d \beta_N^{p^d} \in I^{p^d} \subset I^2 = (0)$ and α_d is invertible), hence $\psi \equiv 1$.

To prove the existence of ψ it is sufficient to prove that, if $\alpha_n = 0$ for all $n > N$ and some integer $N > d$, the coefficient of τ^n in

$$\left(1 - \frac{\alpha_N}{\alpha_d^{p^{N-d}}} \tau^{N-d}\right) \circ \varphi \circ \left(1 - \frac{\alpha_N}{\alpha_d^{p^{N-d}}} \tau^{N-d}\right)^{-1}$$

is zero for all $n \geq N$. But this is an obvious computation as

$$\left(1 - \frac{\alpha_N}{\alpha_d^{p^{N-d}}} \tau^{N-d}\right)^{-1} = 1 + \frac{\alpha_N}{\alpha_d^{p^{N-d}}} \tau^{N-d}.$$

$\square$

REMARK (1.1.3). — If k is a field, we have a map

$$\deg : \mathrm{End}(\mathbb{G}_{a,k}) \to \mathbb{N} \cup \{-\infty\}$$

satisfying the properties

$$\deg(\varphi + \psi) \leq \max(\deg \varphi, \deg \psi),$$
$$\deg(\psi \circ \varphi) = \deg \psi + \deg \varphi,$$

for all $\varphi, \psi \in \mathrm{End}(\mathbb{G}_{a,k})$, and defined by

$$\deg\left(\sum_{n=0}^{d} \alpha_n \tau^n\right) = d$$

if $\alpha_d \neq 0$ and $\deg(0) = -\infty$.

$\square$

We have a ring homomorphism

(1.1.4) $$\partial : \mathrm{End}(\mathbb{G}_{a,k}) \to k$$

defined by

$$\partial\left(\sum_n \alpha_n \tau^n\right) = \alpha_0.$$

(1.2) Drinfeld modules

Let S be a scheme of characteristic p and let d be an integer > 0.

DEFINITION (1.2.1) (Drinfeld). — *A Drinfeld A-module of rank d over S is a pair (E, φ), where E is a commutative group scheme over S and*

$$\varphi : A \to \operatorname{End}(E)$$

is a ring homomorphism from A to the ring $\operatorname{End}(E)$ of endomorphisms of the group scheme E over S such that

(i) *E is locally isomorphic to $\mathbb{G}_{a,S}$ for the Zariski topology on S,*

(ii) *if $U = \operatorname{Spec}(k)$ is an affine open subset of S and $\psi : E_U \xrightarrow{\sim} \mathbb{G}_{a,U}$ is an isomorphism of group schemes over U, for each $a \in A - \{0\}$*

$$\psi \circ \varphi(a) \circ \psi^{-1} = \sum_n \alpha_n(a) \tau^n \in k[\tau]$$

satisfies the following properties :

(a) *$\alpha_n(a)$ is invertible in k if $n = -d \deg(\infty)\infty(a)$;*

(b) *$\alpha_n(a)$ is nilpotent if $n > -d \deg(\infty)\infty(a)$.*

REMARK (1.2.2). — A standard Drinfeld A-module of rank d over S is a pair (E, φ) where E is a line bundle over S and

$$\varphi : A \to \operatorname{End}(E)$$

is a ring homomorphism from A to the ring of endomorphisms $\operatorname{End}(E)$ of the group scheme over S underlying E such that, for each $a \in A - \{0\}$,

$$\varphi(a) = \sum_{n=0}^{-d \deg(\infty)\infty(a)} \alpha_n(a) \tau^n$$

with $\tau^n : E \to E^{\otimes p^n}$ the n-th power of the Frobenius endomorphism,

$$\alpha_n(a) \in H^0(S, E^{\otimes(1-p^n)})$$

for each $n \geq 0$ and $\alpha_n(a)$ never vanishes on S if $n = -d \deg(\infty)\infty(a)$. Each standard Drinfeld A-module gives rise to a Drinfeld A-module if one forgets the $\mathcal{O}_S$-linear structure on E and it follows from lemma (1.1.2) that every Drinfeld A-module is isomorphic to a standard one of the same rank.   □

Let (E, φ) be a Drinfeld A-module over S. It is easy to see that we have a ring homomorphism

$$\partial : \operatorname{End}(E) \to H^0(S, \mathcal{O}_S)$$

such that, if $U = \operatorname{Spec}(k)$ is an affine open subset of S and $\psi : E_U \xrightarrow{\sim} \mathbb{G}_{a,U}$ an isomorphism of group schemes over U, $\partial_U(\psi^{-1} \circ (-) \circ \psi)$ is nothing else than (1.1.4).

DEFINITION (1.2.3) (Drinfeld). — *The characteristic of a Drinfeld A-module* (E, φ) *over* S *is the ring homomorphism* $\partial \circ \varphi : A \to H^0(S, \mathcal{O}_S)$ *or, what is the same, the morphism of schemes*

$$\text{Spec}(\partial \circ \varphi) : S \to X - \{\infty\} = \text{Spec}(A).$$

(1.3) Level structures

Let $I \subset A$ be a non-zero ideal and let (E, φ) be a Drinfeld A-module of rank d over a scheme S of characteristic p. We will denote by $V(I) \subset X - \{\infty\} = \text{Spec}(A)$ the closed subscheme defined by I, and by $\theta : S \to \text{Spec}(A)$ the characteristic of (E, φ).

DEFINITION (1.3.1) (Drinfeld). — *The scheme of* I-*division points for* (E, φ) *is the subscheme*

$$E_I = \bigcap_{a \in I} \text{Ker}(E \xrightarrow{\varphi(a)} E)$$

of E *over* S.

E_I has a natural structure of a scheme of (A/I)-modules induced by φ. If I is generated by $a_1, \ldots, a_s$, we have also

$$E_I = E_{a_1} \cap \cdots \cap E_{a_s}$$

where

$$E_a = \text{Ker}(E \xrightarrow{\varphi(a)} E)$$

for each $a \in A\{0\}$.

If $I = I_1.I_2$ where I_1, I_2 are ideals of A such that $I_1 + I_2 = A$, then E_I is canonically isomorphic to

$$E_{I_1} \oplus_S E_{I_2}$$

as a scheme of modules over

$$A/I = (A/I_1) \oplus (A/I_2).$$

PROPOSITION (1.3.2) (Drinfeld). — *The restriction of the scheme of* (A/I)-*modules* E_I *to* $S - \theta^{-1}(V(I))$ *is locally constant with value* $(A/I)^d$ *for the étale topology on* $S - \theta^{-1}(V(I))$.

Proof : One can always find another ideal J of A such that $I + J = A$ and $I \cdot J = (a)$ is a principal ideal; moreover, if $Z \subset X$ is any finite closed subscheme, one can always choose a J as before such that $Z \cap V(J) = \emptyset$. So it is enough to prove the proposition when $I = (a)$ is principal.

Let $a \in A - \{0\}$, then E_a is clearly finite and étale of constant rank

$$(\operatorname{Card}(A/(a)))^d$$

over $S - \theta^{-1}(V(A/(a)))$. Replacing a by a^2, we get that E_{a^2} is finite and étale of constant rank

$$(\operatorname{Card}(A/(a^2)))^d = ((\operatorname{Card}(A/(a)))^d)^2$$

and

$$E_a = \operatorname{Ker}(E_{a^2} \xrightarrow{\varphi(a)} E_{a^2}).$$

Now the proposition is an easy consequence of the following lemma. □

LEMMA (1.3.3). — *Let $a \in A - \{0\}$ and let M be an A-module of finite length killed by a^2. Let M_a be the kernel of the multiplication by a in M. Then*

(i) $\operatorname{Card}(M) \le \operatorname{Card}(M_a)^2$;

(ii) *the equality holds in* (i) *if and only if M is a free $(A/(a^2))$-module; moreover in this case M_a is a free $(A/(a))$-module.*

Proof: Let x be a closed point of X. It is sufficient to prove the lemma after replacing A by $\mathcal{O}_x$, a by some non-negative power ϖ_x^m of a local uniformizing parameter ϖ_x of $\mathcal{O}_x$ and M by $\mathcal{O}_x \otimes_A M$. But then we can decompose $\mathcal{O}_x \otimes_A M$ as a direct sum of modules isomorphic to $\mathcal{O}_x/(\varpi_x^i)$ with $0 < i \le 2m$ and the lemma follows immediately from the equalities

$$(\mathcal{O}_x/(\varpi_x^i))_{\varpi_x^m} = \begin{cases} \mathcal{O}_x/(\varpi_x^i) & \text{if} \quad 0 < i \le m, \\ (\varpi_x^{i-m})/(\varpi_x^i) & \text{if} \quad m < i \le 2m. \end{cases}$$

□

DEFINITION (1.3.4) (Drinfeld). — *Let us assume that the characteristic of (E, φ) is away from I, i.e. $\theta^{-1}(V(I)) = \emptyset$. A level-$I$ structure on (E, φ) is an isomorphism of schemes of (A/I)-modules over S*

$$\iota : (I^{-1}/A)_S^d \xrightarrow{\sim} E_I$$

where $(I^{-1}/A)_S^d$ is the constant scheme of (A/I)-modules over S with value $(I^{-1}/A)^d$.

Proposition (1.3.2) says that the restriction of (E, φ) to $S - \theta^{-1}(V(I))$ admits level-I structures locally for the étale topology on $S - \theta^{-1}(V(I))$.

REMARK (1.3.5). — In [Dr 1] (§ 5,B), Drinfeld has also extended the notion of level-I structures to cover the case where the characteristic of (E, φ) divides I. □

(1.4) Modular varieties

Let d and I be as before. We will denote by $\mathcal{M}_I^d$ the fibered category over the category of schemes of characteristic p defined by

$$\mathcal{M}_I^d(S) = \begin{cases} \text{the category of triples } (E, \varphi, \iota) \\ \text{where } (E, \varphi) \text{ is a Drinfeld} \\ A\text{-module of rank } d \text{ over } S \\ \text{with characteristic away from } I \\ \text{and where } \iota \text{ is an} \\ I\text{-level structure on } (E, \varphi) \end{cases}$$

(the maps in $\mathcal{M}_I^d(S)$ are the isomorphisms between triples) and the obvious pull-back functors. We get an equivalent fibered category $\mathcal{M}_{I,stand}^d(S)$ by replacing "Drinfeld A-module" by "standard Drinfeld A-module" in the definition of $\mathcal{M}_I^d(S)$. In particular, if $I \neq A$, the choice of a non-zero element $i_0 \in (I^{-1}/A)^d$ defines a trivialization of the line bundle E if $(E, \varphi, \iota) \in \mathrm{ob}\,\mathcal{M}_{I,stand}^d(S)$ and $\mathcal{M}_{I,stand}^d(S)$ (and also $\mathcal{M}_I^d(S)$) is equivalent to a discrete category for each scheme S of characteristic p (the isomorphisms between standard Drinfeld A-modules respect the $\mathcal{O}_S$-linear structure).

The characteristic map of Drinfeld A-modules defines a morphism of fibered categories

$$\Theta : \mathcal{M}_I^d \longrightarrow X - (\{\infty\} \cup V(I))$$

(any $\mathbb{F}_p$-scheme is considered as a fibered category over the category of schemes of characteristic p in the usual way).

THEOREM (1.4.1) (Drinfeld). — *If $I \neq A$, the fibered category $\mathcal{M}_I^d$ is representable by an affine scheme M_I^d of finite type over $\mathbb{F}_p$.*

Proof: It is sufficient to prove that Θ is representable, affine and of finite type. Let us fix a presentation of the $\mathbb{F}_p$-algebra A,

$$A = \mathbb{F}_p[a_1, \ldots, a_\ell]/(f_1(a_1, \ldots, a_\ell), \ldots, f_m(a_1, \ldots, a_\ell)),$$

such that I is generated by $a_1, \ldots, a_s$ ($s \leq \ell$), and let us fix a scheme morphism $\theta : S \to X - (\{\infty\} \cup V(I))$. Rigidifying the line bundle E by the section $\iota(i_0)$ we see that the datum of the isomorphism class of $(E, \varphi, \iota) \in \mathrm{ob}\,\mathcal{M}_{I,stand}^d(S)$ is equivalent to the data

(i) $\varphi(a_\lambda) = \sum_{n=0}^{N_\lambda} \alpha_{\lambda,n} \tau^n \in H^0(S, \mathcal{O}_S)[\tau]$ with

$N_\lambda = -d \deg(\infty)\infty(a_\lambda)$, $\alpha_{\lambda,N_\lambda}$ invertible in $H^0(S, \mathcal{O}_S)$ and $\alpha_{\lambda,0} = \theta^*(a_\lambda)$ $(\lambda = 1, \ldots, \ell)$,

(ii) $\iota(i) \in H^0(S, \mathcal{O}_S)$ for each $i \in (I^{-1}/A)^d$ with $\iota(0) = 0$ and $\iota(i_0) = 1$, subject to the following relations:

(a) $\varphi(a_{\lambda'})\varphi(a_{\lambda''}) = \varphi(a_{\lambda''})\varphi(a_{\lambda'})$, $\forall \lambda', \lambda'' = 1, \ldots, \ell$, and
$f_\mu(\varphi(a_1), \ldots, \varphi(a_\ell)) = 0$, $\forall \mu = 1, \ldots, m$,

(b) $\iota(i' + i'') = \iota(i') + \iota(i'')$, $\forall i', i'' \in (I^{-1}/A)^d$, and $\iota(a_\lambda.i) = \varphi(a_\lambda)(\iota(i))$, $\forall i \in (I^{-1}/A)^d$, $\forall \lambda = 1, \ldots, \ell$,

(c) the polynomial

$$\prod_{i \in (I^{-1}/A)^d} (t - \iota(i)) \in H^0(S, \mathcal{O}_S)[t]$$

is exactly the g.c.d. of the polynomials

$$\sum_{n=0}^{N_\lambda} \alpha_{\lambda,n} t^{p^n}$$

for $\lambda = 1, \ldots, s$.

The theorem is now obvious. $\qquad\square$

If $J \subset I$ are non-zero ideals of A, we have an obvious map of fibered categories

$$r_{I,J} : \mathcal{M}_J^d \to \mathcal{M}_I^d$$

and a commutative diagram

$$
\begin{array}{ccc}
\mathcal{M}_J^d & \xrightarrow{\;\;r\;\;} & \mathcal{M}_I^d \\
\Theta \downarrow & & \downarrow \Theta \\
X - (\{\infty\} \cup V(J)) & \hookrightarrow & X - (\{\infty\} \cup V(I)).
\end{array}
$$

LEMMA (1.4.2). — *The map of fibered categories*

$$r_{I,J} : \mathcal{M}_J^d \to \mathcal{M}_I^d | (X - (\{\infty\} \cup V(J)))$$

is representable, finite, étale and Galois with Galois group the kernel of the reduction modulo I homomorphism

$$GL_d(A/J) \to GL_d(A/I).$$

Proof : We have a right action of $GL_d(A/J)$ on $\mathcal{M}_J^d$: $g \in GL_d(A/J)$ maps (E, φ, ι) onto $(E, \varphi, \iota \circ {}^t g^{-1})$. So the lemma is an obvious consequence of (1.3.2). $\qquad\square$

COROLLARY (1.4.3). — *The fibered category $\mathcal{M}_A^d$ is representable by a Deligne–Mumford algebraic stack of finite type M_A^d over $\mathbb{F}_p$ (see [De–Mu] §4).* □

(1.5) Deformation theory

Let I be an ideal of A and d be an integer ≥ 1. Then we will prove:

THEOREM (1.5.1) (Drinfeld). — *The morphism of Deligne–Mumford algebraic stacks (schemes if $I \neq A$)*

$$\Theta : M_I^d \to X - (\{\infty\} \cup V(I))$$

is smooth of pure relative dimension $d - 1$.

Proof : Thanks to lemma (1.4.2), it is enough to consider the case $I = A$. To prove the theorem in this case we will use deformation theory.

Let k be a ring of characteristic p and let $\varphi : A \to k[\tau]$ be a Drinfeld A-module of rank d over k with characteristic $\theta = \partial \circ \varphi : A \to k$. We will assume for simplicity that φ is standard. Let $\mathcal{O}$ be a thickening of k, i.e. a ring of characteristic p endowed with an ideal $\mathfrak{m}$ such that $\mathfrak{m}^2 = (0)$ and $\mathcal{O}/\mathfrak{m} = k$. We are looking for Drinfeld A-modules (of rank d) over $\mathcal{O}$

$$\widetilde{\varphi} : A \to \mathcal{O}[\tau]$$

such that the reduction modulo $\mathfrak{m}$ of $\widetilde{\varphi}$ is precisely φ; we will denote by $\widetilde{\theta} = \partial \circ \varphi : A \to \mathcal{O}$ the characteristic of $\widetilde{\varphi}$.

Let $\psi : A \to \mathcal{O}[\tau]$ be any lifting of φ as a morphism of $\mathbb{F}_p$-vector spaces. Then ψ will be a Drinfeld A-module of rank d over $\mathcal{O}$ if and only if

$$\psi(a^1)\psi(a^2) = \psi(a^1 a^2)$$

for all $a^1, a^2 \in A$ ($\psi(1)^2 = \psi(1)$ and $\psi(1)$ congruent to 1 modulo $\mathfrak{m}$ imply $\psi(1) = 1$). So let $T : A \times A \to \mathfrak{m}[\tau]$ be the $\mathbb{F}_p$-bilinear map defined by

$$T(a^1, a^2) = \psi(a^1)\psi(a^2) - \psi(a^1 a^2)$$

for each $a^1, a^2 \in A$. Then T satisfies the Hochschild cocycle condition

$$a^1 \cdot T(a^2, a^3) - T(a^1 a^2, a^3) + T(a^1, a^2 a^3) - T(a^1, a^2) \cdot a^3 = 0$$

for all $a^1, a^2, a^3 \in A$ where

$$\mathfrak{m}[\tau] = \{\sum_n \alpha_n \tau^n \in \mathcal{O}[\tau] | \alpha_n \in \mathfrak{m}, \forall n \geq 0\}$$

is viewed as an (A, A)-bimodule in the following way:

$$a \cdot P(\tau) \cdot b = \psi(a)P(\tau)\psi(b)$$

for each $a, b \in A$ and $P(\tau) \in \mathfrak{m}[\tau]$ (as $\mathfrak{m}^2 = (0)$, this (A, A)-bimodule structure is well-defined and independent of the choice of ψ and we can replace $\psi(a)$ by $\partial \circ \psi(a)$ in its definition). Moreover, if we replace ψ by another lifting $\psi + t$ where $t : A \to \mathfrak{m}[\tau]$ is any $\mathbb{F}_p$-linear map, T has to be replaced by

$$T + \delta t$$

where δt is the Hochschild coboundary

$$(\delta t)(a^1, a^2) = a^1 \cdot t(a^2) - t(a^1 a^2) + t(a^1) \cdot a^2$$

for each $a^1, a^2 \in A$. In other words, the obstruction to lifting φ into $\widetilde{\varphi}$ lies in the Hochschild cohomology group

$$H^2(A, \mathfrak{m}[\tau])$$

(see [Ca–Ei] (IX, §6)). Similarly we see that if this obstruction vanishes the isomorphism classes of liftings of φ to $\mathcal{O}$ are parametrized by the Hochschild cohomology group

$$H^1(A, \mathfrak{m}[\tau])$$

and that the automorphism group of a fixed lifting $\widetilde{\varphi}$ of φ to $\mathcal{O}$ is isomorphic to the Hochschild cohomology group

$$H^0(A, \mathfrak{m}[\tau]).$$

Now we have for each integer n (see loc. cit.)

$$H^n(A, \mathfrak{m}[\tau]) = \mathrm{Ext}^n_{A \otimes_{\mathbb{F}_p} A}(A, \mathfrak{m}[\tau])$$

where A is viewed as a (left) $(A \otimes_{\mathbb{F}_p} A)$-module via the augmentation morphism $A \otimes_{\mathbb{F}_p} A \to A$ $(a^1 \otimes a^2 \mapsto a^1 a^2)$ and where the (A, A)-bimodule $\mathfrak{m}[\tau]$ is viewed as a (left) $(A \otimes_{\mathbb{F}_p} A)$-module in the usual way. But by Grothendieck's duality (see [Ha](7.2))

$$\mathrm{Ext}^n_{A \otimes_{\mathbb{F}_p} A}(A, \mathfrak{m}[\tau]) = 0$$

for each integer $n \geq 2$ and in fact

$$R\,\mathrm{Hom}_{A \otimes_{\mathbb{F}_p} A}(A, \mathfrak{m}[\tau]) = (T_{A/\mathbb{F}_p} \overset{L}{\underset{A \otimes_{\mathbb{F}_p} A}{\otimes}} \mathfrak{m}[\tau])[-1]$$

where $T_{A/\mathbb{F}_p} = \mathrm{Hom}_A(\Omega^1_{A/\mathbb{F}_p}, A)$ is viewed as a (right) $(A \otimes_{\mathbb{F}_p} A)$-module via the augmentation map $A \otimes_{\mathbb{F}_p} A \to A$. Already this proves that the Deligne–Mumford stack $\mathcal{M}_A^d$ is smooth over $\mathbb{F}_p$ of pure dimension d. Indeed, let k be a field of characteristic p and let $\varphi : A \to k[\tau]$ be a k-point of $\mathcal{M}_A^d$, then the tangent space of $\mathcal{M}_A^d$ at φ is canonically isomorphic to

$$T_{A/\mathbb{F}_p} \otimes_{A \otimes_{\mathbb{F}_p} A} k[\tau]$$

where the (A, A)-bimodule structure of $k[\tau]$ is given by

$$a \cdot P(\tau) \cdot b = \theta(a) P(\tau) \varphi(b)$$

for each $a, b \in A$ and $P(\tau) \in k[\tau]$ $(\mathcal{O} = k[\varepsilon]/(\varepsilon^2)$, $\mathfrak{m} = (\varepsilon)$, $\mathfrak{m}[\tau] = \varepsilon k[\tau] \cong k[\tau])$. But the $(A \otimes_{\mathbb{F}_p} A)$-module structure of $k[\tau]$ is coming from its $(k \otimes_{\mathbb{F}_p} A)$-module structure via $\theta \otimes id_A$ and we have

LEMMA (1.5.2). — *Let k be a field of characteristic p and let $\varphi : A \to k[\tau]$ be a Drinfeld A-module of rank d over k. Let us consider the $(k \otimes_{\mathbb{F}_p} A)$-module structure on $k[\tau]$ given by*

$$(\alpha \otimes a) \cdot P(\tau) = \alpha P(\tau) \varphi(a)$$

for each $\alpha \in k$, $a \in A$ and $P(\tau) \in k[\tau]$. Then $k[\tau]$ is locally free of constant rank d over $k \otimes_{\mathbb{F}_p} A$.

Proof of the lemma : Let $a \in A$ be a non-constant element (i.e. $\infty(a) < 0$). Then the subring $\mathbb{F}_p[a]$ of A is a polynomial ring in one variable over $\mathbb{F}_p$ and A is finite and locally free of rank

$$- \deg(\infty) \infty(a)$$

as a module over $\mathbb{F}_p[a]$.

Now, using the Euclidean algorithm

$$P(\tau) = P_1(\tau) \varphi(a) + R_1(\tau),$$
$$P_1(\tau) = P_2(\tau) \varphi(a) + R_2(\tau)$$

$$\cdots \cdots$$

where

$$\deg(R_m(\tau)) < \deg(\varphi(a)) = -d \, \deg(\infty) \infty(a)$$

for each m, we can write in a unique way each $P(\tau) \in k[\tau]$ as

$$P(\tau) = \sum_{m \geq 0} \sum_{n=0}^{-d \, \deg(\infty) \infty(a) - 1} \alpha_{m,n} \tau^n \varphi(a)^m$$

and this means exactly that $k[\tau]$ is free of rank

$$-d \, \deg(\infty)\infty(a)$$

over the subring $k[a]$ of $k \otimes_{\mathbb{F}_p} A$.

This finishes the proof of the lemma if $k \otimes_{\mathbb{F}_p} A$ is integral (i.e. $\mathbb{F}_p \subset A$ is exactly the field of constants). Otherwise let $\mathbb{F}_{p^r} \subset A$ be the field of constants, then we have

$$\varphi(A) \subset k[\tau^r] \subset k[\tau]$$

and we have compatible isomorphisms

$$k \otimes_{\mathbb{F}_p} \mathbb{F}_{p^r} \xrightarrow{\sim} \prod_{i=0}^{r-1} k \ , \ \alpha \otimes a \mapsto (\alpha\theta(a)^{p^i})_i,$$

$$k \otimes_{\mathbb{F}_p} A \xrightarrow{\sim} \prod_{i=0}^{r-1} k \otimes_{\theta,\mathbb{F}_{p^r}} A$$

and

$$k[\tau] \xrightarrow{\sim} \bigoplus_{i=0}^{r-1} k[\tau^r]\tau^i.$$

The lemma follows easily from these remarks. □

End of the proof of the theorem : Let us compute the tangent map to $\Theta : M_A^d \to X - \{\infty\}$ at some point $\varphi : A \to k[\tau]$ of M_A^d over a field of characteristic p. It is obviously the map

$$T_{A/\mathbb{F}_p} \otimes_{A \otimes_{\mathbb{F}_p} A} k[\tau] \to T_{A/\mathbb{F}_p} \otimes_{A \otimes_{\mathbb{F}_p} A} k = T_{A/\mathbb{F}_p} \otimes_{A,\theta} k$$

induced by $\partial : k[\tau] \to k$ ($\theta : A \to k$ is the characteristic of φ). But this map is clearly surjective and the theorem is proved. □

(1.6) Hecke algebras, correspondences

Let $\mathbb{A}$ be the ring of adeles of F with its usual topology. Then we have

$$\mathbb{A} = F_\infty \times \mathbb{A}^\infty$$

where

$$\mathbb{A}^\infty = F \otimes_A \widehat{A}$$

and, if we set

$$\mathcal{O}^\infty = \widehat{A},$$

the subring of $\mathbb{A}$

$$\mathcal{O} = \mathcal{O}_\infty \times \mathcal{O}^\infty$$

is a maximal compact subring.

For each non-zero ideal I of A we will denote by K_I^∞ the compact open subgroup

$$\mathrm{Ker}(GL_d(\mathcal{O}^\infty) \longrightarrow GL_d(\mathcal{O}^\infty/\mathcal{O}^\infty I))$$

of $GL_d(\mathbb{A}^\infty)$ (the map is the reduction modulo I map) and we will consider the **Hecke algebra** (over $\mathbb{Q}$)

$$\mathcal{H}_I^\infty = \mathcal{C}_c^\infty(GL_d(\mathbb{A}^\infty)//K_I^\infty)$$

of (locally constant) functions with compact support

$$f^\infty : GL_d(\mathbb{A}^\infty) \to \mathbb{Q}$$

which are invariant by left and right translations by elements of K_I^∞; the product of two functions f_1^∞ and f_2^∞ in $\mathcal{H}_I$ is the convolution product defined by

$$(f_1^\infty * f_2^\infty)(g^\infty) = \int_{GL_d(\mathbb{A}^\infty)} f_1^\infty(h^\infty) f_2^\infty((h^\infty)^{-1} g^\infty) dh^\infty$$

for each $g^\infty \in GL_d(\mathbb{A}^\infty)$, where dh^∞ is the Haar measure on $GL_d(\mathbb{A}^\infty)$ normalized by

$$\mathrm{vol}(K_I^\infty, dh^\infty) = 1.$$

In fact a basis of $\mathcal{H}_I^\infty$ is given by the characteristic functions

$$1_{K_I^\infty g^\infty K_I^\infty}$$

of the double classes $K_I^\infty g^\infty K_I^\infty$ $(g^\infty \in GL_d(\mathbb{A}^\infty))$ and if $g_1^\infty, g_2^\infty \in GL_d(\mathbb{A}^\infty)$ and if

$$K_I^\infty g_1^\infty K_I^\infty g_2^\infty K_I^\infty = \coprod_{n=3}^{N} K_I^\infty g_3^\infty K_I^\infty,$$

we have

$$1_{K_I^\infty g_1^\infty K_I^\infty} * 1_{K_I^\infty g_2^\infty K_I^\infty} = \sum_{n=3}^{N} v_n \cdot 1_{K_I^\infty g_n^\infty K_I^\infty}$$

where

$$v_n = \mathrm{vol}((K_I^\infty g_1^\infty K_I^\infty) \cap (g_n^\infty K_I^\infty (g_2^\infty)^{-1} K_I^\infty), dh^\infty)$$

is a positive integer $(n = 3, \ldots, N)$.

Let $f^\infty \in \mathcal{H}_I^\infty$ and $x \in |X| - (\{\infty\} \cup V(I))$, we will say that x is good for f^∞ if

$$\mathrm{Supp}(f^\infty) \subset GL_n(\mathcal{O}_x) \times GL_n(\mathbb{A}^{\infty,x}) \subset GL_n(\mathbb{A}^\infty).$$

Obviously, for a given $f^\infty \in \mathcal{H}_I^\infty$, almost all the $x \in |X| - (\{\infty\} \cup V(I))$ are good for f^∞ and, for a given $x \in |X| - (\{\infty\} \cup V(I))$, the subset

$$\mathcal{H}_I^{\infty,x} \subset \mathcal{H}_I^\infty$$

with elements the f^∞'s such that x is good for f^∞ is a $\mathbb{Q}$-sub-algebra; moreover as a $\mathbb{Q}$-algebra $\mathcal{H}_I^{\infty,x}$ is isomorphic to

$$\mathcal{C}_c^\infty(GL_d(\mathbb{A}^{\infty,x})/\!/K_I^{\infty,x})$$

(same definition as before but replace $\mathbb{A}^\infty$ by $\mathbb{A}^{\infty,x}$, ...).

Let S be a noetherian scheme and let $Y \xrightarrow{b} S$ be a separated S-scheme of finite type. A geometric correspondence of Y over S is an isomorphism class of commutative diagrams of schemes

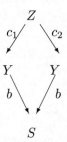

where c_2 is finite and étale and c_1 is proper. If (Z, c) and (Z', c') are two geometric correspondences of Y over S ($c = (c_1, c_2), c' = (c_1', c_2')$) we can form the correspondences

$$(Z, c) + (Z', c') = (Z \amalg Z', (c_1 \amalg c_1', c_2 \amalg c_2'))$$

and

$$(Z, c) \cdot (Z', c') = (Z' \times_{c_2', Y, c_1} Z, (c_1' \circ (Z' \times_Y c_1), c_2 \circ (c_2' \times_Y Z)))$$

of Y on S. Let

$$\mathrm{Corr}(Y/S)_{\mathbb{Z}}$$

be the abelian group generated by the monoid of geometric correspondences of Y on S with the above addition law. The multiplication law obviously induces a ring structure on $\mathrm{Corr}(Y/S)_{\mathbb{Z}}$. We will denote by $[Z, c]$ the class of the geometric correspondence (Z, c) in $\mathrm{Corr}(Y/S)_{\mathbb{Z}}$ (the zero element is $[\emptyset, c]$ and $[Y, (id_Y, id_Y)]$ is a unit for this ring).

The $\mathbf{Q}$-algebra
$$\mathrm{Corr}(Y/S) = \mathbf{Q} \otimes_{\mathbf{Z}} \mathrm{Corr}(Y/S)_{\mathbf{Z}}$$
will be called the $\mathbf{Q}$-algebra of correspondences of Y on S.

(1.7) Hecke operators

For each non-zero ideal I of A, $I \neq A$, and for each $x \in |X| - (\{\infty\} \cup V(I))$ we will define $\mathbf{Q}$-algebra homomorphisms

(1.7.1)
$$\mathcal{H}_I^{\infty} \to \mathrm{Corr}(M_{I,\eta}^d/\eta)^{opp},$$

where $M_{I,\eta}^d$ is the generic fiber of $M_I^d \xrightarrow{\Theta} X - (\{\infty\} \cup V(I))$, and

(1.7.2)
$$\mathcal{H}_I^{\infty,x} \to \mathrm{Corr}(M_{I,X_x}^d/X_x)^{opp},$$

where M_{I,X_x}^d is the restriction of M_I^d to $X_x = \mathrm{Spec}(\mathcal{O}_x) \to X - (\{\infty\} \cup V(I))$, with the following compatibility: the diagram of $\mathbf{Q}$-algebra homomorphisms

(1.7.3)

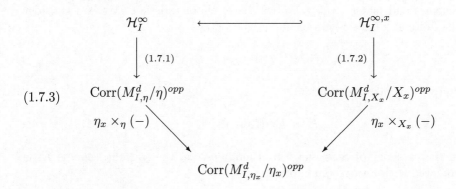

commutes ($\eta \leftarrow \eta_x = \mathrm{Spec}(F_x) \hookrightarrow X_x$ and M_{I,η_x}^d is the fiber of $\Theta : M_I^d \to X - (\{\infty\} \cup V(I))$ at η_x). In particular, we will get by restricting (1.7.2) to x a $\mathbf{Q}$-algebra homomorphism

(1.7.4)
$$\mathcal{H}_I^{\infty,x} \to \mathrm{Corr}(M_{I,x}^d/x)^{opp}$$

where $M_{I,x}^d$ is the fiber of $\Theta : M_I^d \to X - (\{\infty\} \cup V(I))$ at x.

In fact, let $g^{\infty} \in GL_d(\mathbf{A}^{\infty})$ and let $\Delta \subset |X| - \{\infty\}$ be a finite subset containing $V(I)$ and all $x \in |X| - (\{\infty\} \cup V(I))$ such that $(g^{\infty})_x \notin GL_d(\mathcal{O}_x)$. We will construct a geometric correspondence of M_I^d on $X - (\{\infty\} \cup \Delta)$. It will induce the image of $1_{K_I^{\infty} g^{\infty} K_I^{\infty}} \in \mathcal{H}_I^{\infty}$ by (1.7.1) when it is restricted to

η and, for each $x \in |X| - (\{\infty\} \cup \Delta)$, the image of $1_{K_I^\infty g^\infty K_I^\infty} \in \mathcal{H}_I^{\infty,x}$ by (1.7.2) when it is restricted to X_x.

In a first step we assume in addition that $(g^\infty)^{-1}$ is a matrix with coefficients in $\mathcal{O}^\infty$. Then we can choose a non-zero ideal J of A such that $J \subset I$, $V(J) \subset \Delta$, the kernel of the map

$$^t(g^\infty)^{-1} : (F/A)^d \to (F/A)^d$$

is contained in $(J^{-1}/A)^d$ and the image of $(J^{-1}/A)^d$ by $^t(g^\infty)^{-1}$ contains $(I^{-1}/A)^d$. Let us denote by

$$\widetilde{c}_1 : M_J^d|(X - (\{\infty\} \cup \Delta)) \to M_I^d|(X - (\{\infty\} \cup \Delta))$$

the restriction of the morphism $r_{I,J}$ considered in (1.4). We define another morphism

$$\widetilde{c}_2 : M_J^d|(X - (\{\infty\} \cup \Delta)) \to M_I^d|(X - (\{\infty\} \cup \Delta))$$

by the following procedure. Let $S \xrightarrow{\theta} X - (\{\infty\} \cup \Delta)$ be a morphism of schemes and let $(E, \varphi, \iota) \in M_J^d(S)$ be of characteristic θ. We can consider the finite and étale S-subscheme in A-modules

$$\iota(\mathrm{Ker}(^t(g^\infty)^{-1})_S) \subset E_J \subset E$$

and we can take the quotient

$$E' = E/\iota(\mathrm{Ker}(^t(g^\infty)^{-1})_S),$$

in the category of S-schemes in A-modules, i.e. the S-group scheme E' is endowed with a ring homomorphism

$$A \xrightarrow{\varphi'} \mathrm{End}(E').$$

LEMMA (1.7.5) (Drinfeld). — (E', φ') is a Drinfeld A-module of rank d over S with characteristic θ.

Proof: We can assume that $S = \mathrm{Spec}(k)$ is affine, that $E = \mathbb{G}_{a,k}$ and that $\varphi : A \to k[\tau]$ is standard. Then we can view $\iota(\mathrm{Ker}(^t(g^\infty)^{-1})_S)$ as a finite subgroup of k and we can define

$$P(\tau) = \sum_{n=0}^{N} \beta_n \tau^n \in k[\tau]$$

by

$$\sum_{n=0}^{N} \beta_n t^{p^n} = \prod_{\alpha \in \iota(\mathrm{Ker}(^t(g_\infty)^{-1})_S)} (t - \alpha) \; ;$$

we have $\beta_N = 1$, $p^N = \#\iota(\mathrm{Ker}(^t(g^\infty)^{-1})_S)$ and β_0 invertible in k (θ is away from $V(J)$).

Now we can take $E' = \mathbb{G}_{a,k}$ with quotient morphism

$$E \xrightarrow{\;P(\tau)\;} E'$$

and, for each $a \in A - \{0\}$, $\varphi'(a)$ is uniquely determined by the relation

$$P(\tau) \circ \varphi(a) = \varphi'(a) \circ P(\tau).$$

The lemma follows immediately. $\qquad\qquad\qquad\qquad\qquad\qquad\qquad\square$

Now we define a level-I structure ι' on (E', φ') by the following commutative diagram:

$$
\begin{array}{ccc}
(J^{-1}/A)_S^d & \xrightarrow{\;\;\sim\;\;}_{\iota} & E_J \\
\downarrow & & \downarrow \\
(^tg^\infty)^{-1}((J^{-1}/A)^d)_S & \xrightarrow{\;\;\sim\;\;} & E_J/\iota(\mathrm{Ker}(^t(g^\infty)^{-1})_S) \\
\uparrow & & \uparrow \\
(I^{-1}/A)_S^d & \xrightarrow[\iota']{\;\;\sim\;\;} & E_I' \; .
\end{array}
$$

We can set
$$\widetilde{c}_2(E, \varphi, \iota) = (E', \varphi', \iota').$$

Finally, our hypotheses on J imply that

$$K_J^\infty \subset K_I^\infty \cap ((g^\infty)^{-1} K_I^\infty g^\infty).$$

So we have a right action of the finite group

$$K_J^\infty \backslash (K_I^\infty \cap ((g^\infty)^{-1} K_I^\infty g^\infty)) \subset K_J^\infty \backslash K_I^\infty$$

on $M_J^d|(X - (\{\infty\} \cup \Delta))$ (we have

$$K_J^\infty \backslash K_I^\infty = \mathrm{Ker}(GL_d(A/J) \to GL_d(A/I))$$

and a right action of this kernel on M_J^d, see (1.4.2)). The morphism $\tilde{c}_1$ is obviously invariant by the right action of $K_J^\infty \backslash (K_I^\infty \cap ((g^\infty)^{-1} K_I^\infty g^\infty))$ and it is easy to see that this is the same for $\tilde{c}_2$. So by taking the quotient with respect to this right action we get the required geometric correspondence

$$(M_J^d|(X - (\{\infty\} \cup \Delta)))/(K_J^\infty \backslash (K_I^\infty \cap ((g^\infty)^{-1} K_I^\infty g^\infty)))$$

$$c_1 \swarrow \qquad \searrow c_2$$

$$M_I^d|(X - (\{\infty\} \cup \Delta)) \qquad \qquad M_I^d|(X - (\{\infty\} \cup \Delta))$$

$$\Theta \searrow \qquad \swarrow \Theta$$

$$X - (\{\infty\} \cup \Delta) \,.$$

This geometric correspondence is clearly independent of the choice of J. Moreover, if we replace g^∞ by $a^{-1} g^\infty$ where $a \in A - \{0\} \subset F^\times$ is viewed as a central matrix in $GL_d(F) \subset GL_d(\mathbb{A}^\infty)$, we get the same geometric correspondence as for g^∞.

For a general $g^\infty \in GL_d(\mathbb{A}^\infty)$, we can always find some $a \in F^\times$ such that $a(g^\infty)^{-1}$ is a matrix with coefficients in $\mathcal{O}^\infty$. Then we associate to g^∞ the geometric correspondence constructed above for $a^{-1} g^\infty$. It is easy to see that this geometric correspondence depends only on the double class $K_I^\infty g^\infty K_I^\infty$ and the set Δ. We will denote it

$$M_I^d(g^\infty, \Delta)$$

$$c_1 \swarrow \qquad \searrow c_2$$

(1.7.6) $\qquad M_I^d|(X - (\{\infty\} \cup \Delta)) \qquad \qquad M_I^d|(X - (\{\infty\} \cup \Delta))$

$$\Theta \searrow \qquad \swarrow \Theta$$

$$X - (\{\infty\} \cup \Delta) \,.$$

REMARK (1.7.7). — If $g^\infty \in K_A^\infty = GL_d(\mathcal{O}^\infty)$, we can take $\Delta = V(I)$. The morphisms c_1 and c_2 are isomorphisms and $c_1 \circ c_2^{-1}$ is the right action of $K_I^\infty g^\infty \in K_I^\infty \backslash K_A^\infty = GL_d(A/I)$ defined in (1.4) (take $J = I$ so that $\tilde{c}_1$ is the identity map). $\qquad \square$

(1.8) Comments and references

Most of the content of this chapter is due to Drinfeld and taken from [Dr 1]. The proof of (1.5.1) is not exactly Drinfeld's proof but is very similar.

2

Drinfeld A-modules with finite characteristic

(2.0) Notations
In this chapter, we fix an integer $d > 0$, a non-zero ideal I of A, $I \neq A$, a place o of F, $o \neq \infty$, $o \notin V(I)$, and an algebraic closure $\overline{\kappa(o)}$ of $\kappa(o)$; we denote by $\mathrm{Frob}_o \in \mathrm{Gal}(\overline{\kappa(o)}/\kappa(o))$ the geometric Frobenius element.

Our purpose is to describe the set

$$M_{I,o}^d(\overline{\kappa(o)}) = \Theta^{-1}(o)(\overline{\kappa(o)})$$

with the commuting actions of $\mathcal{H}_I^{\infty,o}$ and Frob_o on it.

(2.1) Isogenies
Let k be a field of characteristic p. For any non-zero $u \in k[\tau]$ there exist a unique integer $h \geq 0$ and a unique element $u^{et} \in k[\tau]$ with $\partial(u^{et}) \neq 0$ such that

$$u = u^{et} \cdot \tau^h.$$

If G is the kernel of the finite homomorphism

$$\mathbb{G}_{a,k} \xrightarrow{\ u\ } \mathbb{G}_{a,k}$$

the canonical dévissage of the finite group scheme G over k

$$o \longrightarrow G^c \longrightarrow G \longrightarrow G^{et} \longrightarrow o$$

(G^c connected, G^{et} étale) is given by the following exact sequence:

$$o \longrightarrow \mathrm{Ker}(\tau^h) \longrightarrow G \xrightarrow{\tau^h} \mathrm{Ker}(u^{et}) \longrightarrow o.$$

In particular G^c is isomorphic to

$$\boldsymbol{\alpha}_{p^h} = \mathrm{Spec}(k[t]/(t^{p^h}))$$

The integer h is called the **height** of u and is denoted by $h(u)$ (we have $\deg(u) = h(u) + \deg(u^{et})$).

Conversely:

LEMMA (2.1.1). — *Let $G \subset \mathbf{G}_{a,k}$ be a finite k-subgroup scheme. Then there exists a unique unitary $u \in k[\tau]$ such that $G = \mathrm{Ker}(u)$. Moreover, if $v \in k[\tau]$ and v vanishes identically on G there exists a unique $w \in k[\tau]$ such that $v = w \circ u$.*

Proof : Let $P(t)$ be the unique unitary polynomial in $k[t]$ such that $G = \mathrm{Spec}(k[t]/(P(t)))$. Then there exist $Q_1(t_1,t_2)$, $Q_2(t_1,t_2) \in k[t_1,t_2]$ such that

$$P(t_1 + t_2) = P(t_1)Q_1(t_1,t_2) + P(t_2)Q_2(t_1,t_2)$$

in $k[t_1,t_2]$. But this implies that

$$P(t) = t^{p^N} + \sum_{n=0}^{N-1} \alpha_n t^{p^n}$$

for some $\alpha_0, \ldots, \alpha_{N-1} \in k$ and we can take

$$u(\tau) = \tau^N + \sum_{n=0}^{N-1} \alpha_n \tau^n.$$

If $v \in k[\tau]$ there exist unique w and v' in $k[\tau]$ such that $\deg(v') < \deg(u)$ and

$$v = w \circ u + v'.$$

If $v|\mathrm{Ker}(u) \equiv 0$, then $v'|\mathrm{Ker}(u) \equiv 0$ and $v' = 0$ □

Let (E,φ) and (E',φ') be Drinfeld A-modules of rank d over k.

DEFINITION (2.1.2) (Drinfeld). — *An isogeny from (E,φ) to (E',φ') is a non-zero homomorphism of group schemes over k*

$$E \xrightarrow{u} E'$$

such that

$$\varphi'(a) \circ u = u \circ \varphi(a)$$

for each $a \in A$.

If we choose isomorphisms of group schemes over k, $E \cong \mathbf{G}_{a,k}$, $E' \cong \mathbf{G}_{a,k}$, an isogeny u from (E, φ) to (E', φ') becomes an element of $k[\tau]$ and has a degree and a height. Those two integers are clearly independent of the chosen isomorphisms and are called the **degree of the isogeny**, $\deg(u)$, and the **height of the isogeny**, $h(u)$.

LEMMA (2.1.3). — *For any isogeny u from (E, φ) to (E', φ') there exist an isogeny u' from (E', φ') to (E, φ) and $a \in A - \{0\}$ such that*

$$u' \circ u = \varphi(a).$$

Proof: Let $a \in A - \{0\}$ be such that

$$\varphi(a)(\mathrm{Ker}(u)) = (0)$$

and let $G' \subset E'$ be the group-sub-scheme over k which is the image of $E_a = \mathrm{Ker}(\varphi(a))$ by u. Then we can apply (2.1.1) to $G' \subset E'$ and we get $u' \in \mathrm{Hom}(E', E)$ such that $u' \circ u = \varphi(a)$. If $b \in A$, we have

$$\varphi(b) \circ u' \circ u = u' \circ u \circ \varphi(b) = u' \circ \varphi'(b) \circ u$$

and it follows that

$$\varphi(b) \circ u' = u' \circ \varphi'(b).$$

□

If (E, φ) is a Drinfeld A-module of rank d over k, we will denote by $\mathrm{End}(E, \varphi)$ the subring of $\mathrm{End}(E)$ with elements the u's which commute with $\varphi(a)$ for each $a \in A$ (any non-zero element of $\mathrm{End}(E, \varphi)$ is an isogeny from (E, φ) to itself). Obviously φ factors through $\mathrm{End}(E, \varphi) \subset \mathrm{End}(E)$ and $\mathrm{End}(E, \varphi)$ is a central A-algebra via φ. The degree map on $\mathrm{End}(E, \varphi)$ has a natural extension

$$\deg : F \otimes_A \mathrm{End}(E, \varphi) \longrightarrow \mathbf{Z} \cup \{-\infty\}.$$

We will consider the map

$$\| - \| = p^{\deg(-)/d} : F \otimes_A \mathrm{End}(E, \varphi) \longrightarrow \mathbf{R};$$

it has the following properties:

(i) $\|u\| \geq 0$ and $\|u\| = 0$ if and only if $u = 0$,

(ii) $\|au\| = |a|_\infty \|u\|$,

(iii) $\|u + v\| \leq \max(\|u\|, \|v\|)$,

(iv) $\|u \cdot v\| = \|u\| \cdot \|v\|$,

for any u, $v \in F \otimes_A \mathrm{End}(E, \varphi)$ and any $a \in F$ ($|a|_\infty = p^{-\deg(\infty) \cdot \infty(a)}$).

PROPOSITION (2.1.4) (Drinfeld). — *The A-module $\operatorname{End}(E, \varphi)$ is projective of rank $\leq d^2$. Moreover, $F \otimes_A \operatorname{End}(E, \varphi)$ is a division ring containing F in its center which is fully ramified at the place ∞ of F (i.e. $F_\infty \otimes_A \operatorname{End}(E, \varphi)$ is also a division ring).*

Proof: It follows from the properties (i) and (ii) of $\| \ \|$ that the A-module $\operatorname{End}(E, \varphi)$ is torsion free.

For each finite dimensional F-vector subspace $V \subset F \otimes_A \operatorname{End}(E, \varphi)$, the restriction of $\| \ \|$ to V can be extended to $F_\infty \otimes_F V = V_\infty$ in a natural way and, if we denote this extension again by $\| \ \|$, $\| \ \|$ is a norm on the finite dimensional F_∞-vector space V_∞. The A-submodule

$$V \cap \operatorname{End}(E, \varphi) = V_\infty \cap \operatorname{End}(E, \varphi) \subset V_\infty$$

is discrete. Therefore, $V \cap \operatorname{End}(E, \varphi)$ is finitely generated. But it is torsion free and A is a Dedekind domain, so it is projective.

Now let $a \in A - \{0\}$ be prime to the characteristic of (E, φ). Then the natural map of $(A/(a))$-algebras

$$A/(a) \otimes_A (V \cap \operatorname{End}(E, \varphi)) \longrightarrow \operatorname{End}_{A/(a)}(E_a(\overline{k})),$$

where $\overline{k}$ is an algebraic closure of k, is injective: if $u \in V \cap \operatorname{End}(E, \varphi)$ and $u(\operatorname{Ker} \varphi(a)) = (0)$, thanks to (2.1.1), there exists $v \in \operatorname{End}(E)$ such that

$$u = v\varphi(a);$$

obviously $v \in \operatorname{End}(E, \varphi)$ and $v \in V$. But $E_a(\overline{k})$ is a free $(A/(a))$-module of rank d (see (1.3.2)). So, for each V as above, the rank of the projective A-module $V \cap \operatorname{End}(E, \varphi)$ and consequently $\dim_F(V)$ are bounded by d^2. Taking V larger and larger, we get the first assertion of the proposition.

The fact that $F \otimes_A \operatorname{End}(E, \varphi)$ is a division ring follows immediately from (2.1.3). Now to check that $F_\infty \otimes_A \operatorname{End}(E, \varphi)$ is a division ring it is enough to check that $F_\infty \otimes_A \operatorname{End}(E, \varphi)$ is zero divisor free. But we have a norm

$$\| \ \| : F_\infty \otimes_A \operatorname{End}(E, \varphi) \longrightarrow \mathbf{Q}$$

satisfying the above conditions (i) to (iv). So the proposition follows. $\square$

REMARK (2.1.5). — The norm $\| \ \|$ on $F_\infty \otimes_A \operatorname{End}(E, \varphi)$ is the unique extension of $|\ |_\infty$ on F_∞ and will be denoted $\| \ \|_\infty$ from now on. $\square$

(2.2) Isogeny classes of Drinfeld modules

Let k be a finite extension of $\kappa(o)$ and let $q = p^{r \deg(o)}$ be the number of elements in k. We will denote by θ the composite homomorphism

$$A \longrightarrow\!\!\!\!\!\rightarrow \kappa(o) \hookrightarrow k.$$

Let (E, φ) be a Drinfeld A-module of rank d over k with characteristic θ. Then

(2.2.1) $$\Pi = \Pi_{(E,\varphi)} = \tau^{r \deg(o)} \in \mathrm{End}(E, \varphi)$$

is called the **Frobenius isogeny** of (E, φ).

PROPOSITION (2.2.2) (Drinfeld). — (i) *The center of $D = F \otimes_A \mathrm{End}(E, \varphi)$ is exactly the F-subalgebra $\widetilde{F} = F[\Pi]$ generated by Π.*

(ii) *There are exactly two places $\widetilde{x} = \widetilde{\infty}$ and $\widetilde{x} = \widetilde{o}$ of $\widetilde{F}$ such that*

$$\widetilde{x}(\Pi) \neq 0.$$

The place $\widetilde{\infty}$ is the unique place of $\widetilde{F}$ dividing ∞ and

$$|\Pi|_\infty = q^{1/d}$$

($| \ |_\infty$ is at the same time the natural extension of $| \ |_\infty$ on F_∞ to $\widetilde{F}_{\widetilde{\infty}}$ and the restriction of $\| \ \|_\infty$ on $F_\infty \otimes_F D$ to $\widetilde{F}_{\widetilde{\infty}}$). The place $\widetilde{o}$ divides o.

(iii) *The degree of $\widetilde{F}$ over F divides d and the invariants of D at the places $\widetilde{x}$'s of $\widetilde{F}$ (see (A.3)) are given by*

$$\mathrm{inv}_{\widetilde{x}}(D) = \begin{cases} -[\widetilde{F} : F]/d & \text{if } \widetilde{x} = \widetilde{\infty}, \\ [\widetilde{F} : F]/d & \text{if } \widetilde{x} = \widetilde{o}, \\ 0 & \text{otherwise.} \end{cases}$$

In particular, we have

$$[D : \widetilde{F}] = (d/[\widetilde{F} : F])^2.$$

(iv) *Let $(\widetilde{F}, \Pi)$ be a pair where $\widetilde{F}$ is a finite field extension of F and $\Pi \in \widetilde{F}$. We will say that $(\widetilde{F}, \Pi)$ is a Weil (F, ∞, o)-pair of rank d over k if the following conditions are satisfied:*

(a) *$[\widetilde{F} : F]$ divides d,*

(b) *$F_\infty \otimes_F \widetilde{F}$ is a field,*

(c) *if $| \ |_\infty$ is the natural extension of $| \ |_\infty$ on F_∞ to $F_\infty \otimes_F \widetilde{F}$, we have*

$$|\Pi|_\infty = q^{1/d},$$

(d) *there exists one and only one place $\widetilde{o} \neq \widetilde{\infty}$ of $\widetilde{F}$ such that $\widetilde{o}(\Pi) \neq 0$, and moreover $\widetilde{o}$ divides o (as before, $\widetilde{\infty}$ is the unique place of $\widetilde{F}$ over ∞),*

(e) *$\widetilde{F}$ is generated by Π over F.*

Then the map from the set of isogeny classes of Drinfeld A-modules of rank d over k with characteristic θ to the set of isomorphism classes of Weil (F, ∞, o)-pairs of rank d over k which maps the isogeny class of (E, φ) to the isomorphism class of $(F[\Pi_{(E,\varphi)}], \Pi_{(E,\varphi)})$ is bijective.

Proof : Let $k(\tau)$ be the division ring of fractions of $k[\tau]$. Then $k(\tau)$ is a central division algebra over $\mathbb{F}_p(\Pi)$, where

$$\Pi = \tau^{r\,\deg(o)},$$

whose invariant equals $-1/r\deg(o)$ at $\Pi = \infty$, $1/r\deg(o)$ at $\Pi = o$ and 0 elsewhere. Let $\varphi : A \longrightarrow k[\tau]$ be our Drinfeld A-module of rank d over k with characteristic θ (as we can we assume that $E = \mathbb{G}_{a,k}$). Let $\Phi : F \longrightarrow k(\tau)$ be the natural extension of φ. Then Φ is injective and we will identify F with its image by Φ. Then D is the centralizer of F in $k(\tau)$ or what is the same the centralizer of $F(\Pi)$ in $k(\tau)$. As $F(\Pi)$ contains the center $\mathbb{F}_p(\Pi)$ of $k(\tau)$, $F(\Pi)$ is exactly the center $\widetilde{F}$ of D and

$$[k(\tau) : D] = [F(\Pi) : \mathbb{F}_p(\Pi)]$$

(see (A.1.3)). Moreover, as Π is algebraic over F ($\dim_F k(\tau) < +\infty$), part (i) is now proved, i.e.

$$\widetilde{F} = F(\Pi) = F[\Pi].$$

As $k(\tau)$ is totally ramified over $\mathbb{F}_p(\Pi)$ at $\Pi = \infty$ (resp. $\Pi = o$), the same is true for $\widetilde{F}$ over $\mathbb{F}_p(\Pi)$ and D over $\widetilde{F}$. More precisely there exists a unique place $\widetilde{\infty}$ (resp. $\widetilde{o}$) of $\widetilde{F}$ dividing the place $\Pi = \infty$ (resp. $\Pi = o$) of $\mathbb{F}_p(\Pi)$, and $\widetilde{F}_{\widetilde{\infty}} \otimes_{\widetilde{F}} D$ (resp. $\widetilde{F}_{\widetilde{o}} \otimes_{\widetilde{F}} D$) is a division algebra central over $\widetilde{F}_{\widetilde{\infty}}$ (resp. $\widetilde{F}_{\widetilde{o}}$). In fact, let $\widetilde{\widetilde{\infty}} : k((\tau^{-1}))^{\times} \longrightarrow \mathbb{Z}$ (resp. $\widetilde{\widetilde{o}} : k((\tau))^{\times} \longrightarrow \mathbb{Z}$) be defined by

(resp.
$$\widetilde{\widetilde{\infty}}\left(\sum_{n=N}^{\infty} \alpha_n \tau^{-n}\right) = N$$

$$\widetilde{\widetilde{o}}\left(\sum_{n=N}^{\infty} \alpha_n \tau^{n}\right) = N)$$

if $\alpha_N \neq 0$. Then there exists an integer $e(\widetilde{\widetilde{\infty}}/\widetilde{\infty})$ (resp. $e(\widetilde{\widetilde{o}}/\widetilde{o})) \geq 1$, such that

$$\widetilde{\widetilde{\infty}}(\tilde{a}) = e(\widetilde{\widetilde{\infty}}/\widetilde{\infty})\widetilde{\infty}(\tilde{a}), \ \forall \tilde{a} \in \widetilde{F}_{\widetilde{\infty}},$$

and

(resp.
$$[k(\tau) : \widetilde{F}] = e(\widetilde{\widetilde{\infty}}/\widetilde{\infty})[k : \kappa(\widetilde{\infty})]$$

$$\widetilde{\widetilde{o}}(\tilde{a}) = e(\widetilde{\widetilde{o}}/\widetilde{o})\widetilde{\infty}(\tilde{a}), \ \forall \ \tilde{a} \in \widetilde{F}_{\widetilde{o}},$$

and

$$[k(\tau) : \widetilde{F}] = e(\widetilde{\widetilde{o}}/\widetilde{o})[k : \kappa(\widetilde{o})]).$$

Therefore as

(resp.
$$\widetilde{\widetilde{\infty}}(a) = -d\deg(\infty)\infty(a), \ \forall \ a \in A$$

$$(\widetilde{\widetilde{o}}(a) > o) \Longleftrightarrow (o(a) > o), \ \forall \ a \in A),$$

the place $\widetilde{\infty}$ (resp. $\widetilde{o}$) divides the place ∞ (resp. o) of F.

Obviously, we have
$$\widetilde{x}(\Pi) = 0$$
for each place $\widetilde{x} \neq \widetilde{\infty}$, $\widetilde{o}$ of $\widetilde{F}$. Let us compute $|\Pi|_\infty$. If $e(\widetilde{\infty}/\infty)$ is the ramification index at $\widetilde{\infty}$ of $\widetilde{F}$ over F, we have
$$|\Pi|_\infty = p^{-\deg(\infty)\widetilde{\infty}(\Pi)/e(\widetilde{\infty}/\infty)}.$$

But
$$e(\widetilde{\widetilde{\infty}}/\widetilde{\infty})\widetilde{\infty}(\Pi) = \widetilde{\widetilde{\infty}}\ (\Pi) = -r\deg(o)$$

and
$$e(\widetilde{\widetilde{\infty}}/\widetilde{\infty})e(\widetilde{\infty}/\infty) = d\deg(\infty)$$

(for each $a \in A$ we have $\widetilde{\widetilde{\infty}}\ (a) = -d\deg(\infty)\infty(a)$). So we have
$$\widetilde{\widetilde{\infty}}\ (\Pi) = -r\deg(o)e(\widetilde{\infty}/\infty)/d\deg(\infty)$$

and
$$|\Pi|_\infty = p^{r\deg(o)/d} = q^{1/d}.$$

This finishes the proof of part (ii).

Next let us check that $[\widetilde{F} : F]$ divides d and that
$$[D : \widetilde{F}] = (d/[\widetilde{F} : F])^2.$$

The integer $-\widetilde{\infty}(\Pi)$ is also the ramification index at $\widetilde{\infty}$ of $\widetilde{F}$ over $\mathbb{F}_p(\Pi)$. Consequently, if we set $\deg(\widetilde{\infty}) = [\kappa(\widetilde{\infty}) : \mathbb{F}_p]$, we have
$$[\widetilde{F} : \mathbb{F}_p(\Pi)] = -\widetilde{\infty}(\Pi)\deg(\widetilde{\infty}) = \frac{r\deg(o)}{d}[\widetilde{F} : F]$$

as
$$[\widetilde{F} : F] = e(\widetilde{\infty}/\infty)\deg(\widetilde{\infty})/\deg(\infty).$$

But
$$[\widetilde{F} : \mathbb{F}_p(\Pi)] = [k(\tau) : D]$$

divides
$$[k(\tau) : \mathbb{F}_p(\Pi)]^{1/2} = r\deg(o)$$

and the quotient is $[D : \widetilde{F}]$. Hence the assertion follows.

As $k(\tau)$ is split over $\mathbb{F}_p(\Pi)$ outside $\Pi = \infty$ and $\Pi = o$, D is split over $\widetilde{F}$ outside $\widetilde{\infty}$ and $\widetilde{o}$. The division algebra $\widetilde{F}_{\widetilde{\infty}} \otimes_{\widetilde{F}} D$ is the centralizer of $\widetilde{F}_{\widetilde{\infty}}$ in $k((\tau^{-1}))$. As the invariant of $k((\tau^{-1}))$ over $\mathbb{F}_p((\Pi^{-1}))$ is $-1/[k((\tau^{-1})) : \mathbb{F}_p((\Pi^{-1}))]$, the invariant of $\widetilde{F}_{\widetilde{\infty}} \otimes_{\widetilde{F}} D$ is $-1/[\widetilde{F}_{\widetilde{\infty}} \otimes_{\widetilde{F}} \widetilde{D} : \widetilde{F}_{\widetilde{\infty}}]$ (see (A.2.4)). So part (iii) is proved.

Finally let $(\widetilde{F}, \Pi)$ be a Weil (F, ∞, o)-pair of rank d over k. Let us prove the existence of an embedding $\widetilde{F} \hookrightarrow k(\tau)$ of $\mathbb{F}_p(\Pi)$-algebras. It is sufficient to check that $[\widetilde{F} : \mathbb{F}_p(\Pi)]$ divides $[k(\tau) : \mathbb{F}_p(\Pi)]^{1/2}$ and that $\widetilde{F}$ does not split over $\mathbb{F}_p(\Pi)$ at the places $\Pi = \infty$ and $\Pi = o$, i.e. where $k(\tau)$ is ramified (see (A.3.3)). But $\widetilde{\infty}(\Pi) < 0$ (resp. $\widetilde{o}(\Pi) > 0$) so $\widetilde{\infty}$ (resp. $\widetilde{o}$) divides the place $\Pi = \infty$ (resp. $\Pi = o$) of $\mathbb{F}_p(\Pi)$. Moreover $\widetilde{x}(\Pi) = 0$ for all places $\widetilde{x} \neq \widetilde{\infty}, \widetilde{o}$ of $\widetilde{F}$. So $\widetilde{\infty}$ (resp. $\widetilde{o}$) is the only place of $\widetilde{F}$ dividing the place $\Pi = \infty$ (resp. $\Pi = o$) of $\mathbb{F}_p(\Pi)$ and $\widetilde{F}$ does not split over $\mathbb{F}_p(\Pi)$ at the place $\Pi = \infty$ (resp. $\Pi = o$). As before $|\Pi|_\infty = q^{1/d}$ implies

$$[\widetilde{F} : \mathbb{F}_p(\Pi)] = \frac{r \deg(o)}{d} [\widetilde{F} : F]$$

and, as $[\widetilde{F} : F]$ divides d by hypothesis, $[\widetilde{F} : \mathbb{F}_p(\Pi)]$ divides $r \deg(o) = [k(\tau) : \mathbb{F}_p(\Pi)]^{1/2}$.

Let us choose one embedding $\widetilde{F} \hookrightarrow k(\tau)$. Let $\widetilde{A}$ be the integral closure of A in $\widetilde{F}$, i.e.

$$\widetilde{A} = \{\widetilde{a} \in \widetilde{F} |\ \widetilde{x}(\widetilde{a}) \geq 0 \text{ for each place } \widetilde{x} \neq \widetilde{\infty} \text{ of } \widetilde{F}\}.$$

Using an inner automorphism of $k(\tau)$ we can modify the embedding in such way that it maps A into the maximal order $k[\tau]$ of $k(\tau)$ ($\widetilde{A}$ is contained in some maximal order of $k(\tau)$ and two maximal orders of $k(\tau)$ are conjugate in $k(\tau)$). Let $\varphi : A \longrightarrow k[\tau]$ be the restriction to A of this embedding. It follows easily from the hypotheses on $(\widetilde{F}, \Pi)$ that φ is a Drinfeld A-module of rank d over k with characteristic θ and part (iv) follows (the details are left to the reader). $\qquad\square$

Let $(\widetilde{F}, \widetilde{o})$ be a pair where $\widetilde{F}$ is an finite field extension of F and $\widetilde{o}$ is a place of $\widetilde{F}$ dividing o. We will say that $(\widetilde{F}, \widetilde{o})$ is an (F, ∞, o)-**type of rank** d if the following conditions are satisfied:

(a) $[\widetilde{F} : F]$ divides d,

(b) $F_\infty \otimes_F \widetilde{F}$ is a field,

(c) $\widetilde{F} = F[\Pi]$ for each $\Pi \in \widetilde{F}$ such that $\widetilde{\infty}(\Pi) < 0$, $\widetilde{o}(\Pi) > 0$ and $\widetilde{x}(\Pi) = 0$ for all places $\widetilde{x} \neq \widetilde{\infty}, \widetilde{o}$ of $\widetilde{F}$ ($\widetilde{\infty}$ is the unique place of $\widetilde{F}$ over ∞; if $\widetilde{X}$ is the smooth projective model of $\widetilde{F}$ over $\mathbb{F}_p$, $\text{Pic}^0_{\widetilde{X}/\mathbb{F}_p}(\mathbb{F}_p)$ is a finite group, so some positive power of the line bundle $\mathcal{O}_{\widetilde{X}}(\deg(\widetilde{o})\widetilde{\infty} - \deg(\widetilde{\infty})\widetilde{o})$ is trivial and there exists always a $\Pi \in \widetilde{F}$ as before).

To each Weil (F, ∞, o)-pair $(\widetilde{F}, \Pi)$ of rank d over a finite extension of $\kappa(o)$ we can attach an (F, ∞, o)-type of rank d, $(\widetilde{F}', \widetilde{o}')$, in the following way. Let $\widetilde{F}'$ be the intersection of the intermediate fields

$$F \subset F[\Pi^m] \subset \widetilde{F}$$

($m \in \mathbb{Z}$, $m \geq 1$) and let $\tilde{o}'$ be the place of $\tilde{F}'$ induced by $\tilde{o}$. The pair $(\tilde{F}', \tilde{o}')$ obviously satisfies the above conditions (a) and (b). Let $\Pi' \in \tilde{F}'$ with $\tilde{\infty}'(\Pi') < 0$, $\tilde{o}'(\Pi') > 0$ and $\tilde{x}'(\Pi') = 0$ for all places $\tilde{x}' \neq \tilde{\infty}'$, $\tilde{o}'$ of $\tilde{F}'$ ($\tilde{\infty}'$ is the unique place of $\tilde{F}'$ over ∞, i.e. the place induced by $\tilde{\infty}$). There exists some integer $m \geq 1$ such that $\tilde{F}' = F[\Pi^m]$ and obviously $\tilde{\infty}'(\Pi^m) < 0$, $\tilde{o}'(\Pi^m) > 0$ and $\tilde{x}'(\Pi^m) = 0$ for all places $\tilde{x}' \neq \tilde{\infty}'$, $\tilde{o}'$ of $\tilde{F}'$. So we can find two integers n, $n' \geq 1$ such that

$$\tilde{x}'(\Pi'^{n'}/\Pi^{mn}) = 0$$

for all places $\tilde{x}'$ of $\tilde{F}'$. But this means that $\Pi'^{n'}/\Pi^{mn}$ is a non-zero constant in $\tilde{F}'$, i.e. a root of unity. Increasing n and n' if necessary we can assume that $\Pi'^{n'}/\Pi^{mn} = 1$. Then we have the inclusions

$$F \subset F[\Pi'^{n'}] \subset F[\Pi'] \subset \tilde{F}' \subset F[\Pi^{mn}] \subset \tilde{F}.$$

Therefore $F[\Pi'] = \tilde{F}'$ and $(\tilde{F}', \tilde{o}')$ is an (F, ∞, o)-type of rank d.

To each Drinfeld A-module of rank d over $\overline{\kappa(o)}$ with characteristic $A \longrightarrow \kappa(o) \hookrightarrow \overline{\kappa(o)}$, (E, φ), we can attach an (F, ∞, o)-type of rank d in the following way. There exists a finite extension k of $\kappa(o)$ contained in $\overline{\kappa(o)}$ such that (E, φ) is already defined over k. Let $(F_{(E,\varphi)}, \Pi_{(E,\varphi)})$ be the corresponding Weil (F, ∞, o)-pair of rank d over k. The (F, ∞, o)-type of rank d attached to $(F_{(E,\varphi)}, \Pi_{(E,\varphi)})$ as above, $(\tilde{F}, \tilde{o})$, is clearly independent of the choice of k. Moreover the isomorphism class of $(\tilde{F}, \tilde{o})$ depends only on the isogeny class of (E, φ) (any isogeny between Drinfeld A-modules over $\overline{\kappa(o)}$ is defined over some finite extension of $\kappa(o)$ contained in $\overline{\kappa(o)}$).

COROLLARY (2.2.3). — *The map from the set of isogeny classes of Drinfeld A-modules of rank d over $\overline{\kappa(o)}$ with characteristic $A \longrightarrow \kappa(o) \hookrightarrow \overline{\kappa(o)}$ to the set of isomorphism classes of (F, ∞, o)-types of rank d that we have just constructed is a bijection. Moreover, if $(\tilde{F}, \tilde{o})$ is the (F, ∞, o)-type of rank d corresponding to a Drinfeld A-module (E, φ) of rank d over $\kappa(o)$ with characteristic $A \longrightarrow \kappa(o) \hookrightarrow \overline{\kappa(o)}$, $F \otimes_A \mathrm{End}(E, \varphi)$ is the unique (up to isomorphisms) central division algebra over $\tilde{F}$, whose non-zero invariants are $-[\tilde{F} : F]/d$ at $\tilde{\infty}$ and $[\tilde{F} : F]/d$ at $\tilde{o}$.* $\square$

(2.3) Tate modules of a Drinfeld module

Let k be any field extension of $\kappa(o)$ contained in $\overline{\kappa(o)}$ and let (E, φ) be a Drinfeld A-module of rank d over k with characteristic $A \longrightarrow \kappa(o) \hookrightarrow k$. For each place $x \neq \infty$, o, let $\mathcal{P}_x \subset A$ be the maximal ideal defining $\{x\}$ in

$X - \{\infty\} = \mathrm{Spec}(A)$. For each integer $n \geq 0$, the $\mathcal{O}_x$-module of $\mathcal{P}_x^n$-division points $(\mathcal{O}_x = A_{\mathcal{P}_x})$

$$E_{\mathcal{P}_x^n}(\kappa(o)) \subset E(\kappa(o))$$

is killed by ϖ_x^n, is free of rank d over $\mathcal{O}_x/(\varpi_x^n)$ and is endowed with an action of the Galois group

$$\Gamma = \mathrm{Gal}(\overline{\kappa(o)}/k)$$

(see (1.3.2)). If $n' \geq n$ is another integer, we have the inclusion

$$E_{\mathcal{P}_x^n}(\overline{\kappa(o)}) \subset E_{\mathcal{P}_x^{n'}}(\overline{\kappa(o)}) \subset E(\overline{\kappa(o)})$$

which is compatible with the module structures $(\mathcal{O}_x/(\varpi_x^{n'}) \longrightarrow \mathcal{O}_x/(\varpi_x^n))$ and the actions of Γ. So

$$E_{\mathcal{P}_x^\infty}(\overline{\kappa(o)}) = \varinjlim_n E_{\mathcal{P}_x^n}(\overline{\kappa(o)}) \subset E(\overline{\kappa(o)})$$

is a divisible $\mathcal{O}_x$-module which is non-canonically isomorphic to $(F_x/\mathcal{O}_x)^d$. Moreover it is endowed with a continuous action of Γ.

DEFINITION (2.3.1) (Drinfeld). — *For each place $x \neq \infty$, o of F, the Tate module of (E, φ) at x is the free $\mathcal{O}_x$-module of rank d*

$$T_x(E, \varphi) = \mathrm{Hom}_{\mathcal{O}_x}(F_x/\mathcal{O}_x, E_{\mathcal{P}_x^\infty}(\overline{\kappa(o)}))$$

endowed with its continuous action of Γ.

We have a natural homomorphism of $\mathcal{O}_x$-algebras:

$$(2.3.2) \qquad \mathcal{O}_x \otimes_A \mathrm{End}(E, \varphi) \longrightarrow \mathrm{End}_{\mathcal{O}_x}(T_x(E, \varphi)).$$

THEOREM (2.3.3) (Drinfeld). — (i) *The homomorphism (2.3.2) is injective and its kernel is torsion free.*

(ii) *If k is finite, the image of (2.3.2) is exactly the $\mathcal{O}_x$-submodule of Γ-invariants in $\mathrm{End}_{\mathcal{O}_x}(T_x(E, \varphi))$.*

Proof : Let $a \in A$ be such that $x(a) > 0$ and $y(a) = 0$ for each $y \in |X| - \{\infty, x\}$ (there always exists such an a because $\mathrm{Pic}_X^0(\mathbb{F}_p)$ is finite and hence is a torsional abelian group). For each integer $n \geq 0$, we have a natural map of free $(A/(a^n))$-modules of finite length

$$A/(a^n) \otimes_A \mathrm{End}(E, \varphi) \longrightarrow \mathrm{End}_{A/(a^n)}(E_{a^n}(\overline{\kappa(o)}))$$

To prove part (i), it is enough to prove that this map is injective for all n.

Of course we can assume that $E = \mathbf{G}_{a,k}$. Then part (i) is an obvious consequence of (2.1.1) (see the proof of (2.1.4)).

From now on k is finite. Obviously the image of (2.3.2) is contained in the $\mathcal{O}_x$-submodule

$$\operatorname{End}_{\mathcal{O}_x[\Gamma]}(T_x(E, \varphi)) \subset \operatorname{End}_{\mathcal{O}_x}(T_x(E, \varphi))$$

of Γ-invariants. We can tensor (2.3.2) by F_x over $\mathcal{O}_x$. As F_x is flat over $\mathcal{O}_x$, we get a homomorphism of F_x-algebras

$$F_x \otimes_A \operatorname{End}(E, \varphi) \longrightarrow \operatorname{End}_{F_x[\Gamma]}(F_x \otimes_{\mathcal{O}_x} T_x(E, \varphi)).$$

Thanks to part (i) which is already proved, this last map is injective and it is surjective if and only if (2.3.2) is surjective. Now Γ is topologically generated by the arithmetic Frobenius element. But this element acts on $T_x(E, \varphi)$ as the image of

$$\Pi_x = 1 \otimes \Pi \in \mathcal{O}_x \otimes_A \operatorname{End}(E, \varphi)$$

by (2.3.2). So, if we endow $F_x \otimes_{\mathcal{O}_x} T_x(E, \varphi)$ with the $\widetilde{F}_x$-vector space structure induced by (2.3.2) where

$$\widetilde{F}_x = F_x[\Pi_x] = F_x \otimes_F \widetilde{F} \subset F_x \otimes_F D = D_x$$

($\widetilde{F} = F[\Pi] \subset D = F \otimes_A \operatorname{End}(E, \varphi)$, see (2.2.2)), we have

$$\operatorname{End}_{F_x[\Gamma]}(F_x \otimes_{\mathcal{O}_x} T_x(E, \varphi)) = \operatorname{End}_{\widetilde{F}_x}(F_x \otimes_{\mathcal{O}_x} T_x(E, \varphi)).$$

The F_x-algebra $\widetilde{F}_x$ is semi-simple and in fact a product of field extensions

$$\widetilde{F}_x = \prod_{\widetilde{x}|x} \widetilde{F}_{\widetilde{x}}$$

($\widetilde{x}|x$ is a place of $\widetilde{F}$ dividing x). Similarly we have

$$D_x = \prod_{\widetilde{x}|x} D_{\widetilde{x}}$$

where $D_{\widetilde{x}} = \widetilde{F}_{\widetilde{x}} \otimes_{\widetilde{F}} D$,

$$F_x \otimes_{\mathcal{O}_x} T_x(E, \varphi) = \bigoplus_{\widetilde{x}|x} \widetilde{F}_{\widetilde{x}} \otimes_{\widetilde{F}_x} (F_x \otimes_{\mathcal{O}_x} T_x(E, \varphi))$$

and the above injective F_x-linear map is the product over $\widetilde{x}|x$ of the injective $\widetilde{F}_{\widetilde{x}}$-linear maps

$$D_{\widetilde{x}} \hookrightarrow \operatorname{End}_{\widetilde{F}_{\widetilde{x}}}(\widetilde{F}_{\widetilde{x}} \otimes_{\widetilde{F}_x} (F_x \otimes_{\mathcal{O}_x} T_x(E, \varphi))).$$

Let $\delta_{\tilde{x}}$ be the dimension of the $\widetilde{F}_{\tilde{x}}$-vector space $\widetilde{F}_{\tilde{x}} \otimes_{\widetilde{F}_x} (F_x \otimes_{\mathcal{O}_x} T_x(E, \varphi))$. We obviously have

$$(\delta_{\tilde{x}})^2 \geq \dim_{\widetilde{F}_{\tilde{x}}}(D_{\tilde{x}}),$$

for each $\tilde{x}|x$,

$$\sum_{\tilde{x}|x} \delta_{\tilde{x}}[\widetilde{F}_{\tilde{x}} : F_x] = \dim_{F_x}(F_x \otimes_{\mathcal{O}_x} T_x(E, \varphi)) = d$$

and

$$\sum_{\tilde{x}|x} [\widetilde{F}_{\tilde{x}} : F_x] = [\widetilde{F} : F].$$

But, thanks to (2.2.2),

$$\dim_{\widetilde{F}_{\tilde{x}}}(D_{\tilde{x}}) = (d/[\widetilde{F} : F])^2$$

so we get

$$\delta_{\tilde{x}} = d/[\widetilde{F} : F]$$

for each $\tilde{x}|x$ and part (ii) of the theorem is proved. $\square$

COROLLARY (2.3.4). — *Let (E, φ) be a Drinfeld A-module of rank d over $\overline{\kappa(o)}$ with characteristic map $A \longrightarrow \kappa(o) \hookrightarrow \overline{\kappa(o)}$. Let $(\widetilde{F}, \tilde{o})$ be the (F, ∞, o)-type of its isogeny class. Then, for each $x \in |X| - \{\infty, o\}$, the map (2.3.2) and the inclusion $\widetilde{F} \hookrightarrow F \otimes_A \mathrm{End}(E, \varphi)$ induce a structure of an $\widetilde{F}_x(= F_x \otimes_F \widetilde{F})$-module on $F_x \otimes_{\mathcal{O}_x} T_x(E, \varphi)$ and (2.3.2) induces an isomorphism of $\widetilde{F}_x$-algebras*

$$F_x \otimes_A \mathrm{End}(E, \varphi) \xrightarrow{\sim} \mathrm{End}_{\widetilde{F}_x}(F_x \otimes_{\mathcal{O}_x} T_x(E, \varphi)).$$

$\square$

REMARK (2.3.5). — With the notations of (2.3.4), we have

$$\widetilde{F}_x = \prod_{\tilde{x}|x} \widetilde{F}_{\tilde{x}}$$

and it follows from the proof of (2.3.3) that

$$\dim_{\widetilde{F}_{\tilde{x}}}(\widetilde{F}_{\tilde{x}} \otimes_{\widetilde{F}_x} (F_x \otimes_{\mathcal{O}_x} T_x(E, \varphi))) = d/[\widetilde{F} : F]$$

for each $\tilde{x}|x$. $\square$

Later it will be more convenient for us to replace $T_x(E, \varphi)$ by its $\mathcal{O}_x$-linear dual

$$(2.3.6) \qquad H_x(E, \varphi) = \mathrm{Hom}_{\mathcal{O}_x}(T_x(E, \varphi), \mathcal{O}_x)$$

endowed with the contragredient continuous action of Γ. This free $\mathcal{O}_x$-module of rank d will be called the x-**adic cohomology group** of (E, φ). Similar results hold for $H_x(E, \varphi)$.

(2.4) Dieudonné modules

Let k be a field extension of $\kappa(o)$ contained in $\overline{\kappa(o)}$. We will denote by $\mathcal{O}_k$ the completion of the unramified extension of $\mathcal{O}_o$ with residue field k and by F_k its field of fractions. We will denote by

$$\sigma_o \in \mathrm{Gal}(F_k/F_o)$$

the canonical lifting of the arithmetic Frobenius automorphism of k over $\kappa(o)$ ($p^{\deg(o)}$-th power on k). We fix a uniformizer ϖ_o. We have a canonical isomorphism

$$k[[\varpi_o]] \xrightarrow{\sim} \mathcal{O}_k$$

and

$$\sigma_o\left(\sum_{n=0}^{\infty} \alpha_n \varpi_o^n\right) = \sum_{n=0}^{\infty} \alpha_n^{p^{\deg(o)}} \varpi_o^n$$

(k is perfect).

DEFINITION (2.4.1). — *A Dieudonné $\mathcal{O}_o$-module over k is a free $\mathcal{O}_k$-module of finite rank M endowed with an injective σ_o-linear map $f : M \longrightarrow M$ such that the cokernel of f is of finite length over $\mathcal{O}_k$, i.e. of finite dimension as a k-vector space; the rank of (M, f) is the rank of M as an $\mathcal{O}_k$-module.*

A Dieudonné F_o-module over k is a finite dimensional F_k-vector space N endowed with a bijective σ_o-linear map $f : N \longrightarrow N$; the rank of (N, f) is the dimension of N as an F_k-vector space.

If (M, f) is a Dieudonné $\mathcal{O}_o$-module over k $(F_o \otimes_{\mathcal{O}_o} M, \ F_o \otimes_{\mathcal{O}_o} f)$ is a Dieudonné F_o-module over k of the same rank.

Let r, s be two integers with $r > 0$ and $(r, s) = 1$. Let us set

$$(2.4.2) \qquad N_{r,s} = (F_k)^r$$

and

$$(2.4.3) \qquad f_{r,s}(e_i) = \begin{cases} e_{i+1} & \text{if } i = 1, \ldots, r - 1, \\ \varpi_o^s e_1 & \text{if } i = r, \end{cases}$$

where $(e_1, \ldots, e_r)$ is the canonical basis of $N_{r,s}$ and where we extend $f_{r,s}$ to $N_{r,s}$ by σ_o-linearity; $(N_{r,s}, f_{r,s})$ is obviously a Dieudonné F_o-module over k of rank r. If $s \geq 0$, we set

$$(2.4.4) \qquad M_{r,s} = (\mathcal{O}_k)^r \subset (F_k)^r = N_{r,s}$$

and we have $f_{r,s}(M_{r,s}) \subset M_{r,s}$; $(M_{r,s}, f_{r,s})$ is obviously a Dieudonné $\mathcal{O}_o$-module over k of rank r.

A morphism of Dieudonné $\mathcal{O}_o$-modules (resp. F_o-modules) over k is a linear map between the underlying $\mathcal{O}_k$-modules (resp. F_k-vector spaces) which commutes with the σ_o-linear maps f.

THEOREM (2.4.5). — (i) *The category of Dieudonné F_o-modules over $\overline{\kappa(o)}$ is F_o-linear and semi-simple. Its simple objects (up to isomorphisms) are the above $(N_{r,s}, f_{r,s})$'s ($r, s \in \mathbf{Z}$, $r \geq 1$, $(r,s) = 1$).*

(ii) *The F_o-algebra $\Delta_{r,s}$ of endomorphisms of the simple object $(N_{r,s}, f_{r,s})$ is "the" central division algebra over F_o with invariant $-s/r \in \mathbf{Q}/\mathbf{Z}$.*

□

A proof of (2.4.5) is given in (B.1).

If k is arbitrary the category of Dieudonné F_o-modules over k is obviously F_o-linear, artinian and noetherian. If k' is a field extension of k contained in $\overline{\kappa(o)}$ we have an obvious base change functor from the category of $\mathcal{O}_o$-modules (resp. F_o-modules) over k to the category of those over k'; we will denote it by $k' \otimes_k (-)$.

If (N, f) is an indecomposable Dieudonné F_o-module over k there exist a unique pair of integers (r, s) with $r \geq 1$ and $(r, s) = 1$ and a unique integer $t \geq 1$ such that $\overline{\kappa(o)} \otimes_k (N, f)$ is isomorphic to $(N_{r,s}, f_{r,s})^t$. We will say that the rational number s/r is the **slope** of (N, f). If (N, f) is an arbitrary Dieudonné F_o-module over k, its **slopes** are the slopes of its indecomposable constituents. If $\lambda \in \mathbf{Q}$ is a slope of (N, f), there is a canonical decomposition

$$(N, f) = (N', f') \oplus (N'', f'')$$

such that all the slopes of (N', f') are equal to λ and all the slopes of (N'', f'') are distinct from λ; the **multiplicity of the slope** λ of (N, f) is by definition the dimension of N' over F_k. If (M, f) is a Dieudonné $\mathcal{O}_o$-module over k its slopes (resp. the multiplicity of a given slope) are (resp. is) the slopes (resp. the multiplicity of the given slope) of $(F_o \otimes_{\mathcal{O}_o} M, F_o \otimes_{\mathcal{O}_o} f)$.

PROPOSITION (2.4.6). — *Let (M, f) be a Dieudonné $\mathcal{O}_o$-module over k. Then there exists a unique decomposition*

$$(M, f) = (M^{et}, f^{et}) \oplus (M^c, f^c)$$

where (M^{et}, f^{et}) *and* (M^c, f^c) *are Dieudonné* $\mathcal{O}_o$-*modules over* k *such that*

$$f^{et}(M^{et}) = M^{et}$$

and

$$(f^c)^n(M^c) \subset \varpi_o M^c$$

for all large enough positive integer n. *Moreover* $(F_o \otimes_{\mathcal{O}_o} M^{et}, \; F_o \otimes_{\mathcal{O}_o} f^{et})$ *is the maximal direct factor of* $(F_o \otimes_{\mathcal{O}_o} M, \; F_o \otimes_{\mathcal{O}_o} f)$ *of slope* 0 *and all the slopes of* $(F_o \otimes_{\mathcal{O}_o} M^c, \; F_o \otimes_{\mathcal{O}_o} f^c)$ *are positive.* $\qquad\square$

A proof of (2.4.6) is given in (B.2).

REMARK (2.4.7). — In particular, if (N, f) is a fixed Dieudonné F_o-module over k and $M \subset N$ is an $\mathcal{O}_k$-lattice such that $f(M) \subset M$, then $F_o \otimes_{\mathcal{O}_o} M^{et}$ and $F_o \otimes_{\mathcal{O}_o} M^c$ are independent of M. $\qquad\square$

DEFINITION (2.4.8). — *A* ϖ_o-*divisible* k-*scheme in* $\mathcal{O}_o$-*modules is an inductive system of finite* k-*schemes in* $\mathcal{O}_o$-*modules*

$$G = (G_1 \xrightarrow{i_1} G_2 \xrightarrow{i_2} G_3 \xrightarrow{i_3} \cdots) = \varinjlim_n G_n$$

such that, for each integer $n \geq 1$, *the following properties hold* :

(i) *as a* k-*scheme in* $\kappa(o)$-*vector spaces* $(\kappa(o) \subset \mathcal{O}_o)$, G_n *can be embedded ino* $\mathbb{G}_{a,k}^N$ *for some integer* $N \geq 0$ ($\mathbb{G}_{a,k}$ *has an obvious structure of a* k-*scheme in* $\kappa(o)$-*vector spaces*),

(ii) *as a* k-*scheme in* $\kappa(o)$-*vector spaces, the dimension of* G_n *is nd for some integer* $d \geq 0$ *which is independent of* n (*the order of* G_n *is* $p^{\deg(o)nd}$),

(iii) *the sequence of* k-*schemes in* $\mathcal{O}_o$-*modules*

$$o \longrightarrow G_n \xrightarrow{i_n} G_{n+1} \xrightarrow{\varpi_o^n} G_{n+1}$$

is exact.

The integer d *of* (ii) *is called the rank of* G.

A morphism between two ϖ_o-divisible k-schemes in $\mathcal{O}_o$-modules is a morphism between the corresponding inductive systems of finite k-schemes in $\mathcal{O}_o$-modules.

The category of Dieudonné $\mathcal{O}_o$-modules over k and the category of ϖ_o-divisible k-schemes in $\mathcal{O}_o$-modules are both $\mathcal{O}_o$-linear and exact. We will now construct an equivalence between these two $\mathcal{O}_o$-linear and exact categories.

Let us begin by reviewing a very particular case of Dieudonné's theory. On the one hand, if M is a finite dimensional k-vector space and $M \xrightarrow{f} M$ is a $p^{\deg(o)}$-linear map, we have a functor $G_k(M, f)$ from the category of (commutative, unitary and associative) k-algebras to the category of $\kappa(o)$-vector spaces with

$$(2.4.9) \quad G_k(M, f)(R) = \{g \in \mathrm{Hom}_k(M, R) | g(f(m)) = g(m)^{p^{\deg(o)}}, \forall m \in M\}.$$

It is easy to see that $G_k(M, f)$ is in fact a finite k-scheme in $\kappa(o)$-vector spaces of dimension $\dim_k(M)$ which can be embedded in $\mathbf{G}_{a,k}^N$ for some integer $N \geq 0$ (take a basis of M over k). On the other hand, if G is a finite k-scheme in $\kappa(o)$-vector spaces which can be embedded in $\mathbf{G}_{a,k}^N$ for some integer $N \geq 0$, the group of homomorphisms of k-schemes in $\kappa(o)$-vector spaces

$$(2.4.10) \qquad\qquad M_k(G) = \mathrm{Hom}(G, \mathbf{G}_{a,k})$$

is in fact a finite dimensional k-vector space; moreover, the Frobenius endomorphism of $\mathbf{G}_{a,k} = \mathrm{Spec}(k[t])$ with respect to $\kappa(o)$ $(t \longmapsto t^{p^{\deg(o)}})$ induces a $p^{\deg(o)}$-linear map $f_k(G) : M_k(G) \longrightarrow M_k(G)$.

PROPOSITION (2.4.11). — *The contravariant functors G_k and (M_k, f_k) between the category of pairs (M, f), with M a finite dimensional k-vector space and $M \overset{f}{\longrightarrow} M$ a $p^{\deg(o)}$-linear map, and the category of finite k-schemes in $\kappa(o)$-vector spaces G which can be embedded in $\mathbf{G}_{a,k}^N$ for some integer $N \geq 0$ are exact and quasi-inverse one of the other.*

Moreover, if $G = G_k(M, f)$, the dimension of G (over $\kappa(o)$) is equal to $\dim_k(M)$ and G is connected (resp. étale) if and only if f is nilpotent (resp. an isomorphism). $\qquad\qquad\square$

A proof of (2.4.11) is given in (B.3).

Then we can set

$$(2.4.12) \qquad G_k(M, f) = \varinjlim_n G_k(M/\varpi_o^n M, f \bmod \varpi_o^n M)$$

for any Dieudonné $\mathcal{O}_o$-module (M, f) over k and

$$(2.4.13) \qquad (M_k(G), f_k(G)) = \varprojlim_n (M_k(G_n), f_k(G_n))$$

for any ϖ_o-divisible k-scheme in $\mathcal{O}_o$-modules $G = \varinjlim_n G_n$.

COROLLARY (2.4.14). — *The contravariant functors G_k and (M_k, f_k) between the category of Dieudonné $\mathcal{O}_o$-modules over k and the category of ϖ_o-divisible k-schemes in $\mathcal{O}_o$-modules are $\mathcal{O}_o$-linear, exact and quasi-inverse one of the other.*

Moreover, if $G = G_k(M, f)$, the rank of G is equal to the rank of (M, f) and the canonical decomposition

$$(M, f) = (M^{et}, f^{et}) \oplus (M^c, f^c)$$

(see (2.4.6)) induces the canonical decomposition

$$G = G^c \times_k G^{et}$$

of G into its connected component $G^c = G_k(M^c, f^c)$ and its maximal étale quotient $G^{et} = G_k(M^{et}, f^{et})$. □

(2.5) Dieudonné module of a Drinfeld module

Let k be a field extension of $\kappa(o)$ contained in $\overline{\kappa(o)}$ and let (E, φ) be a Drinfeld A-module of rank d over k with characteristic $A \longrightarrow \kappa(o) \hookrightarrow k$. Let $\mathcal{P}_o \subset A$ be the maximal ideal defining $\{o\}$ in $X - \{\infty\} = \mathrm{Spec}(A)$. For each integer $n \geq 0$, the k-scheme in $(\mathcal{O}_o/(\varpi_o^n))$-modules of $\mathcal{P}_o^n$-division points $(\mathcal{O}_o = A_{\mathcal{P}_o})$

$$E_{\mathcal{P}_o^n} \subset E$$

has the following properties:

LEMMA (2.5.1). — (i) $E_{\mathcal{P}_o^n}$ is finite of dimension nd over $\kappa(o)$.

(ii) *There exists a unique integer $h > 0$ such that $h \deg(o)o(a)$ is the height of $\varphi(a)$ for each $a \in A - \{0\}$ (see (2.1)) and the connected component $E_{\mathcal{P}_o^n}^c$ of $E_{\mathcal{P}_o^n}$ has dimension nh over $\kappa(o)$.*

(iii) $E_{\mathcal{P}_o^n}$ *can be embedded in $\mathbb{G}_{a,k}^N$ for some integer $N \geq 0$.*

Proof: As $E_{\mathcal{P}_o^n} \subset E$, part (iii) is obvious.

Let $h(a)$ be the height of $\varphi(a)$ for $a \in A - \{0\}$. We have

$$h(a_1 a_2) = h(a_1) + h(a_2)$$

for all a_1, $a_2 \in A - \{0\}$ and $h(a) = 0$ if and only if $\partial(\varphi(a)) \neq 0$, i.e. $o(a) = 0$ for all $a \in A - \{0\}$. Therefore there exists an integer $N > 0$ such that $h(a) = No(a)$ for all $a \in A - \{0\}$.

Now, if $a \in A - \{0\}$, the k-scheme in $(A/(a))$-modules of a-division points $E_a = \ker(\varphi(a) : E \longrightarrow E)$ is finite of order $p^{-d \deg(\infty)\infty(a)}$ and its connected component E_a^c is of order $p^{h(a)}$.

Using the same argument as in (1.3.2) we get part (i) of the lemma and that the dimension of $E_{\mathcal{P}_o^n}$ over $\mathbb{F}_p$ is nN. But $E_{\mathcal{P}_o^n}$ is a k-scheme in $\kappa(o)$-vector spaces, so there exists an integer $h > 0$ such that $N = h \deg(o)$ and the lemma is proved. □

We have an inductive system

$$E_{\mathcal{P}_o^\infty} = (E_{\mathcal{P}_o} \hookrightarrow E_{\mathcal{P}_o^2} \hookrightarrow E_{\mathcal{P}_o^3} \hookrightarrow \cdots)$$

of finite k-schemes in $\mathcal{O}_o$-modules. Thanks to (2.5.1), it is in fact a ϖ_o-divisible k-scheme in $\mathcal{O}_o$-module of rank d. We will denote by

(2.5.2) $(H_o(E, \varphi), f_o(E, \varphi)) = (M_k(E_{\mathcal{P}_o^\infty}), f_k(E_{\mathcal{P}_o^\infty}))$

its Dieudonné $\mathcal{O}_o$-module over k and by

(2.5.3) $(H_o(E, \varphi), f_o(E, \varphi)) =$
$$(H_o^{et}(E, \varphi), f_o^{et}(E, \varphi)) \oplus (H_o^c(E, \varphi), f_o^c(E, \varphi))$$

the canonical decomposition of this Dieudonné $\mathcal{O}_o$-module over k (see (2.4.6)). Thanks to (2.5.1), the rank over $\mathcal{O}_k$ of $H_o(E, \varphi)$ (resp. $H_o^c(E, \varphi)$) is d (resp. h).

LEMMA (2.5.4). — *We have the inclusions*

$$\varpi_o H_o(E, \varphi) \subset f_o(E, \varphi)(H_o(E, \varphi)) \subset H_o(E, \varphi)$$

and the equality

$$\dim_k(H_o(E, \varphi)/f_o(E, \varphi)(H_o(E, \varphi))) = 1.$$

Proof : Let $a \in A - \{0\}$ be such that $o(a) = 1$. Then the $\mathcal{P}_o$-primary component of the k-scheme in $(A/(a))$-modules E_a is exactly $E_{\mathcal{P}_o}$ and it follows from (1.3.2) that the connected component of the finite k-group scheme E_a is the same as the connected component $E_{\mathcal{P}_o}^c$ of $E_{\mathcal{P}_o}$.

Now, as $h(a) = h \deg(o) \geq \deg(o)$, the isogeny $\varphi(a)$ of (E, φ) factors through the Frobenius isogeny : if we identify E with $\mathbb{G}_{a,k}$, there exists $u \in k[\tau]$ such that
$$\varphi(a) = u\tau^{p^{\deg(o)}}.$$
But this implies that the kernel of the Frobenius isogeny is contained in E_a and even in $E_{\mathcal{P}_o}^c$ as it is connected. This proves the first part of the lemma. Moreover

$$(H_o(E, \varphi)/f_o(E, \varphi)(H_o(E, \varphi)), 0)$$

is obviously the image by the functor $(M_k(-), f_k(-))$ of the kernel of the Frobenius isogeny and the order of this kernel is $p^{\deg(o)}$. So the lemma is proved. $\square$

We have a natural homomorphism of $\mathcal{O}_o$-algebras

$$(2.5.5) \qquad \mathcal{O}_o \otimes_A \operatorname{End}(E, \varphi)^{opp} \longrightarrow \operatorname{End}(H_o(E, \varphi), f_o(E, \varphi)).$$

THEOREM (2.5.6) (Drinfeld). — (i) *The homomorphism* (2.5.5) *is injective and its cokernel is torsion free.*

(ii) *If k is finite,* (2.5.5) *is an isomorphism.*

To prove this theorem, we will need the following lemma:

LEMMA (2.5.7). — *Let G be a ϖ_o-divisible k-scheme in $\mathcal{O}_o$-modules and let (M, f) be its Dieudonné $\mathcal{O}_o$-module over k. Let d be the rank of G and M. Then*

$$\operatorname{End}(G)^{opp} = \operatorname{End}(M, f)$$

is a free $\mathcal{O}_o$-module of finite rank $\leq d^2$. For each integer $n \geq 0$, the natural homomorphism of $(\mathcal{O}_o/(\varpi_o^n))$-modules from

$$(\mathcal{O}_o/(\varpi_o^n)) \otimes_{\mathcal{O}_o} \operatorname{End}(M, f)$$

to

$$\operatorname{End}(G_n)^{opp} = \operatorname{End}(M/\varpi_o^n M, f \bmod \varpi_o^n M)$$

is injective.

Proof: We have an exact sequence of $\mathcal{O}_o$-modules

$$0 \longrightarrow \operatorname{End}(M, f) \longrightarrow \operatorname{End}_{\mathcal{O}_k}(M) \xrightarrow{\ u \mapsto uf - fu\ } \operatorname{Hom}_{\mathcal{O}_k}(M, M^{(\sigma_o)})$$

where $\operatorname{End}_{\mathcal{O}_k}(M)$ and $\operatorname{Hom}_{\mathcal{O}_k}(M, M^{(\sigma_o)})$ are torsion free as $\mathcal{O}_k$-modules. So $\operatorname{End}(M, f)$ and $\operatorname{End}_{\mathcal{O}_k}(M)/\operatorname{End}(M, f)$ are torsion free as $\mathcal{O}_o$-modules. In particular, for each integer $n \geq 0$, the natural map

$$(\mathcal{O}_o/(\varpi_o^n)) \otimes_{\mathcal{O}_o} \operatorname{End}(M, f) \longrightarrow \operatorname{End}_{\mathcal{O}_k/\varpi_o^n \mathcal{O}_k}(M/\varpi_o^n M)$$

is injective and, consequently, the same is true for the natural map

$$(\mathcal{O}_o/(\varpi_o^n)) \otimes_{\mathcal{O}_o} \operatorname{End}(M, f) \longrightarrow \operatorname{End}(M/\varpi_o^n M, \ f \bmod \varpi_o^n M).$$

Now let $(u_1, \ldots, u_s)$ be a sequence of elements of $\operatorname{End}(M, f)$ which is free over $\mathcal{O}_o$. We will check that $(u_1, \ldots, u_s)$ is also free over $\mathcal{O}_k$ as a sequence of elements of $\operatorname{End}_{\mathcal{O}_k}(M)$. Otherwise, there exist $I \subset \{1, \ldots, s\}$, $I \neq \emptyset$, and a family $(a_i)_{i \in I}$ of non-zero elements of $\mathcal{O}_k$ such that

$$\sum_{i \in I} a_i u_i = 0$$

in $\text{End}_{\mathcal{O}_k}(M)$. Moreover we can assume that the I we choose is minimal among all possible such ones. Let $i_o \in I$ be such that the valuation of a_{i_o} is minimal among the valuations of the a_i's. Dividing the a_i's by a_{i_o} we can assume that $a_{i_o} = 1$. Then

$$0 = f\Big(\sum_{i\in I} a_i u_i\Big) - \Big(\sum_{i\in I} a_i u_i\Big) f = \Big(\sum_{i\in I-\{i_o\}} (a_i^{(\sigma_o)} - a_i) u_i\Big) f.$$

But $M \xrightarrow{f} M$ has a cokernel of finite length over $\mathcal{O}_k$ and M is torsion free over $\mathcal{O}_k$, so it follows that

$$\sum_{i\in I-\{i_o\}} (a_i^{(\sigma_o)} - a_i) u_i = 0.$$

As I is minimal, $a_i^{(\sigma_o)} = a_i$, i.e. $a_i \in \mathcal{O}_o$, for each $i \in I$, and we have a contradiction.

Finally, let $(u_1, \ldots, u_s)$ be a sequence of elements of $\text{End}(M, f)$ which is free over $\mathcal{O}_o$ and of maximal length for this property (s is bounded by the rank over $\mathcal{O}_k$ of $\text{End}_{\mathcal{O}_k}(M)$). Then the $\mathcal{O}_o$-module $\text{End}(M, f)/(\sum_{i=1}^{s} \mathcal{O}_o u_i)$ is a torsion one. As

$$\Big(\sum_{i=1}^{s} \mathcal{O}_k u_i\Big) \cap \text{End}(M, f) = \sum_{i=1}^{s} \mathcal{O}_o u_i,$$

it is contained in the torsion part of the finitely generated $\mathcal{O}_k$-module $\text{End}_{\mathcal{O}_k}(M)/\big(\sum_{i=1}^{s} \mathcal{O}_o \varpi_o^{-N} u_i\big)$ and $\text{End}(M, f)$ is a finitely generated $\mathcal{O}_o$-module. This finishes the proof of the lemma. $\qquad\square$

Proof of (2.5.6) : Firstly let us check part (i). Let $a \in A$ be such that $o(a) > 0$ and $x(a) = 0$ for each $x \in |X| - \{\infty, o\}$. For each integer $n \geq 0$, we have a natural map

$$A/(a^n) \otimes_A \text{End}(E, \varphi)^{opp} \longrightarrow A/(a^n) \otimes_A \text{End}(H_o(E, \varphi), f_o(E, \varphi))$$

and it is enough to check that it is injective. But, thanks to (2.5.7), we have an injective map

$$A/(a^n) \otimes_A \text{End}(H_o(E, \varphi), f_o(E, \varphi)) \hookrightarrow \text{End}(E_{a^n}).$$

So it is enough to check that the composite map

$$A/(a^n) \otimes_A \text{End}(E, \varphi) \longrightarrow \text{End}(E_{a^n})$$

is injective. But this follows from (2.1.1) (see the proof of (2.1.4) for more details).

Secondly let us check part (ii). We are assuming that k is finite; let r be its degree over $\kappa(o)$. The only thing we need to prove is that the injective map of F_o-algebras

$$F_o \otimes_A \mathrm{End}(E,\varphi)^{opp} \longrightarrow F_o \otimes_{\mathcal{O}_o} \mathrm{End}(H_o(E,\varphi), f_o(E,\varphi))$$

is bijective, that is the source and the target have the same dimension. The center of $D = F \otimes_A \mathrm{End}(E,\varphi)$ is $\widetilde{F} = F[\Pi]$ where $(\widetilde{F}, \Pi)$ is the Weil (F, ∞, o)-pair of rank d over k associated to (E,φ) (see (2.2.2)). But the image of Π by the above map is clearly $f_o(E,\varphi)^r$. So, if we endow $F_o \otimes_{\mathcal{O}_o} H_o(E,\varphi)$ with the structure of an $\widetilde{F}_o(= F_o \otimes_F \widetilde{F})$-algebra induced by the above map, $F_o \otimes_{\mathcal{O}_o} f_o(E,\varphi)$ is $\widetilde{F}_o$-linear. In particular we can split $(F_o \otimes_{\mathcal{O}_o} H_o(E,\varphi), \ F_o \otimes_{\mathcal{O}_o} f_o(E,\varphi))$ according to the decomposition

$$\widetilde{F}_o = \prod_{\tilde{x}|o} \widetilde{F}_{\tilde{x}}$$

of the F_o-algebra $\widetilde{F}_o$ as a product of fields. For each place $\tilde{x}$ of $\widetilde{F}$ dividing o, let

$$((F_o \otimes_{\mathcal{O}_o} H_o(E,\varphi))_{\tilde{x}}, (F_o \otimes_{\mathcal{O}_o} f_o(E,\varphi))_{\tilde{x}})$$

be the corresponding piece. Then we have an embedding of $\widetilde{F}_{\tilde{x}}$-algebras

$$D_{\tilde{x}}^{opp} \hookrightarrow \mathrm{End}((F_o \otimes_{\mathcal{O}_o} H_o(E,\varphi))_{\tilde{x}}, (F_o \otimes_{\mathcal{O}_o} f_o(E,\varphi))_{\tilde{x}})$$

where

$$D_{\tilde{x}} = \widetilde{F}_{\tilde{x}} \otimes_F D$$

and it is enough to prove that this embedding is in fact an isomorphism.

Now $(F_o \otimes_{\mathcal{O}_o} H_o(E,\varphi))_{\tilde{x}}$ is an $(F_k \otimes_{F_o} \widetilde{F}_{\tilde{x}})$-module and $(F_o \otimes_{\mathcal{O}_o} f_o(E,\varphi))_{\tilde{x}}$ is $(\sigma_o \otimes_{F_o} \widetilde{F}_{\tilde{x}})$-linear. In general $F_k \otimes_{F_o} \widetilde{F}_{\tilde{x}}$ is not a field. Let $m_{\tilde{x}}$ and $d_{\tilde{x}}$ be the l.c.m. and the g.c.d. of $r = [k : \kappa(o)]$ and $s_{\tilde{x}} = [\kappa(\tilde{x}) : \kappa(o)]$. Then

$$k \otimes_{\kappa(o)} \kappa(\tilde{x}) = \prod_{i=1}^{d_{\tilde{x}}} k_{\tilde{x},i}$$

where $(k_{\tilde{x},1}, \ldots, k_{\tilde{x},d_{\tilde{x}}})$ is a set of representatives of the isomorphism classes of simultaneous extensions of k and $\kappa(\tilde{x})$ over $\kappa(o)$ and accordingly

$$F_k \otimes_{F_o} \widetilde{F}_{\tilde{x}} = \prod_{i=1}^{d_{\tilde{x}}} \widetilde{F}_{\tilde{x},i}.$$

Each $k_{\tilde{x},i}$ has degree $m_{\tilde{x}}$ over $\kappa(o)$ and the $k_{\tilde{x},i}$'s are isomorphic as extensions of $\kappa(o)$. For each $i \in \{1, \ldots, d_{\tilde{x}}\}$, $\widetilde{F}_{\tilde{x},i}$ is "the" unramified extension of $\widetilde{F}_{\tilde{x}}$ with residue field extension $k_{\tilde{x},i}/\kappa(\tilde{x})$ and the $\widetilde{F}_{\tilde{x},i}$'s are isomorphic as extensions of F_o. In fact $(-)^{p^{\deg(o)}} \otimes_{\kappa(o)} \kappa(\tilde{x})$ (resp. $\sigma_o \otimes_{\kappa(o)} \kappa(\tilde{x})$) induces an isomorphism $k_{\tilde{x},i} \xrightarrow{\sim} k_{\tilde{x},i+1}$ (resp. $\widetilde{F}_{\tilde{x},i} \xrightarrow{\sim} \widetilde{F}_{\tilde{x},i+1}$) as extensions of $\kappa(\tilde{x})$ (resp. $\widetilde{F}_{\tilde{x}}$) for each $i \in \{1, \ldots, d_{\tilde{x}}\}$ and the composite isomorphism

(resp.

$$k_{\tilde{x},1} \xrightarrow{\sim} k_{\tilde{x},2} \xrightarrow{\sim} \cdots \xrightarrow{\sim} k_{\tilde{x},d_{\tilde{x}}+1} = k_{\tilde{x},1}$$

$$\widetilde{F}_{\tilde{x},1} \xrightarrow{\sim} \widetilde{F}_{\tilde{x},2} \xrightarrow{\sim} \cdots \xrightarrow{\sim} \widetilde{F}_{\tilde{x},d_{\tilde{x}}+1} = \widetilde{F}_{\tilde{x},1})$$

is $(-)^{p^{s'_{\tilde{x}} s_{\tilde{x}} \deg(o)}}$ (resp. $\sigma_{\tilde{x}}^{s'_{\tilde{x}}}$) where $s'_{\tilde{x}}$ is an integer such that

$$s'_{\tilde{x}} s_{\tilde{x}} \equiv d_{\tilde{x}} \pmod{r}$$

and $\sigma_{\tilde{x}} \in \mathrm{Gal}(\widetilde{F}_{\tilde{x},1}/\widetilde{F}_{\tilde{x}})$ is the canonical lifting of the arithmetic Frobenius automorphism of $k_{\tilde{x},1}$ over $\kappa(\tilde{x})$ ($s'_{\tilde{x}}$ is well-defined up to a multiple of $r/d_{\tilde{x}}$ and $s_{\tilde{x}} r/d_{\tilde{x}} = m_{\tilde{x}}$ is the degree of $k_{\tilde{x},1}$ over $\kappa(o)$). It follows that we have a decomposition

$$(F_o \otimes_{\mathcal{O}_o} H_o(E,\varphi))_{\tilde{x}} = \bigoplus_{i=1}^{d_{\tilde{x}}} (F_o \otimes_{\mathcal{O}_o} H_o(E,\varphi))_{\tilde{x},i}$$

where $(F_o \otimes_{\mathcal{O}_o} H_o(E,\varphi))_{\tilde{x},i}$ is an $\widetilde{F}_{\tilde{x},i}$-vector space for each i and that $(F_o \otimes_{\mathcal{O}_o} f_o(E,\varphi))_{\tilde{x}}$ induces an isomorphism

$$(F_o \otimes_{\mathcal{O}_o} H_o(E,\varphi))_{\tilde{x},i} \xrightarrow{\sim} (F_o \otimes_{\mathcal{O}_o} H_o(E,\varphi))_{\tilde{x},i+1}$$

over $\widetilde{F}_{\tilde{x},i} \xrightarrow{\sim} \widetilde{F}_{\tilde{x},i+1}$ for each i such that the composite isomorphism

$$(F_o \otimes_{\mathcal{O}_o} H_o(E,\varphi))_{\tilde{x},1} \xrightarrow{\sim} (F_o \otimes_{\mathcal{O}_o} H_o(E,\varphi))_{\tilde{x},2} \xrightarrow{\sim} \cdots$$

$$\cdots \xrightarrow{\sim} (F_o \otimes_{\mathcal{O}_o} H_o(E,\varphi))_{\tilde{x},d_{\tilde{x}}+1} = (F_o \otimes_{\mathcal{O}_o} H_o(E,\varphi))_{\tilde{x},1}$$

is $\sigma_{\tilde{x}}^{s'_{\tilde{x}}}$-linear. Then, if we denote by $(F_o \otimes_{\mathcal{O}_o} f_o(E,\varphi))_{\tilde{x}}^{d_{\tilde{x}}}$ this composite isomorphism, it is clear that

$$\mathrm{End}((F_o \otimes_{\mathcal{O}_o} H_o(E,\varphi))_{\tilde{x}}, (F_o \otimes_{\mathcal{O}_o} f_o(E,\varphi))_{\tilde{x}})$$
$$= \mathrm{End}((F_o \otimes_{\mathcal{O}_o} H_o(E,\varphi))_{\tilde{x},1}, (F_o \otimes_{\mathcal{O}_o} f_o(E,\varphi))_{\tilde{x}}^{d_{\tilde{x}}}).$$

If we look at the dimensions over $\widetilde{F}_{\tilde{x}}$, we have

$$\dim_{\widetilde{F}_{\tilde{x}}} D_{\tilde{x}} \leq \dim_{\widetilde{F}_{\tilde{x}}} \mathrm{End}((F_o \otimes_{\mathcal{O}_o} H_o(E,\varphi))_{\tilde{x}}, (F_o \otimes_{\mathcal{O}_o} f_o(E,\varphi))_{\tilde{x}})$$

and we ultimately want to prove that this inequality is in fact an equality. But, thanks to (2.2.2), we have

$$\dim_{\widetilde{F}_{\widetilde{x}}} D_{\widetilde{x}} = (d/[\widetilde{F} : F])^2$$

and, thanks to (2.5.7), we have

$$\dim_{\widetilde{F}_{\widetilde{x}}} \mathrm{End}((F_o \otimes_{\mathcal{O}_o} (E, \varphi))_{\widetilde{x},1}, (F_o \otimes_{\mathcal{O}_o} f_o(E, \varphi))_{\widetilde{x}}^{d_{\widetilde{x}}})$$
$$\leq (\dim_{\widetilde{F}_{\widetilde{x},1}} (F_o \otimes_{\mathcal{O}_o} H_o(E, \varphi))_{\widetilde{x},1})^2.$$

So, for each place $\widetilde{x}$ of $\widetilde{F}$ dividing o, we have

$$(*) \qquad d/[\widetilde{F} : F] \leq \dim_{\widetilde{F}_{\widetilde{x},1}} (F_o \otimes_{\mathcal{O}_o} H_o(E, \varphi))_{\widetilde{x},1}$$

and we want to prove that the equality holds. As we obviously have

$$\dim_{F_k} (F_o \otimes_{\mathcal{O}_o} H_o(E, \varphi)) = \sum_{\widetilde{x}|o} d_{\widetilde{x}} [\widetilde{F}_{\widetilde{x},1} : F_k] \dim_{\widetilde{F}_{\widetilde{x},1}} (F_o \otimes_{\mathcal{O}_o} H_o(E, \varphi))_{\widetilde{x},1},$$
$$\dim_{F_k} (F_o \otimes_{\mathcal{O}_o} H_o(E, \varphi)) = d$$

and

$$d_{\widetilde{x}} [\widetilde{F}_{\widetilde{x},1} : F_k] = [\widetilde{F}_{\widetilde{x}} : F_o],$$

we get the equality

$$d = \sum_{\widetilde{x}|o} [\widetilde{F}_{\widetilde{x}} : F_o] \dim_{\widetilde{F}_{\widetilde{x},1}} (F_o \otimes_{\mathcal{O}_o} H_o(E, \varphi))_{\widetilde{x},1}.$$

Moreover, we have

$$[\widetilde{F} : F] = \sum_{\widetilde{x}|o} [\widetilde{F}_{\widetilde{x}} : F_o].$$

Putting all these things together, we see that the equality holds in $(*)$ for each place $\widetilde{x}$ of $\widetilde{F}$ dividing o. In particular part (ii) of (2.5.6) is proved. $\square$

COROLLARY (2.5.8). — *Let (E, φ) be a Drinfeld A-module of rank d over $\overline{\kappa(o)}$ with characteristic map $A \longrightarrow \kappa(o) \hookrightarrow \overline{\kappa(o)}$. Let $(\widetilde{F}, \tilde{o})$ be the (F, ∞, o)-type of its isogeny class. The map (2.5.5) and the inclusion $\widetilde{F} \hookrightarrow F \otimes_A \mathrm{End}(E, \varphi)$ induce a structure of an $\widetilde{F}_o (= F_o \otimes_F \widetilde{F})$-module on $F_o \otimes_{\mathcal{O}_o} H_o(E, \varphi)$, $F_o \otimes_{\mathcal{O}_o} f_o(E, \varphi)$ is $\widetilde{F}_o$-linear and (2.5.5) induces an isomorphism of $\widetilde{F}_o$-algebras*

$$F_o \otimes_A \mathrm{End}(E, \varphi) \xrightarrow{\sim} \mathrm{End}_{\widetilde{F}_o} (F_o \otimes_{\mathcal{O}_o} H_o(E, \varphi), F_o \otimes_{\mathcal{O}_o} f_o(E, \varphi)).$$

Moreover, the direct sum decomposition

$$F_o \otimes_{\mathcal{O}_o} H_o(E, \varphi) = (F_o \otimes_{\mathcal{O}_o} H_o^{et}(E, \varphi)) \oplus (F_o \otimes_{\mathcal{O}_o} H_o^c(E, \varphi))$$

is the one defined by the decomposition

$$\widetilde{F}_o = \widetilde{F}_o^{\tilde{o}} \times \widetilde{F}_{\tilde{o}}$$

where

$$\widetilde{F}_o^{\tilde{o}} = \prod_{\substack{\tilde{x}|o \\ \tilde{x} \neq \tilde{o}}} \widetilde{F}_{\tilde{x}},$$

the rank of $H_o^{et}(E, \varphi)$ (resp. $H_o^c(E, \varphi)$) over $\mathcal{O}_k$ is equal to $d - h$ (resp. h) where

$$h = [\widetilde{F}_{\tilde{o}} : F_o]d/[\widetilde{F} : F]$$

and $(H_o^c(E, \varphi), f_o^c(E, \varphi))$ has a unique slope $1/h$ (in other words, $F_o \otimes_{\mathcal{O}_o} (H_o^c(E, \varphi), f_o^c(E, \varphi))$ is non-canonically isomorphic to $(N_{h,1}, f_{h,1})$).

Proof: Let k_1 be a large enough finite extension of $\kappa(o)$ contained in $\overline{\kappa(o)}$ such that there exists a Drinfeld A-module (E_1, φ_1) of rank d over k_1 with $(E, \varphi) = \overline{\kappa(o)} \otimes_{k_1} (E_1, \varphi_1)$ and $\widetilde{F} = F[\Pi_1]$ where Π_1 is the Frobenius isogeny of (E_1, φ_1) (see (2.2)). Then $F \otimes_A \operatorname{End}(E, \varphi) = F \otimes_A \operatorname{End}(E_1, \varphi_1) = D_1$ and it follows from the proof of (2.5.6) that

$$(D_{1,\tilde{x}})^{opp} = \operatorname{End}((F_o \otimes_{\mathcal{O}_o} H_o(E_1, \varphi_1))_{\tilde{x}}, (F_o \otimes_{\mathcal{O}_o} f_o(E_1, \varphi))_{\tilde{x}})$$

for each place $\tilde{x}$ of $\widetilde{F}$ dividing o and therefore

$$F_o \otimes_{\mathcal{O}_o} H_o^{et}(E_1, \varphi_1) = \bigoplus_{\substack{\tilde{x}|o \\ \tilde{x} \neq \tilde{o}}} (F_o \otimes_{\mathcal{O}_o} H_o(E_1, \varphi_1))_{\tilde{x}}$$

and

$$F_o \otimes_{\mathcal{O}_o} H_o^c(E_1, \varphi_1) = (F_o \otimes_{\mathcal{O}_o} H_o(E_1, \varphi_1))_{\tilde{o}}$$

($\operatorname{inv}_{\tilde{x}}(D_1) \neq 0$ if and only if $\tilde{x} = \widetilde{\infty}$ or $\tilde{o}$, see (2.2.2)). It also follows from the proof of (2.5.6) that

$$\dim_{F_{k_1}} (F_o \otimes_{\mathcal{O}_o} H_o(E_1, \varphi_1))_{\tilde{x}} = [\widetilde{F}_{\tilde{x}} : F_o]d/[\widetilde{F} : F]$$

for each place $\tilde{x}$ of $\widetilde{F}$ dividing o.

Now we have a canonical isomorphism

$$\overline{\kappa(o)} \otimes_{k_1} (H_o(E_1, \varphi_1), f_o(E_1, \varphi_1)) \xrightarrow{\sim} (H_o(E, \varphi), f_o(E, \varphi))$$

of Dieudonné $\mathcal{O}_o$-modules over $\overline{\kappa(o)}$ and it is clear that $\mathrm{End}((F_o \otimes_{\mathcal{O}_o} H_o(E_1, \varphi_1))_{\tilde{x}}, (F_o \otimes_{\mathcal{O}_o} f_o(E_1, \varphi_1))_{\tilde{x}})$ is exactly the centralizer of the image of Π_1 in $\mathrm{End}((F_o \otimes_{\mathcal{O}_o} H_o(E, \varphi))_{\tilde{x}}, (F_o \otimes_{\mathcal{O}_o} f_o(E, \varphi))_{\tilde{x}})$. But $\widetilde{F}_{\tilde{x}}$ is the field generated by this image of Π_1 over F_o, so we get the equality

$$\mathrm{End}((F_o \otimes_{\mathcal{O}_o} H_o(E_1, \varphi_1))_{\tilde{x}}, (F_o \otimes_{\mathcal{O}_o} f_o(E_1, \varphi_1))_{\tilde{x}})$$
$$= \mathrm{End}_{\widetilde{F}_{\tilde{x}}}((F_o \otimes_{\mathcal{O}_o} H_o(E, \varphi))_{\tilde{x}}, (F_o \otimes_{\mathcal{O}_o} f_o(E, \varphi))_{\tilde{x}})$$

for each place $\tilde{x}$ of $\widetilde{F}$ dividing o. All the assertions of (2.5.8), except the last one about the slopes of $(H_o^c(E, \varphi), F_o^c(R, \varphi))$, follow immediately.

To finish the proof of (2.5.8) it is enough to prove that the semi-simple F_o-algebra $\mathrm{End}(F_o \otimes_{\mathcal{O}_o} H_o^c(E, \varphi), F_o \otimes_{\mathcal{O}_o} f_o^c(E, \varphi))$ is in fact a central simple F_o-algebra with invariant $-1/h$. But this follows from (A.1.2) and (A.2.4) as $(D_{1,\tilde{o}})^{opp}$ is the centralizer in this F_o-algebra of the field $\widetilde{F}_{\tilde{o}}$ and as

$$\mathrm{inv}_{\widetilde{F}_{\tilde{o}}}((D_{1,\tilde{o}})^{opp}) = -[\widetilde{F}_{\tilde{o}} : F_o]/h$$

(see (2.2.2)). $\square$

(2.6) First description of an isogeny class

We will now give a description of the isogeny class

$$(2.6.1) \qquad M_{I,o}^d(\overline{\kappa(o)})_{(\widetilde{F}, \tilde{o})} \subset M_{I,o}^d(\overline{\kappa(o)})$$

associated with a given (F, ∞, o)-type $(\widetilde{F}, \tilde{o})$ of rank k. Such an isogeny class is obviously fixed by the "action" (action by correspondences) of $\mathcal{H}_I^{\infty, o}$ (see (1.7)) and the action of Frob_o (if $(E, \varphi, \iota) \in M_{I,o}^d(\overline{\kappa(o)})$ with $E = \mathbb{G}_{a, \overline{\kappa(o)}}$, then $\mathrm{Frob}_o(E, \varphi, \iota) = (E', \varphi', \iota')$ with $E' = E$,

$$\varphi'(a) = \tau^{\deg(o)} \varphi(a) \tau^{-\deg(o)}$$

for each $a \in A$ and

$$\iota'(i) = \iota(i)^{p^{\deg(o)}}$$

for each $i \in (I^{-1}/A)^d$, so that $\tau^{\deg(o)}$ induces an isogeny from (E, φ) to (E', φ'), the so-called **Frobenius isogeny over** $\kappa(o)$). We will also describe the commuting actions of $\mathcal{H}_I^{\infty, o}$ and Frob_o on $M_{I,o}^d(\overline{\kappa(o)})_{(\widetilde{F}, \tilde{o})}$.

Let (E, φ) be a Drinfeld A-module of rank d over $\overline{\kappa(o)}$ (with characteristic $A \twoheadrightarrow \kappa(o) \hookrightarrow \overline{\kappa(o)}$) in the isogeny class associated with $(\widetilde{F}, \tilde{o})$. If $(E', \varphi', \iota') \in M_{I,o}^d(\overline{\kappa(o)})$ and if $(E, \varphi) \xrightarrow{u} (E', \varphi')$ is an isogeny, for each $x \in |X| - \{\infty\}$, we have an injective map

$$H_x(u) : H_x(E', \varphi') \longleftrightarrow H_x(E, \varphi).$$

Let M_x be the image of $H_x(u)$. If $x \neq o$, M_x is a free $\mathcal{O}_x$-submodule of rank d of $H_x(E, \varphi)$. M_o is a free $\mathcal{O}_{\overline{\kappa(o)}}$-submodule of rank d of $H_o(E, \varphi)$ such that

$$\begin{cases} \varpi_o M_o \subset f_o(E, \varphi)(M_o) \subset M_o, \\ \dim_{\overline{\kappa(o)}}(M_o/f_o(E, \varphi)(M_o)) = 1 \end{cases}$$

(see (2.5.4)). Moreover, if we set

$$M^{\infty, o} = \prod_{x \in |X| - \{\infty, o\}} M_x,$$

the I-level structure ι' induces an isomorphism of $(\mathcal{O}^{\infty, o}/I\mathcal{O}^{\infty, o})$-modules

$$M^{\infty, o}/IM^{\infty, o} \xrightarrow{\sim} (\mathcal{O}^{\infty, o}/I\mathcal{O}^{\infty, o})^d.$$

Let $Y^{\infty, o}$ be the set of free $\mathcal{O}^{\infty, o}$-submodules $M^{\infty, o}$ of rank d of

$$H^{\infty, o}(E, \varphi) = \prod_{x \in |X| - \{\infty, o\}} H_x(E, \varphi)$$

endowed with an isomorphism of $\mathcal{O}^{\infty, o}$-modules

$$M^{\infty, o} \xrightarrow{\sim} (\mathcal{O}^{\infty, o})^d.$$

We have a right action of $K_I^{\infty, o}$ on $Y^{\infty, o}$ by composing the isomorphism $M^{\infty, o} \xrightarrow{\sim} (\mathcal{O}^{\infty, o})^d$ with $^t k^{\infty, o}$ if $k^{\infty, o} \in K_I^{\infty, o}$. Let Y_o be the set of free $\mathcal{O}_{\overline{\kappa(o)}}$-submodules M_o of rank d of $H_o(E, \varphi)$ such that

$$\begin{cases} \varpi_o M_o \subset f_o(E, \varphi)(M_o) \subset M_o, \\ \dim_{\overline{\kappa(o)}}(M_o/f_o(E, \varphi)(M_o)) = 1. \end{cases}$$

Lemma (2.6.2). — *The set of pairs*

$$((E', \varphi', \iota'), u),$$

where $(E', \varphi', \iota') \in M_{I, o}^d(\overline{\kappa(o)})$ *and* $(E, \varphi) \xrightarrow{u} (E', \varphi')$ *is an isogeny, is canonically isomorphic to the set* $(Y^{\infty, o}/K_I^{\infty, o}) \times Y_o$.

Proof : We have already defined a map from the set of pairs $((E', \varphi', \iota'), u)$ to $(Y^{\infty,o}/K_I^{\infty,o}) \times Y_o$ (an isomorphism of $(\mathcal{O}^{\infty,o}/I\mathcal{O}^{\infty,o})$-modules

$$M^{\infty,o}/IM^{\infty,o} \xrightarrow{\sim} (\mathcal{O}^{\infty,o}/I\mathcal{O}^{\infty,o})^d,$$

where $M^{\infty,o}$ is a free $\mathcal{O}^{\infty,o}$-module of rank d, can be lifted to an isomorphism of $\mathcal{O}^{\infty,o}$-modules

$$M^{\infty,o} \xrightarrow{\sim} (\mathcal{O}^{\infty,o})^d$$

and any two such liftings are in the same $K_I^{\infty,o}$-orbit).

Let us define a map in the opposite direction (we let the reader check that these maps are inverse bijections). Let $(M^{\infty,o}, M^{\infty,o} \xrightarrow{\sim} (\mathcal{O}^{\infty,o})^d)$ be an element of $Y^{\infty,o}$ and M_o be an element of Y_o. The quotient $\mathcal{O}^{\infty,o}$-module

$$H^{\infty,o}(E, \varphi)/M^{\infty,o}$$

defines a unique finite étale $\overline{\kappa(o)}$-subscheme in A-modules $G^{\infty,o} \subset E$ such that

$$G^{\infty,o}(\overline{\kappa(o)}) = \mathrm{Hom}_{\mathcal{O}^{\infty,o}}(H^{\infty,o}(E, \varphi)/M^{\infty,o}, \mathbb{A}^{\infty,o}/\mathcal{O}^{\infty,o}).$$

Similarly the quotient $\mathcal{O}_{\overline{\kappa(o)}}$-module

$$H_o(E, \varphi)/M_o$$

endowed with the σ_o-linear map induced by $f_o(E, \varphi)$ defines a unique $\overline{\kappa(o)}$-subscheme in A-modules

$$G_o = G_{\overline{\kappa(o)}}(H_o(E, \varphi)/M_o, f_o(E, \varphi) \bmod M_o)$$

of E (see (2.4.11)). $G^{\infty,o}$ is killed by some ideal of A which is prime to $\mathcal{P}_o$ and G_o is killed by some positive power of $\mathcal{P}_o$ ($\mathcal{P}_o$ is the maximal ideal of A defining o). Then we can take the quotient of (E, φ) by the finite $\overline{\kappa(o)}$-subscheme in A-modules

$$G = G^{\infty,o} \times_{\overline{\kappa(o)}} G_o \subset E$$

(see (2.1.1)) and we get a Drinfeld A-module (E', φ') of rank d over $\overline{\kappa(o)}$ together with an isogeny $(E, \varphi) \xrightarrow{u} (E', \varphi')$. Moreover, the isomorphism $M^{\infty,o} \xrightarrow{\sim} (\mathcal{O}^{\infty,o})^d$ induces a level-I structure ι' on (E', φ'). The pair $((E', \varphi', \iota'), u)$ depends only on the class of $((M^{\infty,o}, M^{\infty,o} \xrightarrow{\sim} (\mathcal{O}^{\infty,o})^d), M_o)$ in $(Y^{\infty,o}/K_I^{\infty,o}) \times Y_o$. $\qquad \square$

We have an action of $A - \{0\} \subset \mathcal{O}^\infty \cap (\mathbb{A}^\infty)^\times$ on $Y^{\infty,o}$ and on Y_o:

$$a \cdot (M^{\infty,o}, M^{\infty,o} \xrightarrow{\sim} (\mathcal{O}^{\infty,o})^d)$$

$$= (aM^{\infty,o}, aM^{\infty,o} \xrightarrow[\sim]{a^{-1}} M^{\infty,o} \xrightarrow{\sim} (\mathcal{O}^{\infty,o})^d)$$

and

$$a \cdot M_o = aM_o.$$

So we can define $F \otimes_A Y^{\infty,o}$ (resp. $F \otimes_A Y_o$) as the sets of pairs (a, y) with $a \in A - \{0\}$ and $y \in Y^{\infty,o}$ (resp. $y \in Y_o$) where two pairs (a', y') and (a'', y'') are identified if $a'' \cdot y' = a' \cdot y''$. We can extend the action of $K_I^{\infty,o}$ to $F \otimes_A Y^{\infty,o}$ by letting $K_I^{\infty,o}$ act trivially on the component a and as before on the component y. If we set

(2.6.3) $$D = F \otimes_A \text{End}(E, \varphi)$$

we can define a left action of $(D^{opp})^\times$ on $F \otimes_A Y^{\infty,o}$ and on $F \otimes_A Y_o$ as follows. We have a left action of $\text{End}(E, \varphi)^{opp}$ on $Y^{\infty,o}$ and on Y_o extending the action of $A - \{0\}$:

$$u \cdot (M^{\infty,o}, M^{\infty,o} \xrightarrow{\sim} (\mathcal{O}^{\infty,o})^d)$$

$$= (H^{\infty,o}(u)(M^{\infty,o}), H^{\infty,o}(u)(M^{\infty,o}) \xrightarrow[\sim]{H^{\infty,o}(u)^{-1}} M^{\infty,o} \xrightarrow{\sim} (\mathcal{O}^{\infty,o})^d)$$

and

$$u \cdot M_o = H_o(u)(M_o).$$

It is obvious that the actions of $(D^{opp})^\times$ and $K_I^{\infty,o}$ on $F \otimes_A Y^{\infty,o}$ commute.

COROLLARY (2.6.4). — *The bijection described in (2.6.2) induces a canonical bijection*

$$(D^{opp})^\times \backslash ((F \otimes_A Y^{\infty,o} / K_I^{\infty,o}) \times F \otimes_A Y_o) \xrightarrow{\sim} M_{I,o}^d(\overline{\kappa(o)})_{(\tilde{F}, \tilde{o})}.$$

□

We have a canonical bijection

(2.6.5) $$\text{Isom}_{\mathbb{A}^{\infty,o}}((\mathbb{A}^{\infty,o})^d, \mathbb{A}^{\infty,o} \otimes_{\mathcal{O}^{\infty,o}} H^{\infty,o}(E, \varphi)) \xrightarrow{\sim} F \otimes_A Y^{\infty,o}$$

which maps g to $(a, (ag((\mathcal{O}^{\infty,o})^d), ag((\mathcal{O}^{\infty,o})^d) \xrightarrow[\sim]{(ag)^{-1}} (\mathcal{O}^{\infty,o})^d))$ if $a \in A - \{0\}$ is divisible enough so that $ag((\mathcal{O}^{\infty,o})^d) \subset H^{\infty,o}(E, \varphi)$. Moreover we have a left action of $(D^{opp})^\times$ and a right action of $GL_d(\mathbb{A}^{\infty,o})$ on the source of (2.6.5) : $\delta \in (D^{opp})^\times$ maps g to $H^{\infty,o}(\delta)g$ and $g' \in GL_d(\mathbb{A}^{\infty,o})$ maps g to gg'. It is clear that (2.6.5) is equivariant for the actions of $(D^{opp})^\times$ and $K_I^{\infty,o} \subset GL_d(\mathbb{A}^{\infty,o})$. In particular, using (2.6.5), we get a right action of $GL_d(\mathbb{A}^{\infty,o})$ on $F \otimes_A Y^{\infty,o}$ extending the action of $K_I^{\infty,o}$ and therefore a right "action" of $\mathcal{H}_I^{\infty,o}$ on $F \otimes_A Y^{\infty,o} / K_I^{\infty,o}$ which commutes with the left action of $(D^{opp})^\times$. We let $\mathcal{H}_I^{\infty,o}$ act trivially on $F \otimes_A Y_o$.

LEMMA (2.6.6). — *The bijection*

$$(D^{opp})^{\times}\backslash((F \otimes_A Y^{\infty,o}/K_I^{\infty,o}) \times F \otimes_A Y_o) \xrightarrow{\sim} M_{I,o}^d(\overline{\kappa(o)})_{(\tilde{F},\tilde{o})}$$

described in (2.6.4) is compatible with the right "actions" of the Hecke algebra $\mathcal{H}_I^{\infty,o}$. □

The set $F \otimes_A Y_o$ (with its left action of $(D^{opp})^{\times}$) can be identified with the set of lattices, i.e. free $\mathcal{O}_{\overline{\kappa(o)}}$-submodules of rank d,

$$M_o \subset F_o \otimes_{\mathcal{O}_o} H_o(E, \varphi)$$

such that

$$\begin{cases} \varpi_o M_o \subset (F_o \otimes_{\mathcal{O}_o} f_o(E,\varphi))(M_o) \subset M_o, \\ \dim_{\overline{\kappa(o)}}(M_o/(F_o \otimes_{\mathcal{O}_o} f_o(E,\varphi))(M_o)) = 1. \end{cases}$$

Let Frob_o act on $F \otimes_A Y_o$ by

$$(2.6.7) \qquad \mathrm{Frob}_o(M_o) = (F_o \otimes_{\mathcal{O}_o} f_o(E,\varphi))(M_o).$$

Obviously, this action commutes with the left action of $(D^{opp})^{\times}$. Let Frob_o act trivially on $F \otimes_A Y^{\infty,o}$.

LEMMA (2.6.8). — *The bijection*

$$(D^{opp})^{\times}\backslash((F \otimes_A Y^{\infty,o}/K_I^{\infty,o}) \times F \otimes_A Y_o) \xrightarrow{\sim} M_{I,o}^d(\overline{\kappa(o)})_{(\tilde{F},\tilde{o})}$$

described in (2.6.4) is compatible with the action of Frob_o. □

(2.7) Isogeny classes as double coset spaces

To get an even more concrete description of the isogeny class $M_{I,o}^d(\overline{\kappa(o)})_{(\tilde{F},\tilde{o})}$ we will fix a basis of the free $\mathbb{A}^{\infty,o}$-module $\mathbb{A}^{\infty,o} \otimes_{\mathcal{O}^{\infty,o}} H^{\infty,o}(E,\varphi)$ and a basis of the $F_{\overline{\kappa(o)}}$-vector space $F_o \otimes_{\mathcal{O}_o} H_o(E,\varphi)$. We will denote by

$$\varepsilon_o \in GL_d(F_{\overline{\kappa(o)}})$$

the matrix of $F_o \otimes_{\mathcal{O}_o} f_o(E,\varphi)$. Corresponding to the choice of these bases, we have inclusions of F-algebras

$$\tilde{F} \longleftrightarrow gl_d(\mathbb{A}^{\infty,o})$$

and

$$\tilde{F} \longleftrightarrow \{h_o \in gl_d(F_{\overline{\kappa(o)}}) | h_o \varepsilon_o = \varepsilon_o \sigma_o(h_o)\};$$

moreover $(\mathbb{A}^{\infty,o} \otimes_F D)^{opp}$ and $(F_o \otimes_F D)^{opp}$ are exactly the centralizers of the images of $\widetilde{F}$ in $gl_d(\mathbb{A}^{\infty,o})$ and in $\{h_o \in gl_d(F_{\overline{\kappa(o)}}) | h_o \varepsilon_o = \varepsilon_o \sigma_o(h_o)\}$ respectively (see (2.3.4) and (2.5.8)).

Thanks to (2.6.5), we can identify $F \otimes_A Y^{\infty,o}$ with $GL_d(\mathbb{A}^{\infty,o})$ and the left (resp. right) action of $(D^{opp})^\times$ (resp. $GL_d(\mathbb{A}^{\infty,o})$) on $F \otimes_A Y^{\infty,o}$ with the action by left translations of $(D^{opp})^\times \subset GL_d(\mathbb{A}^{\infty,o})$ on $GL_d(\mathbb{A}^{\infty,o})$ (resp. the action by right translations of $GL_d(\mathbb{A}^{\infty,o})$ on itself).

If $M_o \subset (F_{\overline{\kappa(o)}})^d$ is a lattice, there exists $h_o \in GL_d(F_{\overline{\kappa(o)}})$ such that

$$M_o = h_o((\mathcal{O}_{\overline{\kappa(o)}})^d)$$

and the class $h_o GL_d(\mathcal{O}_{\overline{\kappa(o)}}) \in GL_d(F_{\overline{\kappa(o)}})/GL_d(\mathcal{O}_{\overline{\kappa(o)}})$ is uniquely determined by M_o. Then $M_o \in F \otimes_A Y_o$ if and only if

$$\begin{cases} \varpi_o((\mathcal{O}_{\overline{\kappa(o)}})^d) \subset h_o^{-1}\varepsilon_o\sigma_o(h_o)((\mathcal{O}_{\overline{\kappa(o)}})^d) \subset (\mathcal{O}_{\overline{\kappa(o)}})^d, \\ \dim_{\overline{\kappa(o)}}((\mathcal{O}_{\overline{\kappa(o)}})^d/(h_o^{-1}\varepsilon_o\sigma_o(h_o)((\mathcal{O}_{\overline{\kappa(o)}})^d))) = 1. \end{cases}$$

But these conditions are equivalent to the following one:

$$h_o^{-1}\varepsilon_o\sigma_o(h_o) \in < \varpi_o >$$

where

$$< \varpi_o > = GL_d(\mathcal{O}_{\overline{\kappa(o)}}) \begin{pmatrix} \varpi_o & & & 0 \\ & 1 & & \\ & & \ddots & \\ 0 & & & 1 \end{pmatrix} GL_d(\mathcal{O}_{\overline{\kappa(o)}}) \subset GL_d(F_{\overline{\kappa(o)}}).$$

So we can identify $F \otimes_A Y_o$ with the subset of $GL_d(F_{\overline{\kappa(o)}})/GL_d(\mathcal{O}_{\overline{\kappa(o)}})$ of classes $h_o GL_d(\mathcal{O}_{\overline{\kappa(o)}})$ such that $h_o^{-1}\varepsilon_o\sigma_o(h_o)$ belong to the double class $< \varpi_o >$. Then the left action of $(D^{opp})^\times$ becomes the action by left translations of $(D^{opp})^\times \subset (F_o \otimes_F D)^\times \subset \{h_o \in GL_d(F_{\overline{\kappa(o)}}) | h_o^{-1}\varepsilon_o\sigma_o(h_o) = \varepsilon_o\}$ on this subset and Frob_o maps $h_o GL_d(\mathcal{O}_{\overline{\kappa(o)}})$ to $\varepsilon_o\sigma_o(h_o)GL_d(\mathcal{O}_{\overline{\kappa(o)}})$.

To sum up we can describe the isogeny class $M_{I,o}^d(\overline{\kappa(o)})_{(\bar{F},\bar{o})}$ in the following way. We arbitrarily fix an embedding of $\mathbb{A}^{\infty,o}$-algebras

$$\mathbb{A}^{\infty,o} \otimes_F \widetilde{F} \longrightarrow gl_d(\mathbb{A}^{\infty,o}),$$

an embedding of F_o-algebras

$$\widetilde{F}_o^{\bar{o}} = \prod_{\substack{\tilde{x}|o \\ \tilde{x}\neq\bar{o}}} \widetilde{F}_{\tilde{x}} \longrightarrow gl_{d-h}(F_o)$$

and an embedding of F_o-algebras

$$\widetilde{F}_{\tilde{o}} \longrightarrow \{h_{\tilde{o}} \in gl_h(F_{\overline{\kappa(o)}})|h_{\tilde{o}}\varepsilon_{\tilde{o}} = \varepsilon_{\tilde{o}}\sigma_o(h_{\tilde{o}})\}$$

where

$$h = [\widetilde{F}_{\tilde{o}} : F_o]d/[\widetilde{F} : F]$$

and

$$\varepsilon_{\tilde{o}} = \begin{pmatrix} 0 & 1 & 0 & \cdots & 0 \\ \vdots & \ddots & \ddots & \ddots & \vdots \\ \vdots & & \ddots & \ddots & 0 \\ 0 & & & \ddots & 1 \\ \varpi_o & 0 & \cdots & & 0 \end{pmatrix} \in GL_h(F_o)$$

(there always exist such embeddings and different choices are conjugate). If D is "the unique" central division $\widetilde{F}$-algebra whose non-zero invariants are $-d/[\widetilde{F} : F]$ at $\widetilde{\infty}$ and $d/[\widetilde{F} : F]$ at $\tilde{o}$, we can identify $(\mathbb{A}^{\infty,o} \otimes_F D)^{opp}$ (resp. $(D_o^{\tilde{o}})^{opp} = \prod_{\substack{\tilde{x}|o \\ \tilde{x}\neq\tilde{o}}} D_{\tilde{x}}^{opp}$, resp. $D_{\tilde{o}}^{opp}$) with the centralizer of $\mathbb{A}^{\infty,o} \otimes_F \widetilde{F}$ in $gl_d(\mathbb{A}^{\infty,o})$ (resp. $\widetilde{F}_o^{\tilde{o}}$ in $gl_{d-h}(F_o)$, resp. $\widetilde{F}_{\tilde{o}}$ in $\{h_{\tilde{o}} \in gl_h(F_{\overline{\kappa(o)}})|h_{\tilde{o}}\varepsilon_{\tilde{o}} = \varepsilon_{\tilde{o}}\sigma_o(h_{\tilde{o}})\}$). We set

$$\varepsilon_o^{\tilde{o}} = \begin{pmatrix} 1 & & 0 \\ & \ddots & \\ 0 & & 1 \end{pmatrix} \in GL_{d-h}(F_o)$$

and

$$\varepsilon_o = \begin{pmatrix} \varepsilon_o^{\tilde{o}} & 0 \\ 0 & \varepsilon_{\tilde{o}} \end{pmatrix} \in GL_d(F_o).$$

It is easy to see that

$$\{h_o \in gl_d(F_{\overline{\kappa(o)}})|h_o\varepsilon_o = \varepsilon_o\sigma_o(h_o)\}$$

is isomorphic to

$$gl_{d-h}(F_o) \times \{h_{\tilde{o}} \in gl_h(F_{\overline{\kappa(o)}})|h_{\tilde{o}}\varepsilon_{\tilde{o}} = \varepsilon_{\tilde{o}}\sigma_o(h_{\tilde{o}})\}$$

(thanks to the diagonal embedding of $gl_{d-h} \times gl_h$ into gl_d). So we have an embedding

$$F_o \otimes_F \widetilde{F} \longrightarrow \{h_o \in gl_d(F_{\overline{\kappa(o)}})|h_o\varepsilon_o = \varepsilon_o\sigma_o(h_o)\}$$

and an identification of $(F_o \otimes_F D)^{opp}$ with the centralizer of the image of this embedding. We can form the coset space

(2.7.1) $(D^{opp})^{\times}\backslash[(GL_d(\mathbb{A}^{\infty,o})/K_I^{\infty,o})$

$\times (\{h_o \in GL_d(F_{\overline{\kappa(o)}})|h_o^{-1}\varepsilon_o\sigma_o(h_o) \in< \varpi_o >\}/GL_d(\mathcal{O}_{\overline{\kappa(o)}}))]$

($< \varpi_o >$ is the double class defined above). On this coset space we have a right "action" of $\mathcal{H}_I^{\infty,o}$ ($\mathcal{H}_I^{\infty,o}$ acts as described before on the first factor $GL_d(\mathbb{A}^{\infty,o})/K_I^{\infty,o}$ and trivially on the second) and an action of Frob_o (Frob_o acts trivially on the first factor and maps $h_o GL_d(\mathcal{O}_{\overline{\kappa(o)}})$ to $\varepsilon_o \sigma_o(h_o) GL_d(\mathcal{O}_{\overline{\kappa(o)}})$).

THEOREM (2.7.2) (Drinfeld). — *We have a* (*non-canonical*) *bijection between the set* (2.7.1) *and* $M_{I,o}^d(\overline{\kappa(o)})_{(\tilde{F},\tilde{o})}$ *which is* $\mathcal{H}_I^{\infty,o}$*-equivariant and* Frob_o*-equivariant.* □

(2.8) Comments and references

Again, most of the material of this chapter is due to Drinfeld and taken from [Dr 1], [Dr 2] and [Dr 3]. Nevertheless, our description (2.7.1) of $M_{I,o}^d(\overline{\kappa(o)})_{(\tilde{F},\tilde{o})}$ is slightly different from the description given by Drinfeld. This difference will become more apparent in the next chapter.

3

The Lefschetz numbers of Hecke operators

(3.0) Introduction

In this chapter, the integer $d > 0$, the non-zero ideal I of A, $I \neq A$, the place o of F, $o \neq \infty$, $o \notin V(I)$, and an algebraic closure $\overline{\kappa(o)}$ of $\kappa(o)$ are fixed as before.

Our purpose is to compute for each $f^{\infty,o} \in \mathcal{H}_I^{\infty,o}$ and each integer $r > 0$ the number of fixed points of the action of $\mathrm{Frob}_o^r \times f^{\infty,o}$ on $M_{I,o}^d(\overline{\kappa(o)})$.

(3.1) The Lefschetz numbers of correspondences

Let k be a finite field of characteristic p with q elements and let $\overline{k}$ be an algebraic closure of k. We will denote by Frob_k the geometric Frobenius element in $\mathrm{Gal}(\overline{k}/k)$.

Let Y be a smooth, separated k-scheme of finite type and let

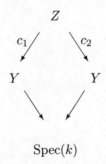

$$Z$$
$$c_1 \swarrow \qquad \searrow c_2$$
$$Y \qquad\qquad Y$$
$$\searrow \qquad \swarrow$$
$$\mathrm{Spec}(k)$$

be a geometric correspondence (see (1.6)). So, Z is also a smooth, separated k-scheme of finite type, c_2 is finite and étale and c_1 is proper.

By a fixed point of $\mathrm{Frob}_k^r \times (Z, c)$, where r is an integer ≥ 0, we mean a point $z \in Z(\bar{k})$ such that

$$\mathrm{Frob}_k^r(c_1(z)) = c_2(z)$$

(Frob_k acts as usual on $Y(\bar{k})$).

LEMMA (3.1.1). — *If $r > 0$, there are only finitely many fixed points of $\mathrm{Frob}_k^r \times (Z, c)$ and they all have multiplicity one.*

Proof : As c_1 is étale, $Z \xrightarrow{c} Y \times_k Y$ is transversal to the graph of Frob_k^r if $r > 0$. □

DEFINITION (3.1.2). — *If $r > 0$, the Lefschetz number $\mathrm{Lef}_r(Z, c)$ is the number of fixed points of $\mathrm{Frob}_k^r \times (Z, c)$ in $Z(\bar{k})$.*

(3.2) Counting of fixed points

We fix some $g^{\infty,o} \in GL_d(\mathbb{A}^{\infty,o})$ and we will denote by $f^{\infty,o} \in \mathcal{H}_I^{\infty,o}$ the characteristic function of the double class

$$K_I^{\infty,o} g^{\infty,o} K_I^{\infty,o} \subset GL_d(\mathbb{A}^{\infty,o}).$$

The image by (1.7.4) of $f^{\infty,o}$ is a geometric correspondence

(3.2.1)

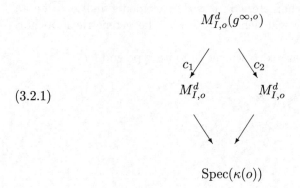

$$M_{I,o}^d(g^{\infty,o})$$

$$c_1 \qquad c_2$$

$$M_{I,o}^d \qquad M_{I,o}^d$$

$$\mathrm{Spec}(\kappa(o))$$

(it follows from (1.7) that

$$\deg(c_2) = |(K_I^{\infty,o} \cap ((g^{\infty,o})^{-1} K_I^{\infty,o} g^{\infty,o})) \backslash K_I^{\infty,o}|.$$

Let r be a positive integer, we will compute the finite number

$$(3.2.2) \qquad \qquad \mathrm{Lef}_r(f^{\infty,o})$$

of fixed points of $\mathrm{Frob}_o^r \times (M_{I,o}^d(g^{\infty,o}), c)$ acting on $M_{I,o}^d(\overline{\kappa(o)})$ (see (3.1.1)). As this action fixes the isogeny classes, we have

$$(3.2.3) \qquad \qquad \mathrm{Lef}_r(f^{\infty,o}) = \sum_{(\widetilde{F},\widetilde{o})} \mathrm{Lef}_r(f^{\infty,o})_{(\widetilde{F},\widetilde{o})}$$

where $(\widetilde{F},\widetilde{o})$ runs over the set of isomorphism classes of (F,∞,o)-types and where $\mathrm{Lef}_r(f^{\infty,o})_{(\widetilde{F},\widetilde{o})}$ is the number of fixed points of $\mathrm{Frob}_o^r \times (M_{I,o}^d(g^{\infty,o}), c)$ in the isogeny class corresponding to the (F,∞,o)-type $(\widetilde{F},\widetilde{o})$ (see (2.2.3)).

Let $(\widetilde{F},\widetilde{o})$ be a (F,∞,o)-type. In (2.7), we have constructed an $\mathcal{H}_I^{\infty,o}$-equivariant and Frob_o-equivariant bijection between

$$(D^{opp})^\times \backslash [GL_d(\mathbf{A}^{\infty,o})/K_I^{\infty,o}$$
$$\times \{h_o \in GL_d(F_{\overline{\kappa(o)}}) | h_o^{-1} \varepsilon_o \sigma_o(h_o) \in < \varpi_o > \}/GL_d(\mathcal{O}_{\overline{\kappa(o)}})]$$

and $M_{I,o}^d(\overline{\kappa(o)})_{(\widetilde{F},\widetilde{o})}$. In particular, we have a bijection between

$$(D^{opp})^\times \backslash [GL_d(\mathbf{A}^{\infty,o})/(K_I^{\infty,o} \cap ((g^{\infty,o})^{-1}K_I^{\infty,o}g^{\infty,o}))$$
$$\times \{h_o \in GL_d(F_{\overline{\kappa(o)}}) | h_o^{-1} \varepsilon_o \sigma_o(h_o) \in < \varpi_o > \}/GL_d(\mathcal{O}_{\overline{\kappa(o)}})]$$

and the restriction to $M_{I,o}^d(\overline{\kappa(o)})_{(\widetilde{F},\widetilde{o})}$ of $M_{I,o}^d(g^{\infty,o})(\overline{\kappa(o)})$. Moreover c_1 (resp. c_2) is induced by the identity on $GL_d(\mathbf{A}^{\infty,o})$ (resp. the right translation by $(g^{\infty,o})^{-1}$ on $GL_d(\mathbf{A}^{\infty,o})$) and the identity on $\{h_o \in GL_d(F_{\overline{\kappa(o)}}) | h_o^{-1} \varepsilon_o \sigma_o(h_o) \in < \varpi_o > \}$. Finally, the action of Frob_k on $M_{I,o}^d(\overline{\kappa(o)})_{(\widetilde{F},\widetilde{o})}$ is induced by the identity on $GL_d(\mathbf{A}^{\infty,o})$ and by

$$h_o \mapsto \varepsilon_o \sigma_o(h_o).$$

So, if $h^{\infty,o} \in GL_d(\mathbf{A}^{\infty,o})$ and $h_o \in GL_d(F_{\overline{\kappa(o)}})$, the double class

$$(D^{opp})^\times (h^{\infty,o}(K_I^{\infty,o} \cap ((g^{\infty,o})^{-1}K_I^{\infty,o}g^{\infty,o})), h_o GL_d(\mathcal{O}_{\overline{\kappa(o)}}))$$

is a fixed point of $\mathrm{Frob}_k^r \times (M_{I,o}^d(g^{\infty,o}), c)$ if and only if there exist $\delta \in (D^{opp})^\times$, $k^{\infty,o} \in K_I^{\infty,o}$ and $k_o \in GL_d(\mathcal{O}_{\overline{\kappa(o)}})$ such that the following conditions are satisfied:

$$\begin{cases} h^{\infty,o}k^{\infty,o} = \delta h^{\infty,o}(g^{\infty,o})^{-1}, \\ N_r(\varepsilon_o)\sigma_o^r(h_o) = \delta h_o k_o; \end{cases}$$

or, what is the same, if and only if there exists $\delta \in (D^{opp})^{\times}$ such that the following conditions are satisfied:

$$\begin{cases} (h^{\infty,o})^{-1}\delta h^{\infty,o} \in K_I^{\infty,o} g^{\infty,o}, \\ (h_o)^{-1}\delta^{-1}N_r(\varepsilon_o)\sigma_o^r(h_o) \in GL_d(\mathcal{O}_{\overline{\kappa(o)}}), \end{cases}$$

where we have set

(3.2.4) $$N_r(h_o') = h_o'\sigma_o(h_o')\cdots\sigma_o^{r-1}(h_o')$$

for all $h_o' \in GL_d(F_{\overline{\kappa(o)}})$.

We will denote by $F_{o,r}$ the unique extension of F_o of degree r which is contained in $F_{\overline{\kappa(o)}}$ and by $\mathcal{O}_{o,r}$ the ring of integers of $F_{o,r}$.

LEMMA (3.2.5). — *The map*

$$GL_d(\mathcal{O}_{\overline{\kappa(o)}}) \to GL_d(\mathcal{O}_{\overline{\kappa(o)}}), k_o \mapsto k_o\sigma^r(k_o)^{-1}$$

is surjective and the inverse image of the identity matrix by this map is $GL_d(\mathcal{O}_{o,r})$.

Proof: See [Gr] (§ 3, Prop. 3). □

LEMMA (3.2.6). — *For any $h^{\infty,o} \in GL_d(\mathbb{A}^{\infty,o})$ and any $h_o \in GL_d(F_{\overline{\kappa(o)}})$, the intersection*

$$(D^{opp})^{\times} \cap [h^{\infty,o}K_I^{\infty,o}(h^{\infty,o})^{-1} \times h_o GL_d(\mathcal{O}_{\overline{\kappa(o)}})h_o^{-1}]$$

inside $GL_d(\mathbb{A}^{\infty,o}) \times GL_d(F_{\overline{\kappa(o)}})$ is reduced to the identity matrix.

Proof: Let δ be in this intersection and let $F' = F[\delta] \subset D^{opp}$ be the field generated by δ over F.

Let x' be a place of F' dividing a place $x \neq \infty$, o of F. As $gl_d(\mathcal{O}_x)$ is compact, the restriction of x' to $F_x' \cap h_x gl_d(\mathcal{O}_x)h_x^{-1}$ is bounded below ($F_x' = F_x \otimes_F F'$ and h_x is the x-component of $h^{\infty,o}$). Therefore the set of integers $\{x'(\delta^n) = nx'(\delta)|n \in \mathbb{Z}\}$ is bounded below and $x'(\delta) = 0$.

Similarly, $x'(\delta) = 0$ for all places of F' dividing o.

As $F_{\infty} \otimes_F F' \subset (F_{\infty} \otimes_F D)^{opp}$ is also a field, we have proved that δ is a constant ($x'(\delta) = 0$ for all places x' of F').

Finally let $x \in V(I)$, then $x(\det(\delta - 1)) > 0$ so that $x(N_{F'/F}(\delta - 1)) > 0$ where $N_{F'/F}$ is the norm map of F' over F and there exists a place x' dividing x such that $x'(\delta - 1) > 0$. Therefore $\delta = 1$. □

Thanks to the lemmas (3.2.5) and (3.2.6), $\mathrm{Lef}_r(f^{\infty,o})_{(\widetilde{F},\widetilde{o})}$ is equal to the number of double classes

$$(D^{opp})^{\times}(h^{\infty,o}(K_I^{\infty,o} \cap ((g^{\infty,o})^{-1}K_I^{\infty,o}g^{\infty,o})), h_o GL_d(\mathcal{O}_{o,r}))$$

with $h^{\infty,o} \in GL_d(\mathbb{A}^{\infty,o})$ and $h_o \in GL_d(F_{\overline{\kappa}(o)})$ such that

$$(h_o)^{-1}\varepsilon_o\sigma_o(h_o) \in <\varpi_o>$$

and there exists $\delta \in (D^{opp})^{\times}$ with the following properties:

$$\begin{cases} (h^{\infty,o})^{-1}\delta h^{\infty,o} \in K_I^{\infty,o}g^{\infty,o}, \\ (h_o)^{-1}\delta^{-1}N_r(\varepsilon_o)\sigma_o^r(h_o) = 1 \,. \end{cases}$$

Moreover, for each such a double class, the conjugacy class of the corresponding $\delta \in (D^{opp})^{\times}$ is well-defined: if we replace $h^{\infty,o}$ by $\delta_1 h^{\infty,o}k^{\infty,o}$ and h_o by $\delta_1 h_o k_r$ with $\delta_1 \in (D^{opp})^{\times}$, $k^{\infty,o} \in K_I^{\infty,o} \cap ((g^{\infty,o})^{-1}K_I^{\infty,o}g^{\infty,o})$ and $k_r \in GL_d(\mathcal{O}_{o,r})$ then δ must be replaced by $\delta_1^{-1}\delta\delta_1$.

Let $(D^{opp})_\natural^{\times}$ be a set of representatives of the conjugacy classes in $(D^{opp})^{\times}$. Let us say that $\delta \in (D^{opp})^{\times}$ is r-**admissible** (**at the place** o) if there exists $h_o \in GL_d(F_{\overline{\kappa(o)}})$ such that

$$(h_o)^{-1}\delta^{-1}N_r(\varepsilon_o)\sigma_o^r(h_o) = 1$$

(the r-admissibility of δ is in fact a property of the conjugacy class of δ in $(D^{opp})^{\times}$: for each $\delta_1 \in (D^{opp})^{\times}$, $\delta_1 N_r(\varepsilon_o) = N_r(\varepsilon_o)\sigma_o^r(\delta_1)$). For each r-admissible $\delta \in (D^{opp})_\natural^{\times}$, let us fix an element $h_o^\delta \in GL_d(F_{\overline{\kappa(o)}})$ such that

$$(h_o^\delta)^{-1}\delta^{-1}N_r(\varepsilon_o)\sigma_o^r(h_o^\delta) = 1$$

and let us set

$$\gamma_r^\delta = (h_o^\delta)^{-1}\varepsilon_o\sigma_o(h_o^\delta).$$

It is easy to see that $\gamma_r^\delta \in GL_d(F_{o,r})$ $(\sigma_o^r(\gamma_r^\delta) = \gamma_r^\delta)$. Then $\mathrm{Lef}_r(f^{\infty,o})_{(\tilde{F},\tilde{o})}$ is equal to the sum over r-admissible $\delta \in (D^{opp})_\natural^{\times}$ of the number of double classes

$$(D^{opp})_\delta^{\times}[h^{\infty,o}(K_I^{\infty,o} \cap ((g^{\infty,o})^{-1}K_I^{\infty,o}g^{\infty,o})), h_o^\delta h_r GL_d(\mathcal{O}_{o,r})]$$

with $h^{\infty,o} \in GL_d(\mathbb{A}^{\infty,o})$ and $h_r \in GL_d(F_{o,r})$ such that

$$\begin{cases} (h^{\infty,o})^{-1}\delta h^{\infty,o} \in K_I^{\infty,o}g^{\infty,o}, \\ (h_r)^{-1}\gamma_r^\delta\sigma_o(h_r) \in <\varpi_o>_r, \end{cases}$$

where
$$(D^{opp})^{\times}_{\delta} = \{\delta' \in (D^{opp})^{\times} | \delta'\delta = \delta\delta'\}$$
is the centralizer of δ in $(D^{opp})^{\times}$ and

$$< \varpi_o >_r = < \varpi_o > \cap GL_d(F_{o,r}) = GL_d(\mathcal{O}_{o,r}) \begin{pmatrix} \varpi_o & & & 0 \\ & 1 & & \\ & & \ddots & \\ 0 & & & 1 \end{pmatrix} GL_d(\mathcal{O}_{o,r}).$$

But the map

$$h^{\infty,o}(K_I^{\infty,o} \cap ((g^{\infty,o})^{-1}K_I^{\infty,o}g^{\infty,o})) \mapsto h^{\infty,o}K_I^{\infty,o}$$

from the set of classes satisfying

$$(h^{\infty,o})^{-1}\delta h^{\infty,o} \in K_I^{\infty,o}g^{\infty,o}$$

to the set of classes satisfying

$$(h^{\infty,o})^{-1}\delta h^{\infty,o} \in K_I^{\infty,o}g^{\infty,o}K_I^{\infty,o}$$

is clearly bijective. So we have proved

PROPOSITION (3.2.7). — *For each (F, ∞, o)-type $(\widetilde{F}, \widetilde{o})$, $\mathrm{Lef}_r(f^{\infty,o})_{(\widetilde{F},\widetilde{o})}$ is equal to the sum over r-admissible δ's in $(D^{opp})^{\times}_{\natural}$ of the number of double classes*
$$(D^{opp})^{\times}_{\delta}[h^{\infty,o}K_I^{\infty,o}, h_o^{\delta}h_rGL_d(\mathcal{O}_{o,r})]$$
with $h^{\infty,o} \in GL_d(\mathbb{A}^{\infty,o})$ and $h_r \in GL_d(F_{o,r})$ such that
$$\begin{cases} (h^{\infty,o})^{-1}\delta h^{\infty,o} \in K_I^{\infty,o}g^{\infty,o}K_I^{\infty,o}, \\ (h_r)^{-1}\gamma_r\sigma_o(h_r) \in < \varpi_o >_r . \end{cases}$$

□

(3.3) Where the orbital integrals come in

We will now give a formula for $\mathrm{Lef}_r(f^{\infty,o})_{(\widetilde{F},\widetilde{o})}$ in terms of orbital integrals and twisted orbital integrals.

For each r-admissible δ in $(D^{opp})^{\times}_{\natural}$, let $GL_d(\mathbb{A}^{\infty,o})_{\delta}$ be the centralizer of δ in $GL_d(\mathbb{A}^{\infty,o})$ and let $GL_{d,\gamma_r^{\delta}}^{\sigma_o}(F_o)$ be the σ_o-centralizer of γ_r^{δ} in $GL_d(F_{o,r})$ (recall that $GL_{d,\gamma_r^{\delta}}^{\sigma_o}$ is an F_o-group and that $GL_{d,\gamma_r^{\delta}}^{\sigma_o}(F_o)$ is the set of $h_r \in GL_d(F_{o,r})$ such that
$$h_r\gamma_r^{\delta} = \gamma_r^{\delta}\sigma_o(h_r)).$$

We will denote by $dh^{\infty,o}$ (resp. dh_r) the Haar measure on $GL_d(\mathbb{A}^{\infty,o})$ (resp. $GL_d(F_{o,r})$) which is normalized by

$$\int_{K_I^{\infty,o}} dh^{\infty,o} = \mathrm{vol}(K_I^{\infty,o}, dh^{\infty,o}) = 1$$

(resp.

$$\int_{GL_d(\mathcal{O}_{o,r})} dh_r = \mathrm{vol}(GL_d(\mathcal{O}_{o,r}), dh_r) = 1).$$

Let $dh_\delta^{\infty,o}$ (resp. $dh_{\delta,r}$) be an arbitrary but fixed Haar measure on $GL_d(\mathbb{A}^{\infty,o})_\delta$ (resp. $GL_{d,\gamma_r^\delta}^{\sigma_o}(F_o)$).

Let us recall that $f^{\infty,o} \in \mathcal{H}_I^{\infty,o}$ is the characteristic function of $K_I^{\infty,o} g^{\infty,o} K_I^{\infty,o} \subset GL_d(\mathbb{A}^{\infty,o})$ and let us denote by

$$f_{o,r} \in \mathcal{C}_c^\infty(GL_d(F_{o,r})//GL_d(\mathcal{O}_{o,r}))$$

the characteristic function of the double class $< \varpi_o >_r \subset GL_d(F_{o,r})$. We can introduce the **orbital integral**

$$(3.3.1) \qquad O_\delta(f^{\infty,o}, dh_\delta^{\infty,o}) = \int_{GL_d(\mathbb{A}^{\infty,o})_\delta \backslash GL_d(\mathbb{A}^{\infty,o})} f^{\infty,o}((h^{\infty,o})^{-1}\delta h^{\infty,o}) \frac{dh^{\infty,o}}{dh_\delta^{\infty,o}}$$

and the **twisted orbital integral**

$$(3.3.2) \qquad TO_{\gamma_r^\delta}(f_{o,r}, dh_{\delta,r}) = \int_{GL_{d,\gamma_r^\delta}^{\sigma_o}(F_o)\backslash GL_d(F_{o,r})} f_{o,r}((h_r)^{-1}\gamma_r^\delta \sigma_o(h_r)) \frac{dh_r}{dh_{\delta,r}}$$

for each r-admissible $\delta \in (D^{opp})_\natural^\times$. As $f^{\infty,o}$ and $f_{o,r}$ are non-negative functions, these integrals make sense even if they are infinite. In fact we will see in (4.8.9) that they are absolutely convergent.

We have the embeddings (see (2.7))

$$(D^{opp})_\delta^\times \subset ((\mathbb{A}^{\infty,o} \otimes_F D)^{opp})_\delta^\times \subset GL_d(\mathbb{A}^{\infty,o})_\delta$$

and

$$(D^{opp})_\delta^\times \subset ((F_o \otimes_F D)^{opp})_\delta^\times \stackrel{i}{\hookrightarrow} GL_{d,\gamma_r^\delta}^{\sigma_o}(F_o)$$

with the obvious notations, where i is the composite of the inclusion

$$((F_o \otimes_F D)^{opp})_\delta^\times \subset \{h_o \in GL_d(F_{\overline{\kappa(o)}}) | \varepsilon_o \sigma_o(h_o) = h_o \varepsilon_o\}_\delta$$

and the isomorphism

$$\{h_o \in GL_d(F_{\overline{\kappa(o)}}) | \varepsilon_o \sigma_o(h_o) = h_o \varepsilon_o\}_\delta \stackrel{\sim}{\to} GL_{d,\gamma_r^\delta}^{\sigma_o}(F_o)$$

given by $h_o \mapsto (h_o^\delta)^{-1} h_o h_o^\delta$ (we let the reader check the details).

In particular, the quotient

$$(D^{opp})_\delta^\times \backslash (GL_d(\mathbb{A}_\delta^{\infty,o}) \times GL_{d,\gamma_r^\delta}^{\sigma_o}(F_o))$$

makes sense and we can endow it with the measure $dh_\delta^{\infty,o} \times dh_{\delta,r}$ divided by the counting measure $d\delta'$ on $(D^{opp})_\delta^\times$.

The following proposition is an obvious corollary of (3.2.7):

PROPOSITION (3.3.3). — *For each (F, ∞, o)-type $(\widetilde{F}, \widetilde{o})$, $\mathrm{Lef}_r(f^{\infty, o})_{(\widetilde{F}, \widetilde{o})}$ is equal to the sum over r-admissible δ's in $(D^{opp})^\times_\natural$ of the product of*

$$\mathrm{vol}((D^{opp})^\times_\delta \backslash (GL_d(\mathbb{A}^{\infty, o}_\delta) \times GL^{\sigma_o}_{d, \gamma^\delta_r}(F_o)), \frac{dh^{\infty, o}_\delta \times dh_{\delta, r}}{d\delta'})$$

by

$$O_\delta(f^{\infty, o}, dh^{\infty, o}_\delta) TO_{\gamma^\delta_r}(f_{o, r}, dh_{\delta, r})$$

(this product is obviously independent of the choices of the Haar measures $dh^{\infty, o}_\delta$ and $dh_{\delta, r}$).   □

As $r > 0$, $\mathrm{Lef}_r(f^{\infty, o})$ is finite and consequently, for each (F, ∞, o)-type $(\widetilde{F}, \widetilde{o})$, $\mathrm{Lef}_r(f^{\infty, o})_{(\widetilde{F}, \widetilde{o})}$ is finite. This implies the following proposition that we will also check directly:

PROPOSITION (3.3.4). — *For each r-admissible $\delta \in (D^{opp})^\times_\natural$ (recall that $r > 0$), the above embeddings*

$$((\mathbb{A}^{\infty, o} \otimes_F D)^{opp})^\times_\delta \hookrightarrow GL_d(\mathbb{A}^{\infty, o})_\delta$$

and

$$((F_o \otimes_F D)^{opp})^\times_\delta \hookrightarrow GL^{\sigma_o}_{d, \gamma^\delta_r}(F_o)$$

are both isomorphisms. In particular, the volume

$$\mathrm{vol}((D^{opp})^\times_\delta \backslash (GL_d(\mathbb{A}^{\infty, o})_\delta \times GL^{\sigma_o}_{d, \gamma^\delta_r}(F_o)), \frac{dh^{\infty, o}_\delta \times dh_{\delta, r}}{d\delta'})$$

is equal to

$$\mathrm{vol}((D^{opp})^\times_\delta \backslash ((\mathbb{A}^\infty \otimes_F D)^{opp})^\times_\delta, \frac{d\delta'^\infty}{d\delta'})$$

where $d\delta'^\infty$ is the Haar measure on $((\mathbb{A}^\infty \otimes_F D)^{opp})^\times_\delta$ induced by $dh^{\infty, o}_\delta \times dh_{\delta, r}$ and the above isomorphisms, and therefore is finite.

REMARK (3.3.5). — It follows from (3.2.6) that $(D^{opp})^\times_\delta$ is discrete in $((\mathbb{A}^{\infty, o} \otimes_F D)^{opp})^\times_\delta$. In fact $((F_\infty \otimes_F D)^{opp})^\times_\delta$ is anisotropic modulo the split component of its center (as an F-group).   □

We will need the following lemma in the proof of (3.3.4):

LEMMA (3.3.6). — *Let $\delta \in (D^{opp})^\times$ and let $F' = F[\delta]$ be the subfield of D^{opp} generated by δ over F. Then the following conditions are equivalent:*

(i) *δ is r-admissible,*

(ii) *F' contains $\widetilde{F}$ and, if o' is the unique place of F' dividing $\widetilde{o}$ $(\widetilde{F_{\widetilde{o}}} \otimes_{\widetilde{F}} F' \subset D^{opp}_{\widetilde{o}}$ is a field), $o'(\delta) = r[F'_{o'} : F_o]/h[\kappa(o') : \kappa(o)]$ and $x'(\delta) = o$ for all other places x' of F' dividing o $(h = [\widetilde{F_{\widetilde{o}}} : F_o]d/[\widetilde{F} : F])$.*

Proof: Let $N = F_{\overline{\kappa}(o)}^d$ and $f : N \to N$ be the σ_o^r-linear map defined by the matrix $\delta^{-1} N_r(\varepsilon_o) \in GL_d(F_{\overline{\kappa(o)}})$. Then (N, f) is a Dieudonné $F_{o,r}$-module over $\overline{\kappa}(o)$ and δ is r-admissible if and only if (N, f) is purely of slope zero, i.e. étale (δ is r-admissible if and only if there exists a basis of N in which the matrix of f is the identity matrix).

Let $\widetilde{F}' = \widetilde{F}[\delta]$ be the subfield of D^{opp} generated by δ over $\widetilde{F}$ and let $\widetilde{o}'$ be the unique place of $\widetilde{F}'$ dividing $\widetilde{o}$ ($\widetilde{F}_{\widetilde{o}} \otimes_{\widetilde{F}} \widetilde{F}' \subset D_{\widetilde{o}}^{opp}$ is a field). We know that

$$N_r(\varepsilon_o) \sigma_o^r(\delta) = \delta N_r(\varepsilon_o)$$

for each $\delta \in (D^{opp})^\times$. So δ induces an automorphism of (N, f) : in fact, we have embeddings of F_o-algebras

$$
\begin{array}{ccc}
F_o \otimes_F \widetilde{F}' & \longrightarrow & \mathrm{End}(N, f) \\
\uparrow & & \uparrow \\
F_o \otimes_F D^{opp} & \longrightarrow & gl_d(F_{\overline{\kappa(o)}})
\end{array}
$$

($\mathrm{End}(N, f)$ is the σ_o^r-centralizer of $\delta^{-1} N_r(\varepsilon_o)$ in $gl_d(F_{\overline{\kappa(o)}})$). We have the decompositions

$$F_o \otimes_F \widetilde{F}' = \prod_{\widetilde{x}|o} \widetilde{F}'_{\widetilde{x}'} = \prod_{\widetilde{x}|o} \prod_{\widetilde{x}'|\widetilde{x}} \widetilde{F}'_{\widetilde{x}'}$$

($\widetilde{x}$ is a place of $\widetilde{F}$ and $\widetilde{x}'$ is a place of $\widetilde{F}'$). Accordingly we have decompositions

$$(N, f) = \bigoplus_{\widetilde{x}|o} (N_{\widetilde{x}}, f_{\widetilde{x}}) = \bigoplus_{\widetilde{x}|o} \bigoplus_{\widetilde{x}'|\widetilde{x}} (N_{\widetilde{x}'}, f_{\widetilde{x}'})$$

of Dieudonné $F_{o,r}$-modules over $\overline{\kappa}(o)$. Now (N, f) is purely of slope zero if and only if each $(N_{\widetilde{x}'}, f_{\widetilde{x}'})$ is purely of slope zero. But $N_{\widetilde{x}} = F_{\overline{\kappa}(o)}^{n_{\widetilde{x}}}$ where $n_{\widetilde{x}} = [\widetilde{F}_{\widetilde{x}} : F_o]d/[\widetilde{F} : F]$ and the matrix of $f_{\widetilde{x}}$ is $\delta_{\widetilde{x}} \in D_{\widetilde{x}}^{opp} \subset gl_{n_{\widetilde{x}}}(F_o)$ if $\widetilde{x} \neq \widetilde{o}$ and $(\delta_{\widetilde{o}})^{-1} N_r(\varepsilon_{\widetilde{o}}) \in D_{\widetilde{o}}^{opp} \subset gl_h(F_{\overline{\kappa}(o)})$ ($n_{\widetilde{o}} = h$) if $\widetilde{x} = \widetilde{o}$. So $(N_{\widetilde{x}'}, f_{\widetilde{x}'})$ is purely of slope zero if and only if $\widetilde{x}'(\delta) = 0$ for each $\widetilde{x}' \neq \widetilde{o}'$ and $(N_{\widetilde{o}'}, f_{\widetilde{o}'})$ is purely of slope zero if and only if $o(\det((\delta_{\widetilde{o}})^{-1} N_r(\varepsilon_o))) = 0$, i.e. $\widetilde{o}'(\delta) = r[\widetilde{F}'_{\widetilde{o}'} : F_o]/h[\kappa(\widetilde{o}') : \kappa(o)]$ (the details are left to the reader).

Finally let us prove that the conditions $\widetilde{x}'(\delta) = 0$ for all places $\widetilde{x}' \neq \widetilde{o}'$ of $\widetilde{F}'$ dividing o and $\widetilde{o}'(\delta) \neq 0$ imply $F' \supset \widetilde{F}$. Let o' be the restriction of $\widetilde{o}'$ to $F' \subset \widetilde{F}'$. Then $o'(\delta) \neq 0$ and $\widetilde{o}'$ is the unique place of $\widetilde{F}'$ dividing

o'. Let $\widetilde{\infty}'$ (resp. ∞') be the unique place of $\widetilde{F}'$ (resp. F') dividing ∞ ($F_\infty \otimes_F F' \subset F_\infty \otimes_F \widetilde{F}' \subset (F_\infty \otimes_F D)^{opp}$ are fields). If X' is the smooth projective model of F' over $\mathbb{F}_p$, $\mathrm{Pic}^0_{X'/\mathbb{F}_p}(\mathbb{F}_p)$ is a finite group. So there exists $\Pi' \in F'$ such that $\infty'(\Pi') \neq 0$, $o'(\Pi') \neq 0$ and $x'(\Pi') = 0$ for all other places x' of F'. We have $\widetilde{\infty}'(\Pi') \neq 0$, $\widetilde{o}'(\Pi') \neq 0$ and $\widetilde{x}'(\Pi') = 0$ for all other places of $\widetilde{F}'$ if we consider Π' as an element of $\widetilde{F}'$. Now let $\Pi \in \widetilde{F}$ be such that $\widetilde{\infty}(\Pi) \neq 0$, $\widetilde{o}(\Pi) \neq 0$ and $\widetilde{x}(\Pi) = 0$ for all other places $\widetilde{x}$ of $\widetilde{F}$. Then $\widetilde{F} = F[\Pi^N]$ for any integer $N \neq 0$ (($\widetilde{F}, \widetilde{o}$) is an ($F, \infty, o$)-type). But we can find non-zero integers N and N' such that

$$\widetilde{x}'(\Pi'^{N'}/\Pi^N) = 0$$

for all places $\widetilde{x}'$ of $\widetilde{F}'$: we have $\widetilde{\infty}'(\Pi) \neq 0$, $\widetilde{o}'(\Pi) \neq 0$ and $\widetilde{x}'(\Pi) = 0$ for all other places of $\widetilde{F}'$ if we view Π as an element of $\widetilde{F}'$. So $\Pi'^{N'}/\Pi^N$ belongs to the finite field of constants in $\widetilde{F}'$ and multiplying N and N' by a positive integer if necessary we can assume that

$$\Pi'^{N'}/\Pi^N = 1.$$

In particular $\Pi^N = \Pi'^{N'} \in F'$ and $\widetilde{F} = F[\Pi^N]$ is contained in F'. $\qquad\square$

Proof of (3.3.4) : As $\delta \in (D^{opp})^\times_\natural$ is r-admissible it follows from (3.3.6) that $F[\delta] \supset \widetilde{F}$ and the desired equalities of centralizers are obvious.

It is well known that the coset space $F_\infty^\times (D^{opp})^\times \backslash ((\mathbb{A} \otimes_F D)^{opp})^\times$ is compact (see [We 1] (3.1.1)). So the coset space $(D^{opp})^\times \backslash ((\mathbb{A}^\infty \otimes_F D)^{opp})^\times$ and its closed subspace $(D^{opp})^\times_\delta \backslash ((\mathbb{A}^\infty \otimes_F D)^{opp})^\times_\delta$ are compact. Therefore the volume of $(D^{opp})^\times_\delta \backslash ((\mathbb{A}^\infty \otimes_F D)^{opp})^\times_\delta$ for any measure is finite. $\qquad\square$

(3.4) Transfer of conjugacy classes

Let $\gamma \in GL_d(F)$. Let us recall that γ is said to be **elliptic** if the F-subalgebra $F[\gamma]$ of $gl_d(F)$ generated by γ is a field, i.e. if the minimal polynomial of γ is irreducible over F.

Let $\gamma \in GL_d(F)$ be elliptic and let $F' = F[\gamma]$. Then γ is said to be **elliptic at the place** ∞ if and only if $F_\infty \otimes_F F'$ is a field, i.e. there exists only one place ∞' of F' dividing the place ∞ of F. On the other hand, γ is said to be r-**admissible (at the place** o) if and only if $o(\det \gamma) = r$ and there exists a place o' of F' dividing the place o of F such that $x'(\gamma) = 0$ for all the places $x' \neq o'$ of F' which divide o.

All these properties depend only on the conjugacy class of γ in $GL_d(F)$.

Let $\gamma \in GL_d(F)$ be elliptic, elliptic at the place ∞ and r-admissible at the place o; let $F' = F[\gamma]$ and let ∞' be the unique place of F' dividing ∞ and o' be the unique place of F' dividing o and such that $o'(\gamma) \neq o$. We can attach to γ an (F, ∞, o)-type $(\widetilde{F}, \widetilde{o})$ of rank d in the following way. There exists $\Pi' \in F'$ such that $\infty'(\Pi') \neq 0$, $o'(\Pi') \neq 0$ and $x'(\Pi') = 0$ for all other places of F' (see the proof of (3.3.6)). We set

$$\widetilde{F} = \bigcap_{n \in \mathbb{Z} - \{0\}} F[\Pi'^n] \subset F'$$

and we let $\widetilde{\infty}$ and $\widetilde{o}$ be the restriction to $\widetilde{F}$ of the places ∞' and o' of F' respectively. Using the same argument as before Corollary (2.2.3), we check that $(\widetilde{F}, \widetilde{o})$ is an (F, ∞, o)-type of rank d. Let D be "the unique" central division algebra over $\widetilde{F}$ with invariants

$$\mathrm{inv}_{\widetilde{x}}(D) = \begin{cases} -[\widetilde{F} : F]/d & \text{if } \widetilde{x} = \widetilde{\infty}, \\ [\widetilde{F} : F]/d & \text{if } \widetilde{x} = \widetilde{o}, \\ 0 & \text{otherwise.} \end{cases}$$

As $\widetilde{F}_{\widetilde{\infty}} \otimes_{\widetilde{F}} F'$ and $\widetilde{F}_{\widetilde{o}} \otimes_{\widetilde{F}} F'$ are fields (there exists only one place of F' dividing $\widetilde{\infty}$ and only one place of F' dividing $\widetilde{o}$: $F_\infty \otimes_F F'$ is a field and if x' is a place of F' dividing $\widetilde{o}$ we obviously have $x'(\gamma) \neq 0$) and as $[F' : \widetilde{F}]$ divides $d/[\widetilde{F} : F]$ ($[F' : F]$ divides d), there exists at least one embedding of $\widetilde{F}$-algebras

$$F' \hookrightarrow D^{opp}$$

and two such embeddings are conjugate in D^{opp} (see (A.3.3)). In particular, we get an element $\delta \in (D^{opp})^\times$ (the image of γ by such an embedding) which is well-defined up to conjugacy in $(D^{opp})^\times$. Moreover, it follows from (3.3.6) that δ is r-admissible at the place o.

Let $GL_d(F)_{\natural,ell}$ be a system of representatives in $GL_d(F)$ of the elliptic conjugacy classes. Then we have constructed a map

(3.4.1) $\{\gamma \in GL_d(F)_{\natural,ell} | \gamma$ is elliptic at

the place ∞ and r-admissible at the place $o\}$

$$\to \coprod_{(\widetilde{F},\widetilde{o})} \{\delta \in (D^{opp})_\natural^\times | \delta \text{ is } r\text{-admissible at the place } o\}$$

where $(\widetilde{F}, \widetilde{o})$ runs over a system of representatives of the isomorphism classes of (F, ∞, o)-type of rank d and where, for each $(\widetilde{F}, \widetilde{o})$, $(D^{opp})_\natural^\times$ has the same meaning as before.

PROPOSITION (3.4.2). — *The map (3.4.1) is bijective.*

Proof: Let us construct an inverse map to (3.4.1). We start with $((\widetilde{F}, \widetilde{o}), \delta)$ and we fix an embedding of F-algebras

$$\widetilde{F} \hookrightarrow gl_d(F)$$

by choosing a basis of $\widetilde{F}$ over F (so that we can identify $gl_d(F)$ with $\mathrm{End}_F(\widetilde{F}^{d/[\widetilde{F}:F]})$). Any two such embeddings are conjugate. Then the centralizer $gl_d(F)_{\widetilde{F}}$ of $\widetilde{F}$ in $gl_d(F)$ can be identified with $gl_{d/[\widetilde{F}:F]}(\widetilde{F})$ and we have an injective map from the set of conjugacy classes in D^{opp} (a central division algebra over $\widetilde{F}$ of dimension $(d/[\widetilde{F}:F])^2$) into the set of conjugacy classes in $gl_d(F)_{\widetilde{F}}$ (if $\delta_1 \in D^{opp}$, $\widetilde{F}' = \widetilde{F}[\delta_1] \subset D^{opp}$ is a field of degree over $\widetilde{F}$ which divides $d/[\widetilde{F}:F]$ and, consequently, it can be embedded into $gl_d(F)_{\widetilde{F}} \cong gl_{d/[\widetilde{F}:F]}(\widetilde{F})$). Let $\gamma_{\widetilde{F}} \in GL_d(F)_{\widetilde{F}}$ (the centralizer of $\widetilde{F}$ in $GL_d(F)$) be a representative of the image of the conjugacy class of δ by the above injective map. We can consider $\gamma_{\widetilde{F}}$ as an element of $GL_d(F)$. Then it is easy to see that $\gamma_{\widetilde{F}}$ is elliptic, elliptic at ∞ and r-admissible at o. So we can map $((\widetilde{F}, \widetilde{o}), \delta)$ to the representative $\gamma \in GL_d(F)_{\natural,ell}$ of the conjugacy class of $\gamma_{\widetilde{F}}$ in $GL_d(F)$. $\qquad\square$

It follows from the definition of (3.4.1) that we have

PROPOSITION (3.4.3). — *Let $\gamma \in GL_d(F)_{\natural,ell}$ be elliptic at ∞ and r-admissible at o. Let $((\widetilde{F}, \widetilde{o}), \delta)$ be the image of γ by (3.4.1). Then $F[\gamma] \subset gl_d(F)$ is a field isomorphic over F to the field $F' = F[\delta] \subset D^{opp}$. We have $F' \supset \widetilde{F}$ and there are unique places ∞' and o' of F' dividing the places $\widetilde{\infty}$ and $\widetilde{o}$ of $\widetilde{F}$ respectively. The centralizer $gl_d(F)_\gamma$ of γ in $gl_d(F)$ is isomorphic to $gl_{d'}(F')$, where $d' = d/[F':F]$, as an F'-algebra and the centralizer $D' = (D^{opp})_\delta$ of δ in D^{opp} is a central division algebra over F' with invariants $1/d'$ at ∞', $-1/d'$ at o' and 0 elsewhere. $\qquad\square$*

Let us recall that for each (F, ∞, o)-type $(\widetilde{F}, \widetilde{o})$ we have fixed embeddings

$$\widetilde{F} \hookrightarrow gl_d(\mathbb{A}^{\infty,o}),$$
$$\widetilde{F} \hookrightarrow \{h_o \in gl_d(F_{\overline{\kappa(o)}}) | h_o \varepsilon_o(h_o) = h_o \varepsilon_o\}$$

in (2.7) and, consequently, identifications between $(\mathbb{A}^{\infty,o} \otimes_F D)^{opp}$ and the centralizer of $\widetilde{F}$ in $gl_d(\mathbb{A}^{\infty,o})$ and between $(F_o \otimes_F D)^{opp}$ and the centralizer of $\widetilde{F}$ in $\{h_o \in gl_d(F_{\overline{\kappa(o)}}) | h_o \varepsilon_o = \varepsilon_o \sigma_o(h_o)\}$. Let us recall also that, for each $\delta \in (D^{opp})_\natural^\times$, we have fixed $h_o^\delta \in GL_d(F_{\overline{\kappa(o)}})$ such that

$$(h_o^\delta)^{-1} \delta^{-1} N_r(\varepsilon_o) \sigma_o^r(h_o^\delta) = 1$$

and we have set

$$\gamma_r^\delta = (h_o^\delta)^{-1} \varepsilon_o \sigma_o(h_o^\delta) \in GL_d(F_{o,r}).$$

COROLLARY (3.4.4). — *Let* $\gamma \in GL_d(F)_{\natural,ell}$ *be elliptic at* ∞ *and* r-*admissible at* o. *Let* $((\widetilde{F}, \widetilde{o}), \delta)$ *be the image of* γ *by* (3.4.1). *Then:*

(i) *the image of* δ *by the embedding*

$$D^{opp} \subset (\mathbb{A}^{\infty,o} \otimes_F D)^{opp} \subset gl_d(\mathbb{A}^{\infty,o})$$

is conjugate to $\gamma \in gl_d(F) \subset gl_d(\mathbb{A}^{\infty,o})$ *in* $gl_d(\mathbb{A}^{\infty,o})$,

(ii) $N_r(\gamma_r^\delta)$ *is conjugate to* $\gamma \in gl_d(F) \subset gl_d(F_{o,r})$ *in* $gl_d(F_{o,r})$.

Proof: Part (i) is obvious. Similarly the image of δ by the embedding

$$D^{opp} \subset (F_o \otimes_F D)^{opp} \subset \{h_o \in gl_d(F_{\overline{\kappa(o)}}) | h_o \varepsilon_o = \varepsilon_o \delta_o(h_o)\}$$

is conjugate to $\gamma \in gl_d(F) \subset gl_d(F_{\overline{\kappa(o)}})$ in $gl_d(F_{\overline{\kappa(o)}})$. But we have

$$(h_o^\delta)^{-1} \delta h_o^\delta = N_r((h_o^\delta)^{-1} \varepsilon_o \sigma_o(h_o^\delta)) = N_r(\gamma_r^\delta).$$

So γ and $N_r(\gamma_r^\delta)$ are conjugate in $gl_d(F_{\overline{\kappa(o)}})$ and therefore in $gl_d(F_{o,r})$. $\quad\square$

REMARK (3.4.5). — From now on we will denote γ_r^δ simply by γ_r and we will view it as a "function" of γ. If we replace h_o^δ by $h_o^\delta h_r$ with $h_r \in GL_d(F_{o,r})$, we have to replace γ_r by $h_r^{-1} \gamma_r \sigma_o(h_r)$. $\quad\square$

At this point, we have obtained the following formula for $\mathrm{Lef}_r(f^{\infty,o})$:

PROPOSITION (3.4.6). — *The number* $\mathrm{Lef}_r(f^{\infty,o})$ *is equal to the sum over the* γ's *in* $GL_d(F)_{\natural,ell}$ *which are elliptic at* ∞ *and* r-*admissible at* o *of the product of*

$$\mathrm{vol}((D^{opp})_\delta^\times \backslash ((\mathbb{A}^\infty \otimes_F D)^{opp})_\delta^\times, \frac{d\delta'^\infty}{d\delta'}),$$

where $((\widetilde{F}, \widetilde{o}), \delta)$ *is the image of* γ *by* (3.4.1) *and* D *is the corresponding central division algebra over* $\widetilde{F}$, *of*

$$O_\gamma(f^{\infty,o}, dh_\gamma^{\infty,o}) = \int_{GL_d(\mathbb{A}^{\infty,o})_\gamma \backslash GL_d(\mathbb{A}^{\infty,o})} f^{\infty,o}((h^{\infty,o})^{-1} \gamma h^{\infty,o}) \frac{dh^{\infty,o}}{dh_\gamma^{\infty,o}},$$

where $dh_\gamma^{\infty,o}$ *is the Haar measure on* $GL_d(\mathbb{A}^{\infty,o})_\gamma$ *induced by* $dh_\delta^{\infty,o}$ (γ *and* δ *are conjugate in* $GL_d(\mathbb{A}^{\infty,o})$), *and of*

$$TO_{\gamma_r}(f_{o,r}, dh_{\gamma,r}) = \int_{GL_{d,\gamma_r}^{\sigma_o}(F_o) \backslash GL_d(F_{o,r})} f_{o,r}((h_r)^{-1} \gamma_r \sigma_o(h_r)) \frac{dh_r}{dh_{\gamma,r}},$$

where $dh_{\gamma,r}$ *is simply a new notation for* $dh_{\delta,r}(\gamma_r = \gamma_r^\delta)$. $\quad\square$

Again the finiteness of $\mathrm{Lef}_r(f^{\infty,o})$ implies the following proposition that we will also check directly:

PROPOSITION (3.4.7). — *There are only finitely many* $\gamma \in GL_d(F)_{\natural,ell}$ *which are elliptic at* ∞, *r-admissible at o and such that there exists* $h^{\infty,o} \in GL_d(\mathbb{A}^{\infty,o})$ *with*

$$f^{\infty,o}((h^{\infty,o})^{-1}\gamma h^{\infty,o}) \neq 0.$$

For the proof of (3.4.7) we will need the following lemma:

LEMMA (3.4.8). — *Let x be a place of F and let $K_x g_x K_x$ be a K_x-double-class $(K_x = GL_d(\mathcal{O}_x))$. Then there exists a constant*

$$C_x = C_x(K_x g_x K_x) \geq 0$$

with the following property: let $d = d_1 + \cdots + d_n$ be any partition of d (n and $d_1, \ldots, d_n$ are positive integers), let P be the corresponding standard parabolic subgroup of GL_d with its standard Levi decomposition

$$P = MN$$

and its canonical isomorphism

$$M \cong GL_{d_1} \times \cdots \times GL_{d_n}$$

(the unipotent radical of P is N and P contains the Borel subgroup of upper triangular matrices) and let

$$p_x = m_x n_x \in P(F_x) = M(F_x)N(F_x)$$

such that there exists $h_x \in GL_d(F_x)$ with

$$h_x^{-1} p_x h_x \in K_x g_x K_x.$$

Then we have

$$|x(\det(m_{x,i}))| \leq C_x$$

for all $i = 1, \ldots, n$ where

$$(m_{x,1}, \ldots, m_{x,n}) \in GL_{d_1}(F_x) \times \cdots \times GL_{d_n}(F_x)$$

is the image of m_x by the above canonical isomorphism.
Moreover, if $K_x g_x K_x = K_x$ (i.e. $g_x \in K_x$), we can take $C_x = 0$.

Proof : Let $d = d_1 + \cdots + d_n$ be a partition of d and $P = MN$ be the corresponding standard parabolic subgroup. We have the Iwasawa decomposition

$$GL_d(F_x) = P(F_x)K_x.$$

So, on the one hand, we can decompose the K_x-double-class $K_x g_x K_x$ into a finite number of K_x-classes

$$K_x g_x K_x = \coprod_{j \in J} p_x^{(j)} K_x$$

where

$$p_x^{(j)} = m_x^{(j)} n_x^{(j)} \in P(F_x) = M(F_x)N(F_x)$$

for each $j \in J$. On the other hand, in the statement of the lemma, we can replace the hypothesis "there exists $h_x \in GL_d(F_x)$ with $h_x^{-1} p_x h_x \in K_x g_x K_x$" by the hypothesis "$p_x \in K_x g_x K_x$" (the $x(\det(m_{x,i}))$'s depend only on the conjugacy class of p_x in $P(F_x)$). Now $P(F_x) \cap K_x = P(\mathcal{O}_x)$, so the hypothesis "$p_x \in K_x g_x K_x$" implies that there exists $j \in J$ such that

$$x(\det(m_{x,i})) = x(\det(m_{x,i}^{(j)}))$$

for all $i = 1, \ldots, n$ ($(m_{x,1}^{(j)}, \ldots, m_{x,n}^{(j)})$ is the image of $m_x^{(j)}$ in $GL_{d_1}(F_x) \times \cdots \times GL_{d_n}(F_x)$ by the canonical isomorphism) and we can take for C_x the maximum of

$$\sup\{|x(\det(m_{x,i}^{(j)}))| \mid 1 \geq i \geq n, j \in J\}$$

when $d = d_1 + \cdots + d_n$ runs over the set of partitions of d. $\qquad\square$

Proof of (3.4.7) : Let $F' = F[\gamma] \in gl_d(F)$. If γ is elliptic, F' is a field. If γ is elliptic at ∞, $F_\infty \otimes_F F'$ is a field and there exists a unique place ∞' of F' dividing ∞. If γ is r-admissible at o, there exists a unique place o' of F' dividing o and such that $o'(\gamma) \neq 0$; moreover

$$o'(\gamma) = r\deg(o)/\deg(o').$$

Finally, if there exists $h^{\infty,o} \in GL_d(\mathbb{A}^{\infty,o})$ with

$$(h^{\infty,o})^{-1}\gamma h^{\infty,o} \in K_I^{\infty,o} g^{\infty,o} K_I^{\infty,o} \subset K_A^{\infty,o} g^{\infty,o} K_A^{\infty,o},$$

thanks to (3.4.8), we have

$$\deg(x')|x'(\gamma)| \leq \deg(x)C_x(K_x g_x K_x),$$

for all places $x \neq \infty, o$ of F and all places x' of F' dividing x, where g_x is the x-component of $g^{\infty,o}$. Indeed we have $F_x \otimes_F F' = F_x' = \prod_{x'|x} F_{x'}'$,

$1 \otimes \gamma = (\gamma_{x'})_{x'|x}$ and a partition $d = \sum\limits_{x'|x} (d/[F' : F])[F'_{x'} : F_x]$; fixing an embedding of F_x-algebras

$$F'_{x'} \hookrightarrow gl_{[F'_{x'}:F_x]}(F_x)$$

for each $x'|x$, we get an element

$$((\gamma_{x'}, \ldots, \gamma_{x'}))_{x'|x} \in \prod_{x'|x} GL_{[F'_{x'}:F_x]}(F_x)^{d/[F':F]} \cong M(F_x),$$

where γ'_x is repeated $d/[F' : F]$ times and M is the standard Levi subgroup of GL_d associated to our partition of d; obviously γ is conjugate to $((\gamma_{x'}, \ldots, \gamma_{x'}))_{x'|x}$ in $GL_d(F_x)$.

Now let

$$P(T) = T^{[F':F]} + a_1 T^{[F':F]-1} + \cdots + a_{[F':F]} \in F[T]$$

be the minimal polynomial of γ. It is not difficult to see that

$$\deg(x) x(a_i) \geq i \ \inf\left\{ \left[\frac{x'(\gamma)}{e(x'/x)} \right] | x' \text{ divides } x \right\}$$

for all places x of F and each $i = 1, \ldots, [F' : F]$, where $e(x'/x)$ is the ramification index of $F'_{x'}$ over F_x.

Therefore, if γ satisfies the hypotheses of the proposition, the degree $[F' : F]$ of its characteristic polynomial $P(T)$ is bounded above by d and the divisors of the coefficients $a_1, \ldots, a_{[F':F]}$ of $P(T)$ are bounded below. So $P(T)$ can take only a finite number of distinct values and the proposition follows. $\qquad\square$

(3.5) Transfer of Haar measures

Let $\gamma \in GL_d(F)_{\natural,ell}$ be elliptic at ∞ and r-admissible at o. Let $((\widetilde{F}, \widetilde{o}), \delta)$ be its image by (3.4.1). We will now give a formula for the volume

$$\text{vol}((D^{opp})^{\times}_{\delta} \backslash ((\mathbb{A}^{\infty} \otimes_F D)^{opp})^{\times}_{\delta}, \frac{d\delta'^{\infty}}{d\delta'})$$

in terms of $GL_d(\mathbb{A})_{\gamma}$ and $((F_{\infty} \otimes_F D)^{opp})^{\times}_{\delta}$.

Let us fix a Haar measure $d\delta'_{\infty}$ on $((F_{\infty} \otimes_F D)^{opp})^{\times}_{\delta}$. Let dz_{∞} be the Haar measure on $F_{\infty}^{\times}$ normalized by $\text{vol}(\mathcal{O}_{\infty}^{\times}, dz_{\infty}) = 1$. We set

$$F' = F[\delta] = F[\gamma]$$

(F' and $F'_\infty = F_\infty \otimes_F F'$ are fields; $F'^\times$ is at the same time the center of $(D^{opp})^\times_\delta$ and the center of $GL_d(F)_\gamma$; see (3.4.3)). As $F^\times_\infty \backslash ((F_\infty \otimes_F D)^{opp})^\times_\delta$ is compact, we have

$$(3.5.1) \qquad \mathrm{vol}((D^{opp})^\times_\delta \backslash ((\mathbf{A}^\infty \otimes_F D)^{opp})^\times_\delta, \frac{d\delta'^\infty}{d\delta'})$$

$$= \frac{\mathrm{vol}(F^\times_\infty (D^{opp})^\times_\delta \backslash ((\mathbf{A} \otimes_F D)^{opp})^\times_\delta, \frac{d\delta'_\mathbf{A}}{dz_\infty d\delta'})}{\mathrm{vol}(F^\times_\infty \backslash ((F_\infty \otimes_F D)^{opp})^\times_\delta, \frac{d\delta'_\infty}{dz_\infty})}$$

where $d\delta'_\mathbf{A} = d\delta'_\infty \times d\delta'^\infty$ and where both the numerator and the denominator are finite.

We want to construct a Haar measure $d\gamma'_\mathbf{A}$ on $GL_d(\mathbf{A})_\gamma$ such that

$$(3.5.2) \qquad \mathrm{vol}(F^\times_\infty (D^{opp})^\times_\delta \backslash ((\mathbf{A} \otimes_F D)^{opp})^\times_\delta, \frac{d\delta'_\mathbf{A}}{dz_\infty d\delta'})$$

$$= \mathrm{vol}(F^\times_\infty GL_d(F)_\gamma \backslash GL_d(\mathbf{A})_\gamma, \frac{d\gamma'_\mathbf{A}}{dz_\infty d\gamma'})$$

where $d\gamma'$ is the counting measure on $GL_d(F)_\gamma$. Thanks to (3.4.3) we can identify $(D^{opp})^\times_\delta$ with the central division algebra D' over F' with invariants $1/d'$ at ∞', $-1/d'$ at o' and 0 elsewhere ($d' = d/[F':F]$) and $gl_d(F)_\gamma$ with $gl_{d'}(F')$. So we have

$$F^\times_\infty (D^{opp})^\times_\delta \backslash ((\mathbf{A} \otimes_F D)^{opp})^\times_\delta = F^\times_\infty D'^\times \backslash (\mathbf{A}' \otimes_{F'} D')^\times$$

and

$$F^\times_\infty GL_d(F)_\gamma \backslash GL_d(\mathbf{A})_\gamma = F^\times_\infty GL_{d'}(F') \backslash GL_{d'}(\mathbf{A}'),$$

where $\mathbf{A}'$ is the ring of adeles of F', we have Haar measures

$$\begin{cases} d\delta'_{\mathbf{A}'} &= d\delta'_\mathbf{A} \text{ on } (\mathbf{A}' \otimes_{F'} D')^\times, \\ dz_\infty & \text{ on } F^\times_\infty, \\ d\delta' & \text{ on } D'^\times, \\ d\gamma' & \text{ on } GL_{d'}(F'), \end{cases}$$

and we want to construct a Haar measure $d\gamma'_{\mathbf{A}'} = d\gamma'_\mathbf{A}$ on $GL_{d'}(\mathbf{A}')$ such that

$$(3.5.3) \qquad \mathrm{vol}(F^\times_\infty D'^\times \backslash (\mathbf{A}' \otimes_{F'} D')^\times, \frac{d\delta'_{\mathbf{A}'}}{dz_\infty d\delta'})$$

$$= \mathrm{vol}(F^\times_\infty GL_{d'}(F') \backslash GL_{d'}(\mathbf{A}'), \frac{d\gamma'_{\mathbf{A}'}}{dz_\infty d\gamma'}).$$

As $GL_{d'}(F')$ is an inner twist of $D'^\times$ as an F'-group scheme there is a general method to construct $d\gamma'_{\mathbb{A}'}$ (see [We 1] and [Ko 1] §1). Let us recall it.

Let k be a field and let k^s be a separable closure of k. Let G and H be two connected reductive k-group schemes. An **inner twisting** between G and H is an isomorphism of k^s-group schemes

$$\psi : k^s \otimes_k G \xrightarrow{\sim} k^s \otimes_k H$$

such that, for each $\sigma \in \mathrm{Gal}(k^s/k)$, there exists at least one $g_\sigma \in G(k_s)$ with

$$\psi^{-1} \circ \sigma(\psi) = \mathrm{Int}(g_\sigma).$$

If there exists an inner twisting between G and H one says that G is an **inner twist** of H.

If we fix such an inner twisting, then it induces an isomorphism of k^s-Lie-algebras

$$\mathrm{Lie}(\psi) : k^s \otimes_k \mathrm{Lie}(G) \xrightarrow{\sim} k^s \otimes_k \mathrm{Lie}(H)$$

with

$$\mathrm{Lie}(\psi)^{-1} \circ \sigma(\mathrm{Lie}(\psi)) = \mathrm{Ad}(g_\sigma)$$

for each $\sigma \in \mathrm{Gal}(k^s/k)$. Therefore, if we set

$$n = \dim_k G = \dim_k H,$$

we get an isomorphism

$$\Lambda^n \mathrm{Lie}(\psi) : k^s \otimes_k \Lambda^n \mathrm{Lie}(G) \xrightarrow{\sim} k^s \otimes_k \Lambda^n \mathrm{Lie}(H)$$

of 1-dimensional k^s-vector spaces with

$$\sigma(\Lambda^n \mathrm{Lie}(\psi)) = \Lambda^n \mathrm{Lie}(\psi),$$

i.e. an isomorphism

$$\Lambda^n \mathrm{Lie}(G) \xrightarrow{\sim} \Lambda^n \mathrm{Lie}(H)$$

of 1-dimensional k-vector spaces. If we replace ψ by $\psi \circ \mathrm{Int}(g)$ for some $g \in G(k^s)$ this last isomorphism is not changed. Moreover we have an exact sequence of groups

$$G(k^s) \xrightarrow{\mathrm{Int}} \mathrm{Aut}(k^s \otimes_k G) \to \Gamma \to 1$$

where $\mathrm{Aut}(k^s \otimes_k G)$ is the group of automorphisms of the k^s-group scheme $k^s \otimes_k G$ and Γ is the group of outer automorphisms of the k^s-group scheme $k^s \otimes_k G$ and is isomorphic to the group of automorphisms of any based root datum associated to $k^s \otimes_k G$ (if $k^s \otimes_k G$ is semi-simple, Γ is finite; if $k^s \otimes_k G$ is an n-dimensional torus, Γ is isomorphic to $GL_n(\mathbb{Z})$). Therefore if we replace η by another inner twisting the isomorphism

$$\Lambda^n \operatorname{Lie}(G) \xrightarrow{\sim} \Lambda^n \operatorname{Lie}(H)$$

associated to ψ is simply multiplied by a root of unity in k.

If k is a non-archimedean local field and if $|dx|$ is the Haar measure on k which gives volume 1 to the ring of integers of k, then any non-zero $\omega_G \in \Lambda^n \operatorname{Lie}(G)^*$ (resp. $\omega_H \in \Lambda^n \operatorname{Lie}(H)^*$) gives rise to a **Haar measure** $|\omega_G|$ (resp. $|\omega_H|$) **on** $G(k)$ (resp. $H(k)$). If

$$\omega_G = \alpha(x_1, \ldots, x_n) dx_1 \wedge \cdots \wedge dx_n$$

(resp.

$$\omega_H = \beta(y_1, \ldots, y_n) dy_1 \wedge \cdots \wedge dy_n)$$

in local analytic coordinates on $G(k)$ (resp. $H(k)$), then

$$|\omega_G| = |\alpha(x_1, \ldots, x_n)||dx_1| \cdots |dx_n|$$

(resp.

$$|\omega_H| = |\beta(y_1, \ldots, y_n)||dy_1| \cdots |dy_n|)$$

in the corresponding local chart. Under the hypothesis that G is an inner twist of H, we can transfer a Haar measure dh on $H(k)$ to $G(k)$ in the following way. We choose a non-zero $\omega_H \in \Lambda^n \operatorname{Lie}(H)^*$. Then there exists a positive real number c such that

$$dh = c|\omega_H|.$$

Let ω_G be the image of ω_H by the transpose of the isomorphism

$$\Lambda^n \operatorname{Lie}(G) \xrightarrow{\sim} \Lambda^n \operatorname{Lie}(H)$$

associated to some inner twisting ψ. Then $|\omega_G|$ is independent of the choice of ψ and we can set

$$dg = c|\omega_G|.$$

Obviously, dg is independent of the choice of ω_H and is called the **transfer of** dh (**from** H **to its inner twist** G).

If k is a function field with completion k_v at the place v and ring of adeles $\mathbb{A}$, we can transfer a Haar measure $dh_{\mathbb{A}}$ on $H(\mathbb{A})$ to an inner twist G of H. We decompose $dh_{\mathbb{A}}$ as an absolutely convergent product

$$dh_{\mathbb{A}} = \prod_v dh_v$$

of Haar measures dh_v on $H(k_v)$. For each v, let dg_v be the transfer of dh_v from $k_v \otimes_k H$ to its inner twist $k_v \otimes_k G$. For almost all v, $k_v \otimes_k G$ is in fact isomorphic to $k_v \otimes_k H$. Therefore the product of Haar measures

$$\prod_v dg_v$$

is also absolutely convergent and defines a Haar measure $dg_\mathbb{A}$ on $G(\mathbb{A})$ which is clearly independent of the decomposition $dh_\mathbb{A} = \prod_v dh_v$. We call $dg_\mathbb{A}$ the **transfer of $dh_\mathbb{A}$ (from H to its inner twist G)**.

Now, let $d\gamma'_\mathbb{A'}$ be the transfer of $d\delta'_\mathbb{A'}$ from $D'^\times$ to its inner twist $GL_{d'}(F')$ (as an F'-group scheme).

PROPOSITION (3.5.4). — *For this choice of $d\gamma'_\mathbb{A'}$ the equality (3.5.3) (and consequently (3.5.2) with $d\gamma'_\mathbb{A} = d\gamma'_\mathbb{A'}$) holds.*

Proof: Let ω_G be the usual volume form on $GL_{d'}$, i.e.

$$\omega_G = \det(\gamma)^{-d'} d\gamma_{11} \wedge d\gamma_{12} \wedge \cdots \wedge d\gamma_{d',d'-1} \wedge d\gamma_{d',d'}.$$

The **Tamagawa measure** of $GL_{d'}(\mathbb{A})$ is defined in the following way (see [We 1] (3.1)):

$$\tau_G = q'^{d'^2(1-g')} \prod_{x'} \lambda_{x'}^{-1} |\omega_G|_{x'}$$

where q' is the number of elements of the field of constants in F', where g' is the genus of F', where

$$\lambda_{x'} = 1 - p^{-\deg(x')}$$

is the convergence factor, where $|\omega_G|_{x'}$ is the Haar measure on $GL_{d'}(F'_{x'})$ associated to ω_G as above and where x' runs over the set of places of F' (if X' is the smooth projective model,

$$q'^{(g'-1)} = \frac{|H^1(X', \mathcal{O}_{X'})|}{|H^0(X', \mathcal{O}_{X'})|} = \text{vol}(F' \backslash \mathbb{A}', \prod_{x'} da'_{x'})$$

where $da'_{x'}$ is the Haar measure on $F'_{x'}$ which is normalized by $\text{vol}(\mathcal{O}'_{x'}, da'_{x'}) = 1$ for each place x' of F'). Let

$$GL_{d'}(\mathbb{A}')^1 = \{(\gamma'_{x'}) \in GL_{d'}(\mathbb{A}') | \sum_{x'} \deg(x')x'(\det(\gamma'_{x'})) = 0\}$$

then the **Tamagawa number** of $GL_{d',F'}$ is defined by

$$\tau(GL_{d',F'}) = \text{vol}(GL_{d'}(F') \backslash GL_{d'}(\mathbb{A}')^1, \frac{\tau_G}{d\gamma'}).$$

Thanks to [We 1] (2.4.3) and (3.3.1), we have

$$\tau(GL_{d',F'}) = \tau(GL_{1,F'})$$

(as

$$\mathrm{vol}(\mathcal{O}'^\times, \prod_{x'} \lambda_{x'}^{-1} \frac{d\gamma'_{x'}}{\gamma'_{x'}}) = 1$$

we have

$$\tau(GL_{1,F'}) = q'^{(1-g')} \frac{|\operatorname{Pic}^0_{X'/\mathbb{F}_p}(\mathbb{F}_p)|}{q'-1}).$$

Let ω_D be the image of ω_G by $\Lambda^{d'^2} \operatorname{Lie}(\psi)^*$ for some inner twisting

$$\psi : (F'^s \otimes_{F'} D')^\times \xrightarrow{\sim} GL_{d'}(F'^s)$$

(F'^s is a separable closure of F'). Then we can start with the F'-rational volume form ω_D to define the Tamagawa measure τ_D on $(\mathbb{A}' \otimes_{F'} D')^\times$. Moreover we can take the same convergent factors (for almost all x', the pair $(D'^\times_{x'}, |\omega_D|_{x'})$ is isomorphic to the pair $(GL_{d'}(F'_{x'}), |\omega_G|_{x'})$). So the Tamagawa measure of $(\mathbb{A}' \otimes_{F'} D')^\times$ is equal to (see [We 1] (3.1))

$$\tau_D = q'^{d'^2(1-g')} \prod_{x'} \lambda_{x'}^{-1} |\omega_D|_{x'}$$

and is the transfer of τ_G from $GL_{d',F'}$ to its inner twist $D'^\times$ (viewed as an F'-group scheme). Let

$$(\mathbb{A}' \otimes_{F'} D')^{\times 1} = \{(\delta'_{x'}) \in (\mathbb{A}' \otimes_{F'} D')^\times \mid \sum_{x'} \deg(x')x'(rn(\delta'_{x'})) = 0\}$$

where rn is the reduced norm, then the Tamagawa number of $D'^\times$ is defined by

$$\tau(D'^\times) = \mathrm{vol}(D'^\times \backslash (\mathbb{A}' \otimes_{F'} D')^{\times 1}, \frac{\tau_D}{d\delta'}).$$

Again thanks to [We 1] (2.4.3) and (3.3.1), we have

$$\tau(D'^\times) = \tau(GL_{1,F'}).$$

But as τ_G is the transfer of τ_D from $D'^\times$ to its inner twist $GL_{d',F'}$, the ratio between the left hand side of (3.5.3) and its right hand side is clearly equal to the ratio $\tau(D'^\times)/\tau(GL_{d',F'})$, so is equal to 1. $\square$

(3.6) The Lefschetz numbers as sums of twisted orbital integrals

Let us summarize the main results of this chapter.

Let $\gamma \in GL_d(F)_{\natural, ell}$ be elliptic at ∞ and r-admissible at o. Let $F' = F[\gamma] \subset gl_d(F)$ be the sub-field generated by γ over F. Let ∞' be the unique place of F' dividing ∞ and let o' be the unique place of F' dividing o and such that $o'(\gamma) \neq 0$. Let D' be "the" central division algebra over F' with invariants $1/d'$ at ∞', $-1/d'$ at o' and 0 elsewhere with $d' = d/[F' : F]$. Then $D'^\times$ is an inner twist of $GL_d(F)_\gamma = GL_{d'}(F')$ as an F'-group scheme. In particular, for each place x of F, we have

$$D'^\times_x = \prod_{x'|x} D'^\times_{x'},$$

$$GL_d(F_x)_\gamma = \prod_{x'|x} GL_{d'}(F'_{x'})$$

and $D'^\times_{x'}$ is an inner twist of $GL_{d'}(F'_{x'})$ for each place x' of F' dividing x. Here we have set

$$\begin{cases} D'_x = F_x \otimes_F D', \\ D'_{x'} = F'_{x'} \otimes_{F'} D'. \end{cases}$$

We arbitrarily choose Haar measures $d\gamma'_\infty$, $d\gamma'^{\infty,o}$ and $d\gamma'_o$ on $GL_d(F_\infty)_\gamma$, $GL_d(\mathbb{A}^{\infty,o})_\gamma$ and $GL_d(F_o)_\gamma$ respectively. We denote by $d\gamma'_\mathbb{A}$ the product Haar measure $d\gamma'_\infty \times d\gamma'^{\infty,o} \times d\gamma'_o$ on $GL_d(\mathbb{A})_\gamma$ and by $d\gamma'$ the counting measure on $GL_d(F)_\gamma$.

We set

$$\overline{GL_d(F_\infty)}_\gamma = D'^\times_\infty.$$

This is an inner twist of $GL_d(F_\infty)_\gamma$ such that $F^\times_\infty \backslash \overline{GL_d(F_\infty)}_\gamma$ is compact and we denote by $d\overline{\gamma}'_\infty$ the transfer of the Haar measure $d\gamma'_\infty$ from $GL_d(F_\infty)_\gamma$ to $\overline{GL_d(F_\infty)}_\gamma$.

Let us recall that we have constructed an element $\gamma_r \in GL_d(F_{o,r})$ such that $N_r(\gamma_r)$ is conjugate to γ in $GL_d(F_{o,r})$ and $GL^{\sigma_o}_{d,\gamma_r}(F_o)$ is isomorphic to $D'^\times_o$ as an F'_o-group scheme (γ_r is not unique but is uniquely determined by γ up to the equivalence relation $\gamma_r \sim h_r^{-1} \gamma_r \sigma_o(h_r)$ for some $h_r \in GL_d(F_{o,r})$). We denote by $d\gamma'_r$ the transfer of the Haar measure $d\gamma'_o$ from $GL_d(F_o)_\gamma$ to its inner twist $GL^{\sigma_o}_{d,\gamma_r}(F_o)$ as an F'_o-group scheme (F'_o is not a field in general but is a product of fields; the definitions of inner twisting and transfer of Haar measures can obviously be extended to cover this case).

Following Kottwitz ([Ko 2]) we introduce the sign

$$\varepsilon(\gamma) = (-1)^{d'-1}$$

(we have

$$d' - 1 = rk_{F'_\infty}(GL_d(F_\infty)_\gamma) - rk_{F'_\infty}(\overline{GL_d(F_\infty)}_\gamma)$$

and
$$d' - 1 = rk_{F'_o}(GL_d(F_o)_\gamma) - rk_{F'_o}(GL^{\sigma_o}_{d,\gamma_r}(F_o)).$$

Here, for a connected reductive group scheme G over a product of fields k, $rk_k(G)$ is the dimension of any maximal k-split torus of G).

Then we can introduce the following four quantities associated to γ:

$$(3.6.1) \qquad \text{vol}(F^\times_\infty GL_d(F)_\gamma \backslash GL_d(\mathbb{A})_\gamma, \frac{d\gamma'_\mathbb{A}}{dz_\infty d\gamma'}),$$

$$(3.6.2) \qquad \frac{\varepsilon(\gamma)}{\text{vol}(F^\times_\infty \backslash \overline{GL_d(F_\infty)}_\gamma, \frac{d\bar{\gamma}'_\infty}{dz_\infty})},$$

$$(3.6.3) \qquad O_\gamma(f^{\infty,o}, d\gamma'^{\infty,o}) = \int_{GL_d(\mathbb{A}^{\infty,o})_\gamma \backslash GL_d(\mathbb{A}^{\infty,o})} f^{\infty,o}((h^{\infty,o})^{-1}\gamma h^{\infty,o}) \frac{dh^{\infty,o}}{d\gamma'^{\infty,o}}$$

(recall that $dh^{\infty,o}$ is the Haar measure on $GL_d(\mathbb{A}^{\infty,o})$ which is normalized by $\text{vol}(K^{\infty,o}_I, dh^{\infty,o}) = 1$), and

$$(3.6.4) \qquad \varepsilon(\gamma)TO_{\gamma_r}(f_{o,r}, d\gamma'_r)$$

$$= \varepsilon(\gamma) \int_{GL^{\sigma_o}_{d,\gamma_r}(F_o) \backslash GL_d(F_{o,r})} f_{o,r}((h_r)^{-1}\gamma_r \sigma_o(h_r)) \frac{dh_r}{d\gamma'_r}$$

(recall that $f_{o,r}$ is the characteristic function of the double class

$$GL_d(\mathcal{O}_{o,r}) \begin{pmatrix} \varpi_o & & & 0 \\ & 1 & & \\ & & \ddots & \\ 0 & & & 1 \end{pmatrix} GL_d(\mathcal{O}_{o,r}) \subset GL_d(F_{o,r})$$

and that dh_r is the Haar measure on $GL_d(F_{o,r})$ which is normalized by $\text{vol}(GL_d(\mathcal{O}_{o,r}), dh_r) = 1$).

Obviously the product of the four quantities (3.6.1) to (3.6.4) is independent of the choices of the Haar measures $d\gamma'_\infty$, $d\gamma'^{\infty,o}$ and $d\gamma'_o$.

THEOREM (3.6.5). — *The number* $\text{Lef}_r(f^{\infty,o})$ *is equal to the sum over the γ's in $GL_d(F)_{\natural,ell}$ which are elliptic at ∞ and r-admissible at o of the products of the four quantities* (3.6.1) *to* (3.6.4) *associated to γ (in fact in this sum only finitely many γ's give a non-zero contribution).* $\qquad \square$

REMARK (3.6.6). — The introduction of the sign $\varepsilon(\gamma)$ will be justified in the next two chapters.

(3.7) Comments and references

In [Dr 2], Drinfeld gives a formula for $\mathrm{Lef}_r(f^{\infty,o})$ directly in terms of orbital integrals (Drinfeld considers only the case $d = 2$, but see [Fl–Ka]). Here, I am following Kottwitz's point of view: the most natural formula for this number of fixed points involves twisted orbital integrals and to obtain Drinfeld's formula one needs a fundamental lemma.

In 1986, in a course at Orsay, Kottwitz explained his point of view in the case of the modular curves associated to GL_2 over $\mathbb{Q}$. The presentation and a large part of the material of this chapter are directly inspired by the notes of this course.

4

The fundamental lemma

(4.0) Introduction

The purpose of this chapter is to replace the twisted orbital integrals $TO_{\gamma_r}(f_{o,r}, dz_r)$ which appear in the formula (3.6.5) for $\mathrm{Lef}_r(f^{\infty,o})$ by ordinary orbital integrals.

The results and the proofs of this chapter are purely local. So we will simply denote by F the local field F_o or more generally any non-archimedean local field. We will denote by $\mathcal{O}$ its ring of integers, by v its discrete valuation, by $\varpi \in \mathcal{O}$ a uniformizer ($v(\varpi) = 1$), by $k = \mathcal{O}/(\varpi)$ its residue field and by q the number of elements of k.

We fix an integer $r \geq 1$ and we denote by k_r "the" finite extension of k of degree r (k_r is a finite field with q^r elements) and by F_r the unramified extension of F with residue field extension k_r over k. We denote by $\mathcal{O}_r$ the ring of integers of F_r and by v_r the discrete valuation of F_r; the restriction of v_r to $\mathcal{O} \subset \mathcal{O}_r$ is v and $\varpi \in \mathcal{O}_r$ is a uniformizer of $\mathcal{O}_r$.

Let $G = GL_d$ for a positive integer d. We set

$$\begin{cases} K &= G(\mathcal{O}) \subset G(F), \\ K_r &= G(\mathcal{O}_r) \subset G(F_r), \end{cases}$$

and we denote by

$$\begin{cases} \mathcal{H} &= \mathcal{C}_c^\infty(G(F)//K), \\ \mathcal{H}_r &= \mathcal{C}_c^\infty(G(F_r)//K_r) \end{cases}$$

the corresponding **Hecke algebras**. K (resp. K_r) is a maximal compact subgroup of the topological group $G(F)$ (resp. $G(F_r)$) and $\mathcal{H}$ (resp. $\mathcal{H}_r$) is the $\mathbb{Q}$-vector space of K-bi-invariant (resp. K_r-bi-invariant) functions with compact supports

$$f : G(F) \to \mathbb{Q}$$

(resp.

$$f_r : G(F_r) \to \mathbb{Q});$$

the product on $\mathcal{H}$ (resp. $\mathcal{H}_r$) is given by the convolution product

$$(f' * f'')(g) = \int_{G(F)} f'(h)f''(h^{-1}g)dh$$

(resp.

$$(f'_r * f''_r)(g_r) = \int_{G(F_r)} f'_r(h_r)f''_r(h_r^{-1}g_r)dh_r)$$

where dh (resp. dh_r) is the Haar measure on $G(F)$ (resp. $G(F_r)$) which is normalized by

$$\int_K dh = 1 \quad (\text{resp.} \int_{K_r} dh_r = 1).$$

(4.1) Satake isomorphism
Following Satake we will analyze the structure of the Hecke algebra $\mathcal{H}$. Similar results will obviously hold for $\mathcal{H}_r$.

It follows from the theorem of elementary divisors ([Bou] Alg. VII.6) that we have the **Cartan decomposition**

(4.1.1) $$G(F) = \coprod_\lambda K\varpi^\lambda K$$

where λ runs over the set

$$\{\lambda = (\lambda_1, \ldots, \lambda_d) \in \mathbb{Z}^d | \lambda_1 \geq \lambda_2 \geq \cdots \geq \lambda_d\}$$

and where we have set

$$\varpi^\lambda = \begin{pmatrix} \varpi^{\lambda_1} & & 0 \\ & \ddots & \\ 0 & & \varpi^{\lambda_d} \end{pmatrix} \in G(F).$$

In particular, we get a basis of the $\mathbb{Q}$-vector space $\mathcal{H}$ by considering the characteristic functions f_λ of the double classes $K\varpi^\lambda K$.

Let $\mathcal{A} = \mathbb{Q}[\sqrt{q}, 1/\sqrt{q}] \subset \mathbb{C}$. We will consider the right K-invariant function

(4.1.2) $$\phi_z : G(F) \to \mathcal{A}[z_1, z_1^{-1}, \ldots, z_d, z_d^{-1}]$$

defined in the following way. Let $B \subset G$ be the Borel subgroup of upper triangular matrices; we have the **Iwasawa decomposition** ([We 2] Ch. II, §2, Thm. 1)

$$(4.1.3) \qquad G(F) = B(F)K \ , \ B(F) \cap K = B(\mathcal{O}).$$

So it is enough to define the right $B(\mathcal{O})$-invariant function $\phi_z|B(F)$. Let

$$(4.1.4) \qquad \delta_{B(F)} : B(F) \to q^{\mathbb{Z}} \subset \mathbb{Q}^{\times}$$

be the modulus character of $B(F)$ (if db is any left or right Haar measure on $B(F)$, then $\delta_{B(F)}(b') = d(b'bb'^{-1})/db$ for each $b' \in B(F)$); in fact

$$\delta_{B(F)}(b) = \prod_{i<j} |b_{ii}/b_{jj}|$$

where

$$|a| = q^{-v(a)}$$

for each $a \in F$. Let

$$(4.1.5) \qquad \chi_z : B(F) \to \mathbb{Q}[z_1, z_1^{-1}, \ldots, z_d, z_d^{-1}]$$

be the quasi-character defined by

$$\chi_z(b) = z_1^{v(b_{11})} \cdots z_d^{v(b_{dd})}.$$

Then we set

$$\phi_z|B(F) = \delta_{B(F)}^{1/2} \chi_z$$

(it is obvious that $\delta_{B(F)}^{1/2}$ and χ_z are right $B(\mathcal{O})$-invariant).

Now the **Satake transform** of $f \in \mathcal{A} \otimes \mathcal{H}$ ($f : G(F) \to \mathcal{A}$) is defined by

$$(4.1.6) \qquad f^{\vee}(z) = \int_{G(F)} f(g)\phi_z(g)dg \in \mathcal{A}[z_1, z_1^{-1}, \ldots, z_d, z_d^{-1}]$$

(recall that dg is the Haar measure on $G(F)$ normalized by $\mathrm{vol}(K, dg) = 1$).

If $P \subset G$ is a standard parabolic subgroup (i.e. $B \subset P \subset G$) and if $P = MN$ is its standard Levi decomposition (N is the unipotent radical of P and $M \subset P$ is the Levi subgroup which contains the torus $T \subset B \subset G$ of diagonal matrices), we will endow $M(F)$ and $N(F)$ with the Haar measures dm and dn normalized by $\mathrm{vol}(M(\mathcal{O}), dm) = 1$ and $\mathrm{vol}(N(\mathcal{O}), dn) = 1$ respectively. Then we have the following integration formula:

$$(4.1.7) \qquad \int_{G(F)} \varphi(g)dg = \int_{M(F)} \int_{N(F)} \int_K \varphi(mnk)dk \ dn \ dm$$

for any locally constant function with compact supports $\varphi : G(F) \to \mathbf{C}$ (dk is the Haar measure on K normalized by $\mathrm{vol}(K, dk) = 1$, i.e. the restriction of dg to K): we have the Iwasawa decomposition

$$G(F) = M(F)N(F)K \ , \ M(F)N(F) \cap K = M(\mathcal{O})N(\mathcal{O}).$$

We can consider the Hecke algebra

$$(4.1.8) \qquad\qquad \mathcal{H}_M = \mathcal{C}_c^\infty(M(F)//M(\mathcal{O}))$$

and, for $f \in \mathcal{A} \otimes \mathcal{H}$, we can define its **constant term along** P, $f^P \in \mathcal{A} \otimes \mathcal{H}_M$, by

$$(4.1.9) \qquad\qquad f^P(m) = \delta_{P(F)}^{1/2}(m) \int_{N(F)} f(mn)dn$$

for each $m \in M(F)$, where

$$(4.1.10) \qquad\qquad \delta_{P(F)} : P(F) \to q^{\mathbf{Z}}$$

is the modulus character of $P(F)$.

We also have a function

$$(4.1.11) \qquad\qquad \phi_{M,z} : M(F) \to \mathcal{A}[z_1, z_1^{-1}, \ldots, z_d, z_d^{-1}]$$

defined by

$$\phi_{M,z}(m) = \delta_{P(F)}^{-1/2}(m)\phi_z(m)$$

for each $m \in M(F)$ and the **Satake transform** of $f \in \mathcal{A} \otimes \mathcal{H}_M$ ($f : M(F) \to \mathcal{A}$) is defined by

$$(4.1.12) \qquad f^\vee(z) = \int_{M(F)} f(m)\phi_{M,z}(m)dm \in \mathcal{A}[z_1, z_1^{-1}, \ldots, z_d, z_d^{-1}].$$

LEMMA (4.1.13). — *The Satake transformations*

$$\mathcal{A} \otimes \mathcal{H} \xrightarrow{\ (-)^\vee\ } \mathcal{A}[z_1, z_1, \ldots, z_d, z_d^{-1}]$$

and

$$\mathcal{A} \otimes \mathcal{H}_M \xrightarrow{\ (-)^\vee\ } \mathcal{A}[z_1, z_1, \ldots, z_d, z_d^{-1}]$$

and the constant term along P

$$\mathcal{A} \otimes \mathcal{H} \xrightarrow{\ (-)^P\ } \mathcal{A} \otimes \mathcal{H}_M$$

are homomorphisms of $\mathcal{A}$-algebras. Moreover the diagram

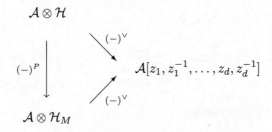

is commutative.

Proof : This is an easy consequence of the integration formula (4.1.7) applied to B and P. $\qquad\square$

REMARK (4.1.14). — For $P = B$ and consequently for $M = T$, we have

$$\mathcal{H}_M = \mathcal{H}_T = \bigoplus_{\lambda \in \mathbf{Z}^d} \mathbf{Q} \cdot f_{T,\lambda}$$

where $f_{T,\lambda}$ is the characteristic function of the double class

$$T(\mathcal{O})\varpi^\lambda T(\mathcal{O}) \subset T(F).$$

Moreover, for each $\lambda, \lambda', \lambda'' \in \mathbf{Z}^d$, we have

$$f_{T,\lambda'} * f_{T,\lambda''} = f_{T,\lambda'+\lambda''}$$

and

$$(f_{T,\lambda})^\vee(z) = z^\lambda = z_1^{\lambda_1} \cdots z_d^{\lambda_d}.$$

So in this case the Satake transformation comes from the isomorphism of $\mathbf{Q}$-algebras

$$\mathcal{H}_T \xrightarrow{\sim} \mathbf{Q}[z_1, z_1^{-1}, \ldots, z_d, z_d^{-1}]$$

which maps $f_{T,\lambda}$ to z^λ.

If P is given by the partition

$$d_1 + \cdots + d_s = d$$

of d so that

$$M = GL_{d_1} \times \cdots \times GL_{d_s}$$

we have obvious $\mathbb{Q}$-algebra isomorphisms

$$\mathcal{H}_M \cong \mathcal{H}_{GL_{d_1}} \otimes \cdots \otimes \mathcal{H}_{GL_{d_s}}$$

and

$$\mathbb{Q}[z_1, z_1^{-1}, \ldots, z_d, z_d^{-1}] \cong \mathbb{Q}[z_1, z_1^{-1}, \ldots, z_{d_1}, z_{d_1}^{-1}] \otimes \cdots$$
$$\cdots \otimes \mathbb{Q}[z_{d_1 + \cdots + d_{s-1} + 1}, z_{d_1 + \cdots + d_{s-1} + 1}^{-1}, \ldots, z_d, z_d^{-1}]$$

and the Satake transformation

$$(-)^\vee : \mathcal{A} \otimes \mathcal{H}_M \to \mathcal{A}[z_1, z_1^{-1}, \ldots, z_d, z_d^{-1}]$$

is the tensor product of the Satake transformations for GL_{d_j} $(j = 1, \ldots, s)$.
$\square$

LEMMA (4.1.15). — *Let $B = TU$ be the Borel subgroup of upper triangular matrices in G with T the maximal torus of diagonal matrices. Let $\gamma \in T(F)$ be regular in $G(F)$, i.e. $\gamma = \mathrm{diag}(\gamma_1, \ldots, \gamma_d)$ with $\gamma_i \in F^\times$ and $\gamma_{i'} \neq \gamma_{i''}$ for any $i' \neq i''$. The centralizer G_γ of γ in G is equal to T. Let $dt = dg_\gamma$ be the Haar measure on $T(F) = G_\gamma(F)$ normalized by $\mathrm{vol}(T(\mathcal{O}), dt) = 1$. Let $f \in \mathcal{H}$ and let*

$$O_\gamma(f, dg_\gamma) = \int_{G_\gamma(F) \backslash G(F)} f(g^{-1}\gamma g) \frac{dg}{dg_\gamma}$$

(this converges absolutely as the orbit $O_G(\gamma)(F)$ of γ in $G(F)$ is obviously closed). Then

$$O_\gamma(f, dg_\gamma) = |D_{T \backslash G}(\gamma)|^{-1/2} f^B(t)$$

where

$$D_{T \backslash G}(t) = \det(1 - \mathrm{Ad}(t^{-1}), \mathrm{Lie}(T(F)) \backslash \mathrm{Lie}(G(F)))$$

for each $t \in T(F)$ $(D_{T \backslash G}(\gamma) \neq 0)$.

Proof: Applying the integration formula (4.1.7) for $P = B$, we get

$$O_\gamma(f, dg_\gamma) = \int_{U(F)} f(u^{-1}\gamma u) du.$$

But, as γ is regular, the morphism of ϖ-adic manifolds

$$U(F) \to U(F) , \ u \mapsto \gamma^{-1} u^{-1} \gamma u$$

is bijective with constant Jacobian

$$J(\gamma) = |\det(1 - \mathrm{Ad}(\gamma^{-1}), \mathrm{Lie}(U(F)))| \neq 0.$$

So by the change of variable

$$u' = \gamma^{-1} u^{-1} \gamma u$$

we get

$$O_\gamma(f, dg_\gamma) = \int_{U(F)} f(\gamma u') \frac{du'}{J(\gamma)}.$$

As

$$|D_{T \backslash G}(\gamma)| = \delta_{B(F)}(\gamma) J(\gamma)^2$$

the lemma is proved.   □

COROLLARY (4.1.16). — *The image of the Satake transformation*

$$\mathcal{A} \otimes \mathcal{H} \xrightarrow{\ (-)^\vee\ } \mathcal{A}[z_1, z_1^{-1}, \ldots, z_d, z_d^{-1}]$$

is contained in the invariants under the natural action of the symmetric group $\mathfrak{S}_d$ *on* $\mathcal{A}[z_1, z_1^{-1}, \ldots, z_d, z_d^{-1}]$.

Proof : If we consider a regular element $\gamma = \mathrm{diag}(\gamma_1, \ldots, \gamma_d) \in T(F) \subset G(F)$ we have

$$f^B(\gamma) = |D_{T \backslash G}(\gamma)|^{1/2} O_\gamma(f, dg_\gamma)$$

thanks to (4.1.15). But if $w \in \mathfrak{S}_d$ and if $\dot{w} \in G(F)$ is the corresponding permutation matrix

$$w \cdot \gamma = \mathrm{diag}(\gamma_{w^{-1}(1)}, \ldots, \gamma_{w^{-1}(d)}) = \dot{w} \, \gamma \, \dot{w}^{-1}.$$

So

$$O_{w \cdot \gamma}(f, dg_{w \cdot \gamma}) = O_\gamma(f, dg_\gamma)$$

$(dg_{\gamma \cdot w} = dt = dg_\gamma)$ and

$$|D_{T \backslash G}(w \cdot \gamma)| = |D_{T \backslash G}(\gamma)|.$$

Therefore we have

$$f^B(w \cdot \gamma) = f^B(\gamma)$$

and the corollary is proved.   □

THEOREM (4.1.17) (Satake). — *The Satake transformation is an isomorphism of* $\mathcal{A}$-*algebras of* $\mathcal{H}$ *onto* $\mathcal{A}[z_1, z_1^{-1}, \ldots, z_d, z_d^{-1}]^{\mathfrak{S}_d}$.

Proof: Let $\lambda, \mu \in \mathbb{Z}^d$ with $\lambda_1 \geq \cdots \geq \lambda_d, \mu_1 \geq \cdots \geq \mu_d$ and let $a(\lambda, \mu) \in \mathcal{A}$ be the coefficient of z^μ in $f_\lambda^\vee(z)$. The theorem follows easily from

CLAIM (4.1.18). — *We have $a(\lambda, \mu) = 0$ unless $\mu \leq \lambda$ for the partial order*

$$(\mu \leq \lambda) \iff \begin{cases} \mu_1 \leq \lambda_1, \\ \mu_1 + \mu_2 \leq \lambda_1 + \lambda_2, \\ \cdots \\ \mu_1 + \cdots + \mu_{d-1} \leq \lambda_1 + \cdots + \lambda_{d-1}, \\ \mu_1 + \cdots + \mu_d = \lambda_1 + \cdots + \lambda_d, \end{cases}$$

and

$$a(\lambda, \lambda) = q^{<\delta, \lambda>}$$

where

$$\delta = (\frac{d-1}{2}, \frac{d-3}{2}, \ldots, \frac{3-d}{2}, \frac{1-d}{2})$$

and

$$< \delta, \lambda > = \sum_{i=1}^{d} \delta_i \lambda_i.$$

Proof of the claim : It follows from (4.1.13) and (4.1.14) that

$$a(\lambda, \mu) = f_\lambda^B(\varpi^\mu)$$

where

$$\varpi^\mu = \mathrm{diag}(\varpi^{\mu_1}, \ldots, \varpi^{\mu_d}) \in T(F).$$

But

$$\delta_{B(F)}^{1/2}(\varpi^\lambda) = q^{<\delta, \lambda>},$$

so the problem is reduced to proving the following assertions:

(1) if there exists $u \in U(F)$ (U is the unipotent radical of B, $B = T.U$) such that

$$\varpi^\mu u \in K \varpi^\lambda K,$$

then $\mu \leq \lambda$;

(2) if $u \in U(F)$, then

$$\varpi^\lambda u \in K \varpi^\lambda K$$

if and only if $u \in U(\mathcal{O}) = U(F) \cap K$.

Let us prove these assertions. For any integer $n \geq 1$ and any $g \in GL_n(F)$, we set

$$v(g) = \min\{v(g_{ij}) \mid 1 \leq i, j \leq n\}.$$

Then it is easy to see that, for any $k_1, k_2 \in GL_n(\mathcal{O})$, we have

$$v(k_1 g k_2) = v(g)$$

and that, for any $\lambda \in \mathbf{Z}^n$ with $\lambda_1 \geq \cdots \geq \lambda_n$, we have

$$\begin{cases} v(g) = \lambda_n, \\ v(\Lambda^2 g) = \lambda_{n-1} + \lambda_n, \\ \cdots \\ v(\Lambda^n g) = \lambda_1 + \cdots + \lambda_n, \end{cases}$$

if

$$g = \begin{pmatrix} \varpi^{\lambda_1} & & 0 \\ & \ddots & \\ 0 & & \varpi^{\lambda_n} \end{pmatrix} \in GL_n(F).$$

So, if $g \in G(F)$, we have

$$g \in K\varpi^\lambda K$$

if and only if

$$\begin{cases} v(g) = \lambda_d, \\ v(\Lambda^2 g) = \lambda_{d-1} + \lambda_d, \\ \cdots \\ v(\Lambda^d g) = \lambda_1 = + \cdots + \lambda_d. \end{cases}$$

Now assertion (1) follows immediatly from the inequalities

$$v(\Lambda^i(\varpi^\mu u)) \leq v(\Lambda^i \varpi^\mu)$$

$(i = 1, 2, \ldots, d)$ with equality if $i = d$ and assertion (2) is left to the reader.
□

Another important consequence of the above claim is

COROLLARY (4.1.19). — *The Satake transform of $f_\lambda \in \mathcal{H}$ where $\lambda = (1, 0, \ldots, 0)$ is*

$$q^{(d-1)/2}(z_1 + \cdots + z_d).$$

Proof: We have $< \delta, \lambda > = (d-1)/2$ and the only $\mu \in \mathbf{Z}^d$ with $\mu_1 \geq \cdots \geq \mu_d$ such that $\mu \leq \lambda$ is λ itself.
□

(4.2) Base change homomorphism

Let $\mathcal{H}$ and $\mathcal{H}_r$ be as in (4.0). We can now define the **base change homomorphism**

(4.2.1) $$b : \mathcal{A}_r \otimes \mathcal{H}_r \to \mathcal{A} \otimes \mathcal{H}$$

by requiring that

$$b|\mathcal{A}_r = \mathbf{Q}[(\sqrt{q})^r, 1/(\sqrt{q})^r] \hookrightarrow \mathbf{Q}[\sqrt{q}, 1/\sqrt{q}] = \mathcal{A}$$

is the inclusion and

$$b(f_r)^\vee(z_1, \ldots, z_d) = f_r^\vee(z_1^r, \ldots, z_d^r)$$

for each $f_r \in \mathcal{H}_r$ (see 4.1.17).

Similarly, for each standard parabolic subgroup $P = MN$ of G we have a base change homomorphism

(4.2.2) $$b_M : \mathcal{A}_r \otimes \mathcal{H}_{M,r} \to \mathcal{A} \otimes \mathcal{H}_M.$$

If P is given by the partition $d_1 + \cdots + d_s = d$, we have

(4.2.3) $$b_M = b_{GL_{d_1}} \otimes \cdots \otimes b_{GL_{d_s}}$$

with the notations of (4.1.14).

LEMMA (4.2.4). — *The diagram*

$$\begin{array}{ccc}
\mathcal{A}_r \otimes \mathcal{H}_r & \xrightarrow{\ b\ } & \mathcal{A} \otimes \mathcal{H} \\
{\scriptstyle (-)^P} \downarrow & & \downarrow {\scriptstyle (-)^P} \\
\mathcal{A}_r \otimes \mathcal{H}_{M,r} & \xrightarrow{\ b_M\ } & \mathcal{A} \otimes \mathcal{H}_M
\end{array}$$

commutes.

Proof: This is an obvious consequence of (4.1.13). □

Following Drinfeld (see [Kaz]) we will now give a complete description of the function

$$f = b(f_r)$$

when

$$f_r^\vee(z) = q^{r(d-1)/2}(z_1 + \cdots + z_d),$$

i.e.

$$f^\vee(z) = q^{r(d-1)/2}(z_1^r + \cdots + z_d^r).$$

PROPOSITION (4.2.5) (Drinfeld). — *For each integer $d \geq 1$, let*

$$\varphi_d : GL_d(F) \to \mathbb{Z}$$

be the function defined by

$$\varphi_d(g) = (1-q)(1-q^2) \cdots (1-q^{\rho-1})$$

if $g \in gl_d(\mathcal{O}) \cap GL_d(F)$ and $v(\det g) = r$, where ρ is the nullity (i.e. the dimension over k of the kernel) of the matrix

$$\overline{g} \in gl_d(k)$$

obtained by reducing g modulo $\varpi gl_d(\mathcal{O})$, and by

$$\varphi_d(g) = 0$$

otherwise (recall that r is a fixed positive integer).
   Then for each standard parabolic subgroup $P = MN$ of $G = GL_d$ given by the partition $d_1 + \cdots + d_s = d$ and for each

$$m = (g_1, \ldots, g_s) \in M(F) = GL_{d_1}(F) \times \cdots \times GL_{d_s}(F)$$

we have

$$(\varphi_d)^P(m) = \sum_{j=1}^{s} q^{r(d-d_j)/2} \varphi_{d_j}(g_j) \prod_{\substack{k=1 \\ k \neq j}}^{s} 1_{GL_{d_k}(\mathcal{O})}(g_k)$$

where $1_{GL_{d_k}(\mathcal{O})}$ is the characteristic function of $GL_{d_k}(\mathcal{O}) \subset GL_{d_k}(F)$.

COROLLARY (4.2.6). — *The Satake transform of φ_d is*

$$(\varphi_d)^{\vee}(z) = q^{r(d-1)/2}(z_1^r + \cdots + z_d^r)$$

and consequently φ_d is our function f.

Proof: We will prove the proposition and its corollary simultaneously.
   Let us assume that the proposition is proved for P corresponding to the partition $s = 2$, $d_1 = 1$ and $d_2 = d - 1$. Then by induction we easily get the proposition for $P = B$ (the Borel subgroup of upper triangular matrices) and therefore we get the corollary (see (4.1.13)). But conversely the corollary and (4.1.13) imply the proposition for a general P. So it is enough to prove the proposition in the particular case where P corresponds to the partition $s = 2$, $d_1 = 1$ and $d_2 = d - 1$.
   The proof of the proposition in this special case is a direct computation. Let us do it. Let

$$m = (g_1, g_2) \in M(F) = F^{\times} \times GL_{d-1}(F)$$

and let

$$n = \begin{pmatrix} 1 & u \\ 0 & 1 \end{pmatrix} \in N(F)$$

with
$$u = (u_1, \ldots, u_{d-1}) \in F^{d-1}.$$

We have
$$\delta_P(m)^{1/2} = q^{[v(\det g_2) - (d-1)v(g_1)]/2}$$

and
$$\varphi_d(mn) = 0$$

unless $g_1 \in \mathcal{O}$, $g_2 \in gl_{d-1}(\mathcal{O})$ and

$$v(g_1) + v(\det g_2) = r.$$

Consequently
$$(\varphi_d)^P(m) = 0$$

unless $g_1 \in \mathcal{O}$, $g_2 \in gl_{d-1}(\mathcal{O})$ and

$$v(g_1) + v(\det g_2) = r.$$

Moreover, if $g_1 \in \mathcal{O}$, $g_2 \in gl_{d-1}(\mathcal{O})$ and $v(g_1) + v(\det g_2) = r$, we have

$$(\varphi_d)^P(m) = q^{[v(\det g_2) - (d-1)v(g_1)]/2} \int_{F^{d-1}} \varphi_d(mn) du$$

with $du = du_1 \cdots du_{d-1}$ and $\mathrm{vol}(\mathcal{O}, du_i) = 1$ $(i = 1, \ldots, d-1)$.

If $g_1 \in \mathcal{O}^\times$, $g_2 \in gl_{d-1}(\mathcal{O})$ and $v(\det g_2) = r$, then obviously $\varphi_d(mn) = 0$ unless $u \in \mathcal{O}^{d-1}$. Moreover, if $u \in \mathcal{O}^{d-1}$, the dimension of the kernel of $\overline{mn}$ is the same as the dimension of the kernel of $\overline{g}_2$ so that

$$\varphi_d(mn) = \varphi_{d-1}(g_2)$$

and we are done in this case $(\varphi_1(g_1) = 0)$.

If $v(g_1) = r$ and $g_2 \in GL_{d-1}(\mathcal{O})$, then obviously $\varphi_d(mn) = 0$ unless $u \in \varpi^{-r}\mathcal{O}^{d-1}$. Moreover, if $u \in \varpi^{-r}\mathcal{O}^{d-1}$, the dimension of the kernel of $\overline{mn}$ is the same as the dimension of the kernel of $\overline{g}_1 = 0$ (i.e. 1) so that

$$\varphi_d(mn) = \varphi_1(g_1)$$

and we are done in this case $(\varphi_{d-1}(g_2) = 0$ and $\mathrm{vol}(\varpi^{-r}\mathcal{O}^{d-1}) = q^{r(d-1)})$.

Finally, if $v(g_1) > 0$, $g_2 \in gl_{d-1}(\mathcal{O})$ and $v(\det g_2) > 0$, $\varphi_d(mn) = 0$ unless $u \in g_1^{-1}\mathcal{O}^{d-1}$. Moreover, if $u \in g_1^{-1}\mathcal{O}^{d-1}$, the dimension of the kernel of $\overline{mn}$ is ρ_2 (resp. $\rho_2 + 1$) if

$$(\overline{g_1 u_1}, \ldots, \overline{g_1 u_{d-1}}) \notin V_2$$

(resp.
$$(\overline{g_1 u_1}, \ldots, \overline{g_1 u_{d-1}}) \in V_2)$$

where ρ_2 is the dimension of the kernel of $\overline{g}_2$ and V_2 is the k-vector subspace of k^{d-1} generated by the lines of the matrix $\overline{g}_2$. So we have

$$\int_{F^{d-1}} \varphi_d(mn)du = (1-q)\cdots(1-q^{\rho_2-1})\,\text{vol}$$

$$+(1-q)\cdots(1-q^{\rho_2-1})(1-q^{\rho_2})\,\text{vol}'$$

where

$$\text{vol} = \text{vol}(\{u \in g_1^{-1}\mathcal{O}^{d-1} \mid (\overline{g_1u_1},\ldots,\overline{g_1u_{d-1}}) \notin V_2\}, du)$$

and

$$\text{vol}' = \text{vol}(\{u \in g_1^{-1}\mathcal{O}^{d-1} \mid (\overline{g_1u_1},\ldots,\overline{g_1u_{d-1}}) \in V_2\}, du).$$

But we have

$$\rho_2 + \dim_k(V_2) = d - 1$$

so that

$$\text{vol}' = q^{d-1-\rho_2}/q^{(v(g_1)+1)(d-1)}$$

and

$$\text{vol} = (q^{d-1} - q^{d-1-\rho_2})/q^{(v(g_1)+1)(d-1)}.$$

Therefore we get

$$\int_{F^{d-1}} \varphi_d(mn)du = 0$$

and we are done in this case too ($\varphi_1(g_1) = 0$ and $\varphi_{d-1}(g_2) = 0$). $\qquad\square$

(4.3) Orbital integrals

We would like to consider orbital integrals of Hecke functions for semi-simple elements of $G(F)$. If the characteristic of F is positive there are two different notions of semi-simple elements (see [Bou] Alg. VII, §5). This can be misleading and, as in [De–Ka–Vi] Appendix 1, we will use other terminology.

DEFINITION (4.3.1). — *An element $\gamma \in G(F)$ is said to be closed if its orbit $O_G(\gamma)(F)$ in $G(F)$ is a closed subset of $G(F)$ for the ϖ-adic topology.*

LEMMA (4.3.2). — (i) *Let $\gamma \in G(F)$ be such that the F-algebra*

$$F' = F[\gamma] \subset gl_d(F)$$

is a product of fields

$$F' = F_1' \times \cdots \times F_s',$$

i.e. such that its minimal polynomial $P(T)$ is a product of distinct irreducible unitary polynomials in $F[T]$,

$$P(T) = P_1'(T)\cdots P_s'(T)$$

($F_j' = F[T]/(P_j'(T))$ for $j = 1,\ldots,s$). Then γ is closed.

(ii) *For any* $\gamma \in G(F)$, *the closure* $\overline{O_G(\gamma)(F)}$ *of* $O_G(\gamma)(F)$ *in* $G(F)$ *for the* ϖ-*adic topology is the disjoint union of finitely many orbits.* $O_G(\gamma)(F)$ *is open in its closure and the other orbits in* $\overline{O_G(\gamma)(F)}$ *are locally closed and of strictly smaller dimension as* ϖ-*adic manifolds. Moreover* $\overline{O_G(\gamma)(F)}$ *contains a unique closed orbit and the minimal polynomial of any element in this closed orbit is the product of the distinct irreducible unitary factors of the minimal polynomial of* γ.

(iii) *The number of orbits which contain a given (closed) orbit in their closure is finite.*

COROLLARY (4.3.3). — *The closed elements of* $G(F)$ *are exactly the* γ's *in* $G(F)$ *such that the* F-*algebra* $F[\gamma] \subset gl_d(F)$ *is a product of fields, i.e. the semi-simple elements in the sense of* [Bou] *Alg. VII,* § 5, *Déf.* 6. □

REMARK (4.3.4). — A closed element $\gamma \in G(F)$ is not necessarily closed in $G(F')$ if F' is a finite extension of F (if char$(F) = 2$ and if $\alpha \in F$ is not a square, $\begin{pmatrix} 0 & 1 \\ \alpha & 0 \end{pmatrix}$ is closed in $GL_2(F)$ but not in $GL_2(F(\sqrt{\alpha}))$). □

Proof of (4.3.2) : Let $\gamma \in G(F)$ with elementary divisors

$$P_1(T)|P_2(T)|\cdots|P_d(T).$$

Then $D_j(T) = P_1(T)\cdots P_j(T)$ is the g.c.d. of the minors of size $j \times j$ of the matrix $T - \gamma \in gl_d(F[T])$. So if $\widetilde{\gamma} \in \overline{O_G(\gamma)(F)}$ and if $\widetilde{P}_1(T)|\widetilde{P}_2(T)|\cdots|\widetilde{P}_d(T)$ are its elementary divisors, $D_j(T)$ divides $\widetilde{D}_j(T) = \widetilde{P}_1(T)\cdots\widetilde{P}_j(T)$ for $j = 1,\ldots,d$ with equality for $j = d$. In other words, the product $\widetilde{P}_{j+1}(T)\cdots\widetilde{P}_d(T)$ divides the product $P_{j+1}(T)\cdots P_d(T)$ for $j = 0,\ldots,d-1$ with equality for $j = 0$.

Part (i), the first assertion of part (ii) and part (iii) of the lemma follow. The dimension of $O_G(\gamma)(F)$ (as a ϖ-adic manifold) coincides with

$$d^2 - \dim_F\{g \in gl_d(F)|g\gamma = \gamma g\}$$

and is equal to

$$2\sum_{j=1}^{d} j(\deg P_j(T) - 1)$$

(see [Bou] Alg. VII, § 5, Ex. 12). Therefore, if $\widetilde{\gamma} \in \overline{O_G(\gamma)(F)} - O_G(\gamma)(F)$,

$$\dim O_G(\gamma)(F) - \dim O_G(\widetilde{\gamma})(F) = 2\sum_{j=0}^{d-1} \deg\Big(\frac{P_{j+1}(T)\cdots P_d(T)}{\widetilde{P}_{j+1}(T)\cdots\widetilde{P}_d(T)}\Big) > 0.$$

Moreover, the union of the orbits $O_G(\tilde{\gamma})(F)$ which are contained in $\overline{O_G(\gamma)(F)}$ and which are of dimension smaller than or equal to a fixed integer is closed in $\overline{O_G(\gamma)(F)}$.

The second assertion of part (ii) of the lemma follows.

Each orbit contained in $\overline{O_G(\gamma)(F)}$ and of minimal dimension for this property is closed. To finish the proof of part (ii) of the lemma it is enough to prove the following assertion. Let $\gamma \in G(F)$ with characteristic polynomial $D_d(T)$, let

$$D_d(T) = Q_1(T)^{d'_1} \cdots Q_s(T)^{d'_s}$$

be the factorization of $D_d(T)$ into irreducibles polynomial $(Q_1(T), \ldots, Q_s(T)$ are distinct irreducible unitary polynomials and $d'_1, \ldots, d'_s$ are positive integers) and let

$$\tilde{P}_j(T) = Q_1(T)^{\varepsilon_{1j}} \cdots Q_s(T)^{\varepsilon_{sj}}$$

where $\varepsilon_{ij} = 0$ if $1 \le j < d-d'_i$ and $\varepsilon_{ij} = 1$ if $d-d'_i \le j \le d$ $(i = 1, \ldots, s)$, then any $\tilde{\gamma} \in G(F)$ with elementary divisors $\tilde{P}_1(T)|\tilde{P}_2(T)|\cdots|\tilde{P}_d(T)$ is contained in $\overline{O_G(\gamma)(F)}$. But it is easy to construct an algebraic map

$$\Gamma : F \to G(F)$$

such that $\Gamma(0) \in O_G(\tilde{\gamma})(F)$ and $\Gamma(x) \in O_G(\gamma)(F)$ for each $x \in F - \{0\}$. Let us do it in the crucial case ([Bou] Alg. VII §5, n° 3) where $P_d(T) = Q(T)^m$ $(s = 1)$ and where $P_j(T) = 1$ for $j = 1, \ldots, d - 1$. We can identify the F-vector space F^d with $F[T]/(Q(T)^m)$ and the action of γ on it with multiplication by T. Now

$$\{T^{n-1}Q(T)^{m-1}, \ldots, TQ(T)^{m-1}, Q(T)^{m-1}, \ldots,$$
$$T^{n-1}Q(T), \ldots, TQ(T), Q(T), T^{n-1}, \ldots, T, 1\}$$

where $n = \deg Q(T)$ $(mn = d)$ is a basis of $F[T]/(Q(T)^m)$ and the matrix of the multiplication by T in this basis is

$$\Gamma(1) = \begin{pmatrix} U & E & 0 & \cdots & 0 \\ 0 & U & \ddots & \ddots & \vdots \\ \vdots & \ddots & \ddots & \ddots & 0 \\ \vdots & & \ddots & U & E \\ 0 & \cdots & \cdots & 0 & U \end{pmatrix} \quad (m \times m \text{ blocks})$$

where

$$U = \begin{pmatrix} -a_{n-1} & 1 & 0 & \cdots & 0 \\ -a_{n-2} & 0 & \ddots & \ddots & \vdots \\ \vdots & \vdots & \ddots & \ddots & 0 \\ \vdots & \vdots & & \ddots & 1 \\ -a_0 & 0 & \cdots & \cdots & 0 \end{pmatrix}, \quad E = \begin{pmatrix} 0 & \cdots & \cdots & 0 \\ \vdots & & & \vdots \\ \vdots & & & \vdots \\ 0 & & & \\ 1 & 0 & \cdots & 0 \end{pmatrix}$$

if we have set

$$Q(T) = T^n + a_{n-1}T^{n-1} + a_{n-2}T^{n-2} + \cdots + a_0.$$

Therefore we can define $\Gamma(x)$ as the matrix obtained by replacing E by xE in the above matrix $\Gamma(1)$ (if $x \neq 0$, $\Gamma(x)$ is conjugate to $\Gamma(1)$ by the diagonal matrix

$$\mathrm{diag}(x^{m-1}1_n, \ldots, x1_n, 1_n)$$

(1_n is the identity matrix in $GL_n(F)$) and the assertion is proved. $\square$

Let $\gamma \in G(F)$ and let $P_d(T)$ be its minimal polynomial. For each irreducible unitary polynomial $Q(T) \in F[T]$ let $m_{Q(T)}$ be the multiplicity of $Q(T)$ as a prime factor of $P_d(T)$ and let

$$V_{Q(T)} = \mathrm{Ker}(Q(\gamma)^{m_{Q(T)}} : F^d \to F^d).$$

Then we have

$$F^d = \bigoplus_{Q(T)} V_{Q(T)}$$

where $Q(T)$ runs through the set of prime factors of $P_d(T)$, i.e. the set of irreducible unitary polynomials in $F[T]$ such that $m_{Q(T)} > 0$ (see [Bou] Alg. VII, §5, Prop. 3). Therefore we have a canonical Levi subgroup of G over F,

$$(4.3.5) \qquad M = \{g \in G \mid g(V_{Q(T)}) \subset V_{Q(T)} \ , \ \forall Q(T)\},$$

associated to γ. It is clear that $\gamma \in M(F)$ and even that $G_\gamma \subset M$.

If $Q_1(T), \ldots, Q_s(T)$ is an ordering on the set of prime factors of $P_d(T)$, we have a canonical parabolic subgroup of G over F,

$$(4.3.6) \qquad P = \{g \in G \mid g(V_{Q_1(T)} \oplus \cdots \oplus V_{Q_i(T)}) \subset V_{Q_1(T)} \oplus \cdots \oplus V_{Q_i(T)},$$
$$\forall i = 1, \ldots, s-1\}$$

associated to γ and to this ordering. If N is the unipotent radical of P, $P = MN$ is a Levi decomposition of P over F.

If $d_{Q(T)} = \dim_F V_{Q(T)}$ for each prime factor $Q(T)$ of $P_d(T)$, M is non-canonically isomorphic to

$$\prod_{Q(T)} GL_{d_{Q(T)}}(F)$$

and

$$d_{Q(T)} = d'_{Q(T)} \deg Q(T)$$

where $d'_{Q(T)}$ is the multiplicity of $Q(T)$ as a prime factor of the characteristic polynomial of γ ($Q(T)$ runs through the set of prime factors of $P_d(T)$).

If γ is closed in $G(F)$, i.e. $m_{Q(T)} = 1$ for each prime factor $Q(T)$ of $P_d(T)$ (see (4.3.3)), $G_\gamma(F)$ is non-canonically isomorphic to

$$\prod_{Q(T)} GL_{d'_{Q(T)}}(F'_{Q(T)}),$$

where

$$F'_{Q(T)} = F[T]/(Q(T))$$

(as before $Q(T)$ runs through the set of prime factors of $P_d(T)$). In particular $G_\gamma(F)$ is unimodular.

If γ is a closed element of $G(F)$ and if dg_γ is a Haar measure on $G_\gamma(F)$ we can define the **orbital integral** of any locally constant function with compact support $f : G(F) \to \mathbb{C}$ at γ by

$$(4.3.7) \qquad O_\gamma(f, dg_\gamma) = \int_{G_\gamma(F)\backslash G(F)} f(g^{-1}\gamma g)\frac{dg}{dg_\gamma}$$

(recall that dg is the Haar measure on $G(F)$ which is normalized by $\mathrm{vol}(K, dg) = 1$). As $O_G(\gamma)(F)$ is closed in $G(F)$, this integral is absolutely convergent and in fact can be reduced to a finite sum (the restriction of f to $O_G(\gamma)(F)$ is locally constant with compact support).

Explicit computation of an orbital integral is a difficult matter in general. But we have several reduction formulas which can be useful. Let us explain the first one (another one will be given in (4.8)).

Let $P = MN$ be a parabolic subgroup of G over F with a Levi decomposition over F. Let K_P be a maximal compact open subgroup of $G(F)$ in **good position with respect to** $P = MN$, i.e. $G(F) = P(F)K_P$ and $P(F) \cap K_P = (M(F) \cap K_P)(N(F) \cap K_P)$ (if $g \in G(F)$ is such that gPg^{-1} and gMg^{-1} are standard we can take $K_P = g^{-1}Kg$). Let dk_P and dn be the Haar measures on K_P and $N(F)$ which are normalized by

$$\mathrm{vol}(K_P, dk_P) = \mathrm{vol}(N(F) \cap K_P, dn) = 1.$$

The K_P-**invariant constant term of** f **along** P of a locally constant function with compact support $f : G(F) \to \mathbb{C}$ is the locally constant function with compact support $f^P : M(F) \to \mathbb{C}$ defined by

$$(4.3.8) \qquad f^P(m) = \delta_{P(F)}^{1/2}(m) \int_{N(F)} \int_{K_P} f(k_P^{-1}mnk_P)dk_Pdn$$

(recall that $\delta_{P(F)} : M(F) \to q^{\mathbb{Z}} \subset \mathbb{Q}^\times$ is the modulus character of P, see (4.1); the integral is absolutely convergent and, in fact, can be reduced to a finite sum).

If dm is the Haar measure on $M(F)$ which is normalized by $\text{vol}(M(F) \cap K_P, dm) = 1$, we have the following integration formula:

$$(4.3.9) \qquad \int_{G(F)} \varphi(g) dg = \int_{M(F)} \int_{N(F)} \int_{K_P} \varphi(mnk_P) dk_P dn \, dm$$

for any locally constant function with compact support $\varphi : G(F) \to \mathbb{C}$.

If $\gamma \in M(F) \subset G(F)$ is closed in $G(F)$ and if $G_\gamma \subset M$, the orbit of γ in $M(F)$ is also closed and we have $G_\gamma(F) = M_\gamma(F)$. Therefore for any locally constant function with compact support $f : G(F) \to \mathbb{C}$ and any Haar measure dg_γ on $G_\gamma(F)$ we can consider the orbital integral

$$(4.3.10) \qquad O_\gamma^M(f^P, dg_\gamma) = \int_{M_\gamma(F) \backslash M(F)} f^P(m^{-1}\gamma m) \frac{dm}{dg_\gamma}$$

(it is absolutely convergent). Now the first reduction formula is the following proposition which generalizes $(4.1.15)$:

PROPOSITION $(4.3.11)$. — *Let* $P = MN$ *and* K_P *be as before. Let* $\gamma \in M(F) \subset G(F)$ *be closed in* $G(F)$ *and such that* $G_\gamma \subset M$. *Then for any locally constant function with compact support* $f : G(F) \to \mathbb{C}$ *and any Haar measure* dg_γ *on* $G_\gamma(F)$, *we have*

$$O_\gamma(f, dg_\gamma) = |D_{M\backslash G}(\gamma)|^{-1/2} O_\gamma^M(f^P, dg_\gamma)$$

where

$$D_{M\backslash G}(m) = \det(1 - Ad(m)^{-1}, \text{Lie}(M(F)) \backslash \text{Lie}(G(F)))$$

for each $m \in M(F)$ $(D_{M\backslash G}(\gamma) \neq 0$ *and the Haar measures* dg, dm, dn *and* dk_P *are normalized as before).*

Proof: Applying the integration formula $(4.3.9)$, we get

$$O_\gamma(f, dg_\gamma) = \int_{M_\gamma(F) \backslash M(F)} \int_{N(F)} \int_{K_P} f(k_P^{-1} n^{-1} m^{-1} \gamma mnk_P) dk_P dn \frac{dm}{dm_\gamma}.$$

But the analytic map of ϖ-adic manifolds

$$N(F) \to N(F) , \quad n \mapsto (m^{-1}\gamma m)^{-1} n^{-1} (m^{-1}\gamma m) n$$

is bijective with constant Jacobian

$$J(\gamma) = J(m^{-1}\gamma m)$$

where

$$J(m) = |\det(1 - Ad(m)^{-1}, \text{Lie}(N(F)))|$$

for each $m \in M(F)$. Therefore by the change of variable

$$n' = (m^{-1}\gamma m)^{-1} n^{-1} (m^{-1}\gamma m) n$$

we get

$$O_\gamma(f, dg_\gamma) = \int_{M_\gamma(F)\backslash M(F)} \int_{N(F)} \int_{K_P} f(k_P^{-1} m^{-1} \gamma m n' k_P) dk_P \frac{dn'}{J(\gamma)} \frac{dm}{dm_\gamma}.$$

The proposition follows if we remark that, for every $m \in M(F)$,

$$|D_{M\backslash G}(m)| = \delta_{P(F)}(m) J(m)^2.$$

$\square$

In particular for any closed γ in $G(F)$ we can apply this proposition to the parabolic subgroup $P = MN$ of G over F with Levi decomposition over F associated to γ and to an ordering of the prime factors of the minimal polynomial of γ. The main advantage that we get is that now γ is elliptic in $M(F)$ (if we identify $M(F)$ with $GL_{d_1}(F) \times \cdots \times GL_{d_s}(F)$ and γ with $(\gamma_1, \ldots, \gamma_s)$, γ_i is elliptic in $GL_{d_i}(F)$ for $i = 1, \ldots, s$).

(4.4) Twisted orbital integrals

Let $\sigma \in \mathrm{Gal}(F_r/F)$ be the lifting of the arithmetic Frobenius element $(\alpha \mapsto \alpha^q)$ in $\mathrm{Gal}(k_r/k)$, so that

$$\mathrm{Gal}(F_r/F) = \sigma^{\mathbb{Z}/(r)}.$$

For $\gamma_r \in G(F_r)$ we define its **norm** by

(4.4.1) $$N_r(\gamma_r) = \gamma_r \sigma(\gamma_r) \cdots \sigma^{r-1}(\gamma_r).$$

We say that $\gamma'_r, \gamma''_r \in G(F_r)$ are σ-**conjugate** if there exists $g_r \in G(F_r)$ such that

$$\gamma''_r = g_r^{-1} \gamma'_r \sigma(g_r).$$

The σ-**centralizer** $G^\sigma_{\gamma_r}$ of $\gamma_r \in G(F_r)$ is the F-group scheme with

$$G^\sigma_{\gamma_r}(R) = \{g_r \in G(F_r \otimes_F R) \mid g_r^{-1} \gamma_r (\sigma \otimes id_R)(g_r) = \gamma_r\}$$

for any commutative F-algebra R.

PROPOSITION (4.4.2) (Saito). — (i) *If* $\gamma_r \in G(F_r)$, $N_r(\gamma_r)$ *is conjugate to at least one element* γ *of* $G(F) \subset G(F_r)$. *Moreover two such* γ's *are conjugate in* $G(F)$ *and the conjugacy class of such a* γ *in* $G(F)$ *depends only on the* σ-*conjugacy class of* γ_r *in* $G(F_r)$.

(ii) *If* γ_r', $\gamma_r'' \in G(F_r)$ *and* $N_r(\gamma_r')$, $N_r(\gamma_r'')$ *are conjugate in* $G(F_r)$, γ_r', γ_r'' *are* σ-*conjugate in* $G(F_r)$.

In other words, we have a well-defined injective map $\mathcal{N}_r$ from the set of σ-conjugacy classes in $G(F_r)$ into the set of conjugacy classes in $G(F)$: we have $\mathcal{N}_r(C_r) = C$ if and only if for each (one) $\gamma_r \in C_r$ and each (one) $\gamma \in C$, $N_r(\gamma_r)$ and γ are conjugate in $G(F_r)$.

Proof: Let $\gamma_r \in G(F_r)$. Then

$$\sigma(N_r(\gamma_r)) = \gamma_r^{-1} N_r(\gamma_r)\gamma_r;$$

so, if $P_1(T)$, $P_2(T), \ldots, P_d(T)$ are the elementary divisors of the matrix $N_r(\gamma_r)$ ($P_j(T) \in F_r[T]$ and $P_j(T)$ divides $P_{j+1}(T)$ for $j = 1, \ldots, d-1$), we have

$$\sigma(P_j(T)) = P_j(T)$$

and

$$P_j(T) \in F[T]$$

for $j = 1, \ldots, d$. Therefore $N_r(\gamma_r)$ is conjugate in $G(F_r)$ to any matrix $\gamma \in G(F)$ with elementary divisors $P_1(T), \ldots, P_d(T)$. Part (i) of the proposition is now obvious.

Let γ_r', γ_r'', $g_r \in G(F_r)$ be such that

$$N_r(\gamma_r'') = g_r^{-1} N_r(\gamma_r')g_r.$$

We want to prove that γ_r' and γ_r'' are σ-conjugate in $G(F_r)$. As

$$g_r^{-1} N_r(\gamma_r')g_r = N_r(g_r^{-1}\gamma_r'\sigma(g_r))$$

we can assume that $g_r = 1$. Let us consider the F-group scheme

$$H = \operatorname{Res}_{F_r/F} G.$$

We have

$$H(F) = G(F_r),$$

$$H(F_r) = G(F_r)^r,$$

with the embedding

$$G(F_r) = H(F) \subset H(F_r) = G(F_r)^r.$$

given by $g_r \mapsto (g_r, \sigma(g_r), \ldots, \sigma^{r-1}(g_r))$ and the natural action of σ on $H(F_r)$ is given by

$$\sigma(g_r^{(0)}, \ldots, g_r^{(r-1)}) = (\sigma(g_r^{(r-1)}), \sigma(g_r^{(0)}), \ldots, \sigma(g_r^{(r-2)})).$$

Let θ be the algebraic automorphism of the F-group scheme H such that

$$\theta(g_r^{(0)}, \ldots, g_r^{(r-1)}) = (g_r^{(1)}, \ldots, g_r^{(r-1)}, g_r^{(0)})$$

on $H(F_r)$. Then θ induces σ on $H(F) = G(F_r)$. If we set

$$g_r^{(0)} = 1$$

and

$$g_r^{(i)} = \sigma^{i-1}(\gamma_r')^{-1} \cdots \sigma(\gamma_r')^{-1} \gamma_r'^{-1} \gamma_r'' \sigma(\gamma_r'') \cdots \sigma^{i-1}(\gamma_r'')$$

for $i = 1, \ldots, r$ ($g_r^{(r)} = 1$ by hypothesis), we have

$$\sigma^i(\gamma_r'') = (g_r^{(i)})^{-1} \sigma^i(\gamma_r') g_r^{(i+1)}$$

for $i = 1, \ldots, r-1$ and the elements γ_r' and γ_r'' in $G(F_r) = H(F)$ are θ-conjugate in $H(F_r)$ (with the obvious definition of θ-conjugacy). Moreover what we want is to prove that γ_r' and γ_r'' are in fact θ-conjugate in $H(F)$. But the equation

$$\gamma_r'' = h_r^{-1} \gamma_r' \theta(h_r)$$

defines in an obvious way a torsor T under the F-group scheme $G_{\gamma_r''}^\sigma = H_{\gamma_r''}^\theta$. We have

$$T(F) = \{g_r \in G(F_r) \mid \gamma_r'' = g_r^{-1} \gamma_r' \sigma(g_r)\},$$
$$T(F_r) = \{h_r \in H(F_r) \mid \gamma_r'' = h_r^{-1} \gamma_r' \theta(h_r)\},$$

we have proved that $T(F_r) \neq \emptyset$ and we want to prove that $T(F) \neq \emptyset$. But if we set

$$R = \{g_r \in gl_d(F_r) \mid \gamma_r'' \sigma(g_r) = g_r \gamma_r''\},$$

R is a finite dimensional F-algebra and $G_{\gamma_r''}^\sigma$ is nothing else than $R^\times$ viewed as an F-group scheme. Therefore ([Se 1], Ex. 2, p. 160)

$$H^1(\mathrm{Gal}(F_r/F), G_{\gamma_r''}^\sigma(F_r)) = 0$$

and the torsor T which is trivial on F_r is also trivial on F. This finishes the proof of the proposition. $\qquad\square$

REMARK (4.4.3). — If $\gamma_r \in G(F_r)$ is such that $N_r(\gamma_r) \in G(F)$, the σ-**orbit** of γ_r,

$$O_G^\sigma(\gamma_r)(F) = \{g_r^{-1}\gamma_r\sigma_r(g_r) \mid g_r \in G(F_r)\}$$

in $G(F_r)$ is nothing else than

$$N_r^{-1}(O_G(N_r(\gamma_r))(F_r))$$

where $O_G(N_r(\gamma_r))(F_r)$ is the orbit of γ_r in $G(F_r)$ and $N_r : G(F_r) \to G(F_r)$ is the norm. Moreover the σ-centralizer $G_{\gamma_r}^\sigma$ is a form of the σ-centralizer $G_{N_r(\gamma_r)}$ (as an F-group scheme). Indeed these two F-group schemes becomes isomorphic over F_r and an F_r-isomorphism is given by

$$\psi : F_r \otimes_F G_{\gamma_r}^\sigma \;\to\; F_r \otimes_F G_{N_r(\gamma_r)},$$
$$(g_r^{(0)}, \ldots, g_r^{(r-1)}) \;\mapsto\; g_r^{(0)},$$

if we identify $G_{\gamma_r}^\sigma$ with $H_{\gamma_r}^\theta$ as in the proof of the proposition. $\qquad\square$

DEFINITION (4.4.4). — *An element $\gamma_r \in G(F_r)$ is said to be σ-closed if $N_r(\gamma_r)$ is conjugate in $G(F_r)$ to a closed element in $G(F)$.*

As F_r is a separable extension of F, $\gamma_r \in G(F_r)$ is σ-closed if and only if $N_r(\gamma_r)$ is closed in $G(F_r)$ (see (4.3.3)). It follows from (4.4.3) that if $\gamma_r \in G(F_r)$ is σ-closed then its σ-orbit $O_G^\sigma(\gamma_r)(F)$ is closed in $G(F_r)$ (for the ϖ-adic topology).

If $\gamma_r \in G(F_r)$ is σ-closed and if $\gamma = N_r(\gamma_r)$ belongs to $G(F)$, $G_{\gamma_r}^\sigma$ has a natural structure of an F'-group scheme where F' is the F-subalgebra of $gl_d(F)$ generated by γ (we have

$$(4.4.5) \qquad F' \subset \{g_r \in gl_d(F_r) \mid g_r\gamma_r = \gamma_r\sigma(g_r)\}).$$

Then it follows from (4.4.3) that $G_{\gamma_r}^\sigma$ is an inner twist of G_γ as an F'-group scheme. In particular $G_{\gamma_r}^\sigma(F)$ is unimodular ($G_{\gamma_r}^\sigma$ is reductive over F') and any Haar measure dg_γ on $G_\gamma(F)$ induces a Haar measure $dg_{\gamma_r}^\sigma$ on $G_{\gamma_r}^\sigma(F)$ (see (3.5)).

Now let γ_r be a σ-closed element of $G(F_r)$, let $dg_{\gamma_r}^\sigma$ be a Haar measure on $G_{\gamma_r}^\sigma(F)$ ($G_{\gamma_r}^\sigma(F)$ is unimodular) and let $f_r : G(F_r) \to \mathbb{C}$ be a locally constant function with compact support. The **twisted orbital integral of f_r at γ_r** is defined by the following absolutely convergent integral:

$$(4.4.6) \qquad TO_{\gamma_r}(f_r, dg_{\gamma_r}^\sigma) = \int_{G_{\gamma_r}^\sigma(F)\backslash G(F_r)} f_r(g_r^{-1}\gamma_r\sigma(g_r)) \frac{dg_r}{dg_{\gamma_r}^\sigma}$$

(recall that the Haar measure dg_r on $G(F_r)$ is normalized by $\mathrm{vol}(K_r, dg_r) = 1$).

If $N_r(\gamma_r) = \gamma \in G(F)$ and $dg^\sigma_{\gamma_r}$ is the transfer of a Haar measure dg_γ on $G_\gamma(F)$ we will also use the notation

$$TO_{\gamma_r}(f_r, dg_\gamma) = TO_{\gamma_r}(f_r, dg^\sigma_{\gamma_r}).$$

As for ordinary orbital integrals we have reduction formulas for twisted orbital integrals.

Let $P = MN$ be a parabolic subgroup of G over F with a Levi decomposition over F. Let $K_{P,r}$ be a maximal compact open subgroup of $G(F_r)$ in good position with respect to $P = MN$ and stable by σ (if $g \in G(F)$ is such that gPg^{-1} and gMg^{-1} are standard we can take $K_{P,r} = g^{-1}K_r g$). Let $dk_{P,r}$, dn_r and dm_r be the Haar measures on $K_{P,r}$, $N(F_r)$ and $M(F_r)$ which are normalized by $\text{vol}(K_{P,r}, dk_{P,r}) = 1$, $\text{vol}(N(F_r) \cap K_{P,r}, dn_r) = 1$ and $\text{vol}(M(F_r) \cap K_{P,r}, dm_r) = 1$ respectively. We can define the (σ-$K_{P,r}$)-**invariant constant term** of any locally constant function with compact support $f_r : G(F_r) \to \mathbb{C}$ **along** P by

(4.4.7) $f_r^{\sigma\text{-}P}(m_r)$

$$= \delta_{P(F_r)}^{1/2}(m_r) \int_{N(F_r)} \int_{K_{P,r}} f_r(k_{P,r}^{-1} m_r n_r \sigma(k_{P,r})) dk_{P,r} dn_r.$$

It is a locally constant function with compact support on $M(F_r)$.

If $\gamma_r \in M(F_r) \subset G(F_r)$ is closed in $G(F_r)$, if $\gamma = N_r(\gamma_r) \in M(F) \subset G(F)$ and if $G_\gamma \subset M$, the σ-orbit of γ_r in $M(F_r)$ is also closed and we have $G^\sigma_{\gamma_r}(F) = M^\sigma_{\gamma_r}(F)$ (obviously we have $G^\sigma_{\gamma_r}(F) \subset G_\gamma(F_r) \subset M(F_r)$). Therefore for any locally constant function with compact support $f_r : G(F_r) \to \mathbb{C}$ and any Haar measure $dg^\sigma_{\gamma_r}$ on $G^\sigma_{\gamma_r}(F)$ we can consider the twisted orbital integral

(4.4.8) $$TO^M_{\gamma_r}(f_r^{\sigma\text{-}P}, dg^\sigma_{\gamma_r}) = \int_{M^\sigma_{\gamma_r}(F) \backslash M(F_r)} f_r^{\sigma\text{-}P}(m_r^{-1}\gamma_r\sigma(m_r)) \frac{dm_r}{dg^\sigma_{\gamma_r}}$$

(this is absolutely convergent). Now our reduction formula is

PROPOSITION (4.4.9). — *Let $P = MN$ and $K_{P,r}$ be as before. Let $\gamma_r \in M(F_r) \subset G(F_r)$ be σ-closed in $G(F_r)$, such that $\gamma = N_r(\gamma_r) \in M(F) \subset G(F)$ and such that $G_\gamma \subset M$. Then for any locally constant function with compact support $f_r : G(F_r) \to \mathbb{C}$ and any Haar measure $dg^\sigma_{\gamma_r}$, on $G^\sigma_{\gamma_r}(F)$, we have*

$$TO_{\gamma_r}(f_r, dg^\sigma_{\gamma_r}) = |D_{M\backslash G}(\gamma)|^{-1/2} TO^M_{\gamma_r}(f_r^{\sigma\text{-}P}, dg^\sigma_{\gamma_r})$$

where $D_{M\backslash G}$ is defined in (4.3.11) $(D_{M\backslash G}(\gamma) \neq 0$ and the Haar measures dg_r, dm_r, dn_r and $dk_{P,r}$ are normalized as before).

Proof: This is the same as the proof of (4.3.11). The details are left to the reader. Let us just remark that

$$|\det(\sigma - \mathrm{Ad}(m_r)^{-1}, \mathrm{Lie}(N(F_r)))| = |\det(1 - \mathrm{Ad}(m_r)^{-1}, \mathrm{Lie}(N(F)))|$$

for each $m_r \in M_r(F_r)$ ($\mathrm{Lie}(N(F_r))$ is considered as an F-vector space and $\sigma - \mathrm{Ad}(m_r)^{-1}$ as an F-linear endomorphism of $\mathrm{Lie}(N(F_r))$; $||$ is the normalized absolute value of F). $\qquad\square$

(4.5) Main theorem

As in chapter 3 we have a notion of r-admissible elements in $G(F)$.

DEFINITION (4.5.1). — *A closed element $\gamma \in G(F)$ is said to be r-admissible if $v(\det \gamma) = r$ and if we can decompose the F-algebra $F' = F[\gamma] \subset gl_d(F)$ into a product of finite field extensions of F,*

$$F' = F'_1 \times \cdots \times F'_s,$$

in such a way that $v'_1(\gamma'_1) \neq 0$ and $v'_j(\gamma'_j) = 0$ for $j = 2, \ldots, s$ where v'_j is the discrete valuation of F'_j and γ'_j is the projection of $\gamma \in F'$ onto F'_j ($j = 1, \ldots, s$).

LEMMA (4.5.2). — *If $\gamma \in G(F)$ is closed and r-admissible there exists $\gamma_r \in G(F_r)$ such that $\gamma = N_r(\gamma_r)$.*

Proof : We will use freely the notations of (4.5.1). Let $d'_1, \ldots, d'_s$ be the positive integers such that the F'-algebra

$$gl_d(F)_\gamma = \{g \in gl_d(F) | g\gamma = \gamma g\}$$

is isomorphic to

$$gl_{d'_1}(F'_1) \times \cdots \times gl_{d'_s}(F'_s).$$

It suffices to find a

$$\gamma'_{j,r} \in gl_{d'_j}(F_r \otimes_F F'_j)$$

such that

$$N_r(\gamma'_{j,r}) = \gamma'_j$$

for each $j = 1, \ldots, s$.

In fact we will prove that

$$\gamma'_j \in F'_j \subset F'_{j,d'_j}$$

is a norm in the étale finite commutative F'_{j,d'_j}-algebra

$$F_r \otimes_F F'_{j,d'_j}$$

where F'_{j,d'_j} is "the" unramified extension of F'_j of degree d'_j $(j = 1, \ldots, s)$ and that will be sufficient. Indeed any embedding of F'_j-algebras

$$F'_{j,d'_j} \hookrightarrow gl_{d'_j}(F'_j)$$

induces an embedding of $(F_r \otimes_F F'_j)$-algebras

$$F_r \otimes_F F'_{j,d'_j} \hookrightarrow gl_{d'_j}(F_r \otimes_F F'_j)$$

such that the restriction of N_r to $F_r \otimes_F F'_{j,d'_j}$ is the norm of $F_r \otimes_F F'_{j,d'_j}$ over F'_{j,d'_j}.

Let δ_j be the g.c.d. of r and $d'_j[k'_j : k]$, let $r_j = r/\delta_j$ and let $F'_{j,d'_j r_j}$ be "the" unramified extension of F'_{j,d'_j} of degree r_j $(F'_{j,d'_j r_j}$ is also "the" unramified extension of F'_j of degree $d'_j r_j)$. Then, the F'_{j,d'_j}-algebra $F_r \otimes_F F'_{j,d'_j}$ is isomorphic to

$$(F'_{j,d'_j r_j})^{\delta_j}$$

and, to prove that γ'_j is a norm in the F'_{j,d'_j}-algebra $F_r \otimes_F F'_{j,d'_j}$, it suffices to prove that γ'_j is a norm in the field extension $F'_{j,d'_j r_j}$ of F'_{j,d'_j}. Thanks to [Se 1] Ch. V, Prop. 3 and its Cor., this will be the case if r_j divides $v'(\gamma'_j)$.

But, if $2 \leq j \leq s$, $v'(\gamma'_j) = 0$ and r_j divides $v'(\gamma'_j)$; if $j = 1$, we have

$$d'_1 v'(\gamma'_1)[k'_1 : k] = r$$

by hypothesis and $r_1 = v'(\gamma'_1)$ (we have $\delta_1 = d'_1[k'_1 : k]$). Therefore the lemma is proved.   □

REMARK (4.5.3). — Let M be the canonical Levi subgroup of G over F associated to γ (see (4.3.5)). Then we have

$$M(F) \cong GL_{d'_1[F'_1:F]}(F) \times \cdots \times GL_{d'_s[F'_s:F]}(F)$$
$$\cup \qquad\qquad \cup \qquad\qquad\qquad \cup$$
$$M_\gamma(F) \cong GL_{d'_1}(F'_1) \times \cdots \times GL_{d'_s}(F'_s)$$

and the γ_r that we have constructed in the proof of (4.5.2) has the following properties. We have

$$\gamma_r = (\gamma'_{1,r}, \ldots, \gamma'_{s,r}) \in M_\gamma(F_r) \subset M(F_r)$$

with

$$\gamma'_{j,r} \in GL_{d'_j}(F_r \otimes_F F'_j) \subset GL_{d'_j[F'_j:F]}(F_r)$$

such that

$$N_r(\gamma'_{j,r}) = \gamma'_j \in GL_{d'_j}(F'_j) \subset GL_{d'_j[F'_j:F]}(F)$$

for $j = 1, \ldots, s$ (recall that $\gamma = (\gamma'_1, \ldots, \gamma'_s)$ as in (4.5.1)). Moreover, the σ-centralizer

$$G^\sigma_{\gamma_r}(F) = M^\sigma_{\gamma_r}(F) \subset M_\gamma(F_r)$$

is isomorphic to

$$D_1'^\times \times GL_{d'_2}(F'_2) \times \cdots \times GL_{d'_s}(F'_s)$$

where

$$D'_1 \subset gl_{d'_1}(F_r \otimes_F F'_j)$$

is the central division algebra over F'_1 with invariant $-1/d'_1$ and where

$$gl_{d'_j}(F'_j) \subset gl_{d'_j}(F_r \otimes_F F'_j)$$

for $j = 2, \ldots, s$.

Indeed, let γ be an elliptic element of $G(F)$ and let us assume for simplicity that the extension $F' = F[\gamma] \subset gl_d(F)$ of F is totally ramified ($F'_1 = F'$, $\gamma'_1 = \gamma$, $k'_1 = k' = k$). Then $F_r \otimes_F F'$ is "the" unramified extension F'_r of degree r of F' and γ is r-admissible if and only if

$$r = v(\det \gamma) = d'v'(\gamma)$$

where $d' = d/[F' : F]$. We set

$$\sigma' = \sigma \otimes 1 \in \mathrm{Gal}(F'_r/F')$$

and we denote by α' the permutation matrix

$$\begin{pmatrix} 0 & 1 & 0 & \cdots & 0 \\ \vdots & \ddots & \ddots & \ddots & \vdots \\ \vdots & & \ddots & \ddots & 0 \\ 0 & & & \ddots & 1 \\ 1 & 0 & \cdots & \cdots & 0 \end{pmatrix} \in gl_{d'}(F').$$

There exists $\beta' \in gl_{d'}(F'_{d'})$ with

$$\alpha' = \beta'\sigma'(\beta')^{-1}$$

where $F'_{d'}$ is the unramified extension of F' of degree d' which is contained in F'_r (see (3.2.5); as $\alpha'^{d'} = 1$ and $\sigma(\alpha') = \alpha'$, we have $\beta'\sigma'^{d'}(\beta')^{-1} = 1$). Now we can embed the torus $(F'_r)^{d'}$ into $gl_{d'}(F'_r)$ by

$$(t'_1, \ldots, t'_{d'}) \mapsto \beta'^{-1}\mathrm{diag}(t'_1, \ldots, t'_{d'})\beta'$$

and we can identify

$$(F'_r)^{d'} \cap gl_{d'}(F') \subset gl_{d'}(F') \subset gl_{d'}(F'_r)$$

with

$$F'_{d'} \hookrightarrow gl_{d'}(F') \subset gl_{d'}(F'_r),$$
$$t'_{d'} \mapsto \beta'^{-1}\mathrm{diag}(\sigma'^{d'-1}(t'_{d'}), \ldots, \sigma'(t'_{d'}), t'_{d'})\beta'.$$

If we consider $\gamma \in F'$ as an element of $F'_{d'}$, there exists $\gamma_{r/d'} \in F'_r$ with norm γ in $F'_{d'}$ $(\gamma_{r/d'}\sigma^{d'}(\gamma_{r/d'}) \cdots \sigma^{r-d'}(\gamma_{r/d'}) = \gamma)$ and if we set

$$\gamma_r = \beta'^{-1}\mathrm{diag}(1, \ldots, 1, \gamma_{r/d'})\beta',$$

we have

$$N'_r(\gamma_r) = \gamma_r\sigma'(\gamma_r) \cdots \sigma'^{r-1}(\gamma_r) = \beta'^{-1}\mathrm{diag}(\gamma, \ldots, \gamma)\beta'.$$

As $v'_r(\gamma_{r/d'}) = 1$, the σ'-centralizer of γ_r in $gl_{d'}(F'_r)$ (which can be identified with the σ'-centralizer of

$$\mathrm{diag}(1, \ldots, 1, \gamma_{r/d'})\alpha'$$

in $gl_{d'}(F'_r)$ by the map $g_r \mapsto \beta'g_r\beta'^{-1}$) is a central division algebra D' over F' with invariant $-1/d'$ as we claimed. $\qquad\square$

DEFINITION (4.5.4). — *A σ-closed element $\gamma_r \in G(F_r)$ is said to be r-admissible if $N_r(\gamma_r)$ is conjugate in $G(F_r)$ to an r-admissible (closed) element $\gamma \in G(F)$ (in particular, $v_r(\det \gamma_r) = 1$).*

Now we are ready to state the particular case that we need of the fundamental lemma for Hecke functions (see [Ar–Cl] (4.5) and (3.13) for the full statement if $\mathrm{char}(F) = 0$).

THEOREM (4.5.5). — *Let*

$$f_r = f_{r,(1,0,\ldots,0)} \in \mathcal{H}_r$$

be the Hecke function with Satake transform

$$f_r^\vee(z) = q^{r(d-1)/2}(z_1 + \cdots + z_d)$$

and let

$$f = b_r(f_r) \in \mathcal{H}$$

be its base change (see (4.2.1)), i.e. the Hecke function with Satake transform

$$f^\vee(z) = q^{r(d-1)/2}(z_1^r + \cdots + z_d^r).$$

Let γ be a closed element of $G(F)$ and let dg_γ be a Haar measure on $G_\gamma(F)$. Then the orbital integral $O_\gamma(f, dg_\gamma)$ is zero unless γ is r-admissible.

Let γ_r be a closed element of $G(F_r)$ and let $dg_{\gamma_r}^\sigma$ be a Haar measure on $G_{\gamma_r}^\sigma(F)$. Then the twisted orbital integral $TO_{\gamma_r}(f_r, dg_{\gamma_r}^\sigma)$ is zero unless γ_r is r-admissible.

Moreover, if γ is an r-admissible closed element of $G(F)$ and if γ_r is an r-admissible σ-closed element of $G(F_r)$ such that

$$N_r(\gamma_r) = \gamma,$$

(see (4.5.2)), then for any Haar measure dg_γ on $G_\gamma(F)$ we have

$$O_\gamma(f, dg_\gamma) = \varepsilon(\gamma) TO_{\gamma_r}(f_r, dg_\gamma)$$

where

$$\varepsilon(\gamma) = (-1)^{rk_{F'}(G_\gamma(F)) - rk_{F'}(G_{\gamma_r}^\sigma(F))}$$

($F' = F[\gamma] \subset gl_d(F)$ is the F-subalgebra generated by γ).

In fact, using the concrete description of the Hecke function f, Drinfeld has computed all the elliptic orbital integrals of f. Following a suggestion of Kottwitz we will deduce (4.5.5) from this computation and a similar one for the twisted orbital integrals of f_r.

(4.6) The elliptic case

In this section we will prove (4.5.5) when γ is elliptic and γ_r is σ-**elliptic** (i.e. $N_r(\gamma_r)$ is conjugate in $G(F_r)$ to an elliptic element γ of $G(F)$).

First of all let us review Drinfeld's computation of elliptic orbital integrals of the function f.

THEOREM (4.6.1) (Drinfeld). — Let γ be an elliptic element of $G(F)$ and let dg_γ be a Haar measure on $G_\gamma(F)$. Let f be the Hecke function on $G(F)$ with Satake transform

$$f^\vee(z) = q^{r(d-1)/2}(z_1^r + \cdots + z_d^r)$$

(recall that $r > 0$).

Then the orbital integral $O_\gamma(f, dg_\gamma)$ is zero unless $v(\det \gamma) = r$. Moreover, if $v(\det \gamma) = r$ and if we normalize dg_γ on $G_\gamma(F) \simeq GL_{d'}(F')$ by $\mathrm{vol}(GL_{d'}(\mathcal{O}'), dg_\gamma) = 1$ ($F' = F[\gamma] \subset gl_d(F)$, $d' = d/[F' : F]$ and $\mathcal{O}'$ is the ring of integers of F'), we have

$$O_\gamma(f, dg_\gamma) = (1 - q') \cdots (1 - q'^{d'-1})[k' : k]$$

(k' is the residue field of $\mathcal{O}'$ and q' is the number of elements of k').

Proof: As the support of f is contained in

$$\{g \in G(F) | v(\det g) = r\}$$

the first assertion is obvious (see (4.2.5), (4.2.6)) and we can assume that $v(\det \gamma) = r$. Then $v'(\gamma) > 0$ (v' is the discrete valuation of F'; recall that $r > 0$); hence $\gamma \in \mathfrak{m}' \subset \mathcal{O}'$, where $\mathfrak{m}'$ is the maximal ideal of $\mathcal{O}'$, and the minimal polynomial of γ belongs to $\mathcal{O}[T]$ and is congruent to $T^{[F':F]}$ modulo $\mathfrak{m}$, where $\mathfrak{m} = \mathcal{O}\varpi$ is the maximal ideal of $\mathcal{O}$. In particular the set

$$\{g \in G(F) | g^{-1}\gamma g \in gl_d(\mathcal{O})\}$$

is non-empty (see [Bou] Alg. VII). Let

$$\Gamma \subset \{g \in G(F) | g^{-1}\gamma g \in gl_d(\mathcal{O})\}$$

be a system of representatives of the double classes in

$$G_\gamma(F) \backslash \{g \in G(F) | g^{-1}\gamma g \in gl_d(\mathcal{O})\} / K.$$

Then we have

$$O_\gamma(f, dg_\gamma) = \sum_{g_0 \in \Gamma} \mathrm{vol}((G_\gamma(F) \cap g_0 K g_0^{-1}) \backslash (g_0 K g_0^{-1}), \frac{dg}{dg_\gamma}) f(g_0^{-1}\gamma g_0)$$

$$= \sum_{g_0 \in \Gamma} \frac{f(g_0^{-1}\gamma g_0)}{\mathrm{vol}(G_\gamma(F) \cap g_0 K g_0^{-1}, dg_\gamma)}$$

(recall that $\mathrm{vol}(K, dg) = 1$).

Now let us identify $G(F)/K$ with the set of lattices L in F^d (free $\mathcal{O}$-submodules of rank d) as usual:

$$gK \mapsto g(\mathcal{O}^d).$$

We get an identification of

$$\{g \in G(F) | g^{-1}\gamma g \in gl_d(\mathcal{O})\}/K \subset G(F)/K$$

with the set of lattices L in F^d such that $\gamma(L) \subset L$ and an identification of Γ with a system of representatives of the $G_\gamma(F)$-orbits in the set of lattices L in F^d such that $\gamma(L) \subset L$. If $L = g_0(\mathcal{O}^d) \in \Gamma$, we have

$$g_0 K g_0^{-1} = \{g \in G(F) | g(L) = L\}$$

and the nullity of $\overline{g_0^{-1}\gamma g_0} \in gl_d(k)$ is equal to

$$\rho(L) = \dim_k \operatorname{Ker}(L/\varpi L \xrightarrow{\gamma} L/\varpi L) = \dim_k(L/(\gamma(L) + \varpi L)).$$

Therefore we can rewrite the above formula for $O_\gamma(f, dg_\gamma)$ as

$$O_\gamma(f, dg_\gamma) = \sum_{L \in \Gamma} \frac{(1-q)(1-q^2)\cdots(1-q^{\rho(L)-1})}{\operatorname{vol}(\{g \in G_\gamma(F)|g(L) = L\}, dg_\gamma)}$$

(see (4.2.5), (4.2.6)).

Let us also fix a basis of F^d as an F'-vector space, i.e. an isomorphism of F'-vector spaces $F'^{d'} \cong F^d$. For each lattice $L \subset F^d$ such that $\gamma(L) \subset L$, $\mathcal{O}'L$ is an $\mathcal{O}'$-lattice in $F'^{d'}$ (a free $\mathcal{O}'$-submodule of rank d'). We have $G_\gamma(F) = GL_{d'}(F')$ and we can choose Γ in such way that $\mathcal{O}'L = \mathcal{O}'^{d'}$ for each $L \in \Gamma$. In other words, we can choose Γ to be a system of representatives of the $GL_{d'}(\mathcal{O}')$-orbits in the set of lattices L in F^d such that

$$L \subset \mathcal{O}'^{d'},$$

$$\mathcal{O}'L = \mathcal{O}'^{d'}$$

and

$$\gamma(L) \subset L.$$

For such a choice of Γ and for any $L \in \Gamma$, we have

$$\operatorname{vol}(\{g \in G_\gamma(F)|g(L) = L\}, dg_\gamma) = 1/i(L)$$

where $i(L)$ is the index of the subgroup

$$\{g \in GL_{d'}(F')|g(L) = L\}$$

of $GL_{d'}(\mathcal{O}')$ (thanks to our normalization of the Haar measure dg_γ on $G_\gamma(F) = GL_{d'}(F')$ we have $\operatorname{vol}(GL_{d'}(\mathcal{O}'), dg_\gamma) = 1$). But $i(L)$ is also the number of elements of the $GL_{d'}(\mathcal{O}')$-orbit of L and we can rewrite the above formula for $O_\gamma(f, dg_\gamma)$ as

$$O_\gamma(f, dg_\gamma) = \sum_L (1-q)(1-q^2)\cdots(1-q^{\rho(L)-1})$$

where L runs through the set of all lattices of F^d such that $L \subset \mathcal{O}'^{d'}$, $\mathcal{O}'L = \mathcal{O}'^{d'}$ and $\gamma(L) \subset L$.

The $\mathcal{O}$-subalgebra $\mathcal{O}[\gamma] \subset \mathcal{O}'$ is free of rank $[F' : F]$ as an $\mathcal{O}$-module, is local with maximal ideal $\mathcal{O}[\gamma] \cap \mathfrak{m}' = (\varpi, \gamma)$ and has residue field $\mathcal{O}[\gamma]/(\varpi, \gamma) = k$. More generally let $R \subset \mathcal{O}'$ be any $\mathcal{O}$-subalgebra which is free of rank $[F' : F]$ as an $\mathcal{O}$-module, let $\mathfrak{m}_R = R \cap \mathfrak{m}'$ be its unique maximal ideal and let us assume that its residue field $R/\mathfrak{m}_R$ is equal to k. We set

$$O_R = \sum_L (1 - q)(1 - q^2) \cdots (1 - q^{\rho_R(L)-1})$$

where L runs through the set of lattices L in F^d such that $L \subset \mathcal{O}'^{d'}$, $\mathcal{O}'L = \mathcal{O}'^{d'}$ and $RL \subset L$ and where

$$\rho_R(L) = \dim_k(L/\mathfrak{m}_R L)$$

(there are only finitely many L's : for a large enough positive integer n, $\mathfrak{m}'^n \subset R$ and $\mathfrak{m}'^n(\mathcal{O}'^{d'}) = \mathfrak{m}'^n\mathcal{O}'L = \mathfrak{m}'^n L \subset RL \subset L \subset \mathcal{O}'^{d'}$). Obviously we have

$$O_{\mathcal{O}[\gamma]} = O_\gamma(f, dg_\gamma).$$

Let n be the minimal positive integer such that $\mathfrak{m}'^n \subset R$. By induction on n we will now prove that

$$O_R = (1 - q')(1 - q'^2) \cdots (1 - q'^{d'-1})[k' : k]$$

for all R and the theorem (4.6.1) will follow. Let us begin with the case $n = 1$. We have

$$\mathfrak{m}'(\mathcal{O}'^{d'}) \subset L \subset \mathcal{O}'^{d'}$$

for each L occurring in the definition of O_R and if we set

$$V = L/\mathfrak{m}'L \subset k'^{d'}$$

we have

$$O_R = \sum_V (1 - q)(1 - q^2) \cdots (1 - q^{\dim_k(V)-1})$$

where V runs through the set of k-vector subspaces of $k'^{d'}$ such that $k'V = k'^{d'}$. For each non-zero k-vector space of finite dimension W let us set

$$\delta_W = \sum_V (1 - q)(1 - q^2) \cdots (1 - q^{\dim_k(V)-1})$$

where V runs through the set of all non-zero k-vector subspaces of W. As the grassmannian of s-dimensional k-vector subspaces in a fixed r-dimensional k-vector space has

$$\frac{(1 - q^{r-s+1})(1 - q^{r-s+2}) \cdots (1 - q^r)}{(1 - q)(1 - q^2) \cdots (1 - q^s)}$$

elements, we have

$$\delta_W = \sum_{s=1}^{r} \frac{(1 - q^{r-s+1})(1 - q^{r-s+2}) \cdots (1 - q^r)}{1 - q^s}$$

where $r = \dim_k(W)$ and it follows that

$$\delta_W = r = \dim_k(W)$$

(see (C.1.1)).

Now, for each k'-vector subspace $V' \subset k'^{d'}$, we have

$$\sum_{W'}(-1)^{d'-r'} q'^{(d'-r')(d'-r'-1)/2} = \begin{cases} 1 & \text{if } V' = k'^{d'}, \\ 0 & \text{otherwise,} \end{cases}$$

where W' runs through the set of k'-vector spaces such that $V' \subset W' \subset k'^{d'}$ and where $r' = \dim_{k'}(W')$ (see (C.1.4)). Therefore we have

$$O_R = \sum_{W'}(-1)^{d'-r'} q'^{(d'-r')(d'-r'-1)/2} \delta_{W'},$$

i.e.

$$O_R = \sum_{W'}(-1)^{d'-r'} r' q'^{(d'-r')(d'-r'-1)/2}[k' : k],$$

where W' runs through the set of all non-zero k'-vector subspaces of $k'^{d'}$ and where $r' = \dim_{k'}(W')$. Finally we get

$$O_R = \sum_{r'=1}^{d'}(-1)^{d'-r'} r' q'^{(d'-r')(d'-r'-1)/2} \frac{(1 - q'^{d'-r'+1}) \cdots (1 - q'^{d'})}{(1 - q')\cdots(1 - q'^{r'})}[k' : k]$$

and

$$O_R = (1 - q') \cdots (1 - q'^{d'-1})[k' : k]$$

(see (C.1.3)).

To finish the proof of the theorem (4.6.1) it is now sufficient to prove that

$$O_R = O_{R_1}$$

where we have set

$$R_1 = R + \mathfrak{m}'^{n-1}$$

($\mathfrak{m}'^n \subset R$, $\mathfrak{m}'^{n-1} \not\subset R$ and $n \geq 2$). First of all let us remark that $R_1 \subset \mathcal{O}'$ is an $\mathcal{O}$-subalgebra which is free of rank $[F' : F]$ as an $\mathcal{O}$-module, that

$$\mathfrak{m}_{R_1} = \mathfrak{m}_R + \mathfrak{m}'^{n-1} = R_1 \cap \mathfrak{m}'$$

is the unique maximal ideal of R_1 and that the residue field of R_1 is equal to the residue field of R, i.e. is equal to k. In particular O_{R_1} is well-defined. Secondly we have a map

$$L \mapsto L_1 = R_1 L = L + \mathfrak{m}'^{n-1} L$$

from the set of lattices L in F^d such that $L \subset O'^{d'}, O'L = O'^{d'}$ and $RL \subset L$ to the set of lattices L_1 in F^d such that $L_1 \subset O'^{d'}, O'L_1 = O'^{d'}$ and $R_1 L_1 \subset L_1$. If $L_1 = L + \mathfrak{m}'^{n-1} L$ we have $\mathfrak{m}_R L_1 = \mathfrak{m}_R L$ ($\mathfrak{m}'^n \subset \mathfrak{m}_R \subset \mathfrak{m}'$) and we have a commutative diagram of k-vector spaces

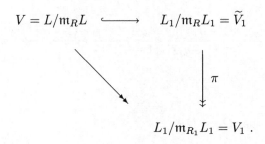

The above map $L \mapsto L_1$ is surjective and its fiber at L_1 is isomorphic to the set of k-vector subspaces V of $\widetilde{V}_1 = L_1/\mathfrak{m}_R L_1$ such that $\pi(V) = V_1 = L_1/\mathfrak{m}_{R_1} L_1$ (the bijection maps a point L in this fiber to $V = L/\mathfrak{m}_R L$). Indeed, if $L_1 \supset L \supset \mathfrak{m}_R L_1$ and $L_1 = L + \mathfrak{m}_{R_1} L_1$, we have $L_1 = L + \mathfrak{m}'^{n-1} L_1$ and consequently $L_1 = L + \mathfrak{m}'^{n-1} L + \mathfrak{m}'^{2n-2} L_1$; but $\mathfrak{m}'^{2n-2} \subset \mathfrak{m}'^n \subset \mathfrak{m}_R$ and we get $L_1 = L + \mathfrak{m}'^{n-1} L + \mathfrak{m}_R L_1$; thanks to Nakayama's lemma it follows that $L_1 = L + \mathfrak{m}'^{n-1} L$. Therefore

$$O_R = \sum_{L_1} (1 - q)(1 - q^2) \cdots (1 - q^{\dim_k(V_1)-1}) \varepsilon_{\widetilde{V}_1 \xrightarrow{\pi} V_1}$$

where L_1 runs through the set of lattices in F^d such that $L_1 \subset O'^{d'}$, $O'L_1 = O'^{d'}$ and $RL_1 \subset L_1$, where $\widetilde{V}_1 = L_1/\mathfrak{m}_R L_1$, $V_1 = L_1/\mathfrak{m}_{R_1} L_1$ and $\pi : \widetilde{V}_1 \longrightarrow V_1$ is the canonical projection and where

$$\varepsilon_{\widetilde{V}_1 \xrightarrow{\pi} V_1} = \sum_V (1 - q^{\dim_k(V_1)})(1 - q^{\dim_k(V_1)+1}) \cdots (1 - q^{\dim_k(V)-1}).$$

In the last sum, V runs through the set of k-vector subspaces of $\widetilde{V}_1$ such that $\pi(V) = V_1$. Now let $r_1 = \dim_k(V_1)$ and $\widetilde{r}_1 = \dim_k(\widetilde{V}_1)$, let $W_1 = \text{Ker}(\pi)$ and let r be an integer with $r_1 \le r \le \widetilde{r}_1$. The map

$$V \mapsto (W_1 \cap V, \widetilde{V}_1 \longrightarrow \widetilde{V}_1/V \xleftarrow{\sim} W_1/(W_1 \cap V))$$

is a bijection from the set of k-vector subspaces $V \subset \tilde{V}_1$ such that $\pi(V) = V_1$ onto the set of pairs (W, f) where W is a k-vector subspace of W_1 and $\tilde{V}_1 \xrightarrow{f} W_1/W$ is a k-linear map such that its restriction to $W_1 \subset \tilde{V}_1$ is the canonical projection $W_1 \longrightarrow\!\!\!\!\!\rightarrow W_1/W$. Hence the number of such V's of dimension r is equal to

$$\frac{(1 - q^{\tilde{r}_1 - r + 1})(1 - q^{\tilde{r}_1 - r + 2}) \cdots (1 - q^{\tilde{r}_1 - r_1})}{(1 - q)(1 - q^2) \cdots (1 - q^{r - r_1})} q^{r_1(\tilde{r}_1 - r)}$$

and we have

$$\varepsilon_{\tilde{V}_1 \xrightarrow{\pi} V_1} =$$
$$\sum_{r = r_1}^{\tilde{r}_1} \frac{(1 - q^{r_1}) \cdots (1 - q^{r-1})(1 - q^{\tilde{r}_1 - r + 1}) \cdots (1 - q^{\tilde{r}_1 - r_1})}{(1 - q) \cdots (1 - q^{r - r_1})} q^{r_1(\tilde{r}_1 - r)}.$$

It follows that

$$\varepsilon_{\tilde{V}_1 \xrightarrow{\pi} V_1} = 1$$

(see (C.1.1)) and that $O_R = O_{R_1}$. $\hspace{2cm}$ □

The corresponding result for twisted orbital integrals of f_r is

THEOREM (4.6.2). — *Let $\gamma_r \in G(F_r)$ be such that $N_r(\gamma_r) = \gamma$ belongs to $G(F)$ and is elliptic in $G(F)$. Let $dg^\sigma_{\gamma_r}$ be a Haar measure on $G^\sigma_{\gamma_r}(F)$. Let f_r be the Hecke function on $G(F_r)$ with Satake transform*

$$f_r^\vee(z) = q^{r(d-1)/2}(z_1 + \cdots + z_d).$$

Then the twisted orbital integral $TO_{\gamma_r}(f_r, dg^\sigma_{\gamma_r})$ is zero unless $v(\det \gamma) = r$. Moreover, if $v(\det \gamma) = r$ and if we normalize $dg^\sigma_{\gamma_r}$ on $G^\sigma_{\gamma_r}(F) \simeq D'^\times$ by $\mathrm{vol}(\mathcal{D}'^\times, dg^\sigma_{\gamma_r}) = 1$ (D' is the central division algebra with invariant $-1/d'$ over F' where $F' = F[\gamma] \subset gl_d(F)$ and $d' = d/[F' : F]$, see (4.5.3) ; $\mathcal{D}'$ is the maximal order of D'), we have

$$TO_{\gamma_r}(f_r, dg^\sigma_{\gamma_r}) = [k' : k]$$

(k' *is the residue field of $\mathcal{O}'$).*

The theorem will be an easy consequence of the following proposition:

PROPOSITION (4.6.3) (Drinfeld). — *Let $\gamma_r \in G(F_r)$. Let us assume that*

$$(\gamma_r \sigma)(\mathcal{O}_r^d) \subset \mathcal{O}_r^d,$$

that there exists a positive integer N with

$$(\gamma_r \sigma)^N(\mathcal{O}_r^d) \subset \varpi \mathcal{O}_r^d$$

where ϖ is a uniformizer of $\mathcal{O}$ and that

$$v_r(\det \gamma_r) = 1.$$

Then each lattice L_r in F_r^d such that

$$(\gamma_r \sigma)(L_r) \subset L_r$$

has the form
$$L_r = (\gamma_r \sigma)^n (\mathcal{O}_r^d)$$

for some integer n.

Proof: We will follow [Dr 3] (2.7). Since

$$(\gamma_r \sigma)^{mN}(\mathcal{O}_r^d) \subset \varpi^m \mathcal{O}_r^d$$

and

$$(\gamma_r \sigma)^{-mN}(\mathcal{O}_r^d) \supset \varpi^{-m} \mathcal{O}_r^d$$

for any integer $m \geq 0$, there exists a unique $n \in \mathbb{Z}$ such that

$$L_r \subset (\gamma_r \sigma)^n (\mathcal{O}_r^d)$$

and

$$L_r \not\subset (\gamma_r \sigma)^{n+1}(\mathcal{O}_r^d).$$

As $v_r(\det \gamma_r) = 1$, the length of the finite $\mathcal{O}_r$-module $\mathcal{O}_r^d/(\gamma_r\sigma)(\mathcal{O}_r^d)$ is 1. Therefore, if $x \in L_r$ and $x \notin (\gamma_r\sigma)^{n+1}(\mathcal{O}_r^d)$, x generates the $\mathcal{O}_r$-module of length 1

$$(\gamma_r \sigma)^n(\mathcal{O}_r^d)/(\gamma_r\sigma)^{n+1}(\mathcal{O}_r^d).$$

So $\{x, (\gamma_r\sigma)(x), \ldots, (\gamma_r\sigma)^{s-1}(x)\}$ generate the $\mathcal{O}_r$-module of length s

$$(\gamma_r\sigma)^n(\mathcal{O}_r^d)/(\gamma_r\sigma)^{n+s}(\mathcal{O}_r^d)$$

for each integer $s \geq 0$. In particular, we have

$$(\gamma_r\sigma)^n(\mathcal{O}_r^d) = L_r + (\gamma_r\sigma)^{n+s}(\mathcal{O}_r^d)$$

for each integer $s \geq 0$. But, if s is large enough, $(\gamma_r\sigma)^{n+s}(\mathcal{O}_r^d)$ is contained in L_r and we get

$$(\gamma_r\sigma)^n(\mathcal{O}_r^d) = L_r$$

as desired. $\qquad\square$

Proof of (4.6.2) : If $g_r^{-1}\gamma_r\sigma(g_r) \in K_r\varpi^{(1,0,\dots,0)}K_r$ for some $g_r \in G(F_r)$, then

$$v_r(\det(g_r^{-1}\gamma_r\sigma(g_r))) = v_r(\det\gamma_r) = 1$$

and

$$v(\det\gamma) = v_r(\det N_r(\gamma_r)) = r.$$

Hence, the first assertion of the theorem is obvious.

Let us assume that $v(\det\gamma) = r$ or what is the same that $v_r(\det\gamma_r) = 1$. The subset $gl_d(\mathcal{O}_r) \cap G(F_r)$ of $G(F_r)$ is the disjoint union of the double classes $K_r\varpi^\lambda K_r$ with $\lambda = (\lambda_1,\dots,\lambda_d) \in \mathbf{Z}^d$ and $\lambda_1 \geq \lambda_2 \geq \cdots \geq \lambda_d \geq 0$. On each double class $K_r\varpi^\lambda K_r$ the function $v_r \circ \det$ is constant with value $\lambda_1 + \cdots + \lambda_d$. But, for any λ as before, we have $\lambda_1 + \cdots + \lambda_d = 1$ if and only if $\lambda = (1,0,\dots,0)$. Therefore, for any $g_r \in G(F_r)$ we have

$$g_r^{-1}\gamma_r\sigma(g_r) \in gl_d(\mathcal{O}_r)$$

if and only if

$$g_r^{-1}\gamma_r\sigma(g_r) \in K_r\varpi^{(1,0,\dots,0)}K_r.$$

Now, by similar arguments to those in the proof of (4.6.1) we get the formula

$$TO_{\gamma_r}(f_r, dg_{\gamma_r}^\sigma) = \sum_{L_r \in \Gamma_r} \frac{1}{\mathrm{vol}(\{g_r \in D'^\times | g_r(L_r) = L_r\}, dg_{\gamma_r}^\sigma)}$$

where Γ_r is a system of representatives for the $D'^\times$-orbits in the set of lattices L_r in F_r^d such that

$$(\gamma_r\sigma)(L_r) \subset L_r$$

(recall that $\mathrm{vol}(K_r, dg_r) = 1$).

Up to σ-conjugacy we can assume that $\gamma_r \in gl_d(\mathcal{O}_r)$ so that the hypotheses of the proposition (4.6.3) are satisfied : we have

$$(\gamma_r\sigma)^r = \gamma,$$

γ is elliptic in $G(F)$ and $v'(\gamma) > 0$ where v' is the discrete valuation of the field $F' = F[\gamma] \subset gl_d(F)$, so that

$$(\gamma_r\sigma)^{rs}(\mathcal{O}_r^d) = \gamma^s(\mathcal{O}_r^d) \subset \varpi\mathcal{O}_r^d$$

as long as $sv'(\gamma)$ is larger than the ramification index of F' over F. Therefore, each lattice L_r in F_r^d such that $(\gamma_r\sigma)(L_r) \subset L_r$ has the form $L_r = (\gamma_r\sigma)^n(\mathcal{O}_r^d)$ for some integer n. But, as $(\gamma_r\sigma)^r = \gamma \in F'^\times \subset D'^\times$, the lattices $(\gamma_r\sigma)^{n'}(\mathcal{O}_r^d)$ and $(\gamma_r\sigma)^{n''}(\mathcal{O}_r^d)$ are in the same $D'^\times$-orbit for all integers n', n'' such that $n' - n'' \equiv 0 \pmod{r}$ and

$$\sum_{L_r \in \Gamma_r} [D'^\times/\gamma^{\mathbf{Z}} : \{g_r \in D'^\times | g_r(L_r) = L_r\}] = r.$$

As $\{g_r \in D'^{\times}|g_r(L_r) = L_r\} \subset D'^{\times} \subset D'^{\times}/\gamma^{\mathbb{Z}}$ for each $L_r \in \Gamma_r$ and as $\mathrm{vol}(D'^{\times}, dg^{\sigma}_{\gamma_r}) = 1$, we have

$$TO_{\gamma_r}(f_r, dg^{\sigma}_{\gamma_r}) = \sum_{L_r \in \Gamma_r} [D'^{\times} : \{g_r \in D'^{\times}|g_r(L_r) = L_r\}]$$

and to finish the proof of the theorem it suffices to check that

$$[D'^{\times}/\gamma^{\mathbb{Z}} : D'^{\times}] = r/[k' : k].$$

But we have

$$r = v(\det \gamma) = [k' : k]d'v'(\gamma)$$

and the last assertion follows. $\qquad\qquad\qquad\qquad\qquad\qquad\qquad\square$

LEMMA (4.6.4). — *Let d' be a positive integer, let F' be a non-archimedean local field with ring of integers $\mathcal{O}'$ and let D' be a central division algebra of dimension d'^2 over F' with maximal order $\mathcal{D}'$. Let dg' be a Haar measure on $GL_{d'}(F')$ and let $d\bar{g}'$ be the transfer of dg' from $GL_{d'}(F')$ to its inner twist $D'^{\times}$ (see (3.5)). Then we have*

$$\frac{\mathrm{vol}(GL_{d'}(\mathcal{O}'), dg')}{\mathrm{vol}(D'^{\times}, d\bar{g}')} = (q' - 1) \cdots (q'^{d'-1} - 1)$$

where q' is the number of elements of the residue field k' of $\mathcal{O}'$.

Proof: We will follow [Ko 1] § 1 (see also [Ro] 3). It is enough to prove the lemma for one particular Haar measure dg'. Let $\mathcal{G}'$ be the following smooth group scheme over $\mathrm{Spec}(\mathcal{O}')$:

$$\mathcal{G}' = \{g' \in gl_{d',\mathcal{O}'}| \det \tilde{g}' \text{ is invertible}\}$$

where $\tilde{g}' = (\tilde{g}'_{ij})_{1 \le i,j \le d'}$ with

$$\tilde{g}'_{ij} = \begin{cases} g'_{ij} & \text{if } i \le j, \\ \varpi' g'_{ij} & \text{if } i > j, \end{cases}$$

if $g' = (g'_{ij})_{1 \le i,j \le d'}$ (ϖ' is a uniformizer of $\mathcal{O}'$), and the group law is given by

$$g' = g'_1 \cdot g'_2$$

if and only if

$$\tilde{g}' = \tilde{g}'_1 \tilde{g}'_2$$

(ordinary matrix product). Then the general fiber $F' \otimes_{\mathcal{O}'} \mathcal{G}'$ of $\mathcal{G}'$ is isomorphic to $GL_{d',F'}$ but its special fiber $k' \otimes_{\mathcal{O}'} \mathcal{G}'$ is a connected solvable algebraic group of dimension d'^2 over k' (for example, if $d' = 2$, $k' \otimes_{\mathcal{O}'} \mathcal{G}'$ is the k'-group scheme

$$k' \otimes_{\mathcal{O}'} \mathcal{G}' = \left\{ \begin{pmatrix} g'_{11} & g'_{12} \\ g'_{21} & g'_{22} \end{pmatrix} \in gl_{2,k'} \,|\, g'_{11}g'_{22} \neq 0 \right\}$$

and the group law is given by

$$\begin{pmatrix} g'_{11} & g'_{12} \\ g'_{21} & g'_{22} \end{pmatrix} \cdot \begin{pmatrix} h'_{11} & h'_{12} \\ h'_{21} & h'_{22} \end{pmatrix} = \begin{pmatrix} g'_{11}h'_{11} & g'_{11}h'_{12} + g'_{12}h'_{22} \\ g'_{12}h'_{11} + g'_{22}h'_{21} & g'_{22}h'_{22} \end{pmatrix}.$$

Let ω' be a left invariant volume form on $\mathcal{G}'$ such that the restriction of ω' to the special fiber $k' \otimes_{\mathcal{O}'} \mathcal{G}'$ is non-zero. For example, we can take

$$\omega' = \frac{dg'_{11} \wedge \cdots \wedge dg'_{1d'} \wedge \cdots \wedge dg'_{d',1} \wedge \cdots \wedge dg'_{d'd'}}{(\det \widetilde{g}')^{d'}}.$$

Such a volume form is uniquely determined up to the product by a unit of $\mathcal{O}'$. Therefore the Haar measure $|\omega'|'$ on $\mathcal{G}'(F') \cong GL_{d'}(F')$ which is induced by ω' (see (3.5)) is independent of the choice of ω'. Moreover it is easy to see that

$$\mathrm{vol}(\mathcal{G}'(\mathcal{O}'), |\omega'|') = q'^{-d'^2} |\mathcal{G}'(k')|,$$

i.e.

$$\mathrm{vol}(\mathcal{G}'(\mathcal{O}'), |\omega'|') = q'^{-d'}(q'-1)^{d'}.$$

Now let $F'_{d'}$ be the unramified extension of degree d' of F' and let $\mathcal{O}'_{d'}$ be its ring of integers. We will denote by $k'_{d'}$ the residue field of $\mathcal{O}'_{d'}$ and by $\sigma' \in \mathrm{Gal}(F'_{d'}/F')$ the lifting of the arithmetic Frobenius element (q'-th power) in $\mathrm{Gal}(k'_{d'}/k')$. There exists an integer e', e' prime to d', such that D' is isomorphic to

$$F'_{d'}[\tau']/(\tau'^{d'} - \varpi'),$$

with commutation relation

$$\tau' a' = \sigma'^{e'}(a')\tau'$$

for each $a' \in F'_{d'}$ and D' is isomorphic to

$$\sum_{i=0}^{d'-1} \mathcal{O}'_{d'}\tau'^i \subset F'_{d'}[\tau']/(\tau'^{d'} - \varpi')$$

(see (A.2.5), (A.2.6)). Moreover, we can embed D' into $gl_{d'}(F'_{d'})$ by

$$\tau' \mapsto \begin{pmatrix} 0 & 1 & \cdots & 0 \\ \vdots & \ddots & \ddots & \vdots \\ 0 & & \ddots & 1 \\ \varpi' & 0 & \cdots & 0 \end{pmatrix} = \varepsilon'$$

and

$$a' \mapsto \begin{pmatrix} a' & & & 0 \\ & \sigma'^{e'}(a') & & \\ & & \ddots & \\ 0 & & & \sigma'^{e'(d'-1)}(a') \end{pmatrix}$$

and the image of D' is the F'-algebra of invariants under the automorphism

$$g' \mapsto \varepsilon'^{-1}\sigma'^{e'}(g')\varepsilon'$$

of $gl_{d'}(F'_{d'})$. In other words, let θ' be the automorphism of the $\mathcal{O}'_{d'}$-group scheme $\mathcal{O}'_{d'} \otimes_{\mathcal{O}'} \mathcal{G}'$ defined by

$$\theta'(g') = h'$$

if

$$\widetilde{h}' = \varepsilon'^{-1}\sigma'^{e'}(\widetilde{g}')\varepsilon'$$

and let $\mathrm{Gal}(F'_{d'}/F')$ act on $\mathcal{O}'_{d'} \otimes_{\mathcal{O}'} \mathcal{G}'$ by $\sigma' \mapsto \theta'$. By descent we get a new smooth $\mathcal{O}'$-group scheme $\overline{\mathcal{G}}'$ with an inner twisting

$$\psi' : \mathcal{O}'_{d'} \otimes_{\mathcal{O}'} \mathcal{G}' \xrightarrow{\sim} \mathcal{O}'_{d'} \otimes_{\mathcal{O}'} \overline{\mathcal{G}}'$$

over $\mathcal{O}'_{d'}$ such that

$$\overline{\mathcal{G}}'(F') \cong D',$$
$$\overline{\mathcal{G}}'(\mathcal{O}') \cong \mathcal{D}',$$

(for more details see [Re] §14). Obviously the transfer of ω' to $\overline{\mathcal{G}}'$ by ψ' is a volume form on $\overline{\mathcal{G}}'$ such that its restriction to the special fiber $k' \otimes_{\mathcal{O}'} \overline{\mathcal{G}}'$ is non-zero. Therefore, if we denote by this transfer $\overline{\omega}'$, we have

$$\mathrm{vol}(\overline{\mathcal{G}}'(\mathcal{O}'), |\overline{\omega}'|') = q'^{-d'^2}|\overline{\mathcal{G}}'(k')|,$$

i.e.

$$\mathrm{vol}(\overline{\mathcal{G}}'(\mathcal{O}'), |\overline{\omega}'|') = q'^{-d'}(q'^{d'} - 1).$$

Finally, we have

$$\mathcal{G}'(\mathcal{O}') \subset GL_{d'}(\mathcal{O}') \subset GL_{d'}(F') = \mathcal{G}'(F')$$

and the index of $\mathcal{G}'(\mathcal{O}')$ in $GL_{d'}(\mathcal{O}')$ is equal to

$$[GL_{d'}(k') : B'(k')] = \frac{(q'-1)\cdots(q'^{d'} - 1)}{(q'-1)^{d'}}$$

where $B' \subset GL_{d'}$ is the Borel subgroup over $\mathcal{O}'$ of upper triangular matrices. The lemma follows. $\qquad\square$

Proof of (4.5.5) *when γ is elliptic and γ_r is σ-elliptic* : If $v(\det \gamma) \neq r$, it follows from (4.6.1) that $O_\gamma(f, dg_\gamma) = 0$ and from (4.6.2) that $TO_{\gamma_r}(f_r, dg_{\gamma_r}^\sigma) = 0$ for any $\gamma_r \in G(F_r)$ with norm γ. Therefore, we can assume that $v(\det \gamma) = r$. But in this case, the theorem follows from (4.6.1), (4.6.2) and (4.6.4). $\square$

(4.7) The general case

In this section we will deduce the general case of (4.5.5) from the elliptic case.

First of all, let us review the following result of Kottwitz (fundamental lemma for the unit in the Hecke algebra, see [Ko 3]).

PROPOSITION (4.7.1) (Kottwitz). — *Let 1_K and 1_{K_r} be the characteristic functions of K in $G(F)$ and K_r in $G(F_r)$ respectively (1_K is the unit in $\mathcal{H}$, 1_{K_r} is the unit in $\mathcal{H}_r$ and $1_K = b(1_{K_r})$). Let γ be elliptic in $G(F)$ and let dg_γ be a Haar measure on $G_\gamma(F)$.*
Then $O_\gamma(1_K, dg_\gamma) = 0$ if γ is not a norm in $G(F_r)$.
Moreover, if $\gamma = N_r(\gamma_r)$ for some $\gamma_r \in G(F_r)$, we have

$$O_\gamma(1_K, dg_\gamma) = TO_{\gamma_r}(1_{K_r}, dg_\gamma).$$

REMARK (4.7.2). — The same results holds if we assume only that γ is closed (see (4.3.11) and (4.4.9)). $\square$

Proof : Obviously $O_\gamma(1_K, dg_\gamma) = 0$ unless $v(\det \gamma) = 0$ (resp. $TO_{\gamma_r}(1_{K_r}, dg_{\gamma_r}^\sigma) = 0$ unless $v_r(\det \gamma_r) = 0$) for each elliptic (resp. σ-elliptic) element γ in $G(F)$ (resp. γ_r in $G(F_r)$) and for each Haar measure dg_γ on $G_\gamma(F)$ (resp. $dg_{\gamma_r}^\sigma$ on $G_{\gamma_r}^\sigma(F)$). So we can assume that $v(\det \gamma) = 0$. In this case γ is a norm in $G(F_r)$ (see the proof of (4.5.2)) and the first assertion of the proposition is proved.

Now let $\gamma_r \in G(F_r)$ be such that $N_r(\gamma_r) = \gamma$ belongs to $G(F)$, is elliptic and $v(\det \gamma) = 0$. Let F_∞ be the completion of the maximal unramified extension of F ($F_\infty \supset F_r \supset F$) with residue field k_∞ (k_∞ is an algebraic closure of k containing k_r) and let $\sigma \in \text{Gal}(F_\infty/F)$ be the lifting of the arithmetic Frobenius element in $\text{Gal}(k_\infty/k)$ (σ induces σ in $\text{Gal}(F_r/F)$). There exists $g_\infty \in G(F_\infty)$ such that

$$\gamma_r = g_\infty \sigma(g_\infty)^{-1}$$

(see (3.2.5)). Let us fix such a g_∞. As $\sigma^r(\gamma_r) = \gamma_r$, we have

$$\sigma^r(g_\infty)\sigma^{r+1}(g_\infty)^{-1} = g_\infty \sigma(g_\infty)^{-1}$$

and, if we set

$$\tilde{\gamma} = \sigma^r(g_\infty)^{-1} g_\infty,$$

we have $\sigma(\widetilde{\gamma}) = \widetilde{\gamma}$ and $\widetilde{\gamma} \in G(F)$. Moreover, we have

$$\gamma = N_r(\gamma_r) = g_\infty \sigma^r (g_\infty)^{-1}.$$

Therefore, we have

$$\widetilde{\gamma} = g_\infty^{-1} \gamma g_\infty$$

and, as $\gamma, \widetilde{\gamma} \in G(F)$, there exists $g_1 \in G(F)$ such that

$$\widetilde{\gamma} = g_1^{-1} \gamma g_1.$$

Let us fix such a g_1. We get group isomorphisms

$$G_\gamma(F) \xrightarrow{\sim} G_{\widetilde{\gamma}}(F) \;,\;\; g \mapsto g_1^{-1} g g_1$$

and

$$G_{\widetilde{\gamma}}(F) \xrightarrow{\sim} G_{\gamma_r}^\sigma(F) \;,\;\; g \mapsto g_\infty g g_\infty^{-1}$$

(we have

$$\sigma^r(g_\infty g g_\infty^{-1}) = g_\infty \widetilde{\gamma}^{-1} g \widetilde{\gamma} g_\infty^{-1} = g_\infty g g_\infty^{-1}$$

and

$$g_\infty g g_\infty^{-1} \gamma_r = g_\infty g \sigma(g_\infty)^{-1} = \gamma_r \sigma(g_\infty g g_\infty^{-1})$$

for any $g \in G_{\widetilde{\gamma}}(F)$) and bijections

$$X^\gamma \xrightarrow{\sim} X^{\widetilde{\gamma}} \;,\;\; L \mapsto g_1^{-1}(L)$$

and

$$X^{\widetilde{\gamma}} \xrightarrow{\sim} X_r^{\gamma_r \sigma} \;,\;\; L \mapsto g_\infty(L)$$

(we have

$$\sigma^r(g_\infty L) = g_\infty \widetilde{\gamma}^{-1}(L) = g_\infty(L)$$

and

$$\gamma_r \sigma(g_\infty(L)) = g_\infty(L)$$

for any $L \in X^{\widetilde{\gamma}}$). Here X (resp. X_r) is the set of lattices L in F^d (resp. L_r in F_r^d) and

$$X^\gamma = \{L \in X | \gamma(L) = L\},$$
$$X^{\widetilde{\gamma}} = \{L \in X | \widetilde{\gamma}(L) = L\}$$

(resp.

$$X_r^{\gamma_r \sigma} = \{L_r \in X_r | \gamma_r \sigma(L_r) = L_r\}).$$

As the two bijections are equivariant with respect to the two group homomorphisms in the obvious way, we get bijections

$$G_\gamma(F) \backslash X^\gamma \xrightarrow{\sim} G_{\widetilde{\gamma}}(F) \backslash X^{\widetilde{\gamma}}$$

and

$$G_{\tilde{\gamma}}(F)\backslash X^{\tilde{\gamma}} \stackrel{\sim}{\to} G^{\sigma}_{\gamma_r}(F)\backslash X^{\gamma_r\sigma}_r.$$

But, as in the proof of (4.6.1) (resp. (4.6.2)), we have

$$O_{\gamma}(1_K, dg_{\gamma}) = \sum_{L \in \Gamma} \frac{1}{\text{vol}(\{g \in G_{\gamma}(F)|g(L) = L\}, dg_{\gamma})}$$

(resp.

$$TO_{\gamma_r}(1_{K_r}, dg^{\sigma}_{\gamma_r}) = \sum_{L_r \in \Gamma_r} \frac{1}{\text{vol}(\{g_r \in G^{\sigma}_{\gamma_r}(F)|g_r\sigma(L_r) = L_r\}, dg^{\sigma}_{\gamma_r})}),$$

where Γ (resp. Γ_r) is a system of representatives of $G_{\gamma}(F)$-orbits (resp. $G^{\sigma}_{\gamma_r}(F)$-orbits) in X^{γ} (resp. $X^{\gamma_r\sigma}_r$), for any Haar measure dg_{γ} (resp. $dg^{\sigma}_{\gamma_r}$) on $G_{\gamma}(F)$ (resp. $G^{\sigma}_{\gamma_r}(F)$) and the proposition follows. □

End of the proof of (4.5.5) : Let γ be a closed element in $G(F)$ and let $P = MN$ be the parabolic subgroup of G over F canonically associated to γ and to an ordering of the set of prime factors of the minimal polynomial of γ (see (4.3.6)). We have

$$M(F) \cong GL_{d_1}(F) \times \cdots \times GL_{d_s}(F),$$
$$\gamma = (\gamma_1, \ldots, \gamma_s)$$

for some partition $(d_1, \ldots, d_s)$ of d and γ_j is elliptic in $GL_{d_j}(F)$ for $j = 1, \ldots, s$. Thanks to (4.3.11) and (4.2.5), (4.2.6) we have

$$O_{\gamma}(f, dg_{\gamma}) = |D_{M\backslash G}(\gamma)|^{-1/2} \sum_{j=1}^{s} q^{r(d-d_j)/2} O^{GL_{d_j}}_{\gamma_j}(\varphi_{d_j}, dg_{\gamma_j})$$

$$\times \prod_{\substack{k=1 \\ k \neq j}} O^{GL_{d_k}}_{\gamma_k}(1_{GL_{d_k}(\mathcal{O})}, dg_{\gamma_k})$$

$(G_{\gamma}(F) = M_{\gamma}(F) \cong \prod_{j=1}^{s} GL_{d_j}(F)_{\gamma_j}$ and $dg_{\gamma} = \prod_{j=1}^{s} dg_{\gamma_j})$. But the j-th term of this sum is zero unless $v(\det \gamma_j) = r$ and $v(\det \gamma_k) = 0$ for all $k = 1, \ldots, s$, $k \neq j$. Therefore, $O_{\gamma}(f, dg_{\gamma})$ is zero unless γ is r-admissible. Moreover, if γ is r-admissible, we can choose the above ordering in such way that

$$v(\det \gamma_j) = \begin{cases} r & \text{if } j = 1, \\ 0 & \text{if } j = 2, \ldots, s, \end{cases}$$

and we have

$$O_\gamma(f, dg_\gamma) = |D_{M\backslash G}(\gamma)|^{-1/2} q^{r(d-d_1)/2} O_{\gamma_1}^{GL_{d_1}}(\varphi_{d_1}, dg_{\gamma_1})$$
$$\times \prod_{j=2}^{s} O_{\gamma_j}^{GL_{d_j}}(1_{GL_{d_j}(\mathcal{O}_j)}, dg_{\gamma_j}).$$

Similarly, let $\gamma_r \in G(F_r)$ be such that $N_r(\gamma_r) = \gamma \in G(F)$ and is closed in $G(F)$. Up to σ-conjugacy, we can assume that $\gamma_r \in M(F)$ where $P = MN$ is associated to γ as before. In particular, we have

$$\gamma_r = (\gamma_{r,1}, \ldots, \gamma_{r,s})$$

with $\gamma_{r,j} \in GL_{d_j}(F_r)$ and $N_r(\gamma_{r,j}) = \gamma_j$ ($j = 1, \ldots, s$). From (4.4.9) and (4.2.5), (4.2.6), we get as before that $TO_{\gamma_r}(f_r, dg_\gamma) = 0$ unless γ_r is r-admissible and that

$$TO_{\gamma_r}(f, dg_\gamma) = |D_{M\backslash G}(\gamma)|^{-1/2} q^{r(d-d_1)/2} TO_{\gamma_{r,1}}^{GL_{d_1}}(\psi_{d_1}, dg_{\gamma_1})$$
$$\times \prod_{j=2}^{s} TO_{\gamma_{r,j}}^{GL_{d_j}}(1_{GL_{d_j}(\mathcal{O}_{r,j})}, dg_{\gamma_j}).$$

for any r-admissible γ_r such that

$$v_r(\det \gamma_{r,j}) = \begin{cases} 1 & \text{if } j = 1, \\ 0 & \text{if } j = 2, \ldots, s, \end{cases}$$

where ψ_{d_1} is the Hecke function on $GL_{d_1}(F_r)$ with Satake transform

$$q^{r(d_1-1)/2}(z_1 + \cdots + z_{d_1}).$$

But $\varphi_{d_1} = b(\psi_{d_1})$ and to finish the proof of the theorem (4.5.5) it suffices to apply the elliptic case of (4.5.5) which has been proved in (4.6) and (4.7.1) ($\varepsilon(\gamma) = \varepsilon(\gamma_1)$). $\qquad\square$

(4.8) Non-closed orbital integrals

In the previous sections we have considered only closed orbital integrals of the Hecke functions f. Now we would like to consider more general ones. There are two difficulties. On the one hand it is not clear that $G_\gamma(F)$ is unimodular for a non-closed element γ in $G(F)$. On the other hand, if the orbit $O_G(\gamma)(F)$ is not closed, it is not clear if the orbital integral $O_\gamma(f, dg_\gamma)$ is convergent. First of all we will review how to solve these difficulties (see [Ra] if $\text{char}(F) = 0$ and [De–Ka–Vi] Appendice 1 if $\text{char}(F) > 0$).

Let $\gamma \in G(F)$ and let $Q_1, \ldots, Q_s(T)$ be an ordering of the set of prime factors of the minimal polynomial $P_d(T)$ of γ. For each prime factor $Q(T)$ of $P_d(T)$ we have defined the F-vector subspace of F^d,

$$V_{Q(T)} = \mathrm{Ker}(Q(\gamma)^{m_{Q(T)}} : F^d \to F^d),$$

where $m_{Q(T)}$ is the multiplicity of $Q(T)$ in $P_d(T)$ (see (4.3)). Let us now introduce the flag of F-vector subspaces

(4.8.1) $(0) = V_{Q(T),0} \subsetneq V_{Q(T),1} \subsetneq \cdots \subsetneq V_{Q(T),m_{Q(T)}} = V_{Q(T)}$

where

$$V_{Q(T),k} = \mathrm{Ker}(Q(T)^k : V_{Q(T)} \to V_{Q(T)})$$

for $k = 0, \ldots, m_{Q(T)}$. We have a canonical parabolic subgroup of $GL(V_{Q(T)})$ (over F)

$$P''_{Q(T)} = \{g \in GL(V_{Q(T)}) | g(V_{Q(T),k}) \subset V_{Q(T),k} \ , \ k = 0, \ldots, m_{Q(T)}\}$$

and therefore a canonical parabolic subgroup

(4.8.2) $P'' = \displaystyle\prod_{Q(T)} P''_{Q(T)} \subset \prod_{Q(T)} GL(V_{Q(T)}) = M$

(over F) where $Q(T)$ runs through the set of prime factors of $P_d(T)$ (see (4.3.5)).

If $P = MN$ is the parabolic subgroup of G with Levi decomposition (over F) which is associated to the ordering $Q_1(T), \ldots, Q_s(T)$ (see (4.3.6)), we set

(4.8.3) $P' = P''N \subset MN = P \subset G.$

Then P' is a parabolic subgroup of G over F which depends on the ordering $Q_1(T), \ldots, Q_s(T)$.

We will denote by $N'' \subset P''$ the unipotent radical of P'' and by

$$\pi'' : P'' \longrightarrow P''/N'' = \overline{P}''$$

its maximal reductive quotient ($N' = N''N$ is the unipotent radical of P' and $P' \longrightarrow P'/N = P'' \overset{\pi''}{\longrightarrow} \overline{P}''$ is its maximal reductive quotient).

LEMMA (4.8.4). — *For any* $\gamma \in G(F)$, *the centralizer* G_γ *of* γ *in* G *is contained in* P'' *(in particular* $\gamma \in P''(F)$*) and* $\overline{\gamma} = \pi''(\gamma)$ *is elliptic (and therefore closed) in* $\overline{P}''(F)$. *Moreover*

$$\overline{O_{P''}(\gamma)(F)} = \pi''^{-1}(O_{\overline{P}''}(\overline{\gamma})(F))$$

(*closure for the* ϖ-*adic topology in* $P''(F)$).

An element of

$$\overline{P}''(F) = \prod_{Q(T)} \prod_{k=1}^{m_{Q(T)}} GL(V_{Q(T),k}/V_{Q(T),k-1})$$

is elliptic (resp. closed) if and only if each of its components is elliptic (resp. closed).

Proof : We can assume that $P_d(T) = Q(T)^m$ $(m = m_{Q(T)})$ for some irreducible unitary polynomial $Q(T)$ in $F[T]$ $(M = G)$.

The inclusion $G_\gamma \subset P''$ is obvious.

Each component of $\overline{\gamma}$ in $GL(V_k/V_{k-1})$ $(V_k = V_{Q(T),k}$ for $k = 0, \ldots, m)$ has a minimal polynomial equal to $Q(T)$. Therefore $\overline{\gamma}$ is elliptic in $\overline{P}''(F)$.

Let $O_{\overline{P}''}(\overline{\gamma}) \subset \overline{P}''$ and $\Omega^\circ = O_{P''}(\gamma) \subset P''$ be the algebraic orbits (they are locally closed F-subschemes of $\overline{P}''$ and P'' respectively, smooth and geometrically connected over F, and they are isomorphic to the homogeneous schemes $\overline{P}''_{\overline{\gamma}}\backslash\overline{P}''$ and $P''_\gamma\backslash P''$ respectively). Let $\Omega = \pi''^{-1}(O_{\overline{P}''}(\overline{\gamma})) \subset P''$. Ω is also a locally closed F-subscheme of P'', smooth and geometrically connected over F, and we have $\Omega^\circ \subset \Omega$.

Let R be a finite dimensional F-algebra and let H be the F-group scheme of invertible elements in R. Then H is smooth over F and the fppf cohomology group

$$H^1_{\mathrm{fppf}}(\mathrm{Spec}(F), H)$$

coincides with the étale one

$$H^1_{\mathrm{ét}}(\mathrm{Spec}(F), H) = H^1(\mathrm{Gal}(F^s/F), H(F^s))$$

where F^s is a separable closure of F (see [SGA 3] XXIV, 8.1 (i) and [Gro] Rq. (11.8)). But the last cohomology group is trivial (see [Se 1] Ch. X, § 1, Ex. 2). In particular

$$H^1_{\mathrm{fppf}}(\mathrm{Spec}(F), \overline{P}''_{\overline{\gamma}}) = \{1\}$$

and

$$H^1_{\mathrm{fppf}}(\mathrm{Spec}(F), P''_\gamma) = \{1\}.$$

Therefore, the sets of F-rational points $\Omega(F)$ and $\Omega^\circ(F)$ with their induced structures of ϖ-adic manifold coincide with $\pi''^{-1}(O_{\overline{P}''}(\overline{\gamma})(F))$ and $O_{P''}(\gamma)(F)$ respectively.

To finish the proof of the lemma it suffices to check that Ω° is a Zariski dense open subset of Ω. In fact we will prove that the inclusion $\Omega^\circ \subset \Omega$ is étale ($\Omega^\circ \neq \emptyset$ and Ω is irreducible). As Ω° is homogeneous under the action of P'' by conjugacy, it is enough to prove that the tangent map at $\gamma \in \Omega^\circ(F)$ to this inclusion is bijective. But this tangent map is nothing else than

$$\mathfrak{p}''_\gamma\backslash\mathfrak{p}'' \to \mathfrak{p}'' \times_{\overline{\mathfrak{p}}''} (\overline{\mathfrak{p}}''_{\overline{\gamma}}\backslash\overline{\mathfrak{p}}''),$$
$$\mathfrak{p}''_\gamma + \xi \mapsto ([\gamma, \xi], \overline{\mathfrak{p}}''_{\overline{\gamma}} + \mathrm{Lie}(\pi'')(\xi)),$$

where $\mathfrak{p}''$, $\mathfrak{p}''_\gamma$, $\overline{\mathfrak{p}}''$, $\overline{\mathfrak{p}}''_{\overline{\gamma}}$ are the Lie algebras of P'', P''_γ, $\overline{P}''$, $\overline{P}''_{\overline{\gamma}}$ respectively, where $\mathrm{Lie}(\pi'') : \mathfrak{p}'' \to \overline{\mathfrak{p}}''$ is the map induced by π'' and where the fiber product is for the maps $\mathrm{Lie}(\pi'')$ and $[\overline{\gamma}, -] : \overline{\mathfrak{p}}''_{\overline{\gamma}}\backslash\overline{\mathfrak{p}}'' \to \overline{\mathfrak{p}}''$.

As $\mathfrak{p}''_\gamma = \mathrm{Ker}([\gamma, -])$, this tangent map is injective. Moreover, the dimension over F of the source (resp. the target) is

$$\dim_F(\mathfrak{p}'') - \dim_F(\mathfrak{p}''_\gamma)$$

(resp.

$$\dim_F(\mathfrak{p}'') - \dim_F(\overline{\mathfrak{p}}''_{\overline{\gamma}})).$$

So the problem is reduced to proving that

$$\dim_F(\mathfrak{p}''_\gamma) = \dim_F(\overline{\mathfrak{p}}''_{\overline{\gamma}}).$$

But, if we denote by $\mathfrak{n}''$ the Lie algebra of N'', we have a commutative diagram

$$
\begin{array}{ccccccccc}
& & 0 & & 0 & & 0 & & \\
& & \downarrow & & \downarrow & & \downarrow & & \\
0 & \longrightarrow & \mathrm{Ker}([\gamma, -] : \mathfrak{n}'' \to \mathfrak{n}'') & \longrightarrow & \mathfrak{p}''_\gamma & \longrightarrow & \overline{\mathfrak{p}}''_{\overline{\gamma}} & & \\
& & \downarrow & & \downarrow & & \downarrow & & \\
0 & \longrightarrow & \mathfrak{n}'' & \longrightarrow & \mathfrak{p}'' & \longrightarrow & \overline{\mathfrak{p}}'' & \longrightarrow & 0 \\
& & {\scriptstyle [\gamma,-]}\downarrow & & {\scriptstyle [\gamma,-]}\downarrow & & {\scriptstyle [\overline{\gamma},-]}\downarrow & & \\
0 & \longrightarrow & \mathfrak{n}'' & \longrightarrow & \mathfrak{p}'' & \longrightarrow & \overline{\mathfrak{p}}'' & \longrightarrow & 0 \\
& & \downarrow & & \downarrow & & \downarrow & & \\
& & \mathfrak{n}''/[\gamma, \mathfrak{n}''] & \overset{(*)}{\longrightarrow} & \mathfrak{p}''/[\gamma, \mathfrak{p}''] & \longrightarrow & \overline{\mathfrak{p}}''/[\overline{\gamma}, \overline{\mathfrak{p}}''] & \longrightarrow & 0 \\
& & \downarrow & & \downarrow & & \downarrow & & \\
& & 0 & & 0 & & 0 & &
\end{array}
$$

where the rows and the columns are exact. Applying the snake lemma, we see that the equality

$$\dim_F(\mathfrak{p}''_\gamma) = \dim_F(\overline{\mathfrak{p}}''_{\overline{\gamma}})$$

is equivalent to the nullity of the map $(*)$ or what is the same to the inclusion

$$\mathfrak{n}'' \subset [\gamma, \mathfrak{p}''].$$

To prove the last inclusion, we will follow [Ho] Lemma 2 (b). We can choose a sequence of F-vector subspaces $W_1, \ldots, W_m$ of F^d such that

$$
\begin{cases}
V_k (= V_{Q(T),k}) = W_1 \oplus \cdots \oplus W_k, \\
Q(\gamma)(W_k) \subset W_{k-1},
\end{cases}
$$

for $k = 0, \ldots, m$ ($W_0 = (0)$). For each $k, \ell \in \{1, \ldots, m\}$ we denote by

$$Q(\gamma)_k : W_k \to W_{k-1}$$

the map induced by $Q(\gamma)$ and we set

$$\mathfrak{n}''_{k\ell} = \mathrm{Hom}_F(W_\ell, W_k).$$

Then $Q(\gamma)_k$ in injective if $k \geq 2$ and we have

$$\mathfrak{n}'' = \bigoplus_{1 \leq k < \ell \leq m} \mathfrak{n}''_{k\ell} \subset \bigoplus_{1 \leq k \leq \ell \leq m} \mathfrak{n}''_{k\ell} = \mathfrak{p}''.$$

We want to prove that

$$\mathfrak{n}'' \subset [\gamma, \mathfrak{p}''].$$

But we have

$$[R(\gamma), \mathfrak{p}''] \subset [\gamma, \mathfrak{p}'']$$

for any $R(T) \in F[T]$ and it suffices to prove that

$$\mathfrak{n}'' \subset [Q(\gamma), \mathfrak{p}''].$$

Let us begin by proving that

$$\bigoplus_{2 \leq \ell \leq m} \mathfrak{n}''_{1\ell} \subset [Q(\gamma), \bigoplus_{1 \leq \ell \leq m} \mathfrak{n}''_{1\ell}].$$

We have

$$[Q(\gamma), n_{1\ell}] = -n_{1\ell} Q(\gamma)_{\ell+1}$$

for each $n_{1\ell} \in \mathfrak{n}''_{1\ell}$. As $Q(\gamma)_{\ell+1}$ is injective, it follows that

$$[Q(\gamma), -] : \mathfrak{n}''_{1\ell} \to \mathfrak{n}''_{1,\ell+1}$$

is surjective for $\ell = 1, \ldots, m$ and we are done. Now let us assume that

$$\bigoplus_{k+1 \leq \ell \leq m} \mathfrak{n}''_{k\ell} \subset [Q(\gamma), \bigoplus_{k \leq \ell \leq m} \mathfrak{n}''_{k\ell}]$$

for some $k \in \{1, \ldots, m-1\}$ and let us prove the same statement for $k+1$. We have

$$[Q(\gamma), n_{k+1,\ell}] = Q(\gamma)_{k+1} n_{k+1,\ell} - n_{k+1,\ell} Q(\gamma)_{\ell+1}$$

for each $n_{k+1,\ell} \in \mathfrak{n}''_{k+1,\ell}$. As $Q(\gamma)_{\ell+1}$ is injective,

$$[Q(\gamma), -] : \mathfrak{n}''_{k+1,\ell} \to \mathfrak{n}''_{k,\ell} \oplus \mathfrak{n}''_{k+1,\ell+1}$$

is surjective modulo $\mathfrak{n}''_{k,\ell}$ for $\ell = k+1, \ldots, m$ and we are done. By induction we get the inclusion $\mathfrak{n}'' \subset [Q(\gamma), \mathfrak{p}'']$ and the lemma is proved. $\square$

REMARK (4.8.5). — It follows from the proof of (4.8.4) that we have a canonical exact sequence

$$0 \to \mathrm{Ker}(\mathfrak{n}'' \xrightarrow{[\gamma,-]} \mathfrak{n}'') \to \mathfrak{p}_\gamma'' \to \overline{\mathfrak{p}}_{\overline{\gamma}}'' \to \mathfrak{n}''/[\gamma, \mathfrak{n}''] \to 0.$$

In particular, we have a canonical isomorphism of 1-dimensional F-vector spaces

$$\bigwedge^{\dim_F(\mathfrak{p}_\gamma'')} \mathfrak{p}_\gamma'' \cong \bigwedge^{\dim_F(\overline{\mathfrak{p}}_{\overline{\gamma}}'')} \overline{\mathfrak{p}}_{\overline{\gamma}}''.$$

$\square$

LEMMA (4.8.6). — For each $\gamma \in G(F)$, the centralizer $G_\gamma(F)$ of γ in $G(F)$ is unimodular.

Proof: Let

$$\mathrm{Ad}_{G_\gamma} : G_\gamma(F) \to \mathfrak{g}_\gamma$$

be the adjoint action of $G_\gamma(F)$ on its Lie algebra $\mathfrak{g}_\gamma$. We want to prove that

$$v(\det(\mathrm{Ad}_{G_\gamma}(g_\gamma))) = 0$$

for each $g_\gamma \in G_\gamma(F)$ (see [Bou] Lie II, § 3, Cor. of Prop. 55). But we have

$$G_\gamma(F) = P_\gamma''(F),$$

where P'' is defined in (4.8.2), and, thanks to (4.8.5),

$$\det(\mathrm{Ad}_{P_\gamma''}(p_\gamma)) = \det(\mathrm{Ad}_{\overline{P}_{\overline{\gamma}}''}(\pi''(p_\gamma)))$$

for each $p_\gamma \in P_\gamma''(F)$ ($\mathrm{Ad}_{P_\gamma''}$ and $\mathrm{Ad}_{\overline{P}_{\overline{\gamma}}''}$ are the adjoint actions of $P_\gamma''(F)$ on $\mathfrak{p}_\gamma''$ and of $\overline{P}_{\overline{\gamma}}''(F)$ on $\overline{\mathfrak{p}}_{\overline{\gamma}}''$ respectively). The problem is reduced to checking that $\overline{P}_{\overline{\gamma}}''(F)$ is unimodular. Therefore the lemma follows from (4.8.4) ($\overline{\gamma}$ is elliptic in $\overline{P}''(F)$). $\square$

REMARK (4.8.7). — We have a canonical bijection from the set of Haar measures on $G_\gamma(F)$ onto the set of Haar measures on $\overline{P}_{\overline{\gamma}}''(F)$. Indeed, we have $G_\gamma(F) = P_\gamma''(F)$. So, if dg_γ is a Haar measure on $G_\gamma(F)$, $dg_\gamma = c|\omega|$ for some invariant volume form

$$\omega \in \left(\bigwedge^{\dim_F(\mathfrak{p}_\gamma'')} \mathfrak{p}_\gamma'' \right)^*$$

and some constant $c > 0$ (see (3.5)). The corresponding Haar measure $d\overline{p}_{\overline{\gamma}}''$ on $\overline{P}_\gamma''(F)$ is $c|\overline{\omega}|$ where $\overline{\omega}$ is the image of ω by the canonical isomorphism

$$\left(\bigwedge^{\dim_F(\mathfrak{p}_\gamma'')} \mathfrak{p}_\gamma'' \right)^* \cong \left(\bigwedge^{\dim_F(\overline{\mathfrak{p}}_{\overline{\gamma}}'')} \overline{\mathfrak{p}}_{\overline{\gamma}}'' \right)^*$$

(see (4.8.5)). $\square$

Let $\gamma \in G(F)$, let dg_γ be a Haar measure on $G_\gamma(F)$ and let $f : G(F) \to \mathbb{C}$ be a locally constant function with compact support. At least formally we can consider the orbital integral

$$(4.8.8) \qquad O_\gamma(f, dg_\gamma) = \int_{G_\gamma(F)\backslash G(F)} f(g^{-1}\gamma g) \frac{dg}{dg_\gamma}$$

(recall that dg is the Haar measure on $G(F)$ which is normalized by $\mathrm{vol}(K, dg) = 1$).

PROPOSITION (4.8.9). — *The orbital integral* $O_\gamma(f, dg_\gamma)$ *is absolutely convergent.*

To prove this proposition we will need two lemmas.

Let $M \subset G$ be the Levi subgroup of G (over F) canonically associated to γ and let $P = MN$ be the parabolic subgroup (over F) with Levi subgroup M associated to an ordering on the set of prime factors of the minimal polynomial of γ (see (4.3.5) and (4.3.6)). Let K_P be a maximal compact open subgroup of $G(F)$ in good position with respect to $P = MN$ (see (4.3)). We can form the K_P-invariant constant term of f along P, $f^P : M(F) \to \mathbb{C}$ (see (4.3.8)) and, at least formally, we can consider the orbital integral

$$O_\gamma^M(f^P, dg_\gamma) = \int_{M_\gamma(F)\backslash M(F)} f^P(m^{-1}\gamma m) \frac{dm}{dg_\gamma}$$

(recall that dk_P, dn, dm are the Haar measures on K_P, $N(F)$, $M(F)$ which are normalized by $\mathrm{vol}(K_P, dk_P) = 1$, $\mathrm{vol}(N(F) \cap K_P, dn) = 1$, $\mathrm{vol}(M(F) \cap K_P, dm) = 1$ respectively and that $M_\gamma(F) = G_\gamma(F)$).

LEMMA (4.8.10). — *The orbital integral* $O_\gamma(f, dg_\gamma)$ *is absolutely convergent if and only if the orbital integral* $O_\gamma^M(f^P, dg_\gamma)$ *is absolutely convergent. Moreover, if this is the case, we have*

$$O_\gamma(f, dg_\gamma) = |D_{M\backslash G}(\gamma)|^{-1/2} O_\gamma^M(f^P, dg_\gamma)$$

where $D_{M\backslash G}(\gamma)$ *is defined in* (4.3.11).

Proof : The proof is the same as the proof of (4.3.11). $\qquad\qquad\square$

Let $P'' \subset M$ be the parabolic subgroup constructed in (4.8.2). Let $P'' = M''N''$ be a Levi decomposition of P'' (over F) and let $K_{P''}^M \subset M(F)$ be a maximal compact open subgroup of $M(F)$ in good position with respect to P''. We can form the $K_{P''}^M$-invariant constant term of f^P along P'',

$$(f^P)^{P''}(m'') = \delta_{P''(F)}^{M(F)}(m'')^{1/2} \int_{N''(F)} \int_{K_{P''}^M} f^P((k_{P''}^M)^{-1} m'' \mathfrak{n}'' k_{P''}^M) dk_{P''}^M d\mathfrak{n}''$$

where $\delta_{P''(F)}^{M(F)} : M''(F) \to q^{\mathbb{Z}}$ is the modulus character of $P''(F)$ in $M(F)$. If we choose K_P and $K_{P''}^M$ in such way that

$$K_{P''}^M = M(F) \cap K_P$$

we have

$$(f^P)^{P''} = f^{P'}$$

where

$$P' = P''N = M''N''N$$

(see (4.8.3)). The Haar measure dg_γ on $G_\gamma(F)$ induces a Haar measure $d\overline{p}_\gamma''$ on $\overline{P}_{\overline{\gamma}}''(F)$ (see (4.8.7)). But, if $\gamma = \gamma_{M''}\gamma_{N''}$ with $\gamma_{M''} \in M''(F)$ and $\gamma_{N''} \in N''(F)$, we have $\pi''(\gamma_{M''}) = \overline{\gamma}$ and π'' induces an isomorphism of $M_{\gamma_{M''}}''(F)$ onto $\overline{P}_{\overline{\gamma}}''(F)$. So dg_γ induces a Haar measure $dm_{\gamma_{M''}}''$ on $M_{\gamma_{M''}}''(F)$. By construction of P'', $\gamma_{M''}$ is closed (and even elliptic) in $M''(F)$. In particular, the orbital integral

$$O_{\gamma_{M''}}((f^P)^{P''}, dm_{\gamma_{M''}}'') = \int_{M_{\gamma_{M''}}''(F)\backslash M''(F)} (f^P)^{P''}(m''^{-1}\gamma_{M''}m'')\frac{dm''}{dm_{\gamma_{M''}}''}$$

is absolutely convergent (the Haar measure dm'' on $M''(F)$ is normalized by $\mathrm{vol}(M''(F) \cap K_{P''}^M, dm'') = 1$).

LEMMA (4.8.11). — *The orbital integral $O_\gamma^M(f^P, dg_\gamma)$ is absolutely convergent and there exists a positive constant $c(\gamma)$ (independent of f) such that*

$$O_\gamma^M(f^P, dg_\gamma) = c(\gamma)^{-1}\delta_{P''(F)}^{-1/2}(\gamma_{M''})O_{\gamma_{M''}}^{M''}((f^P)^{P''}, dm_{\gamma_{M''}}'').$$

Proof : On

$$\pi''^{-1}(O_{\overline{P}''}(\overline{\gamma})(F)) = O_{M''}(\gamma_{M''})(F)N''(F)$$

we have the measure

$$dx = \frac{dm''}{dm_{\gamma_{M''}}''}d\mathfrak{n}''$$

$(\mathrm{vol}(N''(F) \cap K_P^M, d\mathfrak{n}'') = 1)$. The group $P''(F)$ acts by conjugacy on $\pi''^{-1}(O_{\overline{P}''}(\overline{\gamma})(F))$ and it is easy to see that

$$d(p''xp''^{-1}) = \delta_{P''(F)}(p'')dx$$

for each $p'' \in P''(F)$. In particular, dx induces a measure dx° on the open dense $P''(F)$-invariant subset

$$O_{P''}(\gamma)(F) \subset \pi''^{-1}(O_{\overline{P}''}(\overline{\gamma})(F))$$

(see (4.8)) which satisfies

$$d(p''x^\circ p''^{-1}) = \delta_{P''(F)}(p'')dx^\circ$$

for each $p'' \in P''(F)$.

But on $O_{P''}(\gamma)(F) \simeq P''_\gamma(F)\backslash P''(F)$ we also have the measure

$$dy = \frac{dm''dn''}{dg_\gamma}$$

$(P''_\gamma(F) = G_\gamma(F))$ which satisfies

$$d(p''yp''^{-1}) = \delta_{P''(F)}(p'')dy$$

for each $p'' \in P''(F)$.

Therefore, there exists a constant $c(\gamma) \in \mathbb{R}$, $c(\gamma) > 0$, such that

$$dx^\circ = c(\gamma)dy$$

(see [We 3] Ch. II, §9).

As the closed subset

$$\pi''^{-1}(O_{\overline{P}''}(\overline{\gamma})(F)) - O_{P''}(\gamma)(F) \subset \pi''^{-1}(O_{\overline{P}''}(\overline{\gamma})(F))$$

has measure 0 for dx, the lemma follows (for any locally constant function with compact support $\varphi : M(F) \to \mathbb{C}$ we have the integration formula

$$\int_{M(F)} \varphi(m)dm = \int_{M''(F)} \int_{N''(F)} \int_{K_{P''}^M} \varphi(m''n''k_{P''}^M)dk_{P''}^M dn''dm''$$

where $\mathrm{vol}(K_{P''}^M, dk_{P''}^M) = 1$).   □

Proof of (4.8.9) : This is a direct consequence of (4.8.10) and (4.8.11).   □

REMARK (4.8.12). — If we choose K_P and $K_{P''}^M$ in such a way that

$$K_{P''}^M = M(F) \cap K_P$$

and if we consider

$$f^{P'} = (f^P)^{P''}$$

as a function on $\overline{P}''(F)$, $f^{P'}$ is independent of the choice of the Levi decomposition $P'' = M''N''$ and we have the following reduction formula :

$$O_\gamma(f, dg_\gamma) = |D_{G/M}(\gamma)|^{-1/2}c(\gamma)^{-1}\delta_{P''(F)}(\overline{\gamma})^{-1/2} \times O_{\overline{\gamma}}^{\overline{P}''}(f^{P'}, d\overline{p}''_{\overline{\gamma}})$$

where $d\overline{p}''_{\overline{\gamma}}$ is the Haar measure on $\overline{P}''_{\overline{\gamma}}(F)$ which corresponds to dg_γ (see (4.8.7)).   □

For the Hecke function f on $G(F)$ with Satake transform

$$f^\vee(z) = q^{r(d-1)/2}(z_1^r + \cdots + z_d^r)$$

$(r > 0)$ we have the following vanishing theorem for its orbital integrals:

THEOREM (4.8.13). — *Let* $\gamma \in G(F)$ *be such that* $O_\gamma(f, dg_\gamma) \neq 0$ *for some Haar measure* dg_γ *on* $G_\gamma(F)$. *Then there exists a partition* $d = d_1 + d_2$ *with* $d_1 > 0$, $d_2 \geq 0$, *an elliptic element* $\gamma_1 \in GL_{d_1}(F)$ *with* $v(\det \gamma_1) = r$ *and an element* $\gamma_2 \in GL_{d_2}(\mathcal{O})$ *such that* γ *is conjugate to*

$$\begin{pmatrix} \gamma_1 & 0 \\ 0 & \gamma_2 \end{pmatrix}$$

in $G(F)$. *Moreover, if we assume for simplicity that*

$$\gamma = \begin{pmatrix} \gamma_1 & 0 \\ 0 & \gamma_2 \end{pmatrix},$$

if we identify $G_\gamma(F)$ *with*

$$GL_{d_1}(F)_{\gamma_1} \times GL_{d_2}(F)_{\gamma_2}$$

and if we take

$$dg_\gamma = dg_{1,\gamma_1} \times dg_{2,\gamma_2}$$

where dg_{j,γ_j} *is a Haar measure on* $GL_{d_j}(F)_{\gamma_j}$ *for* $j = 1, 2$, *we have*

$$O_\gamma(f, dg_\gamma) = O_{\gamma_1}(f_1, dg_{1,\gamma_1}) O_{\gamma_2}(f_2, dg_{2,\gamma_2})$$

where f_1 *is the Hecke function on* $GL_{d_1}(F)$ *with Satake transform*

$$f_1^\vee(z) = q^{r(d_1-1)/2}(z_1^r + \cdots + z_{d_1}^r)$$

and where f_2 *is the characteristic function of* $GL_{d_2}(\mathcal{O}) \subset GL_{d_2}(F)$.

Proof: As the support of f is contained in

$$\{g \in G(F) \mid v(\det g) = r\}$$

we can assume that $v(\det \gamma) = r$. Then there exists a partition $d = d_1 + d_2$ with $d_1 > 0$, $d_2 \geq 0$ and there exist $\gamma_1 \in GL_{d_1}(F)$, $\gamma_2 \in GL_{d_2}(F)$ such that γ is conjugate to

$$\tilde{\gamma} = \begin{pmatrix} \gamma_1 & 0 \\ 0 & \gamma_2 \end{pmatrix}$$

in $G(F)$, $v(\det \gamma_1) > 0$, the minimal polynomial of γ_1 is equal to $Q_1(T)^{m_1}$ for some irreducible unitary polynomial $Q_1(T)$ and some integer $m_1 \geq 1$ and the minimal polynomial of γ_2 is prime to $Q_1(T)$. The partition $d = d_1 + d_2$ defines a standard parabolic subgroup $P \subset G$ (over F) with a standard Levi decomposition $P = MN$ (over F). Let $dg_{\tilde{\gamma}}$ be the Haar measure on $M_{\tilde{\gamma}}(F) = G_{\tilde{\gamma}}(F)$ which is induced by dg_γ. It follows easily from (4.3.11) (or its proof) that

$$O_\gamma(f, dg_\gamma) = O_{\tilde{\gamma}}(f, dg_{\tilde{\gamma}}) = |D_{M \backslash G}(\gamma)|^{-1/2} O_{\tilde{\gamma}}^M(f^P, dg_{\tilde{\gamma}}).$$

But, thanks to (4.2.5), (4.2.6), we have

$$f^P(m) = q^{rd_2/2} \varphi_{d_1}(g_1) 1_{GL_{d_2}(\mathcal{O})}(g_2) + q^{rd_1/2} 1_{GL_{d_1}(\mathcal{O})}(g_1) \varphi_{d_2}(g_2)$$

for all

$$m = \begin{pmatrix} g_1 & 0 \\ 0 & g_2 \end{pmatrix} \in M(F)$$

(with the notations of loc. cit.). Let us identify $M(F)$ with $GL_{d_1}(F) \times GL_{d_2}(F)$, $M_{\tilde{\gamma}}(F)$ with $GL_{d_1}(F)_{\gamma_1} \times GL_{d_2}(F)_{\gamma_2}$ and let us split $dg_{\tilde{\gamma}}$ into $dg_{1,\gamma_1} \times dg_{2,\gamma_2}$ where dg_{j,γ_j} is a Haar measure on $GL_{d_j}(F)_{\gamma_j}$ $(j = 1, 2)$. We get

$$O_\gamma(f, dg_\gamma)$$

$$= |D_{M \backslash G}(\gamma)|^{-1/2} [q^{rd_2/2} O_{\gamma_1}^{GL_{d_1}}(\varphi_{d_1}, dg_{1,\gamma_1}) O_{\gamma_2}^{GL_{d_2}}(1_{GL_{d_2}(\mathcal{O})}, dg_{2,\gamma_2})$$

$$+ q^{rd_1/2} O_{\gamma_1}^{GL_{d_1}}(1_{GL_{d_1}(\mathcal{O})}, dg_{1,\gamma_1}) O_{\gamma_2}^{GL_{d_2}}(\varphi_{d_2}, dg_{2,\gamma_2})].$$

Therefore, if $O_\gamma(f, dg_\gamma) \neq 0$, we necessarily have $v(\det \gamma_1) = r$, $v(\det \gamma_2) = 0$ and

$$O_\gamma(f, dg_\gamma) =$$

$$|D_{M \backslash G}(\gamma)|^{-1/2} q^{rd_2/2} O_{\gamma_1}^{GL_{d_1}}(\varphi_{d_1}, dg_{1,\gamma_1}) O_{\gamma_2}^{GL_{d_2}}(1_{GL_{d_2}(\mathcal{O})}, dg_{2,\gamma_2}).$$

Moreover, in this case we have

$$|D_{M \backslash G}(\gamma)| = |\det \gamma_1|^{d_2} = q^{rd_2}$$

and to finish the proof of the theorem we can assume that $d_2 = 0$, i.e. that the minimal polynomial of γ is equal to $Q(T)^m$ for some unitary irreducible polynomial $Q(T)$ and some integer $m \geq 1$.

We want to prove that $O_\gamma(f, dg_\gamma) = 0$ unless $m = 1$. Let P'' $(= P')$ be the parabolic subgroup of G $(= M)$ (over F) associated to γ in (4.8.2). Thanks to (4.8.12), we have

$$O_\gamma(f, dg_\gamma) = c(\gamma)^{-1} \delta_{P''(F)}(\overline{\gamma})^{-1/2} O_{\overline{\gamma}}^{\overline{P}''}(f^{P''}, d\overline{p}_{\overline{\gamma}}'')$$

(with the notations of loc. cit.). But we are assuming that $v(\det \overline{\gamma}) = v(\det \gamma) = r$. Therefore, if P'' is proper $(P'' \underset{\neq}{\subset} G)$, $\overline{\gamma}$ cannot belong to the support of $f^{P''}$ (use (4.2.5) and (4.2.6) to compute $f^{P''}$; the details are left to the reader). The theorem is now completely proved. $\square$

(4.9) Comments and references

The main result of this chapter is the theorem (4.5.5). It is a particular case of the fundamental lemma for Hecke functions on $GL_d(F)$. This fundamental lemma says that the conclusion of (4.5.5) are valid for any $f_r \in \mathcal{H}_r$ and $f = b_r(f_r)$. At least if $\mathrm{char}(F) = 0$, it has been completely proved by Arthur and Clozel using global methods ([Ar-Cl]). Earlier work on this subject was done by Saito, Shintani, Langlands ($d = 2$; see [Sa], [Sh] and [Lan 1]) and Kottwitz ($d = 3$, $f_r = 1_{K_r} \in \mathcal{H}_r$ for general d; see [Ko 3]). The concept of fundamental lemma has been introduced by Langlands (see [Lan 2]).

The proof of (4.5.5) is essentially due to Drinfeld. It has not yet been published. The references are [Kaz] Thm. 9 and a letter of Drinfeld which is quoted and partly reproduced in [Fl] (8.5). The statement of Theorem 9 in [Kaz] is not completely correct. Our theorem (4.8.13) is hopefully the correct one.

In the sections (4.1) and (4.2) we have closely followed the notes of Kottwitz's course at Orsay, in 1986. Another reference is [Car].

A lot of complications in this chapter are due to the fact that we need a non-archimedean local field of positive characteristic and that such a field is not perfect. In particular an element of $GL_d(F)$ which is semi-simple can become unipotent over the algebraic closure of F and we don't have Jordan decomposition. In (4.3) we explain how to work over such a non-archimedean local field using ideas of Howe (see [Ho]). We are following [De-Ka-Vi] Appendice 1.

5

Very cuspidal Euler–Poincaré functions

(5.0) Introduction

The purpose of this chapter is to introduce a function f_∞ on $GL_d(F_\infty)$ such that its orbital integral on any elliptic element γ in $GL_d(F_\infty)$ is given by (3.6.2) (for suitable choices of Haar measures) and such that its orbital integral on any other element of $GL_d(F_\infty)$ is zero. In [Ko 1] §2, Kottwitz has given a whole family of examples of such functions. By taking a suitable linear combination of Kottwitz's functions we can require an important extra property for f_∞: our f_∞ will be very cuspidal.

The results and the proofs of this chapter are purely local. So we will use the same notations as in the previous chapter: F is a non-archimedean local field with ring of integers $\mathcal{O}$, with discrete valuation v and with residue field k, ϖ is a uniformizer of $\mathcal{O}$, q is the number of elements in k, $G = GL_d$, $F^\times$ is the center of $G(F)$ and $K = G(\mathcal{O}) \subset G(F)$.

As before $B \subset G$ is the (standard) Borel subgroup of upper triangular matrices. Let

$$\mathcal{B}^\circ \subset K \subset G(F)$$

be the corresponding **Iwahori subgroup** (the inverse image of $B(k) \subset G(k)$ by the reduction modulo ϖ homomorphism $K \to G(k)$).

The **Hecke algebra**

$$\widetilde{\mathcal{H}}^{ad} = \mathcal{C}_c^\infty(F^\times \backslash G(F) // \mathcal{B}^\circ)$$

is the $\mathbb{Q}$-vector space of $\mathcal{B}^\circ$-bi-invariant functions with compact supports

$$f : F^\times \backslash G(F) \to \mathbb{Q}$$

endowed with the convolution product

$$(f' * f'')(g) = \int_{F^\times \backslash G(F)} f'(h) f''(h^{-1}g) \frac{dh}{dz}$$

where dh (resp. dz) is the Haar measure on $G(F)$ (resp. $F^\times$) normalized by $\mathrm{vol}(K, dh) = 1$ (resp. $\mathrm{vol}(\mathcal{O}^\times, dz) = 1$).

(5.1) The function f

As before $T \subset B$ is the maximal torus of diagonal matrices. We will write the elements of T in the following way:

$$t = \mathrm{diag}(t_1, \ldots, t_d);$$

and if $\lambda \in \mathbf{Z}^d$ we set

$$t^\lambda = t_1^{\lambda_1} \cdots t_d^{\lambda_d}.$$

This gives us an identification

$$\mathbf{Z}^d \xrightarrow{\sim} \mathrm{Hom}(T, GL_1).$$

If $(\varepsilon_i)_{i=1,\ldots,d}$ is the canonical basis of $\mathbf{Z}^d$, then the set

$$R \subset \mathbf{Z}^d$$

of roots of (G, T) is equal to

$$\{\varepsilon_i - \varepsilon_j | 1 \le i, j \le d \text{ and } i \ne j\}$$

and the set

$$\Delta = \{\alpha_i = \varepsilon_i - \varepsilon_{i+1} | 1 \le i \le d-1\} \subset R$$

of simple roots of (G, T, B) is identified with $\{1, \ldots, d-1\}$ by $i \mapsto \alpha_i$.

The symmetric group

$$W = \mathfrak{S}_d$$

is the Weyl group of (G, T) and

$$S = \{s_1, \ldots, s_{d-1}\},$$

where

$$s_i = (i, i+1),$$

is the set of simple reflexions associated to Δ (s_i is associated to $\varepsilon_i - \varepsilon_{i+1}$). W acts on $\mathbf{Z}^d$ by

$$(w, \lambda) \mapsto (\lambda_{w^{-1}(1)}, \ldots, \lambda_{w^{-1}(d)})$$

so that
$$(\dot{w}t\dot{w}^{-1})^\lambda = t^{w^{-1}(\lambda)}$$
for all $w \in W, \lambda \in \mathbb{Z}^d$ and $t \in T$, and we have

$$w(\varepsilon_i) = \varepsilon_{w(i)} \quad (i = 1, \ldots, d).$$

For each subset $I \subset \Delta$ we can define a partition of d,

$$d_I = (d_1, \ldots, d_{|\Delta-I|+1}),$$

with
$$\Delta - I = \{d_1, d_1 + d_2, \ldots, d_1 + \cdots + d_{|\Delta-I|}\},$$
a standard parabolic subgroup

$$P_I \subset G,$$

a standard Levi decomposition of P_I,

$$P_I = M_I N_I,$$

with M_I isomorphic to $GL_{d_1} \times \cdots \times GL_{d_{|\Delta-I|+1}}$, and a subgroup of W,

$$W_I \subset W,$$

generated by the $(s_i)_{i \in I}$. We have

$$\begin{cases} d_\emptyset = (1, \ldots, 1), P_\emptyset = B, W_\emptyset = \{1\}, \\ d_\Delta = (d), P_\Delta = G, W_\Delta = W. \end{cases}$$

For each $I \subset \Delta$, we denote by $\mathcal{P}_I^\circ$ the **standard parahoric subgroup corresponding to** P_I : $\mathcal{P}_I^\circ$ is the inverse image of $P_I(k) \subset G(k)$ by the reduction modulo ϖ homomorphism $K \to G(k)$. We have

$$\begin{cases} \mathcal{P}_\emptyset^\circ = \mathcal{B}^\circ, \\ \mathcal{P}_\Delta^\circ = K. \end{cases}$$

For each $I \subset \Delta$, let $\mathcal{P}_I$ be the normalizer of $\mathcal{P}_I^\circ$ in $G(F)$. If s', s'' are positive integers such that

$$\begin{cases} |\Delta - I| + 1 = s's'', \\ d_j = d_{j+s''} = \cdots = d_{j+(s'-1)s''} \quad (j = 1, \ldots, s'') \end{cases}$$

and if s' is maximal for these properties, we have

$$\mathcal{P}_I = \mathcal{P}_I^\circ \rtimes \varepsilon^{(d_1 + \cdots + d_{s''})\mathbb{Z}}$$

where

$$\varepsilon = \begin{pmatrix} 0 & 1 & 0 & \cdots & 0 \\ \vdots & \ddots & \ddots & \ddots & \vdots \\ \vdots & & \ddots & \ddots & 0 \\ 0 & & & \ddots & 1 \\ \varpi & 0 & \cdots & \cdots & 0 \end{pmatrix} \in G(F)$$

$(d = s'(d_1 + \cdots + d_{s''}), \ \varepsilon^d = \varpi \in F^\times \subset G(F))$.

For each $I \subset \Delta$, we denote by

(5.1.1) $\chi_I : F^\times \backslash G(F) \to \mathbb{Q}$

the function supported by $F^\times \backslash \mathcal{P}_I$ which is equal to $(-1)^{(s'-1)s''n}$ on

$$F^\times \backslash \mathcal{P}_I^\circ \varepsilon^{(d_1 + \cdots + d_{s''})n} \varpi^{\mathbb{Z}} \subset F^\times \backslash \mathcal{P}_I$$

for all $n \in \mathbb{Z}$ (s' and s'' being as before). Obviously χ_I belongs to $\tilde{\mathcal{H}}^{ad}$.

We can now introduce the **Euler–Poincaré function**

(5.1.2) $f = \sum_{I \subset \Delta} (-1)^{|\Delta - I|} \dfrac{\chi_I}{(|\Delta - I|) + 1) \operatorname{vol}(\mathcal{P}_I^\circ, dg)} \in \tilde{\mathcal{H}}^{ad}$.

The main purpose of this chapter is to prove the following theorem:

THEOREM (5.1.3). — (i) (Kottwitz) *Let* $\gamma \in G(F)$ *be elliptic, i.e.* $F' = F[\gamma] \subset gl_d(F)$ *is a field. Then the centralizer* G_γ *of* γ *in* G *has a natural structure of an* F'-*group scheme and as such is isomorphic to* $GL_{d',F'}$ *where* $d' = d/[F' : F]$. *Let* D' *be "the" central division algebra over* F' *with invariant* $1/d'$ *and let* $\overline{G}_\gamma$ *be the restriction "à la Weil" from* F' *to* F *of the* F'-*group scheme* $D'^\times (\overline{G}_\gamma(F) = D'^\times)$. *Then for any Haar measure* dg_γ *on* $G_\gamma(F)$ *we have a corresponding Haar measure* $d\bar{g}_\gamma$ *on* $\overline{G}_\gamma(F)$ *($d\bar{g}_\gamma$ is the transfer of* dg_γ *from* $G_\gamma(F)$ *to its inner twist* $\overline{G}_\gamma(F)$ *as an* F'-*group scheme) and the orbital integral*

$$O_\gamma(f, dg_\gamma) = \int_{G_\gamma(F) \backslash G(F)} f(g^{-1} \gamma g) \frac{dg}{dg_\gamma}$$

is equal to

$$\frac{\varepsilon(\gamma)}{\operatorname{vol}(F^\times \backslash \overline{G}_\gamma(F), \frac{d\bar{g}_\gamma}{dz})}$$

where $\varepsilon(\gamma) = (-1)^{d'-1}$.

(ii) *For each proper standard parabolic subgroup $P \underset{\neq}{\subset} G$ with standard Levi decomposition $P = MN$ the K-invariant constant term along P*

$$f^P(m) = \delta_{P(F)}^{1/2}(m) \int_{N(F)} \int_K f(k^{-1}mnk) dk dn,$$

where dk (resp. dn) is the Haar measure on K (resp. $N(F)$) normalized by $\mathrm{vol}(K, dk) = 1$ (resp. $\mathrm{vol}(N(\mathcal{O}), dn) = 1$), is identically zero on $M(F)$ ($f^P(m) = 0$, $\forall m \in M(F)$).

(iii) *For any $\gamma \in G(F)$ which is not elliptic and any Haar measure dg_γ on the centralizer $G_\gamma(F)$ of γ in $G(F)$ (see (4.8.6)), the orbital integral $O_\gamma(f, dg_\gamma)$ (see (4.8.9)) is zero.*

DEFINITION (5.1.4). — *A function in $\widetilde{\mathcal{H}}^{ad}$ is said to be very cuspidal if, for each proper standard parabolic subgroup $P \underset{\neq}{\subset} G$ with standard Levi decomposition $P = MN$, its K-invariant constant term along P, f^P, is identically zero on $M(F)$.*

(5.2) Kottwitz's functions

Let $\mathcal{I}$ be the **Bruhat–Tits building** of $F^\times \backslash G(F)$, let $\mathcal{F}$ be its set of **facets** and let d its canonical metric (see [Br–Ti] or [Ti]). Let us recall that $(\mathcal{I}, d)$ is a complete metric space, that $\mathcal{F}$ is a set of subsets of $\mathcal{I}$, that each $\sigma \in \mathcal{F}$ is an open simplex of dimension $\dim(\sigma) \leq d - 1$, that $\mathcal{F}$ is endowed with an ordering such that $\sigma' \leq \sigma$ if and only if $\sigma' \subset \bar{\sigma}$ where $\bar{\sigma}$ is the closure of σ in $\mathcal{I}$ and that in fact $\mathcal{I}$ is a simplicial complex of dimension $d - 1$ (see loc. cit.). Let $\mathcal{F}_s \subset \mathcal{F}$ be the subset of facets of dimension s for each $s = 0, \ldots, d - 1$ ($\mathcal{F} = \mathcal{F}_0 \amalg \cdots \amalg \mathcal{F}_{d-1}$). The elements of $\mathcal{F}_0$ (resp. $\mathcal{F}_{d-2}, \mathcal{F}_{d-1}$) are called the **vertices** (resp. the **walls**, the **chambers**) of $\mathcal{I}$. The vertices of $\mathcal{I}$ are the dilatation classes of lattices $L \subset F^d$ (L is a free $\mathcal{O}$-submodule of rank d and L is equivalent to aL for all $a \in F^\times$). A subset $\Sigma \subset \mathcal{F}_0$ with $s + 1$ elements ($0 \leq s \leq d - 1$) is the set of vertices of the simplex $\bar{\sigma} \subset \mathcal{I}$ for some $\sigma \in \mathcal{F}_s$ if and only if there exists an ordering of Σ,

$$\Sigma = \{\sigma_0^0, \ldots, \sigma_0^s\},$$

and lattices $L_0, \ldots, L_s$ in the dilatation classes $\sigma_0^0, \ldots, \sigma_0^s$ respectively such that

$$\varpi L_s = L_0 \underset{\neq}{\subset} L_1 \underset{\neq}{\subset} \cdots \underset{\neq}{\subset} L_s \subset F^d$$

(such an ordering is unique up to a cyclic permutation of the indices if it exists). If L and L' are lattices in F^d and if

$$L \supset L' \supset \varpi^r L$$

but $\varpi L \not\supset L'$ and $L' \not\supset \varpi^{r-1}L$ for some integer $r \geq 0$, then the distance between the two corresponding vertices of $\mathcal{I}$ is 1 if and only if $r = 1$. Therefore a finite subset $\Sigma \subset \mathcal{F}_0$ is the set of vertices of the simplex $\bar{\sigma} \subset \mathcal{I}$ for some $\sigma \in \mathcal{F}$ if and only if $d(\sigma_0, \sigma_0') = 1$ for any two distinct elements σ_0, σ_0' of Σ.

Let us recall that between any two points x, x' of $\mathcal{I}$ there exists a unique geodesic $[x, x']$ and that $\mathcal{I}$ is contractible (see loc. cit.).

The group $F^\times \backslash G(F)$ (and also $G(F)$) acts simplicially and isometrically on $\mathcal{I}$ (if $g \in G(F)$ and $L \subset F^d$ is a lattice, $g(L) \subset F^d$ is another lattice) and therefore acts on $\mathcal{F}$. If $\sigma \in \mathcal{F}$, we denote by $G(F)_\sigma$ (resp. $F^\times \backslash G(F)_\sigma$) the stabilizer of σ in $G(F)$ (resp. $F^\times \backslash G(F)$) and by

$$sgn_\sigma : G(F)_\sigma \to \{\pm 1\}$$

the sign of the permutation action of $G(F)_\sigma$ on the set of vertices of $\bar{\sigma}$. The subgroup $F^\times \backslash G(F)_\sigma \subset F^\times \backslash G(F)$ is compact open; in fact, if $G(F)_\sigma^\circ$ is the intersection of the stabilizers of the vertices of σ, $F^\times \backslash G(F)_\sigma^\circ$ is a parahoric subgroup of $F^\times \backslash G(F)$ and is a subgroup of finite index in $F^\times \backslash G(F)_\sigma$.

For each subset $I \subset \Delta$ we have a corresponding facet σ_I with $\dim(\sigma_I) = |\Delta - I|$: let us write

$$\Delta - I = \{d_1, d_1 + d_2, \ldots, d_1 + \cdots + d_{s-1}\}$$

for the partition $d_I = (d_1, \ldots, d_s)$ of d and let us set

$$L_j = (\varpi^{-1}\mathcal{O})^{d_1 + \cdots + d_j} \oplus \mathcal{O}^{d_{j+1} + \cdots + d_s}$$

for $j = 0, \ldots, |\Delta - I| + 1 = s$; then

$$\varpi L_s = L_0 \underset{\neq}{\subset} L_1 \underset{\neq}{\subset} \cdots \underset{\neq}{\subset} L_s \subset F^d$$

and $\bar{\sigma}_I$ is the simplex with vertices the similarity classes of $L_0, L_1, \ldots, L_{s-1}$. Moreover

$$G(F)_{\sigma_I} = \mathcal{P}_I,$$
$$G(F)_{\sigma_I}^\circ = \mathcal{P}_I^\circ \varpi^{\mathbb{Z}},$$

and

$$sgn_{\sigma_I} = \chi_I | \mathcal{P}_I$$

with the notations of (5.1).

Each orbit of $F^\times \backslash G(F)$ in $\mathcal{F}$ meets the set $\{\sigma_I | I \subset \Delta\}$ and $\sigma_{I'}$, $\sigma_{I''}$ ($I', I'' \subset \Delta$) are in the same orbit if and only if

$$|\Delta - I'| = |\Delta - I''|$$

and the corresponding partitions of d,

$$d = d'_1 + \cdots + d'_{|\Delta - I'| + 1},$$

$$d = d''_1 + \cdots + d''_{|\Delta - I''| + 1}$$

(see (5.1)), are the same up to a cyclic permutation of the indices.

For each system of representatives

$$\mathcal{S} \subset \mathcal{F}$$

for the orbits of $F^\times \backslash G(F)$ on $\mathcal{F}$ we have the following function introduced by Kottwitz (see [Ko 1]§ 2):

$$(5.2.1) \qquad f_{\mathcal{S}} = \sum_{\sigma \in \mathcal{S}} (-1)^{\dim(\sigma)} \frac{sgn_\sigma}{\mathrm{vol}(F^\times \backslash G(F)_\sigma, \frac{dg}{dz})} \in \widetilde{\mathcal{H}}^{ad}$$

where sgn_σ is extended to all $G(F)$ by putting $sgn_\sigma(g) = 0$ for $g \notin G(F)_\sigma$.

LEMMA (5.2.2). — *The function* $f \in \widetilde{\mathcal{H}}^{ad}$ *defined in* (5.1.2) *is equal to*

$$\frac{1}{|\{\mathcal{S} | \mathcal{S} \subset \{\sigma_I | I \subset \Delta\}\}|} \sum_{\mathcal{S} \subset \{\sigma_I | I \subset \Delta\}} f_{\mathcal{S}}$$

where $\mathcal{S}$ *runs over the systems of representatives of the orbits of* $F^\times \backslash G(F)$ *on* $\mathcal{F}$.

Proof : Let us fix $I_0 \subset \Delta$ and let N_{I_0} be the number of the facets σ_I $(I \subset \Delta)$ which are in the same orbit under $F^\times \backslash G(F)$ than σ_{I_0}. Then we obviously have

$$\frac{1}{N_{I_0}} = \frac{|\{\mathcal{S} | \sigma_{I_0} \in \mathcal{S} \subset \{\sigma_I | I \subset \Delta\}\}|}{|\{\mathcal{S} | \mathcal{S} \subset \{\sigma_I | I \subset \Delta\}\}|}$$

and it is enough to prove that

$$N_{I_0} \mathrm{vol}(F^\times \backslash G(F)_{\sigma_{I_0}}, \frac{dg}{dz}) = (|\Delta - I_0| + 1) \mathrm{vol}(\mathcal{P}^\circ_{I_0}, dg).$$

We let the reader check the last equality.   □

(5.3) Elliptic orbital integrals of f

For each $\gamma \in G(F)$ we denote by $\mathcal{I}(\gamma)$ (resp. $\mathcal{F}(\gamma)$) the set of fixed points of γ in $\mathcal{I}$ (resp. $\mathcal{F}$). For any $\tau \in \mathcal{F}(\gamma)$, the intersection

$$\tau(\gamma) = \tau \cap \mathcal{I}(\gamma)$$

is non-empty (it contains at least the barycenter of $\overline{\tau}$). $G_\gamma(F)$ acts on $\mathcal{I}(\gamma)$ (resp. $\mathcal{F}(\gamma)$).

LEMMA (5.3.1). — *For each $\gamma \in G(F)$ and each $\tau \in \mathcal{F}(\gamma)$, $\tau(\gamma)$ is an open simplex and*

$$sgn_\tau(\gamma) = (-1)^{\dim(\tau) - \dim(\tau(\gamma))}.$$

Proof: We can assume that $\tau = \sigma_I$ for some $I \subset \Delta$ and therefore that $\gamma \in \mathcal{P}_I$. Let s', s'' be the positive integers with $|\Delta - I| + 1 = s's''$, with

$$d_j = d_{j+s''} = \cdots = d_{j+(s'-1)s''}$$

and such that s' is maximal for these two properties for all $j = 1, \ldots, s''$. Then there exists $n \in \{0, \ldots, s' - 1\}$ such that $\gamma \in \mathcal{P}_I^\circ \varepsilon^{(d_1 + \cdots + d_{s''})n} \varpi^{\mathbf{Z}}$. The set of vertices of σ_I can be identified with

$$\{1, \ldots, s'', \ldots, 1 + (s' - 1)s'', \ldots, s's''\}$$

and the permutation γ of this set can be described in the following way. Let m be the g.c.d. of s' and n and let $\overline{s}' = s'/m$ and $\overline{n} = n/m$. For each $j = 1, \ldots, s''$ and each $k = 0, \ldots, m - 1$, we have a cyclic permutation

$$C_{j,k} = \begin{pmatrix} j + ks'' & j + (k+m)s'' \cdots j + (k + (\overline{s}' - 1)m)s'' \\ j + (k + (\overline{s}' - 1)m)s'' \ j + ks'' & \cdots j + (k + (\overline{s}' - 2)m)s'' \end{pmatrix}.$$

The permutation γ is clearly

$$\prod_{j=1}^{s''} \prod_{k=0}^{m-1} (C_{j,k})^{\overline{n}}$$

(any two $C_{j,k}$'s commute as their supports are equal or disjoint).

Now, let $\tau_0^{j,k}$ be the barycenter of the facet $\tau^{j,k} \subset \overline{\tau}$ with set of vertices (for $\overline{\tau^{j,k}} \subset \overline{\tau}$)

$$\{j + (k + \ell m)s'' | \ell = 0, \ldots, \overline{s}' - 1\}$$

($j = 1, \ldots, s''$ and $k = 0, \ldots, m - 1$). Then $\overline{\tau(\gamma)}$ is the convex hull of the subset

$$\{\tau_0^{j,k} | j = 1, \ldots, s'' \text{ and } k = 0, \ldots, m - 1\}$$

in the simplex $\overline{\tau}$. Therefore $\tau(\gamma)$ is a simplex of dimension $ms'' - 1$.

Finally, the signature of the permutation γ of the set of vertices of $\overline{\tau}$ is equal to

$$(-1)^{(\overline{s}' - 1)s'' m \overline{n}}$$

and it is easy to check that

$$(\overline{s}' - 1)s'' m \overline{n} \equiv \dim(\tau) - \dim(\tau(\gamma)) \pmod 2$$

($\dim(\tau) = s's'' - 1$, $\dim(\tau(\gamma)) = ms'' - 1$, $s' - m = m(\overline{s}' - 1)$ and $(\overline{s}' - 1)(\overline{n} - 1)$ is even as $(\overline{s}', \overline{n}) = 1$). $\qquad\square$

For each $\tau \in \mathcal{F}(\gamma)$, we set

$$G_\gamma(F)_\tau = G_\gamma(F) \cap G(F)_\tau;$$

this is an open compact subgroup of $G_\gamma(F)$.

LEMMA (5.3.2). — *For each $\sigma \in \mathcal{F}$, each closed $\gamma \in G(F)$ (see (4.3)) and each Haar measure dg_γ on $G_\gamma(F)$, there are finitely many $G_\gamma(F)$-orbits in $(G(F).\sigma) \cap \mathcal{F}(\gamma)$ and the orbital integral*

$$O_\gamma\left(\frac{sgn_\sigma}{\mathrm{vol}(F^\times \backslash G(F)_\sigma, \frac{dg}{dz})}, dg_\gamma\right)$$

is equal to

$$\sum_\tau \frac{sgn_\tau(\gamma)}{\mathrm{vol}(F^\times \backslash G_\gamma(F)_\tau, \frac{dg_\gamma}{dz})}$$

where τ runs through a set of representatives for the $G_\gamma(F)$-orbits in $(G(F).\sigma) \cap \mathcal{F}(\gamma)$.

Proof: As $\gamma \in G(F)$ is closed, the quotient

$$G_\gamma(F) \backslash \{g \in G(F) | sgn_\sigma(g^{-1}\gamma g) \neq 0\}$$

is compact. But $sgn_\sigma(g^{-1}\gamma g) \neq 0$ if and only if $\gamma \cdot (g \cdot \sigma) = g \cdot \sigma$. Therefore the first assertion of the lemma is proved ($G_\gamma(F)_\sigma$ is open) and our orbital integral is equal to

$$\sum_\tau \frac{sgn_\sigma(g_\tau^{-1}\gamma g_\tau)}{\mathrm{vol}(F^\times \backslash G_\gamma(F)_\tau, \frac{dg_\gamma}{dz})}$$

where τ runs over a set of representatives for the $G_\gamma(F)$-orbits in $(G(F) \cdot \sigma) \cap \mathcal{F}(\gamma)$ and where g_τ is a fixed element of $G(F)$ such that $g_\tau \cdot \sigma = \tau$. But we obviously have

$$sgn_\sigma(g_\tau^{-1}\gamma g_\tau) = sgn_\tau(\gamma)$$

and the lemma is proved. $\qquad\square$

COROLLARY (5.3.3). — *For each closed $\gamma \in G(F)$ (see (4.3.1)) and each Haar measure dg_γ on $G_\gamma(F)$, there are finitely many $G_\gamma(F)$-orbits in $\mathcal{F}(\gamma)$ and the orbital integral*

$$O_\gamma(f, dg_\gamma)$$

is equal to

$$\sum_\tau (-1)^{\dim(\tau)} \mathrm{vol}(F^\times \backslash G_\gamma(F)_\tau, \frac{dg_\gamma}{dz})^{-1}$$

where τ runs through a set of representatives for the $G_\gamma(F)$-orbits in $\mathcal{F}(\gamma)$.

Proof : This is a direct consequence of the lemmas (5.2.2), (5.3.1) and (5.3.2). $\qquad\square$

If we endow $\mathcal{I}(\gamma) \subset \mathcal{I}$ with the induced topology, $\mathcal{I}(\gamma)$ is a CW-complex (see [Wh] §5) with (open) cells $(\tau(\gamma))_{\tau \in \mathcal{F}(\gamma)}$ and $G_\gamma(F)$ acts in a cellular way on $\mathcal{I}(\gamma)$.

LEMMA (5.3.4). — *If $\mathcal{I}(\gamma)$ is non-empty, $\mathcal{I}(\gamma)$ is contractible.*

Proof : Let $x_0 \in \mathcal{I}(\gamma)$ and let $x \in \mathcal{I}(\gamma)$. There exists a unique geodesic $[x_0, x]$ in $\mathcal{I}$ from x_0 to x. But $\gamma[x_0, x]$ is also a geodesic in $\mathcal{I}$ from x_0 to x (γ acts isometrically on $\mathcal{I}$). Therefore $[x_0, x] \subset \mathcal{I}(\gamma)$ and the lemma follows. $\qquad\square$

LEMMA (5.3.5). — *If γ is elliptic in $G(F)$, $\mathcal{I}(\gamma)$ is non-empty.*

Proof : Let $F' = F[\gamma] \subset gl_d(F)$, let e be the ramification index of F' over F and let $d' = d/[F' : F]$. We identify $F'^{d'}$ to F^d by choosing a basis of F' over F and we consider the lattices

$$\varpi L_e = L_0 \underset{\neq}{\subset} L_1 \underset{\neq}{\subset} \cdots \underset{\neq}{\subset} L_e \subset F'^{d'} = F^d$$

where

$$L_j = (\varpi'^{-j}\mathcal{O}')^{d'}$$

(ϖ (resp. ϖ') is a uniformizer in $\mathcal{O}$ (resp. $\mathcal{O}'$); ϖ'^e/ϖ is a unit in $\mathcal{O}'$). Now, for each $j = 0, 1, \ldots, e - 1$, the lattice $\gamma(L_j)$ is in the same dilatation class as L_k where $0 \leq k \leq e - 1$ and

$$k \equiv j + v'(\gamma) \pmod{e}.$$

Therefore, γ fixes the e-dimensional facet of $\mathcal{I}$ given by the lattices $(L_j)_{j=0,\ldots,e-1}$ and $\mathcal{I}(\gamma)$ is non-empty (it contains at least the barycenter of this facet). $\qquad\square$

We will now review some results of Serre about **Euler–Poincaré measures** (see [Se 2] (3.3)). Let H be a unimodular, locally compact group and let $\mathcal{J}$ be a CW-complex on which H acts in a cellular way. We denote by $\mathcal{C}$ the set of (open) cells of $\mathcal{J}$. Let us assume that :

(i) $\mathcal{J}$ is non-empty and contractible;

(ii) $\mathcal{J}$ is locally compact;

(iii) for each $\tau \in \mathcal{C}$, the stabilizer H_τ of τ in H is a compact, open subgroup of H ;

(iv) any compact subgroup of H is contained in H_τ for at least one $\tau \in \mathcal{C}$;

(v) $H \backslash \mathcal{C}$ is finite

(it follows from these conditions that $\mathcal{J}$ is of finite dimension).

If dh is a Haar measure on H, we can set

$$\chi_H(dh) = \sum_{\tau \in \Sigma} (-1)^{\dim(\tau)} \operatorname{vol}(H_\tau, dh)^{-1}$$

where Σ is a system of representatives for the H-orbits in $\mathcal{C}$ ($\chi_H(dh)$ is independent of the choice of Σ and if we replace dh by $c \cdot dh$ for some positive constant c, $\chi_H(dh)$ is multiplied by $1/c$).

PROPOSITION (5.3.6) (Serre). — *Let Δ be a discrete, torsion free, cocompact subgroup of H. Then the cohomology groups*

$$H^q(\Delta, \mathbb{Q})$$

($q \in \mathbb{Z}$) of Δ with coefficients in the trivial Δ-module $\mathbb{Q}$ are finite dimensional $\mathbb{Q}$-vector spaces and vanish for $q < 0$ or $q > \dim(\mathcal{J})$.

Moreover, the Euler–Poincaré characteristic

$$\chi(\Delta) = \sum_q (-1)^q \dim_{\mathbb{Q}}(H^q(\Delta, \mathbb{Q}))$$

is equal to

$$\chi_H(dh) \operatorname{vol}(\Delta \backslash H, \frac{dh}{d\delta})$$

where dh is a Haar measure on H and $d\delta$ is the counting measure on Δ.

COROLLARY (5.3.7). — *If H admits a least one discrete, torsion free, cocompact subgroup, $\chi_H(dh)$ is independent of the particular CW-complex $\mathcal{J}$ satisfying the conditions (i) to (v) that we used to define it.*   □

Proof of (5.3.6) : As Δ is discrete and torsion free, $\Delta \cap H_\tau = \{1\}$ for each $\tau \in \mathcal{C}$ (H_τ is compact open in H) and Δ acts freely on $\mathcal{J}$. As $H \backslash \mathcal{C}$ is finite and $\Delta \backslash H$ is compact, Δ acts properly on $\mathcal{J}$ and the quotient space $\Delta \backslash \mathcal{J}$ is compact. As $\mathcal{J}$ is contractible

$$H^q(\Delta, \mathbb{Q}) = H^q(\Delta \backslash \mathcal{J}, \mathbb{Q})$$

(singular cohomology) for each $q \in \mathbb{Z}$ and the first assertion is obvious. Moreover

$$\chi(\Delta) = \chi(\Delta \backslash \mathcal{J}) = \sum_\tau (-1)^{\dim(\tau)}$$

where τ runs through a set of representatives of the Δ-orbits in $\mathcal{C}$. But, on the one hand, if Σ is a set of representatives of the H-orbits in $\mathcal{C}$, for each $\tau \in \Sigma$, $H \cdot \tau$ is the union of $|\Delta \backslash H / H_\tau|$ distinct Δ-orbits and we have

$$\chi(\Delta) = \sum_{\tau \in \Sigma} |\Delta \backslash H / H_\tau| (-1)^{\dim(\tau)}.$$

On the other hand, for each $\tau \in \Sigma$, we have

$$\mathrm{vol}(\Delta \backslash H, \frac{dh}{d\delta}) = |\Delta \backslash H / H_\tau| \, \mathrm{vol}(H_\tau, dh).$$

Therefore the proposition follows. $\square$

Let $\gamma \in G(F)$ be elliptic. Then $G_\gamma(F) \cong GL_{d'}(F')$ where $F' = F[\gamma]$ is the F-subalgebra of $gl_d(F)$ generated by γ and $d' = d/[F' : F]$. Let $H = F'^\times \backslash GL_{d'}(F')$. Then the above conditions (i) to (v) are satisfied by $\mathcal{J} = \mathcal{I}(\gamma)$ and $\mathcal{J} = \mathcal{I}'$ where $\mathcal{I}'$ is the Bruhat–Tits building of $F'^\times \backslash GL_{d'}(F')$ (see (5.3.3), (5.3.4) and (5.3.5)). For $\mathcal{J} = \mathcal{I}(\gamma)$, it follows from (5.3.3) that

$$O_\gamma(f, dg_\gamma) = \chi_H(\frac{dg_\gamma}{dz'}) / \mathrm{vol}(F^\times \backslash F'^\times, \frac{dz'}{dz}),$$

for any Haar measures dg_γ on $G_\gamma(F) \cong GL_{d'}(F')$ and dz' on $F'^\times$.

PROPOSITION (5.3.8). — *For any positive integer d' and any non-archimedean local field F', $F'^\times \backslash GL_{d'}(F')$ has discrete, torsion free, cocompact subgroups.*

Proof when $\mathrm{char}(F') > 0$ (*see* [Bo-Ha] *for the general case*): Let us change our notations. We will now denote by F' a function field in one variable over a finite field and our non-archimedean local field will be $F'_{\infty'}$ for some place ∞' of F' (any non-archimedean local field of positive characteristic is obtained in this way). Let D' be a central division algebra of dimension d'^2 over F' which is split at ∞' ($D'_{\infty'} = F'_{\infty'} \otimes_{F'} D'$ is isomorphic to the $F'_{\infty'}$-algebra $gl_{d'}(F'_{\infty'})$). Let

$$A' = \{\alpha' \in F'^\times | v'(\alpha') \geq 0, \forall v' \neq \infty'\} \cup \{0\}$$

(v' place of F') and let $\mathcal{D}' \subset D'$ be a maximal A'-order of D'. Then

$$A'^\times \backslash \mathcal{D}'^\times \subset F'^\times \backslash \mathcal{D}'^\times \subset F'^\times_{\infty'} \backslash \mathcal{D}'^\times_{\infty'} \cong F'^\times_{\infty'} \backslash GL_{d'}(F'_{\infty'})$$

is a discrete, cocompact subgroup. Replacing $A'^\times \backslash \mathcal{D}'^\times$ by a subgroup $\Delta \subset A'^\times \backslash \mathcal{D}'^\times$ of finite index if necessary, we get a discrete, torsion free, cocompact subgroup of $F'^\times_{\infty'} \backslash GL_{d'}(F'_{\infty'})$ (see [Bo-Ha] for more details). $\square$

Thanks to this proposition we can compute $\chi_H(dh)$ using $\mathcal{J} = \mathcal{I}'$. But we have (see [Se 2] Thm. 7)

PROPOSITION (5.3.9). — Let $H = F'^\times \backslash GL_{d'}(F')$ and $\mathcal{J} = \mathcal{I}'$ be as before. Let dg' (resp. dz') be the Haar measure on $GL_{d'}(F')$ (resp. $F'^\times$) which is normalized by $\mathrm{vol}(GL_{d'}(\mathcal{O}'), dg') = 1$ (resp. $\mathrm{vol}(\mathcal{O}'^\times, dz') = 1$) where $\mathcal{O}'$ is the ring of integers of F'. Then we have

$$\chi_H\left(\frac{dg'}{dz'}\right) = \frac{1}{d'} \prod_{i=1}^{d'-1} (1 - q'^i)$$

where q' is the number of elements of the residue field k' of $\mathcal{O}'$.

Proof : The same argument as in (5.2.2) shows that

$$\chi_H\left(\frac{dg'}{dz'}\right) = \sum_{I' \subset \Delta'} \frac{(-1)^{|\Delta'-I'|}}{(|\Delta' - I'| + 1)\,\mathrm{vol}(\mathcal{P}_{I'}^\circ, dg')}$$

where $\Delta' = \{1, 2, \ldots, d' - 1\}$ and $\mathcal{P}_{I'}^\circ \subset GL_{d'}(\mathcal{O}')$ is the inverse image of the parabolic subgroup $P_{I'}(k') \subset GL_{d'}(k')$ by the reduction modulo ϖ' homomorphism $GL_{d'}(\mathcal{O}') \longrightarrow GL_{d'}(k')$ for each $I' \subset \Delta'$ (recall that $\mathrm{vol}(\mathcal{O}'^\times, dz') = 1$). But

$$\mathrm{vol}(\mathcal{P}_{I'}^\circ, dg') = 1/[GL_{d'}(k') : P_{I'}(k')]$$

thanks to our normalization of dg' and therefore

$$\chi_H\left(\frac{dg'}{dz'}\right) = \sum_{(d_1',\ldots,d_{s'}')} \frac{(-1)^{s'-1} \prod_{i=1}^{d'} (1 - q'^i)}{s' \prod_{j=1}^{s'} \prod_{i=1}^{d_j'} (1 - q'^i)}$$

where $(d_1', \ldots, d_{s'}')$ runs through the set of partitions of d' ($s', d_1', \ldots, d_{s'}'$ are positive integers and $d' = d_1' + \cdots + d_{s'}'$).

Now, in the ring of formal power series $\mathbb{Q}[[x, y]]$, we have the equalities

$$\prod_{i=1}^{\infty} \frac{1}{1 - xy^i} = 1 + \sum_{n=1}^{\infty} \sum_{m=1}^{n} P(n, m) x^m y^n$$

$$= 1 + \sum_{j=1}^{\infty} \frac{x^j y^j}{(1 - y) \cdots (1 - y^j)}$$

where $P(n, m)$ is the number of partitions of n in m positive parts. Taking the logarithm, we get the equality

$$\sum_{d'=1}^{\infty} \frac{x^{d'} y^{d'}}{d'(1 - y^{d'})} = \sum_{d'=1}^{\infty} \sum_{(d'_1, \ldots, d'_{s'})} \frac{(-1)^{s'-1} x^{d'} y^{d'}}{s' \prod_{j=1}^{s'} \prod_{i=1}^{d'_j} (1 - y^i)}$$

and, therefore, the equality

$$\frac{1}{d'(1 - y^{d'})} = \sum_{(d'_1, \ldots, d'_{s'})} \frac{(-1)^{s'-1}}{s' \prod_{j=1}^{s'} \prod_{i=1}^{d'_j} (1 - y^i)}$$

for each integer $d' \geq 1$. The proposition follows. $\qquad\square$

Proof of (5.1.3) (i) : If $dg_\gamma = dg'$ and dz' are normalized as in (5.3.9), we have shown that

$$O_\gamma(f, dg_\gamma) = \chi_H(\frac{dg_\gamma}{dz'})/\operatorname{vol}(F^\times \backslash F'^\times, \frac{dz'}{dz})$$

$$= \frac{1}{d'} \prod_{i=1}^{d'-1} (1 - q'^i)/\operatorname{vol}(F^\times \backslash F'^\times, \frac{dz'}{dz})$$

(γ elliptic in $G(F)$, $F' = F[\gamma] \subset gl_d(F)$, $d' = d/[F' : F]$ and $G_\gamma(F) \cong GL_{d'}(F')$). But we have $\varepsilon(\gamma) = (-1)^{d'-1}$ and

$$\operatorname{vol}(F^\times \backslash \overline{G}_\gamma(F), \frac{dg_\gamma}{dz}) = d' \operatorname{vol}(\mathcal{D}'^\times, d\overline{g}_\gamma) \operatorname{vol}(F^\times \backslash F'^\times, \frac{dz'}{dz})$$

where $\mathcal{D}'$ is the maximal order of D' (D' is the central division algebra over F' with invariant $1/d'$, $\overline{G}_\gamma(F) \cong D'^\times$ and $\operatorname{vol}(\mathcal{O}'^\times, dz') = 1$). As

$$\frac{\operatorname{vol}(GL_{d'}(\mathcal{O}'), dg_\gamma)}{\operatorname{vol}(\mathcal{D}'^\times, d\overline{g}_\gamma)} = (q' - 1) \cdots (q'^{d'-1} - 1)$$

(see (4.6.4)), part (i) of (5.1.3) is proved. $\qquad\square$

(5.4) K-invariant constant terms of f

The proof of (5.1.3) (ii) that we will give in this section and the next one is due to Waldspurger. Let us begin with the computation of the K-invariant constant terms of the functions $\chi_I (I \subset \Delta)$.

Let $P = MN$ be a standard parabolic subgroup of G (over F) with its standard Levi decomposition (over F). We have

$$P = P_{\Delta^M},$$

$$M = M_{\Delta^M},$$

$$N = N_{\Delta^M},$$

for some subset $\Delta^M \subset \Delta$ (see (5.1)).

If J is a subset of Δ^M,

$$P_J^M = P_J \cap M \subset M$$

is a standard parabolic subgroup of M (over F) with standard Levi decomposition

$$P_J^M = M_J(N_J \cap M).$$

In fact, if

$$\Delta - \Delta^M = \{d_1, d_1 + d_2, \ldots, d_1 + \cdots + d_{s-1}\}$$

where $(d_1, \ldots, d_s)$ is a partition of d, we have

$$M \cong GL_{d_1} \times \cdots \times GL_{d_s}$$

and

$$P_J^M \cong P_{J_1} \times \cdots \times P_{J_s}$$

where

$$J_j = J \cap [d_1 + \cdots + d_{j-1} + 1, d_1 + \cdots + d_j - 1]$$

and where

$$P_{J_j} \subset GL_{d_j}$$

is the corresponding parabolic subgroup for $j = 1, \ldots, s$.

The standard parahoric subgroup of $M(F)$ corresponding to P_J^M is nothing else than

$$\mathcal{P}_J^{M,\circ} = \mathcal{P}_J^\circ \cap M(F).$$

With the above notations,

$$\mathcal{P}_J^{M,\circ} = \mathcal{P}_{J_1}^\circ \times \cdots \times \mathcal{P}_{J_s}^\circ$$

where

$$\mathcal{P}_{J_j}^\circ \subset GL_{d_j}(F)$$

is the standard parahoric subgroup corresponding to P_{J_j} for $j = 1, \ldots, s$. Let $\mathcal{P}_J^M$ be the normalizer of $\mathcal{P}_J^{M,\circ}$ in $M(F)$.

Let $W^M = W_{\Delta^M} \subset W$ be the Weyl group of (M, T). With the above notations, we have

$$W^M = \mathfrak{S}_{d_1} \times \cdots \times \mathfrak{S}_{d_s}.$$

For any $I \subset \Delta$, we consider the subset

(5.4.1) $\qquad D_{M,I} = \{w \in W | w^{-1}(\Delta^M) \subset R^+ \text{ and } w(I) \subset R^+\}$

of W where

$$R^+ = \{\varepsilon_i - \varepsilon_j | 1 \leq i < j \leq d\} \subset R$$

is the set of positive roots of (G, T, B). If

$$\Delta - I = \{e_1, e_1 + e_2, \ldots, e_1 + \cdots + e_{t-1}\}$$

we denote by $A_{M,I}$ the set of tableaux

$$a = (a_{jk})_{\substack{j=1,\ldots,s \\ k=1,\ldots,t}}$$

such that $a_{jk} \in \mathbb{Z}$ and $a_{jk} \geq 0$ for all j, k, that

$$\sum_{k=1}^{t} a_{jk} = d_j$$

for all j and that

$$\sum_{j=1}^{s} a_{jk} = e_k$$

for all k. To each $w \in D_{M,I}$ we can associate a tableau $a \in A_{M,I}$ in the following way: let $\dot{w} \in GL_d(\mathbb{Z})$ be the permutation matrix corresponding to w, let $(\overline{v}_1, \ldots, \overline{v}_d)$ be the canonical basis of k^d over k, let

$$\overline{V}_j = k\overline{v}_{d_1+\cdots+d_{j-1}+1} \oplus \cdots \oplus k\overline{v}_{d_1+\cdots+d_j}$$

for $j = 1, \ldots, s$ and let

$$\overline{L}_k = k\overline{v}_{e_1+\cdots+e_{k-1}+1} \oplus \cdots \oplus k\overline{v}_{e_1+\cdots+e_k}$$

for $k = 1, \ldots, t$, then

$$a_{jk} = \dim_k(\overline{V}_j \cap \dot{w}(\overline{L}_k))$$

for all j, k (we obviously have

$$\bigoplus_{k=1}^{t} \overline{V}_j \cap \dot{w}(\overline{L}_k) = \overline{V}_j$$

for all j and

$$\bigoplus_{j=1}^{s} \overline{V}_j \cap \dot{w}(\overline{L}_k) = \dot{w}(\overline{L}_k)$$

for all k). It is well known that this map

(5.4.2) $\qquad D_{M,I} \xrightarrow{\sim} A_{M,I}$

is bijective.

LEMMA (5.4.3). — (i) $D_{M,I}$ is a set of representatives of $W^M \backslash W/W_I$ in W. In fact, each double coset in $W^M \backslash W/W_I$ contains a unique element of minimal length and this is the representative of the coset in $D_{M,I}$.

(ii) The map

$$W \to P(k) \backslash G(k)/P_I(k), \ w \mapsto P(k) \, \dot{w} \, P_I(k),$$

induces a bijection

$$W^M \backslash W/W_I \xrightarrow{\sim} P(k) \backslash G(k)/P_I(k).$$

Proof : For example, see [Ca] (2.7.3) and (2.8.1). □

COROLLARY (5.4.4). — The map

$$W \to P(\mathcal{O}) \backslash K/\mathcal{P}_I^\circ, \ w \mapsto P(\mathcal{O}) \, \dot{w} \, \mathcal{P}_I^\circ$$

induces a bijection

$$W^M \backslash W/W_I \xrightarrow{\sim} P(\mathcal{O}) \backslash K/\mathcal{P}_I^\circ.$$

In particular, K is the disjoint union of the double cosets $P(\mathcal{O}) \, \dot{w} \, \mathcal{P}_I^\circ$ where w runs through $D_{M,I}$. □

LEMMA (5.4.5). — For each $w \in D_{M,I}$, we have

$$P(F) \cap (\dot{w} \, \mathcal{P}_I \, \dot{w}^{-1}) = (M(F) \cap (\dot{w} \, \mathcal{P}_I \, \dot{w}^{-1}))(N(F) \cap (\dot{w} \, \mathcal{P}_I \, \dot{w}^{-1})).$$

Proof : Let $(v_1, \cdots, v_d)$ be the canonical basis of F^d over F, let

$$V_j = F v_{d_1 + \cdots + d_{j-1} + 1} \oplus \cdots \oplus F v_{d_1 + \cdots + d_j}$$

for $j = 1, \ldots, s$ and let

$$L_k = (\varpi^{-1} \mathcal{O})^{e_1 + \cdots + e_k} \oplus \mathcal{O}^{e_{k+1} + \cdots + e_t}$$

for $k = 0, 1, \ldots, t$. Then $P(F)$ (resp. $M(F)$, resp. $N(F)$) is the subgroup of g's in $G(F)$ such that

$$g(V_1 \oplus \cdots \oplus V_j) = V_1 \oplus \cdots \oplus V_j$$

(resp.

$$g(V_j) = V_j,$$

resp.

$$(1 - g)(V_1 \oplus \cdots \oplus V_j) \subset V_1 \oplus \cdots \oplus V_{j-1})$$

for all j and $\dot{w}\,\mathcal{P}_I\,\dot{w}^{-1}$ is the subgroup of g's in $G(F)$ such that there exists an integer n (depending on g) with

$$g(\dot{w}\,(L_k)) = \dot{w}\,(L_{k+n})$$

for all $k \in \mathbf{Z}$ (if $k = n_0 t - m_0$ with $m_0, n_0 \in \mathbf{Z}$ and $0 \le m_0 \le t - 1$, we have set

$$L_k = \varpi^{-n_0} L_{m_0}).$$

But it is clear that

$$(V_1 \oplus \cdots \oplus V_j) \cap \dot{w}\,(L_k) = (V_1 \cap \dot{w}\,(L_k)) \oplus \cdots \oplus (V_j \cap \dot{w}\,(L_k))$$

for each $j = 1, \ldots, s$ and each $k \in \mathbf{Z}$. The lemma follows.   □

LEMMA (5.4.6). — *Let $w \in D_{M,I}$ and let $J = \Delta^M \cap w(I) \subset \Delta^M$. Then we have*

$$M(F) \cap (\dot{w}\,\mathcal{P}_I^{\circ}\,\dot{w}^{-1}) = M(\mathcal{O}) \cap (\dot{w}\,\mathcal{P}_I^{\circ}\,\dot{w}^{-1}) = \mathcal{P}_J^{M,\circ}$$

and

$$N(F) \cap (\dot{w}\,\mathcal{P}_I\,\dot{w}^{-1}) = N(\mathcal{O}) \cap (\dot{w}\,\mathcal{P}_I^{\circ}\,\dot{w}^{-1}).$$

Proof: As $\dot{w}\,\mathcal{P}_I^{\circ}\,\dot{w}^{-1} \subset K$, we have

$$M(F) \cap (\dot{w}\,\mathcal{P}_I^{\circ}\,\dot{w}^{-1}) = M(\mathcal{O}) \cap (\dot{w}\,\mathcal{P}_I^{\circ}\,\dot{w}^{-1}).$$

But we have

$$M(k) \cap (\dot{w}\,P_I(k)\,\dot{w}^{-1}) = P_J^M(k)$$

for each $w \in D_{M,I}$ (see [Ca] (2.8.9) for example). So

$$M(\mathcal{O}) \cap (\dot{w}\,\mathcal{P}_I^{\circ}\,\dot{w}^{-1}) = \mathcal{P}_J^{M,\circ}.$$

As

$$\dot{w}\,\mathcal{P}_I^{\circ}\,\dot{w}^{-1} = \{g \in \dot{w}\,\mathcal{P}_I\,\dot{w}^{-1} \mid v(\det g) = 0\},$$

we have

$$N(F) \cap (\dot{w}\,\mathcal{P}_I\,\dot{w}^{-1}) = N(F) \cap (\dot{w}\,\mathcal{P}_I^{\circ}\,\dot{w}^{-1}).$$

But $\dot{w}\,\mathcal{P}_I^{\circ}\,\dot{w}^{-1} \subset K$ and the lemma is completely proved.   □

REMARK (5.4.7). — As $M(F) \cap (\dot{w}\,\mathcal{P}_I\,\dot{w}^{-1})$ normalizes $M(F) \cap (\dot{w}\,\mathcal{P}_I^{\circ}\,\dot{w}^{-1}) = \mathcal{P}_J^{M,\circ}$, we have

$$\mathcal{P}_J^{M,\circ} \subset M(F) \cap (\dot{w}\,\mathcal{P}_I\,\dot{w}^{-1}) \subset \mathcal{P}_J^{M}.$$

□

For each locally constant function f on $G(F)$ (resp. f^M on $M(F)$) we set

$$f_K(g) = \int_K f(k^{-1}gk)dk$$

for each $g \in G(F)$ (resp.

$$(f^M)_{K^M}(m) = \int_{K^M} f^M((k^M)^{-1}mk^M)dk^M$$

for each $m \in M(F)$) where dk (resp. dk^M) is the Haar measure on K (resp. $K^M = M(\mathcal{O})$) which is normalized by $\mathrm{vol}(K, dk) = 1$ (resp. $\mathrm{vol}(K^M, dk^M) = 1$).

For each $w \in D_{M,I}$ we denote by

$$\chi_I^{M,w} : M(F) \to \mathbf{Q}$$

the restriction to $M(F)$ of the function

$$g \mapsto \chi_I(\dot{w}^{-1} g \, \dot{w})$$

on $G(F)$. The support of this function $\chi_I^{M,w}$ is exactly $M(F) \cap (\dot{w} \, \mathcal{P}_I \, \dot{w}^{-1})$ and $\chi_I^{M,w}$ is bi-invariant under $M(F) \cap (\dot{w} \, \mathcal{P}_I^{\circ} \, \dot{w}^{-1}) = \mathcal{P}_J^{M,\circ}$.

LEMMA (5.4.8). — *The K-invariant constant term of χ_I along P is given by the following formula:*

$$\frac{(\chi_I)^P}{\mathrm{vol}(\mathcal{P}_I^{\circ}, dg)} = \delta_{P(F)}^{1/2} \left(\sum_{w \in D_{M,I}} \frac{\chi_I^{M,w}}{\mathrm{vol}(\mathcal{P}_{\Delta^M \cap w(I)}^{M,\circ}, dm)} \right)_{K^M}.$$

Proof: Thanks to (5.4.2), for each $g \in G(F)$ we have

$$(\chi_I)_K(g) = \sum_{w \in D_{M,I}} \int_{P(\mathcal{O})\dot{w}\mathcal{P}_I^{\circ}} \chi_I(k^{-1}gk)dk.$$

As χ_I is bi-invariant under $\mathcal{P}_I^{\circ}$, for any Haar measure dk^P on $P(\mathcal{O})$ this formula can be rewritten

$$\frac{(\chi_I)_K(g)}{\mathrm{vol}(\mathcal{P}_I^{\circ}, dk)} = \sum_{w \in D_{M,I}} \frac{\int_{P(\mathcal{O})} \chi_I(\dot{w}^{-1} (k^P)^{-1} g k^P \, \dot{w})dk^P}{\mathrm{vol}(P(\mathcal{O}) \cap (\dot{w} \, \mathcal{P}_I^{\circ} \, \dot{w}^{-1}), dk^P)}.$$

Let us take $dk^P = dk^M dk^N = dk^N dk^M$ where dk^M (resp. dk^N) is the Haar measure on $M(\mathcal{O})$ (resp. $N(\mathcal{O})$) which is normalized by $\mathrm{vol}(M(\mathcal{O}), dk^M) = 1$ (resp. $\mathrm{vol}(N(\mathcal{O}), dk^N) = 1$). Thanks to (5.4.5) and (5.4.6), we have

$$P(\mathcal{O}) \cap (\dot{w} \, \mathcal{P}_I^\circ \, \dot{w}^{-1}) = \mathcal{P}_J^{M,\circ}(N(\mathcal{O}) \cap (\dot{w} \, \mathcal{P}_I^\circ \, \dot{w}^{-1}))$$

and

$$\mathrm{vol}(P(\mathcal{O}) \cap (\dot{w} \, \mathcal{P}_I^\circ \, \dot{w}^{-1}), dk^P) = \mathrm{vol}(\mathcal{P}_J^{M,\circ}, dk^M) \mathrm{vol}(N(\mathcal{O}) \cap (\dot{w} \, \mathcal{P}_I^\circ \, \dot{w}^{-1}), dk^N).$$

Therefore it suffices to check that

$$\int_{N(F)} \int_{P(\mathcal{O})} \chi_I(\dot{w}^{-1} \, (k^P)^{-1} m n k^P \, \dot{w}) dk^P \, dn$$
$$= \mathrm{vol}(N(\mathcal{O}) \cap (\dot{w} \, \mathcal{P}_I^\circ \, \dot{w}^{-1}), dk^N)(\chi_I^{M,w})_{K^M}(m)$$

for each $m \in M(F)$ to finish the proof of the lemma.

Clearly we can change the order of integrations. Moreover, if $k^P = k^M k^N$ ($k^M \in M(\mathcal{O}), k^N \in M(\mathcal{O})$), we have

$$(k^P)^{-1} m n k^P = m' n'$$

where

$$m' = (k^M)^{-1} m k^M$$

and

$$n' = m'^{-1}(k^N)^{-1} m'((k^M)^{-1} n k^M) k^N.$$

So if we make the change of variables

$$(k^N, n) \mapsto (k^N, n')$$

we get the formula

$$\int_{N(\mathcal{O})} \int_{N(F)} \chi_I(\dot{w}^{-1} \, (k^M k^N)^{-1} m n k^M k^N \, \dot{w}) dn \, dk^N$$
$$= \int_{N(F)} \chi_I(\dot{w}^{-1} \, m' n' \, \dot{w}) dn'$$

(the absolute value of the Jacobian is 1 and $\mathrm{vol}(N(\mathcal{O}), dk^N) = 1$). In particular, to finish the proof of the lemma it suffices to check that

$$\int_{N(F)} \chi_I(\dot{w}^{-1} \, m' n' \, \dot{w}) dn' = \mathrm{vol}(N(\mathcal{O}) \cap (\dot{w} \, \mathcal{P}_I^\circ \, \dot{w}^{-1}), dn')(\chi_I^{M,w})(m')$$

for each $m' \in M(F)$ (recall that $\mathrm{vol}(N(\mathcal{O}), dn') = 1$). But thanks to (5.4.5) and to (5.4.6) we have

$$P(F) \cap (\dot{w} \, \mathcal{P}_I \, \dot{w}^{-1}) = (M(F) \cap \dot{w} \, \mathcal{P}_I \, \dot{w}^{-1})(N(\mathcal{O}) \cap (\dot{w} \, \mathcal{P}_I^{\circ} \, \dot{w}^{-1})).$$

Therefore, either

$$n' \notin N(\mathcal{O}) \cap (\dot{w} \, \mathcal{P}_I^{\circ} \, \dot{w}^{-1})$$

and we have

$$\chi_I(\dot{w}^{-1} \, m'n' \, \dot{w}) = 0,$$

or

$$n' \in N(\mathcal{O}) \cap (\dot{w} \, \mathcal{P}_I^{\circ} \, \dot{w}^{-1})$$

and we have

$$\chi_I(\dot{w}^{-1} \, m'n' \, \dot{w}) = \chi_I(\dot{w}^{-1} \, m' \, \dot{w}) = (\chi_I^{M,w})(m').$$

The lemma follows. $\qquad\qquad\qquad\qquad\qquad\qquad\qquad\qquad\qquad\qquad\square$

To go further, we need to compute the intersection $M(F) \cap (\dot{w} \, \mathcal{P}_I \, \dot{w}^{-1})$ and the function $\chi_I^{M,w}$ for each $w \in D_{M,I}$. As before, let $(e_1, \ldots, e_t)$ be the partition of d such that

$$\Delta - I = \{e_1, e_1 + e_2, \ldots, e_1 + \cdots + e_{t-1}\}.$$

Let t', t'' be the positive integers such that

$$t = t't'',$$
$$e_k = e_{k+t''} = \cdots = e_{k+(t'-1)t''}$$

for each $k = 1, \ldots, t''$ and such that t'' is minimal for these two first properties. Then we have

$$d = t'e'',$$
$$\mathcal{P}_I = \mathcal{P}_I^{\circ} \rtimes \varepsilon^{e''\mathbf{Z}}$$

and

$$\chi_I(\mathcal{P}_I^{\circ} \varepsilon^{e''n}) = \{(-1)^{(t'-1)t''n}\}$$

for each $n \in \mathbf{Z}$, where we have set

$$e'' = e_1 + \cdots + e_{t''}.$$

Let $w \in D_{M,I}$ and let $a \in A_{M,I}$ be the corresponding tableau. Let

$$J = \Delta^M \cap w(I).$$

We identify M with $GL_{d_1} \times \cdots \times GL_{d_s}$, Δ^M with

$$\coprod_{j=1}^{s} \Delta^{GL_{d_j}}$$

where

$$\Delta^{GL_{d_j}} = \{1, \ldots, d_j - 1\}$$

for $j = 1, \ldots, s$, J with

$$\coprod_{j=1}^{s} J_j$$

where

$$\Delta^{GL_{d_j}} - J_j = \{a_{j1}, a_{j1} + a_{j2}, \ldots, a_{j1} + \cdots + a_{j,t-1}\}$$

for $j = 1, \ldots, s$, $\mathcal{P}_J^{M,\circ}$ with

$$\prod_{j=1}^{s} \mathcal{P}_{J_j}^{GL_{d_j},\circ}$$

and $\mathcal{P}_J^M$ with

$$\prod_{j=1}^{s} (\mathcal{P}_{J_j}^{GL_{d_j},\circ} \rtimes \varepsilon_j^{e_j''\mathbf{Z}})$$

where

$$\varepsilon_j = \begin{pmatrix} 0 & 1 & 0 & \cdots & 0 \\ \vdots & \ddots & \ddots & \ddots & \vdots \\ \vdots & & \ddots & \ddots & 0 \\ 0 & & & \ddots & 1 \\ \varpi & 0 & \cdots & \cdots & 0 \end{pmatrix} \in GL_{d_j}(F)$$

and e_j'' is some positive integer which divides d_j for $j = 1, \ldots, s$.

LEMMA (5.4.9). — *Let $n \in \mathbf{Z}$ and let u' be the g.c.d. of n and t'. Then:*

(i) *the intersection $M(F) \cap (\dot{w} \, \mathcal{P}_I^{\circ} \varepsilon^{e''n} \, \dot{w}^{-1})$ is non-empty if and only if*

$$a_{jk} = a_{j,k+t''u'} = \cdots = a_{j,k+(\bar{t}'-1)t''u'}$$

for all $j = 1, \ldots, s$ and $k = 1, \ldots, t''u'$ where we have set $\bar{t}' = t'/u'$ (in particular, if this intersection is non empty, $\bar{t}'$ divides d_j for each $j = 1, \ldots, s$);

(ii) *if the equivalent conditions of part (i) are satisfied, the intersection $M(F) \cap (\dot{w} \, \mathcal{P}_I^{\circ} \varepsilon^{e''n} \, \dot{w}^{-1})$ is equal to*

$$\prod_{j=1}^{s} (\mathcal{P}_{J_j}^{GL_{d_j},\circ} \varepsilon_j^{f_j''\bar{n}})$$

where we have set

$$f_j'' = a_{j1} + \cdots + a_{j,t''u'} = d_j/\bar{t}'$$

for all $j = 1, \ldots, s$ and

$$\bar{n} = n/u'.$$

Proof: For each $j = 1, \ldots, s$ and each $k = 0, 1, \ldots, t$ we have

$$V_j = F^{d_j}$$

and

$$V_j \cap \dot{w}(L_k) = L_{jk} = (\varpi^{-1}\mathcal{O})^{a_{j1} + \cdots + a_{jk}} \oplus \mathcal{O}^{a_{j,k+1} + \cdots + a_{jt}}.$$

Moreover, we have seen that, for all $k \in \mathbf{Z}$,

$$\dot{w}(L_k) = L_{1k} \oplus \cdots \oplus L_{sk}.$$

Therefore, $M(F) \cap (\dot{w}\,\mathcal{P}_I^\circ \varepsilon^{e''n}\,\dot{w}^{-1})$ is equal to

$$\prod_{j=1}^{s} \{g_j \in GL_{d_j}(F) | g_j(L_{jk}) = L_{j,k+t''n}, \forall k \in \mathbf{Z}\}$$

and it is non-empty if and only if

$$a_{jk} = a_{j,k+t''n}$$

for all $j = 1, \ldots, s$ and $k \in \mathbf{Z}/t\mathbf{Z} \cong \{1, \ldots, t\}$. The lemma follows. $\qquad \square$

We denote by δ the g.c.d. of $d_1, \ldots, d_s$ and we set

(5.4.10) $$\bar{d}_j = d_j/\delta$$

for all $j = 1, \ldots, s$.

COROLLARY (5.4.11). — *Let $m = (g_1, \ldots, g_s) \in M(F)$ ($g_j \in GL_{d_j}(F)$ for $j = 1, \ldots, s$) and let us set*

$$\mu_j = v(\det g_j) \in \mathbf{Z}$$

for all $j = 1, \ldots, s$. Then we have

$$m \in M(F) \cap (\dot{w}\,\mathcal{P}_I\,\dot{w}^{-1})$$

if and only if the following conditions are satisfied:

(1) $g_j \varepsilon_j^{-\mu_j} \in \mathcal{P}_{J_j}^{GL_{d_j},\circ}$ for all $j = 1, \ldots, s$,

(2) there exists $\nu \in \mathbb{Z}$ such that

$$\mu_j = \nu \bar{d}_j$$

for all $j = 1, \ldots, s$,

(3) δ divides $\nu t'$,

(4) if u' is the g.c.d. of t' and $\nu t'/\delta$, we have

$$a_{jk} = a_{j,k+t''u'} = \cdots = a_{j,k+(\bar{t}'-1)t''u'}$$

for all $j = 1, \ldots, s$ and all $k = 1, \ldots, t''u'$ where we have set $\bar{t}' = t'/u'$.

Moreover, if $m \in M(F) \cap (\dot{w} \, \mathcal{P}_I \, \dot{w}^{-1})$, we have

$$\chi_I^{M,w}(m) = (-1)^{(t'-1)t''u'}$$

with the above notations. $\hfill\square$

(5.5) The function f is very cuspidal

It follows from (5.4.8) that the K-invariant constant term of our Euler–Poincaré function f (see (5.1.2)) along the standard parabolic subgroup $P = MN$ is given by

$$(5.5.1) \qquad\qquad f^P = \delta_{P(F)}^{1/2} (\varphi_M)_{K^M}$$

where φ_M is the function on $M(F)$ which is given by

$$(5.5.2) \qquad \varphi_M = \sum_{I \subset \Delta} \sum_{w \in D_{M,I}} \frac{(-1)^{|\Delta - I|} \chi_I^{M,w}}{(|\Delta - I| + 1) \operatorname{vol}(\mathcal{P}_{\Delta^M \cap w(I)}^{M,\circ}, dm)}.$$

Let $m = (g_1, \ldots, g_s) \in M(F)$ $(g_j \in GL_{d_j}(F), \ j = 1, \ldots, s)$ and let

$$\mu_j = v(\det g_j)$$

for $j = 1, \ldots, s$. Following (5.4.11), $\varphi_M(m) = 0$ if

$$\mu_{j'}/\bar{d}_{j'} \neq \mu_{j''}/\bar{d}_{j''}$$

for some $j', j'' \in \{1, \ldots, s\}$, $j' \neq j''$. Moreover, if there exists $\nu \in \mathbb{Z}$ such that

$$\mu_j/\bar{d}_j = \nu$$

for all $j = 1, \ldots, s$, we have

$$(5.5.3) \qquad \varphi_M(m) = \sum_{J \subset \Delta^M} c_J \frac{1_{\mathcal{P}_J^{M,\circ}}(g_1 \varepsilon_1^{-\mu_1}, \ldots, g_s \varepsilon_s^{-\mu_s})}{\text{vol}(\mathcal{P}_J^{M,\circ}, dm)}$$

where $1_{\mathcal{P}_J^{M,\circ}}$ is the characteristic function of $\mathcal{P}_J^{M,\circ}$ inside $M(F)$ and where

$$(5.5.4) \qquad c_J = \sum_{\substack{I \subset \Delta \\ \delta \text{ divides } \nu t'}} \frac{(-1)^{t-1}}{t} \sum_{\substack{w \in D_{M,I} \\ J = \Delta^M \cap w(I) \\ a_{jk} = a_{j,k+t''u'}, \forall j,k}} (-1)^{(t'-1)t''u'}$$

for all $J \subset \Delta^M$ with $(e_1, \ldots, e_t)$ the partition of d corresponding to I, t'' the smallest positive integer such that $e_k = e_{k+t''}$ for each $k \in \mathbb{Z}/t\mathbb{Z} \cong \{1, \ldots, t\}$, $t' = t/t''$, u' the g.c.d. of t' and $\nu t'/\delta$ and $a \in A_{M,I}$ the tableau corresponding to w (recall that δ is the g.c.d. of $d_1, \ldots, d_s$).

We will now prove the following combinatorial statement which immediately implies part (ii) of (5.1.3).

PROPOSITION (5.5.5) (Waldspurger). — *If P is proper ($s \geq 2$), then $c_J = 0$ for all $J \subset \Delta^M$. Therefore φ_M is identically zero on $M(F)$.*

Proof: First of all let us write

$$\frac{\nu}{\delta} = \frac{\bar{\nu}}{\bar{\delta}}$$

where $\bar{\nu}$ and $\bar{\delta}$ are relatively prime integers ($\bar{\delta} > 0$). Then δ divides $\nu t'$ if and only if $\bar{\delta}$ divides t' and, if this is the case, the g.c.d. u' of t' and $\nu t'/\delta$ is equal to $t'/\bar{\delta}$. Moreover, we have

$$t''\bar{\delta}u' - 1 + (\bar{\delta}u' - 1)t''u' \equiv t''u' - 1 \pmod 2.$$

So we can rewrite the formula for c_J as

$$c_J = \sum_{\substack{I \subset \Delta \\ \bar{\delta} \text{ divides } t' \\ w \in D_{M,I} \\ J = \Delta^M \cap w(I) \\ a_{jk} = a_{j,k+(t/\bar{\delta})}, \forall j,k}} \frac{(-1)^{(t/\bar{\delta})-1}}{t}.$$

In particular $c_J = 0$ if $\bar{\delta}$ doesn't divide one of the t_j's ($j = 1, \ldots, s$) or if $\bar{\delta}$ divides t_j and

$$e_{jk} \neq e_{j,k+(t_j/\bar{\delta})}$$

for some $j \in \{1, \ldots, s\}$ and some $k \in \mathbb{Z}/t_j\mathbb{Z} \cong \{1, \ldots, t_j\}$. Here, $t_j = |\Delta^{GL_{d_j}} - J_j| + 1$ and $d_{J_j} = (e_{j1}, \ldots, e_{jt_j})$ is the partition of d_j such that

$$|\Delta^{GL_{d_j}} - J_j| = \{e_{j1}, e_{j1} + e_{j2}, \ldots, e_{j1} + \cdots + e_{j,t_j-1}\}$$

(in fact $(e_{j1}, \ldots, e_{jt_j})$ can be deduced from the sequence $(a_{j1}, \ldots, a_{jt})$ by canceling the zero terms). So we can and we will assume that $\bar{\delta}$ divides t_j for all $j = 1, \ldots, s$ and that

$$e_{jk} = e_{j,k+(t_j/\bar{\delta})}$$

for all $j = 1, \ldots, s$ and all $k \in \mathbb{Z}/t_j\mathbb{Z} \cong \{1, \ldots, t_j\}$.

We have a natural bijection between the set of pairs (I, w) where $I \subset \Delta$ and $w \in D_{M,I}$ such that

$$\Delta^M \cap w(I) = J$$

and the set of pairs $(t, (K_j)_{j=1,\ldots,s})$ where $t \in \{1, \ldots, d\}$ and $(K_j)_{j=1,\ldots,s}$ is a family of subsets of $\{1, \ldots, t\}$ such that $|K_j| = t_j$ for $j = 1, \ldots, s$ and

$$K_1 \cup \cdots \cup K_s = \{1, \ldots, t\}.$$

To (I, w) one associates $t = |\Delta - I| + 1$ and

$$K_j = \{k \in \{1, \ldots, t\} | a_{jk} \neq 0\}$$

$(j = 1, \ldots, s)$ where $a \in A_{M,I}$ is the tableau corresponding to w; if we set

$$K_j = \{k_1, \ldots, k_{t_j}\}$$

with $k_1 < \cdots < k_{t_j}$, we have

$$a_{jk_\ell} = e_{j\ell}$$

for each $j = 1, \ldots, s$ and each $\ell = 1, \ldots, t_j$.

If (I, w) corresponds to $(t, (K_j)_{j=1,\ldots,s})$, the conditions

$$a_{jk} = a_{j,k+(t/\bar{\delta})}$$

for all $j = 1, \ldots, s$ and all $k \in \mathbb{Z}/t\mathbb{Z} \cong \{1, \ldots, t\}$ on the tableau $a \in A_{M,I}$ corresponding to w are now equivalent to the conditions

$$K_j = \overline{K}_j \amalg (\overline{K}_j + (t/\bar{\delta})) \amalg \cdots \amalg (\overline{K}_j + (t/\bar{\delta})(\bar{\delta} - 1))$$

for some subsets

$$\overline{K}_j \subset \{1, \ldots, t/\bar{\delta}\}$$

with $t_j/\bar{\delta}$ elements for $j = 1, \ldots, s$ (recall that we are assuming that $\bar{\delta}$ divides t_j and that $e_{jk} = e_{j,k+(t_j/\bar{\delta})}$ for all $j = 1, \ldots, s$ and all $k \in \mathbb{Z}/t_j\mathbb{Z} \cong \{1, \ldots, t_j\}$). Therefore, we get the formula

$$c_J = \frac{1}{\bar{\delta}} c_{\bar{t}_1, \ldots, \bar{t}_s}.$$

where $\bar{t}_j = t_j/\bar{\delta}$ $(j = 1, \ldots, s)$ and

$$c_{\bar{t}_1,\ldots,\bar{t}_s} = \sum_{\bar{t}=1}^{d/\bar{\delta}} \frac{(-1)^{\bar{t}}}{\bar{t}} c_{\bar{t}_1,\ldots,\bar{t}_s}^{\bar{t}}$$

with $c_{\bar{t}_1,\ldots,\bar{t}_s}^{\bar{t}}$ equal to the number of families $(\overline{K}_j)_{j=1,\ldots,s}$ of subsets $\overline{K}_j \subset \{1,\ldots,\bar{t}\}$ such that $|\overline{K}_j| = \bar{t}_j$ $(j = 1,\ldots,s)$ and

$$\overline{K}_1 \cup \cdots \cup \overline{K}_s = \{1,\ldots,\bar{t}\}$$

for all integers $\bar{t} \geq 1$ and $\bar{t}_1,\ldots,\bar{t}_s \geq 0$.

Let $X_1,\ldots,X_s$ be indeterminates over $\mathbb{Q}$. Then

$$(-1)^{\bar{t}} c_{\bar{t}_1,\ldots,\bar{t}_s}^{\bar{t}}$$

is the coefficient of $X_1^{\bar{t}_1} \cdots X_s^{\bar{t}_s}$ in the polynomial

$$\left(1 - \prod_{j=1}^{s} (1 + X_j)\right)^{\bar{t}}$$

for all $\bar{t}$. Indeed this polynomial is equal to

$$\sum_{\substack{\bar{K}_1,\ldots,\bar{K}_s,\bar{L} \subset \{1,\ldots,\bar{t}\} \\ \bar{K}_1 \cup \cdots \cup \bar{K}_s \subset \bar{L}}} (-1)^{|\bar{L}|} X_1^{|\bar{K}_1|} \cdots X_s^{|\bar{K}_s|}$$

and

$$\sum_{\bar{K} \subset \bar{L} \subset \{1,\ldots,\bar{t}\}} (-1)^{|\bar{L}|} = \begin{cases} (-1)^{\bar{t}} & \text{if } \bar{K} = \{1,\ldots,\bar{t}\}, \\ 0 & \text{otherwise}, \end{cases}$$

for each $\overline{K} \subset \{1,\ldots,\bar{t}\}$. Therefore $c_{\bar{t}_1,\ldots,\bar{t}_s}$ is the coefficient of $X_1^{\bar{t}_1} \cdots X_s^{\bar{t}_s}$ in the polynomial

$$-\sum_{\bar{t}=1}^{d/\bar{\delta}} \frac{1}{\bar{t}} \left(1 - \prod_{j=1}^{s} (1 + X_j)\right)^{\bar{t}}.$$

Finally, the formal power series

$$\sum_{j=1}^{s} \log(1 + X_j) = -\sum_{\bar{t} \geq 1} \frac{(-1)^{\bar{t}}}{\bar{t}} (X_1^{\bar{t}} + \cdots + X_s^{\bar{t}})$$

is equal to

$$\log\left(\prod_{j=1}^{s} (1 + X_j)\right) = -\sum_{\bar{t} \geq 1} \frac{1}{\bar{t}} \left(1 - \prod_{j=1}^{s} (1 + X_j)\right)^{\bar{t}}$$

and differs from the above polynomial by terms of total degree $> d/\bar{\delta}$. Therefore, if $\bar{t}_1 + \cdots + \bar{t}_s \leq d/\bar{\delta}$, $c_{\bar{t}_1,\dots,\bar{t}_s}$ is also the coefficient of $X_1^{\bar{t}_1} \cdots X_s^{\bar{t}_s}$ in

$$-\sum_{\bar{t} \geq 1} \frac{(-1)^{\bar{t}}}{\bar{t}}(X_1^{\bar{t}} + \cdots + X_s^{\bar{t}}).$$

In particular, if $s \geq 2$, $\bar{t}_1, \dots, \bar{t}_s \geq 1$ and $\bar{t}_1 + \cdots \bar{t}_s \leq d/\bar{\delta}$, we have

$$c_{\bar{t}_1,\dots,\bar{t}_s} = 0$$

and the proposition follows. $\square$

(5.6) Non-elliptic orbital integrals of f

Let us finish the proof of (5.1.3). Let $\gamma \in G(F)$ be non-elliptic. We want to prove that $O_\gamma(f, dg_\gamma)$ is zero for any Haar measure dg_γ on $G_\gamma(F)$.

But, thanks to (4.8.10) and (4.8.11), we have

$$O_\gamma(f, dg_\gamma) = |D_{M\backslash G}(\gamma)|^{-1/2} c(\gamma)^{-1} \delta_{P''(F)}^{-1/2}(\gamma_{M''}) O_{\gamma_{M''}}^{M''}((f^P)^{P''}, dm''_{\gamma_{M''}})$$

with the notations of loc. cit. Moreover, up to conjugacy, one can assume that the parabolic subgroups $P \subset G$, $P'' \subset M$ and their Levi subgroups $M \subset P$, $M'' \subset P''$ are standard and that $K_P = G(\mathcal{O})$ and $K_{P''}^M = M(\mathcal{O})$.

As γ is non-elliptic, P or P'' is proper ($P \underset{\neq}{\subset} G$ or $P'' \underset{\neq}{\subset} M$). Therefore, part (ii) of (5.1.3) which is already proved implies that $(f^P)^{P''}$ is identically zero on $M''(F)$ and part (iii) of (5.1.3) follows. $\square$

(5.7) Comments and references

In [Ko 1] § 2, Kottwitz introduces a whole family of functions for a connected reductive group G over a non-archimedean local field of characteristic 0, the so-called Euler–Poincaré functions (in fact, he assumes moreover that the connected center of G is anisotropic over F). The orbital integrals of these functions for a given orbit are all the same. If the orbit is elliptic (semi-simple), these orbital integrals are all equal to 1 for a natural normalization of Haar measures. Otherwise, the orbital integrals vanish.

My main goal in this chapter has been to extend Kottwitz's results for $G = PGL_d$ over a non-archimedean local field of positive characteristic. At least, I have given one function f over $PGL_d(F)$ with the required orbital integrals (see (5.1.2)). It is a suitable linear combination of Kottwitz's functions.

The function f (which makes sense over any non-archimedean local field) has an important property: it is very cuspidal. This property is non-invariant (it depends on the choices of a maximal compact subgroup and a maximal split torus in good position) but it is useful for two purposes. First of all, it easily implies the vanishing of all orbital integrals of f except for elliptic orbits (an invariant property). Secondly, if we plug this very cuspidal Euler–Poincaré function in the non-invariant Arthur trace formula the geometric side simplifies a lot as we shall see in the second volume of this book.

I introduced the notion of very cuspidal function and the function f in a series of letters to Kottwitz and Arthur at the end of 1988 (see [Lau 1]). At this time, I was not able to prove that f is very cuspidal in general. In fact, for a given $J \underset{\neq}{\subset} \Delta$, it is easy to reduce the statement

$$f^{P_J}|M_J(\mathcal{O}) \equiv 0$$

to the following combinatorial formulas: for any $I \subset J$, we have

$$c_{I,J} \overset{\text{dfn}}{=} \sum_{\substack{w \in W \\ w^{-1}(J) \subset R^+ \\ w^{-1}(I) \subset \Delta}} \frac{(-1)^{|I|+t_w^J}}{(n - |I| - t_w^J) \binom{n - |I|}{t_w^J}} = 0$$

with

$$t_w^J = |\{\alpha \in \Delta \mid w(\alpha) \in R^+ - J\}|$$

and the notations of (5.1) and (5.4). More precisely, for each $m \in M_J(\mathcal{O})$ with reduction $\overline{m} \in M_J(k)$ modulo ϖ, we have

$$f^{P_J}(m) = \sum_{I \subset J} c_{I,J} \mathrm{tr}(\mathrm{Ind}_{P_I^J(k)}^{M_J(k)}(1)(\overline{m}))$$

where 1 is the trivial character of the finite group $P_I^J(k)$. But I was unable to prove these combinatorial formulas except for small values of d ($d = 2, 3, 4$). So I asked Waldspurger to help me and he kindly did. In less than one week (!) he found the proof that I have included in this chapter with his permission. One year later, Kassel pointed out that, at least for $I = J = \emptyset$, the above combinatorial formula is classical. In fact the vanishing of $c_{\emptyset,\emptyset}$ for all d is a direct consequence of the Worpitzky formula (see [Fo–Sc] Ch. 2, §6, for example) and it is easy to generalize the Worpitzky formula in order to obtain $c_{\emptyset,J} = 0$ for all d and all $J \underset{\neq}{\subset} \Delta$. There is little doubt that it is possible to prove $c_{I,J} = 0$ in this way in general. This would give a slightly different proof of (5.1.3) (ii) than Waldspurger's one.

For reductive groups over $\mathbb{R}$ one can also find very cuspidal Euler–Poincaré functions (see [Lau 2] and [La]).

6

The Lefschetz numbers as sums of global elliptic orbital integrals

We fix a positive integer d, a non-zero ideal $I \subset_{\neq} A$ and a place o of F, $o \neq \infty$ and $o \notin V(I)$. Then we have a smooth affine scheme $M_{I,o}^d$ of pure dimension $d - 1$ over $\kappa(o)$.

Let $\overline{\kappa(o)}$ be an algebraic closure of $\kappa(o)$ and let $\mathrm{Frob}_o \in \mathrm{Gal}(\overline{\kappa(o)}/\kappa(o))$ be the geometric Frobenius element. Let $f^{\infty,o} \in \mathcal{H}_I^{\infty,o}$ be a Hecke operator and let r be a positive integer.

Then $f^{\infty,o}$ acts on $M_{I,o}^d$ (by a correspondence) and we can consider the number of fixed points

$$\mathrm{Lef}_r(f^{\infty,o})$$

of $\mathrm{Frob}_o^r \times f^{\infty,o}$ acting on $M_{I,o}^d(\overline{\kappa(o)})$ (strictly speaking, we have considered only the case where $f^{\infty,o}$ is the characteristic function of a double class $K_I^{\infty,o} g^{\infty,o} K_I^{\infty,o}$ in $GL_d(\mathbb{A}^{\infty,o})$; in general $f^{\infty,o}$ is a $\mathbb{Q}$-linear combination of such characteristic functions and $\mathrm{Lef}_r(f^{\infty,o})$ is defined by linearity).

Let $K_o = GL_d(\mathcal{O}_o)$, $K_\infty = GL_d(\mathcal{O}_\infty)$ and let $\mathcal{B}_\infty^\circ \subset K_\infty$ be the standard Iwahori subgroup. Let dg_o, dg_∞, dz_∞ be the Haar measures on $GL_d(F_o)$, $GL_d(F_\infty)$, $F_\infty^\times$ which are normalized by $\mathrm{vol}(K_o, dg_o) = 1$, $\mathrm{vol}(K_\infty, dg_\infty) = 1$, $\mathrm{vol}(\mathcal{O}_\infty^\times, dz_\infty) = 1$ respectively.

Let

$$f_o \in \mathcal{H}_o = \mathcal{C}_c^\infty(GL_d(F_o)//K_o)$$

be the Hecke function with Satake transform

$$f_o^\vee(z) = p^{\deg(o)r(d-1)/2}(z_1^r + \cdots + z_d^r)$$

(see (4.2.6)). Let

$$f_\infty \in \tilde{\mathcal{H}}_\infty^{ad} = \mathcal{C}_c^\infty(F_\infty^\times \backslash GL_d(F_\infty) // \mathcal{B}_\infty^\circ)$$

be the very cuspidal Euler–Poincaré function that we define in (5.1.2).

Let us recall that $dg^{\infty,o}$ is the Haar measure on $GL_d(\mathbb{A}^{\infty,o})$ which is normalized by $\mathrm{vol}(K_I^{\infty,o}, dg^{\infty,o}) = 1$. We get a Haar measure $dg_\mathbb{A} = dg_\infty dg^{\infty,o} dg_o$ on $GL_d(\mathbb{A})$. Let

$$f_\mathbb{A} = f_\infty f^{\infty,o} f_o \in \mathcal{C}_c^\infty(F_\infty^\times \backslash GL_d(\mathbb{A}) // \mathcal{B}_\infty^\circ K_I^{\infty,o} K_o).$$

For each elliptic element $\gamma \in G(F)$ let us choose arbitrarily a Haar measure $dg_{\gamma,\mathbb{A}}$ on its centralizer $GL_d(\mathbb{A})_\gamma$ in $GL_d(\mathbb{A})$. Let $d\gamma'$ be the counting measure on $GL_d(F)_\gamma$, the centralizer of γ in $GL_d(F)$. Then the orbital integral

$$O_\gamma(f_\mathbb{A}, dg_{\gamma,\mathbb{A}}) = \int_{GL_d(\mathbb{A})_\gamma \backslash GL_d(\mathbb{A})} f_\mathbb{A}(g_\mathbb{A}^{-1} \gamma g_\mathbb{A}) \frac{dg_\mathbb{A}}{dg_{\gamma,\mathbb{A}}}$$

is absolutely convergent (in fact can be reduced to a finite sum), depends only on the conjugacy class of γ in $GL_d(F)$ and vanishes outside a finite number of elliptic conjugacy classes in $GL_d(F)$. Moreover, the product of $O_\gamma(f_\mathbb{A}, dg_{\gamma,\mathbb{A}})$ by the volume

$$a(\gamma, dg_{\gamma,\mathbb{A}}) = \mathrm{vol}(F_\infty^\times GL_d(F)_\gamma \backslash GL_d(\mathbb{A})_\gamma, \frac{dg_{\gamma,\mathbb{A}}}{dz_\infty d\gamma'})$$

is independent of the choice of $dg_{\gamma,\mathbb{A}}$ and is a rational number.

Let $GL_d(F)_{\natural,ell}$ be a system of representatives of the conjugacy classes of elliptic elements in $GL_d(F)$. Then, combining (3.6.5), (4.5.5) and (5.1.3)(i), we get

THEOREM (6.1). — *The number of fixed points* $\mathrm{Lef}_r(f^{\infty,o})$ *is equal to*

$$\sum_{\gamma \in GL_d(F)_{\natural,ell}} a(\gamma, dg_{\gamma,\mathbb{A}}) O_\gamma(f_\mathbb{A}, dg_{\gamma,\mathbb{A}})$$

(*this sum is well-defined, independent of the choices of* $GL_d(F)_{\natural,ell}$ *and* $dg_{\gamma,\mathbb{A}}$ *and is a rational number*). □

7

Unramified principal series representations

(7.0) Introduction

We will use the same notations as in chapters 4 and 5: F is a non-archimedean local field with ring of integers $\mathcal{O}$, $G = GL_d, \ldots$.

The purpose of this chapter is to review the construction and the main properties of the unramified principal series representations of $G(F)$ (see [Be–Ze 1], [Be–Ze 2], [Cas 1]).

(7.1) Parabolic induction and restriction

As in (5.1), $B \subset G$ is the (standard) Borel subgroup of upper triangular matrices, $T \subset B$ is the maximal torus of diagonal matrices and $\Delta \cong \{1, \ldots, d-1\}$ is the set of simple roots of (G, T, B). For each subset $I \subset \Delta$, we have a corresponding standard parabolic subgroup $P_I \subset G$ with a standard Levi decomposition $P_I = M_I N_I$. For each $J \subset I \subset \Delta$, $P_J^I = P_J \cap M_I$ is a standard parabolic subgroup of M_I and $P_J^I = M_J N_J^I$, where $N_J^I = N_J \cap M_I$, is its standard Levi decomposition. For each $I \subset \Delta$, we denote by Z_I the center of M_I.

For each $I \subset \Delta$, $M_I(F)$ is a unimodular, locally compact, totally discontinuous, separated, topological group. Let $\text{Rep}_s(M_I(F))$ be the abelian category of **smooth representations** of $M_I(F)$ and let $\text{Rep}_a(M_I(F))$ be its full subcategory of **admissible representations** (see (D.1)).

Let $J \subset I \subset \Delta$ and let P be a parabolic subgroup of M_I which admits M_J as a Levi component, i.e. such that $P = M_J N$ where N is the unipotent radical of P (for example, $P = P_J^I$). Then, $N(F) \subset P(F) \subset M_I(F)$ are closed subgroups such that

(A) $N(F)$ is normal in $P(F)$,

(B) any compact subset of $N(F)$ is contained in a compact subgroup of $N(F)$ (in fact, is contained in $z_J N(\mathcal{O}) z_J^{-1}$ for some $z_J \in Z_J(F)$),

(C) $M_I(F) = P(F) M_I(\mathcal{O})$,

(D) for any Haar measure dn on $N(F)$ and any $p \in P(F)$, we have

$$d(pnp^{-1}) = \delta_{P(F)}(p) dn$$

where $\delta_{P(F)} : P(F) \to \mathbb{R}_{>0}$ is the modulus character of $P(F)$ (thanks to (B), $N(F)$ is unimodular and $\delta_{P(F)}$ is trivial on $N(F)$).

The **induction functor**

(7.1.1) $$i_{M_J(F),P(F)}^{M_I(F)} : \mathrm{Rep}_s(M_J(F)) \to \mathrm{Rep}_s(M_I(F))$$

is defined in the following way (see also (D.3)). If $(\mathcal{W}, \rho) \in \mathrm{ob}\, \mathrm{Rep}_s(M_J(F))$, then $i_{M_J(F),P(F)}^{M_I(F)}(\mathcal{W}, \rho)$ is the object $(\mathcal{V}, \pi)$ of $\mathrm{Rep}_s(M_I(F))$ where $\mathcal{V}$ is the $\mathbb{C}$-vector space of locally constant functions

$$v : M_I(F) \to \mathcal{W}$$

such that

$$v(m_J n m_I) = \delta_{P(F)}^{1/2}(m_J) \rho(m_J)(v(m_I)),$$

for any $m_I \in M_I(F)$, any $m_J \in M_J(F)$ and any $n \in N(F)$, and where π is the left action of $M_I(F)$ on such functions which is induced by the right translation on $M_I(F)$.

The **restriction functor** (also called the **modified Jacquet functor**)

(7.1.2) $$r_{M_I(F)}^{M_J(F),P(F)} : \mathrm{Rep}_s(M_I(F)) \to \mathrm{Rep}_s(M_J(F))$$

is defined in the following way (see also (D.3)). If $(\mathcal{V}, \pi) \in \mathrm{ob}\, \mathrm{Rep}_s(M_I(F))$, then $r_{M_I(F)}^{M_J(F),P(F)}(\mathcal{V}, \pi)$ is the object $(\mathcal{W}, \rho)$ of $\mathrm{Rep}_s(M_J(F))$ where

$$\mathcal{W} = \mathcal{V}/\mathcal{V}(N(F))$$

with $\mathcal{V}(N(F))$ the $\mathbb{C}$-vector subspace of $\mathcal{V}$ generated by the elements

$$\pi(n)(v) - v \quad (n \in N(F), v \in \mathcal{V})$$

and where ρ is induced by

$$\delta_{P(F)}^{-1/2} \otimes (\pi | P(F)) : P(F) \to \mathrm{Aut}_{\mathbb{C}}(\mathcal{V}).$$

If $P = P_J^I$, we will also use the notations

$$i_J^I = i_{M_J(F),P(F)}^{M_I(F)}$$

and

$$r_I^J = r_{M_I(F)}^{M_J(F),P(F)}$$

for these functors.

The first properties of these functors are

PROPOSITION (7.1.3). — (i) *The functors* $i_{M_J(F),P(F)}^{M_I(F)}$ *and* $r_{M_I(F)}^{M_J(F),P(F)}$ *are exact and* $r_{M_I(F)}^{M_J(F),P(F)}$ *is left adjoint to* $i_{M_J(F),P(F)}^{M_I(F)}$.

(ii) *For any* $(\mathcal{W},\rho) \in \text{ob Rep}_s(M_J(F))$, *the contragredient representation of* $i_{M_J(F),P(F)}^{M_I(F)}(\mathcal{W},\rho)$ *is canonically isomorphic to* $i_{M_J(F),P(F)}^{M_I(F)}(\widetilde{\mathcal{W}},\tilde{\rho})$ *where* $(\widetilde{\mathcal{W}},\tilde{\rho})$ *is the contragredient representation of* $(\mathcal{W},\rho)$.

(iii) *The functor* $i_{M_J(F),P(F)}^{M_I(F)}$ *carries admissible representations into admissible ones.*

(iv) *If* $K \subset J \subset I$, *the functors* i_K^I *and* $i_J^I \circ i_K^J$ (*resp.* r_I^K *and* $r_J^K \circ r_I^J$) *are canonically isomorphic.*   □

For the convenience of the reader, a detailed proof of the proposition is reproduced in appendix D (see (D.3.3), (D.3.5), (D.3.6) and (D.3.14)). It is based on the properties (A) to (D) of the pair $N(F) \subset P(F)(\subset M_I(F))$.

The properties of $r_{M_I(F)}^{M_J(F),P(F)}$ which are analogous to the properties (7.1.3) (ii) and (iii) of $i_{M_J(F),P(F)}^{M_I(F)}$ are more subtle and more difficult to prove. They depend on other properties of the pair $N(F) \subset P(F)(\subset M_I(F))$ that we will now state. For simplicity, we will restrict ourself to the case $P = P_J^I$.

Let $\widetilde{B} \subset G$ be the Borel subgroup of lower triangular matrices. We have $T = B \cap \widetilde{B}$ and a Levi decomposition $\widetilde{B} = T\widetilde{U}$, where $\widetilde{U}$ is the unipotent radical of $\widetilde{B}$. For each $I \subset \Delta$ let $\widetilde{P}_I$ be the parabolic subgroup of G generated by $\widetilde{U}$ and M_I. It has a Levi decomposition $\widetilde{P}_I = M_I\widetilde{N}_I$, where $\widetilde{N}_I$ is the unipotent radical of $\widetilde{P}_I$. For each $J \subset I \subset \Delta$, let $\widetilde{P}_J^I$ be the parabolic subgroup $\widetilde{P}_J \cap M_I$ of M_I. It has a Levi decomposition $\widetilde{P}_J^I = M_J\widetilde{N}_J^I$, where $\widetilde{N}_J^I = \widetilde{N}_J \cap M_I$.

For each non-negative integer n, let

$$K(n) = 1 + \varpi^{n+1}gl_d(\mathcal{O}) \subset K \subset G(F)$$

be the **principal congruence subgroup of level** n. For each $n \in \mathbb{Z}_{\geq 0}$, $K(n+1)$ is a normal subgroup of $K(n)$ and the decreasing sequence $(K(n))_{n\in\mathbb{Z}_{\geq 0}}$ of compact open subgroups of $G(F)$ form a basis of neighborhoods of the identity.

Let $J \subset I \subset \Delta$. We set

$$\mathcal{Z}_J^I(1) = \{z_J \in Z_J(F)| \, |\alpha(z_J)| \leq 1, \forall \alpha \in I - J\}.$$

We have

$$\mathcal{Z}_J^I(1) = Z_J(F) \cap \mathcal{Z}_\emptyset^I(1).$$

For each $\varepsilon \in]0, 1]$, we set

$$\mathcal{Z}^I_{\emptyset, J}(\varepsilon) = \{z_\emptyset \in \mathcal{Z}^I_\emptyset(1) | \, |\alpha(z_\emptyset)| \leq \varepsilon, \forall \alpha \in I - J\}$$

and

$$\mathcal{Z}^I_J(\varepsilon) = Z_J(F) \cap \mathcal{Z}^I_{\emptyset, J}(\varepsilon).$$

Then, we have the following properties:

(E) $N^I_J(F) \cap \widetilde{N}^I_J(F) = \{1\}$ and $P^I_J(F) \cap \widetilde{P}^I_J(F) = M_J(F)$,

(F) $\delta_{\widetilde{P}^I_J(F)}(m_J) = \delta_{P^I_J(F)}(m_J^{-1})$ for any $m_J \in M_J(F)$,

(G) $K(n) \cap M_I(F)$ **is in good position with respect to** the triple $(N^I_J(F), M_J(F), \widetilde{N}^I_J(F))$, i.e.

$$K(n) \cap M_I(F) = (K(n) \cap N^I_J(F))(K(n) \cap M_J(F))(K(n) \cap \widetilde{N}^I_J(F))$$

for each $n \in \mathbb{Z}_{\geq 0}$,

(H) for any $z_\emptyset \in \mathcal{Z}^I_\emptyset(1)$ and, therefore, for any $z_\emptyset \in \mathcal{Z}^I_J(1)$ we have

$$z_\emptyset(K(n) \cap N^I_J(F))z_\emptyset^{-1} \subset K(n) \cap N^I_J(F)$$

and

$$z_\emptyset(K(n) \cap \widetilde{N}^I_J(F))z_\emptyset^{-1} \supset K(n) \cap \widetilde{N}^I_J(F)$$

for each $n \in \mathbb{Z}_{\geq 0}$,

(I) for each pair of compact open subgroups (Γ_1, Γ_2) of $N^I_J(F)$ and each pair of compact open subgroups $(\widetilde{\Gamma}_1, \widetilde{\Gamma}_2)$ of $\widetilde{N}^I_J(F)$ there exists $\varepsilon \in]0, 1]$ such that

$$z_\emptyset \Gamma_1 z_\emptyset^{-1} \subset \Gamma_2$$

and

$$z_\emptyset \widetilde{\Gamma}_1 z_\emptyset^{-1} \supset \widetilde{\Gamma}_2$$

for any $z_\emptyset \in \mathcal{Z}^I_{\emptyset, J}(\varepsilon)$ and, therefore, for any $z_\emptyset \in \mathcal{Z}^I_J(\varepsilon)$.

Thanks to these properties, we have

THEOREM (7.1.4). — (i) (Jacquet) *The functor r^J_I carries admissible representations into admissible ones.*

(ii) (Casselman) *Let $\widetilde{r}^J_I$ be the functor $r^{M_J(F), \widetilde{P}^I_J(F)}_{M_I(F)}$ from $\mathrm{Rep}_s(M_I(F))$ to $\mathrm{Rep}_s(M_J(F))$. Then, for each $(\mathcal{V}, \pi) \in \mathrm{ob}\,\mathrm{Rep}_s(M_I(F))$ with contragredient representation $(\widetilde{\mathcal{V}}, \widetilde{\pi})$, there is a canonical isomorphism ι from $\widetilde{r}^J_I(\widetilde{\mathcal{V}}, \widetilde{\pi})$ onto the contragredient representation of $r^J_I(\mathcal{V}, \pi)$.*

Moreover, let us set $r_I^J(\mathcal{V}, \pi) = (\mathcal{W}, \rho)$ and $\tilde{r}_I^J(\tilde{\mathcal{V}}, \tilde{\pi}) = (\check{\mathcal{W}}, \check{\rho})$ and let us denote by

$$p : \mathcal{V} \longrightarrow \mathcal{V}/\mathcal{V}(N_J^I(F)) = \mathcal{W}$$

and

$$\check{p} : \tilde{\mathcal{V}} \longrightarrow \tilde{\mathcal{V}}/\tilde{\mathcal{V}}(\tilde{N}_J^I(F)) = \check{\mathcal{W}}$$

the canonical projections. Then, for each non-negative integer n, there exists $\varepsilon \in]0,1]$ such that

$$< \iota(\check{p}(\tilde{v})), \rho(z_J)(p(v)) > = \delta_{P_J^I(F)}^{-1/2}(z_J) < \tilde{v}, \pi(z_J)(v) >$$

for any $v \in \mathcal{V}^{K(n) \cap M_I(F)}$, any $\tilde{v} \in \tilde{\mathcal{V}}^{K(n) \cap M_I(F)}$ and any $z_J \in \mathcal{Z}_J^I(\varepsilon)$. $\square$

A proof of this theorem is given in (D.3.13).

COROLLARY (7.1.5) (Casselman). — *We keep the notations of (7.1.4) (ii). Then, for each non-negative integer n, there exists $\varepsilon' \in]0,1]$ such that*

$$< \iota(\check{p}(\tilde{v})), \rho(z_\emptyset')(p(v)) > = \delta_{P_J^I(F)}^{-1/2}(z_\emptyset') < \tilde{v}, \pi(z_\emptyset')(v) >$$

for any $v \in \mathcal{V}^{K(n) \cap M_I(F)}$, any $\tilde{v} \in \tilde{\mathcal{V}}^{K(n) \cap M_I(F)}$ and any $z_\emptyset' \in \mathcal{Z}_{\emptyset,J}^I(\varepsilon')$.

Proof : Let us fix a non-negative integer n and let $\varepsilon \in]0,1]$ be as in (7.1.4) (ii) for this fixed n. Obviously $\mathcal{Z}_J^I(\varepsilon) \neq \emptyset$. Let us arbitrarily fix $z_J \in \mathcal{Z}_J^I(\varepsilon)$ and let us set

$$\varepsilon' = \inf\{|\alpha(z_J)| \,|\, \alpha \in I - J\}.$$

We have $\varepsilon' \in]0,\varepsilon] \subset]0,1]$ and

$$\mathcal{Z}_{\emptyset,J}^I(\varepsilon') \subset z_J \mathcal{Z}_\emptyset^I(1) \subset \mathcal{Z}_\emptyset^I(1).$$

Now, let us fix $v \in \mathcal{V}^{K(n) \cap M_I(F)}$, $\tilde{v} \in \tilde{\mathcal{V}}^{K(n) \cap M_I(F)}$ and $z_\emptyset' = z_J z_\emptyset \in \mathcal{Z}_{\emptyset,J}^I(\varepsilon')$ with $z_\emptyset \in \mathcal{Z}_\emptyset^I(1)$. We have

$$\pi(z_\emptyset)(v) \in \mathcal{V}^{K(n) \cap \check{P}_\emptyset^I(F)},$$

so that

$$\pi(e_{K(n) \cap M_I(F)})(\pi(z_\emptyset)(v)) = \pi_{K(n) \cap N_\emptyset^I(F)}(\pi(z_\emptyset)(v))$$

(see (D.3.7)). Therefore, we obtain

$$\pi(z_J)(\pi(e_{K(n) \cap M_I(F)})(\pi(z_\emptyset)(v)))$$
$$= \pi_{z_J(K(n) \cap N_\emptyset^I(F))z_J^{-1}}(\pi(z_\emptyset')(v))$$

and

$$\rho(z_J)(p(\pi(e_{K(n)\cap M_I(F)})(\pi(z_\emptyset(v)))))$$
$$= \delta^{1/2}_{P^I_J(F)}(z_\emptyset)\rho_{K(n)\cap N^J_\emptyset(F)}(\rho(z'_\emptyset)(p(v))).$$

Thanks to (7.1.4) (ii), we have

$$< \iota(\check{p}(\tilde{v})), \rho(z_J)(\pi(e_{K(n)\cap M_I(F)})(\pi(z_\emptyset)(v))) >$$
$$= \delta^{-1/2}_{P^I_J(F)}(z_J) < \tilde{v}, \pi(e_{K(n)\cap M_I(F)})(\pi(z_\emptyset)(v)) >,$$

so that

$$< \iota(\check{p}(\tilde{v})), \rho_{K(n)\cap N^J_\emptyset(F)}(\rho(z'_\emptyset)(p(v))) >$$
$$= \delta^{-1/2}_{P^I_J(F)}(z'_\emptyset) < \tilde{v}, \pi_{z_J(K(n)\cap N_\emptyset I(F))z^{-1}_J}(\pi(z'_\emptyset)(v)) > .$$

But we have

$$\tilde{\pi}_{z_J(K(n)\cap N^I_\emptyset(F))z^{-1}_J}(\tilde{v}) = \tilde{v}$$

and

$$\tilde{\rho}_{K(n)\cap N^J_\emptyset(F)}(\iota(\check{p}(\tilde{v}))) = \iota(\check{p}(\tilde{v}))$$

as $\tilde{v}$ (resp. $\iota(\check{p}(\tilde{v}))$) is fixed by $K(n) \cap M_I(F)$ (resp. $K(n) \cap M_J(F)$) and the proof of the corollary is completed. $\square$

(7.2) Cuspidal representations

Let $I \subset \Delta$ and let $d_I = (d_1, \ldots, d_s)$ be the corresponding partition of d. We can identify M_I with $GL_{d_1} \times \cdots \times GL_{d_s}$. Obviously, $M_I(F)$ is countable at infinity. Let

$$(7.2.1) \qquad M_I(F)^1 = \{m_I = (g_1, \ldots, g_s) \in M_I(F) | \, |\det g_j| = 1,$$
$$\forall j = 1, \ldots, s\}.$$

Then $M_I(F)^1$ is an open, normal subgroup of $M_I(F)$, the quotient group $M_I(F)/M_I(F)^1$ is commutative (it is isomorphic to $\mathbb{Z}^s$), $Z_I(F) \cap M_I(F)^1$ is compact (it is isomorphic to $(\mathcal{O}^\times)^s$) and the quotient group $M_I(F)/Z_I(F)M_I(F)^1$ is finite (it is isomorphic to $(\mathbb{Z}/d_1\mathbb{Z}) \times \cdots \times (\mathbb{Z}/d_s\mathbb{Z})$).

A **quasi-cuspidal** (resp. **cuspidal**) representation of $M_I(F)$ is a smooth (resp. admissible) representation of $M_I(F)$ such that all its matrix coefficients have compact supports modulo $Z_I(F)$ (see (D.4)).

LEMMA (7.2.2). — *A smooth representation* $(\mathcal{V}, \pi)$ *of* $M_I(F)$, *with contra-gredient representation* $(\widetilde{\mathcal{V}}, \tilde{\pi})$, *is quasi-cuspidal if and only if, for each* $v \in \mathcal{V}$ *and each* $\tilde{v} \in \widetilde{\mathcal{V}}$, *there exists* $\varepsilon \in]0, 1]$ *such that*

$$< \tilde{v}, \pi(z_\emptyset)(v) > = 0$$

for all $z_\emptyset \in \mathcal{Z}_\emptyset^I(1)$ *with*

$$|\alpha(z_\emptyset)| \leq \varepsilon$$

for at least one $\alpha \in I$.

Proof: We have the Cartan decomposition

$$M_I(F) = M_I(\mathcal{O}) \mathcal{Z}_\emptyset^I(1) M_I(\mathcal{O})$$

(see (4.1)) and, for each $v \in \mathcal{V}$ and each $\tilde{v} \in \widetilde{\mathcal{V}}$, $\pi(M_I(\mathcal{O}))(v)$ and $\tilde{\pi}(M_I(\mathcal{O}))(\tilde{v})$ are finite sets. Therefore, $(\mathcal{V}, \pi)$ is quasi-cuspidal if and only if, for each $v \in \mathcal{V}$ and each $\tilde{v} \in \widetilde{\mathcal{V}}$, the subset

$$\{z_\emptyset \in \mathcal{Z}_\emptyset^I(1)| < \tilde{v}, \pi(z_\emptyset)(v) > \neq 0\}$$

of $\mathcal{Z}_\emptyset^I(1)$ is compact modulo $Z_I(F)$.

But a closed subset Ω of $\mathcal{Z}_\emptyset^I(1)$ is compact modulo $Z_I(F)$ if and only if there exists $\varepsilon \in]0, 1]$ such that

$$|\alpha(\Omega)| \subset]\varepsilon, 1]$$

for each $\alpha \in I$. The lemma follows. □

THEOREM (7.2.3) (Jacquet). — *A smooth representation* $(\mathcal{V}, \pi)$ *of* $M_I(F)$ *is quasi-cuspidal if and only if* $r_I^J(\mathcal{V}, \pi) = (0)$ *for every* $J \subsetneq I$.

Proof: Thanks to (7.1.3) (iv), we have $r_I^J(\mathcal{V}, \pi) = (0)$ for every $J \subsetneq I$ if and only if $r_I^{I-\{\alpha\}}(\mathcal{V}, \pi) = (0)$ for every $\alpha \in I$.

Let $\alpha \in I$ and let us assume that $r_I^{I-\{\alpha\}}(\mathcal{V}, \pi) = (0)$, i.e. $\mathcal{V} = \mathcal{V}(N_{I-\{\alpha\}}^I(F))$. If $v \in \mathcal{V}$ and $\tilde{v} \in \widetilde{\mathcal{V}}$, there exist compact open subgroups Γ_1 and Γ_2 in $N_{I-\{\alpha\}}^I(F)$ such that $v \in \text{Ker}(\pi_{\Gamma_1})$ (see (D.3)) and $\tilde{v} \in \widetilde{\mathcal{V}}^{\Gamma_2}$. But, thanks to the property (I) of (7.1), there exists $\varepsilon > 0$ such that

$$z_\emptyset \Gamma_1 z_\emptyset^{-1} \subset \Gamma_2$$

for any $z_\emptyset \in \mathcal{Z}_\emptyset^I(1)$ with $|\alpha(z_\emptyset)| \leq \varepsilon$. Therefore, we have

$$
\begin{aligned}
< \tilde{v}, \pi(z_\emptyset)(v) > &= < \tilde{\pi}_{\Gamma_2}(\tilde{v}), \pi(z_\emptyset)(v) > \\
&= < \tilde{v}, \pi_{\Gamma_2}(\pi(z_\emptyset)(v)) > \\
&= < \tilde{v}, \pi(z_\emptyset)(\pi_{z_\emptyset^{-1}\Gamma_2 z_\emptyset}(v)) > \\
&= 0
\end{aligned}
$$

for any $z_\emptyset \in \mathcal{Z}_\emptyset^I(1)$ with $|\alpha(z_\emptyset)| \leq \varepsilon$, as $\tilde{\pi}_{\Gamma_2}(\tilde{v}) = \tilde{v}$, $\pi_{\Gamma_1}(v) = 0$ and $z_\emptyset^{-1}\Gamma_2 z_\emptyset \supset \Gamma_1$. Thanks to (7.2.2) this completes the proof of the "if" part of the theorem.

Conversely, let us assume that $(\mathcal{V}, \pi)$ is quasi-cuspidal and let $J \underset{\neq}{\subset} I$. We want to prove that $\mathcal{V} = \mathcal{V}(N_J^I(F))$. Let $v \in \mathcal{V}$ and let $n \in \mathbf{Z}_{\geq 0}$ be such that $v \in \mathcal{V}^{K(n) \cap M_I(F)}$. Then

$$
\{m_I \in M_I(F) | \pi(e_{K(n) \cap M_I(F)})(\pi(m_I)(v)) \neq 0\}
$$

is compact modulo Z_I (see (D.4.1)). Therefore, there exists $z_J \in \mathcal{Z}_J^I(1)$ such that

$$
\pi(e_{K(n) \cap M_I(F)})(\pi(z_J)(v)) = 0 .
$$

But we have

$$
K(n) \cap M_I(F) = (K(n) \cap N_J^I(F))(K(n) \cap \widetilde{P}_J^I(F))
$$

and

$$
z_J^{-1}(K(n) \cap \widetilde{P}_J(F))z_J \subset K(n) \cap \widetilde{P}_J^I(F) \subset K(n) \cap M_I(F),
$$

so that

$$
\pi(e_{K(n) \cap M_I(F)})(\pi(z_J)(v)) = \pi(z_J)(\pi_{z_J^{-1}(K(n) \cap N_J^I(F))z_J}(v)).
$$

It follows that

$$
\pi_{z_J^{-1}(K(n) \cap N_J^I(F))z_J}(v) = 0
$$

and that

$$
v \in \mathcal{V}(N_J^I(F))
$$

as required. $\qquad\qquad\square$

COROLLARY (7.2.4). — *Let $(\mathcal{V}, \pi)$ be a smooth irreducible representation of $M_I(F)$. Then*

(i) *there exist $J \subset I$, a cuspidal irreducible representation $(\mathcal{W}, \rho)$ of $M_J(F)$ and an embedding $(\mathcal{V}, \pi) \hookrightarrow i_J^I(\mathcal{W}, \rho)$ in $\mathrm{Rep}_s(M_I(F))$,*

(ii) *$(\mathcal{V}, \pi)$ is admissible.*

Proof : Assertion (ii) follows immediately from assertion (i) and (7.1.3)
(iii).

Let us choose $J \subset I$ such that $r_I^J(\mathcal{V}, \pi) \neq (0)$ and $r_I^K(\mathcal{V}, \pi) = (0)$ for all
$K \subsetneq J$. Then, thanks to (7.1.3) (iv) and (7.2.3), $r_I^J(\mathcal{V}, \pi)$ is quasi-cuspidal.
If $(\mathcal{W}, \rho)$ is an irreducible quotient of $r_I^J(\mathcal{V}, \pi)$ in $\mathrm{Rep}_s(M_J(F))$, $(\mathcal{W}, \rho)$ is
again quasi-cuspidal (see (7.1.3) (i) and (7.2.3)) and, therefore, cuspidal (see
(D.4.2)). Since

$$\mathrm{Hom}_{\mathrm{Rep}_s(M_J(F))}(r_I^J(\mathcal{V}, \pi), (\mathcal{W}, \rho)) \neq (0)$$

and since r_I^J is left adjoint to i_J^I (see (7.1.3) (i)), we have

$$\mathrm{Hom}_{\mathrm{Rep}_s(M_I(F))}((\mathcal{V}, \pi), i_J^I(\mathcal{W}, \rho)) \neq (0).$$

Since $(\mathcal{V}, \pi)$ is irreducible, any non-zero morphism from $(\mathcal{V}, \pi)$ to $i_J^I(\mathcal{W}, \rho)$ in
$\mathrm{Rep}_s(M_I(F))$ is an embedding.

Now, to finish the proof of assertion (i) and of the corollary, it suffices to
check that $r_I^J(\mathcal{V}, \pi)$ has at least one irreducible quotient in $\mathrm{Rep}_s(M_J(F))$.
Equivalently (see (D.1.2)), let us check that the corresponding $(\mathbb{C} \otimes
\mathcal{C}_c^\infty(M_J(F)))$-module $\mathcal{V}/\mathcal{V}(N_J^I(F))$ has at least one irreducible quotient.
Thanks to Zorn's lemma, this will be the case if this module is finitely
generated. But the $(\mathbb{C} \otimes \mathcal{C}_c^\infty(M_I(F)))$-module $\mathcal{V}$ corresponding to $(\mathcal{V}, \pi)$
is irreducible, so that it is generated by any non-zero v in $\mathcal{V}$. Let us fix
$v \in \mathcal{V} - \{0\}$. The subset $\pi(M_I(\mathcal{O}))(v)$ of $\mathcal{V}$ is finite ($M_I(\mathcal{O})$ is compact and
$(\mathcal{V}, \pi)$ is smooth) and, as

$$M_I(F) = P_J^I(F)M_I(\mathcal{O}),$$

its image in $\mathcal{V}/\mathcal{V}(N_J^I(F))$ generates this $(\mathbb{C} \otimes \mathcal{C}_c^\infty(M_J(F)))$-module.   □

(7.3) Principal series representations

For each smooth character

$$\chi : T(F) \to \mathbb{C}^\times$$

(i.e. a group homomorphism with an open kernel), we can form the **principal
series representation**

(7.3.1) $$(I(\chi), i(\chi)) = i_\emptyset^\Delta(\mathbb{C}, \chi) \in \mathrm{ob} \, \mathrm{Rep}_s(G(F)).$$

Its space $I(\chi)$ is the $\mathbb{C}$-vector space of locally constant functions

$$v : G(F) \to \mathbb{C}$$

such that
$$v(tug) = (\delta_{B(F)}^{1/2}\chi)(t)v(g)$$
for all $t \in T(F)$, $u \in U(F)$ and $g \in G(F)$ and
$$(i(\chi)(g')(v))(g) = v(gg')$$
for all $g, g' \in G(F)$. In particular, it admits a central character
$$(7.3.2) \qquad \omega_{i(\chi)} = \chi|Z(F)$$
($\delta_{B(F)}$ is trivial on $Z(F)$; $Z \stackrel{\text{dfn}}{=} Z_\Delta$). It follows from (7.1.3) (iii) that $(I(\chi), i(\chi))$ is admissible.

THEOREM (7.3.3) (Casselman; Bernstein and Zelevinsky). — (i) $(I(\chi), i(\chi))$ has finite length and its length is at most $|W| = d!$.

(ii) Let χ' be another smooth character of $T(F)$. Then:

(a) either $\chi' \notin W(\chi) = \{w(\chi) = \chi \circ w^{-1}|w \in W\}$ (recall that $w \in W$ acts on $T(F)$ by $t \mapsto \dot{w} t \dot{w}^{-1}$ where $\dot{w} \in G(F)$ is the corresponding permutation matrix) and there is no irreducible representation of $G(F)$ which is isomorphic at the same time to a subquotient of $(I(\chi), i(\chi))$ and to a subquotient of $(I(\chi'), i(\chi'))$,

(b) or $\chi' \in W(\chi)$ and any irreducible representation of $G(F)$ occurs (up to isomorphism) as a Jordan-Hölder subquotient with the same finite multiplicity in $(I(\chi), i(\chi))$ and in $(I(\chi'), i(\chi'))$ (in other words, the semi-simplifications of $(I(\chi), i(\chi))$ and $(I(\chi'), i(\chi'))$ are isomorphic).

(iii) Let $\chi' \in W(\chi)$ and let us set
$$W(\chi, \chi') = \{w \in W|w(\chi) = \chi'\}.$$
Then the $\mathbb{C}$-vector space
$$\operatorname{Hom}_{\operatorname{Rep}_s(G(F))}((I(\chi), i(\chi)), (I(\chi'), i(\chi')))$$
is non-zero and its dimension is finite and at most $|W(\chi, \chi')|$.

The proof of this theorem is based on the following key lemma.

KEY LEMMA (7.3.4). — Let $I \subset \Delta$ and let $\chi : T(F) \to \mathbb{C}^\times$ be a smooth character. Then there exists a numbering $\{w_1, \ldots, w_N\}$ of
$$D_{\emptyset, I} = \{w \in W|w(I) \subset R^+\}$$
satisfying the following condition: the smooth representation $r_\Delta^I(I(\chi), i(\chi))$ of $M_I(F)$ admits a filtration
$$F_n = F_n(r_\Delta^I(i_\emptyset^\Delta(\mathbb{C}, \chi))) \quad (n = 0, 1, \ldots, N),$$
$$(0) = F_0 \subset F_1 \subset \cdots \subset F_N = r_\Delta^I(I(\chi), i(\chi))$$
in $\operatorname{Rep}_s(M_I(F))$ such that, for each $n = 1, \ldots, N$,
$$F_n/F_{n-1} = \operatorname{gr}_n^F = \operatorname{gr}_n^F(r_\Delta^I(I(\chi), i(\chi)))$$
is isomorphic to $i_\emptyset^I(\mathbb{C}, w_n^{-1}(\chi))$.

Proof: The set $D_{\emptyset,I}$ is a system of representatives of the classes in W/W_I (see (5.4.3) (i)) and, therefore, G is the disjoint union of the double classes $B\,\dot{w}\,P_I$ ($w \in D_{\emptyset,I}$) (see [Ca] (2.8.1)). There exists at least one numbering. $\{w_1,\ldots,w_N\}$ of $D_{\emptyset,I}$ such that, for each $n \in \{1,\ldots,N\}$,

$$B\,\dot{w}_1\,P_I \cup B\,\dot{w}_2\,P_I \cup \cdots \cup B\,\dot{w}_n\,P_I$$

is a Zariski open subset of G; in particular, $B\,\dot{w}\,P_I$ is locally closed for the Zariski topology in G for each $w \in D_{\emptyset,I}$ (see [Sp] (10.2.13) in the case $I = \emptyset$). Let us fix such a numbering. It follows that $G(F)$ is the disjoint union of the double classes $B(F)\,\dot{w}\,P_I(F)$ ($w \in D_{\emptyset,I}$) and that, for each $n \in \{1,\ldots,N\}$,

$$B(F)\,\dot{w}_1\,P_I(F) \cup B(F)\,\dot{w}_2\,P_I(F) \cup \cdots \cup B(F)\,\dot{w}_n\,P_I(F)$$

is open in $G(F)$ for the ϖ-adic topology (for each $w \in D_{\emptyset,I}$, $B(F)\,\dot{w}\,P_I(F) \subset G(F)$ is the subset of F-rational points in $B\,\dot{w}\,P_I \subset G$).

Let us set $(\mathcal{V},\pi) = (I(\chi), i(\chi))$. For each $n = 0,\ldots,N$, let $\mathcal{V}_n$ be the $\mathbb{C}$-vector space of locally constant functions

$$v : G(F) \to \mathcal{C}$$

such that

$$v(tug) = (\delta_{B(F)}^{1/2}\chi)(t)v(g)$$

for any $t \in T(F)$, any $u \in U(F)$ and any $g \in G(F)$ and such that

$$\mathrm{Supp}(v) \subset \bigcup_{i=1}^{n} B(F)\,\dot{w}_i\,P_I(F).$$

Obviously, we have

$$(0) = \mathcal{V}_0 \subset \mathcal{V}_1 \subset \mathcal{V}_2 \subset \cdots \subset \mathcal{V}_N = \mathcal{V}$$

and

$$\pi(P_I(F))(\mathcal{V}_n) \subset \mathcal{V}_n \quad (n = 0, 1, \ldots, N).$$

For each $n \in \{0, \ldots, N\}$, let

$$\mathcal{V}_n(N_I(F)) \subset \mathcal{V}_n$$

be the $\mathbb{C}$-vector subspace generated by the elements $\pi(n_I)(v) - v$ ($n_I \in N_I(F), v \in \mathcal{V}_n$). As in (D.3), one sees easily that

$$\mathcal{V}_n(N_I(F)) = \bigcup_{\Gamma} \mathrm{Ker}(\pi_\Gamma : \mathcal{V}_n \to \mathcal{V}_n)$$

where Γ runs through the set of all compact open subgroups of $N_I(F)$, so that

$$\mathcal{V}_n(N_I(F)) = \mathcal{V}_n \cap \mathcal{V}(N_I(F))$$

and

$$\mathcal{V}_n/\mathcal{V}_n(N_I(F)) \subset \mathcal{V}/\mathcal{V}(N_I(F))$$

$(n = 0, 1, \ldots, N)$. Let us set

$$(\mathcal{W}, \rho) = r_\Delta^I(\mathcal{V}, \pi) = r_\Delta^I(I(\chi), i(\chi))$$

and

$$\mathcal{W}_n = \mathcal{V}_n/\mathcal{V}_n(N_I(F)) \subset \mathcal{V}/\mathcal{V}(N_I(F)) = \mathcal{W}$$

$(n = 0, 1, \ldots, N)$. We have

$$(0) = \mathcal{W}_0 \subset \mathcal{W}_1 \subset \cdots \subset \mathcal{W}_N = \mathcal{W}$$

and

$$\rho(M_I(F))(\mathcal{W}_n) \subset \mathcal{W}_n \quad (n = 0, 1, \ldots, N).$$

In other words, we have a filtration

$$F_n = (\mathcal{W}_n, \rho|\mathcal{W}_n) \quad (n = 0, 1, \ldots, N)$$

of $(\mathcal{W}, \rho)$ in $\mathrm{Rep}_s(M_I(F))$. We want to prove that, for each $n = 1, \ldots, N$, gr_n^F is isomorphic to $i_\emptyset^I(\mathbb{C}, w_n^{-1}(\chi))$.

Let us fix $n \in \{1, \ldots, N\}$ and let us denote by $\overline{\mathcal{V}}_n$ the $\mathbb{C}$-vector space of locally constant functions

$$\overline{v} : P_I(F) \to \mathbb{C}$$

such that

$$\overline{v}(tn_\emptyset^I \nu_I p_I) = w_n^{-1}(\delta_{B(F)}^{1/2}\chi)(t)\overline{v}(p_I)$$

for any $t \in T(F)$, $n_\emptyset^I \in N_\emptyset^I(F)$, $\nu_I \in (\dot{w}_n^{-1} U(F) \dot{w}_n) \cap N_I(F)$ and $p_I \in P_I(F)$ and such that the support of $\overline{v}$ is compact modulo $(\dot{w}_n^{-1} B(F) \dot{w}_n) \cap P_I(F)$ (recall that

$$(\dot{w}_n^{-1} B(F) \dot{w}_n) \cap P_I(F) = T(F)N_\emptyset^I(F)((\dot{w}_n^{-1} U(F) \dot{w}_n) \cap N_I(F));$$

see [Ca] (2.8.7)). We have an exact sequence

$$0 \to \mathcal{V}_{n-1} \overset{\alpha}{\hookrightarrow} \mathcal{V}_n \overset{\beta}{\to} \overline{\mathcal{V}}_n$$

where α is the inclusion and

$$\beta(v)(p_I) = v(\dot{w}_n p_I)$$

for each $v \in \mathcal{V}_n$ and each $p_I \in P_I(F)$ (if $v \in \mathcal{V}_n$, $\mathrm{Supp}(v) \cap (B(F) \, \dot{w}_n \, P_I(F))$ is compact modulo $B(F)$ so that $\mathrm{Supp}(\beta(v))$ is compact modulo $(\dot{w}_n^{-1} \, B(F) \dot{w}_n) \cap P_I(F)$). Moreover, β is surjective. Indeed, we have a commutative square of $\mathbb{C}$-linear maps

$$
\begin{array}{ccc}
\mathbb{C} \otimes \mathcal{C}_c^\infty (\overset{n}{\underset{i=1}{\bigcup}} B(F) \, \dot{w}_i \, P_I(F)) & \longrightarrow & \mathbb{C} \otimes \mathcal{C}_c^\infty (B(F) \, \dot{w}_n \, P_I(F)) \\
\mathcal{P} \Big\downarrow & & \overline{\mathcal{P}} \Big\downarrow \\
\mathcal{V}_n & \xrightarrow{\quad\quad \beta \quad\quad} & \overline{\mathcal{V}}_n
\end{array}
$$

where

$$
\mathcal{P}(\varphi)(g) = \int_{T(F)} (\delta_{B(F)}^{-1/2} \chi^{-1})(t) \int_{U(F)} \varphi(utg) \, du \, dt
$$

and

$$
\overline{\mathcal{P}}(\overline{\varphi})(p_I) = \int_{T(F)} (\delta_{B(F)}^{-1/2} \chi^{-1})(t) \int_{U(F)} \overline{\varphi}(ut \, \dot{w}_n \, p_I) \, du \, dt
$$

and where the top horizontal arrow is the restriction; $\mathcal{P}$ and $\overline{\mathcal{P}}$ are surjective (see (D.3.4)); therefore, it suffices to check that the top horizontal arrow is also surjective; but any covering of a compact subset of a locally compact, totally discontinuous, separated, topological space by open subsets has a finite refinement of pairwise disjoint compact open subsets and that restriction is clearly surjective. The action $\pi|P_I(F)$ of $P_I(F)$ on $\mathcal{V}_n$ induces an action $\overline{\pi}_n$ of $P_I(F)$ on $\overline{\mathcal{V}}_n$. Then gr_n^F is nothing else than the $\mathbb{C}$-vector space

$$
\overline{\mathcal{V}}_n / \overline{\mathcal{V}}_n(N_I(F))
$$

together with the action of $M_I(F)$ which is induced by

$$
\delta_{P_I(F)}^{-1/2} \otimes (\overline{\pi}_n | M_I(F)).
$$

Now let us choose a Haar measure $d\nu_I$ on $(\dot{w}_n^{-1} \, U(F) \, \dot{w}_n) \cap N_I(F)$ (for example, we can take the Haar measure which gives the volume 1 to $(\dot{w}_n^{-1} \, U(\mathcal{O}) \, \dot{w}_n) \cap N_I(\mathcal{O})$). For each $\overline{v} \in \overline{\mathcal{V}}_n$, let us set

$$
\mathcal{T}_n(\overline{v})(m_I) = \int_{((\dot{w}_n^{-1} U(F) \dot{w}_n) \cap N_I(F)) \backslash N_I(F)} \delta_{P_I(F)}^{-1/2} (m_I) \overline{v}(n_I m_I) \frac{dn_I}{d\nu_I}
$$

($m_I \in M_I(F)$) (this integral is absolutely convergent as the function $n_I \mapsto \overline{v}(n_I m_I)$ is compactly supported modulo $(\dot{w}_n^{-1} \, U(F) \, \dot{w}_n) \cap N_I(F)$). It is easy to see that the function $\mathcal{T}_n(\bar{v})$ on $M_I(F)$ belongs to the space of $i_\emptyset^I(\mathbb{C}, w_n^{-1}(\chi))$, that $\mathcal{T}_n$ vanishes identically on $\bar{\mathcal{V}}_n(N_I(F))$ and that $\mathcal{T}_n$ induces an intertwining operator

$$\tau_n : \mathrm{gr}_n^F \to i_\emptyset^I(\mathbb{C}, w_n^{-1}(\chi))$$

in $\mathrm{Rep}_s(M_I(F))$ (we have

$$\frac{\delta_{B(F)}^{1/2}(t)}{w_n^{-1}(\delta_{B(F)}^{1/2})(t)} = \left| \det\left(\mathrm{Ad}(t), \frac{\mathrm{Lie}(N_I(F))}{\mathrm{Lie}((\dot{w}_n^{-1} U(F) \dot{w}_n) \cap N_I(F))} \right) \right|$$

and

$$\delta_{B(F)}(t) = \delta_{P_\emptyset^I(F)}(t)\delta_{P_I(F)}(t)$$

for any $t \in T(F)$). The operator τ_n is an epimorphism. Indeed, if $\mathcal{V}_n^I$ is the space of $i_\emptyset^I(\mathbb{C}, \chi)$, we have a commutative square of $\mathbb{C}$-linear maps

$$
\begin{array}{ccc}
\mathbb{C} \otimes \mathcal{C}_c^\infty(P_I(F)) & \xrightarrow{\;\mathcal{S}\;} & \mathbb{C} \otimes \mathcal{C}_c^\infty(M_I(F)) \\
\Big\downarrow{\scriptstyle Q_n} & & \Big\downarrow{\scriptstyle \mathcal{P}_n^I} \\
\mathcal{V}_n & \xrightarrow[\;\mathcal{T}_n\;]{} & \mathcal{V}_n^I
\end{array}
$$

where

$$\mathcal{S}(\varphi)(m_I) = \int_{N_I(F)} \delta_{P_I(F)}^{-1/2}(m_I)\varphi(n_I m_I)dn_I,$$

$$Q_n(\varphi)(p_I) = \int_{T(F)} w_n^{-1}(\delta_{B(F)}^{-1/2}\chi^{-1})(t)$$

$$\times \int_{N_\emptyset^I(F)} \int_{(\dot{w}_n^{-1}U(F)\dot{w}_n)\cap N_I(F)} \varphi(n_\emptyset^I \nu_I t p_I)d\nu_I dn_\emptyset^I dt$$

and

$$\mathcal{P}_n^I(\psi)(m_I) = \int_{T(F)} (\delta_{P_\emptyset^I(F)}^{-1/2} w_n^{-1}(\chi^{-1}))(t) \int_{N_\emptyset^I(F)} \psi(n_\emptyset^I t m_I)dn_\emptyset^I dt;$$

as $\mathcal{P}_n^I$ and $\mathcal{S}$ are surjective (see (D.3.4) for $\mathcal{P}_n^I$ and remark that

$$\psi \mapsto (n_I m_I \mapsto \delta_{P_I(F)}^{1/2}(m_I)1_{N_I(\mathcal{O})}(n_I)\varphi(m_I))$$

is a section of $\mathcal{S}$), it follows that $\mathcal{T}_n$ is surjective. Finally, the operator $\mathcal{T}_n$ is a monomorphism, i.e. $\mathrm{Ker}(\mathcal{T}_n) \subset \overline{\mathcal{V}}_n(N_I(F))$. Indeed, let $\bar{v} \in \mathrm{Ker}(\mathcal{T}_n)$ and let us check that there exists a compact open subgroup Γ of $N_I(F)$ with

$$\int_\Gamma \bar{v}(p_I n_I) dn_I = 0$$

for all $p_I \in P_I(F)$, so that $\bar{v} \in \overline{\mathcal{V}}_n(N_I(F))$. There exists a finite family $(p_{I,1}, \ldots, p_{I,A})$ of elements of $P_I(F)$ such that

$$\mathrm{Supp}(\bar{v}) \subset \bigcup_{a=1}^{A} ((\dot{w}_n^{-1} B(F) \dot{w}_n) \cap P_I(F)) p_{I,a} P_I(\mathcal{O}).$$

Thanks to the Iwasawa decomposition

$$M_I(F) = P_\emptyset^I(F) M_I(\mathcal{O})$$

and the inclusion

$$P_\emptyset^I(F) \subset (\dot{w}_n^{-1} B(F) \dot{w}_n) \cap P_I(F),$$

we can (and we will) assume that

$$p_{I,a} = n_{I,a} \in N_I(F) \quad (a = 1, \ldots, A).$$

Then there exists a compact open subgroup of $N_I(F)$ such that

$$m_I^{-1} n_{I,a} m_I \in \Gamma$$

for all $m_I \in M_I(\mathcal{O})$ and all $a \in \{1, \ldots, A\}$ and such that

$$\Gamma \supset M_I(\mathcal{O}).$$

In fact, let us take Γ of the form $z_I N_I(\mathcal{O}) z_I^{-1}$ for some $z_I \in Z_I(F)$ (see (7.1)), so that

$$m_I \Gamma m_I^{-1} = \Gamma$$

for each $m_I \in M_I(\mathcal{O})$. We have

$$\mathrm{Supp}(\bar{v}) \subset ((\dot{w}_n^{-1} B(F) \dot{w}_n) \cap P_I(F)) M_I(\mathcal{O}) \Gamma$$

and, for each $p_I \in P_I(F)$, either

$$p_I \notin ((\dot{w}_n^{-1} B(F) \dot{w}_n) \cap P_I(F)) M_I(\mathcal{O}) \Gamma$$

and
$$\int_\Gamma \bar{v}(p_I n_I) dn_I = 0 \ ,$$

or
$$p_I = tn_\emptyset^I \nu_I m_I n_I' \ ,$$

with $t \in T(F)$, $n_\emptyset^I \in N_\emptyset^I(F)$, $\nu_I \in (\dot{w}_n^{-1} U(F) \dot{w}_n) \cap N_I(F)$, $m_I \in M_I(\mathcal{O})$ and $n_I' \in \Gamma$, and

$$\int_\Gamma \bar{v}(p_I n_I) dn_I = w_n^{-1}(\delta_{B(F)}^{1/2} \chi)(t) \int_\Gamma \bar{v}(m_I n_I'') dn_I''$$

$$= w_n^{-1}(\delta_{B(F)}^{1/2} \chi)(t) \int_\Gamma \bar{v}(n_I''' m_I) dn_I'''$$

$(n_I'' = n_I' n_I, \ n_I''' = m_I n_I'' m_I^{-1})$. But this last integral is zero as the support of the function

$$N_I(F) \to \mathbb{C}, \ n_I''' \mapsto \bar{v}(n_I''' m_I)$$

is contained in

$$((\dot{w}_n^{-1} U(F) \dot{w}_n) \cap N_I(F))\Gamma \subset N_I(F)$$

and as

$$\int_{((\dot{w}_n^{-1} U(F) \dot{w}_n) \cap N_I(F)) \backslash N_I(F)} \bar{v}(n_I''' m_I) \frac{dn_I'''}{d\nu_I} = 0$$

for all $m_I \in M_I(F)$ by hypothesis. $\qquad\square$

Proof of (7.3.3) : Let us begin with the following remark. Let $(\mathcal{V}, \pi)$ be a non-zero subquotient of $(I(\chi), i(\chi))$, let $I \subset \Delta$ be such that $r_\Delta^I(\mathcal{V}, \pi)$ is non-zero and quasi-cuspidal (see (7.2.3) and (7.1.3) (iv)) and let $(\mathcal{W}, \rho)$ be an irreducible subquotient of $r_\Delta^I(\mathcal{V}, \pi)$ in $\text{Rep}_s(M_I(F))$ (there always exists at least one irreducible subquotient of any smooth representation of $M_I(F)$, i.e. any non-degenerate $(\mathbb{C} \otimes \mathcal{C}_c^\infty(M_I(F)))$-module has at least one irreducible subquotient (see (D.1.2)): indeed replacing the module by a submodule we can assume that it is finitely generated and then we apply Zorn's lemma). Then $(\mathcal{W}, \rho)$ is cuspidal (see (7.2.3), (7.1.3) (i) and (D.4.2)) and there exists $w \in D_{\emptyset, I}$ such that $(\mathcal{W}, \rho)$ is an irreducible subquotient of $i_\emptyset^I(\mathbb{C}, w^{-1}(\chi))$ (see (7.1.3) (i) and (7.3.4)). So there exists a subrepresentation of $i_\emptyset^I(\mathbb{C}, w^{-1}(\chi))$, for some $w \in D_{\emptyset, I}$, which is isomorphic to $(\mathcal{W}, \rho)$ (see (D.4.11): $i_\emptyset^I(\mathbb{C}, w^{-1}(\chi))$ admits a central character). In particular, we have

$$\text{Hom}_{\text{Rep}_s(M_I(F))}((\mathcal{W}, \rho), i_\emptyset^I(\mathbb{C}, w^{-1}(\chi))) \neq (0)$$

for some $w \in D_{\emptyset, I}$ and we get (see (7.1.3) (i))

$$\text{Hom}_{\text{Rep}_s(T(F))}(r_I^\emptyset(\mathcal{W}, \rho), (\mathbb{C}, w^{-1}(\chi))) \neq (0)$$

for some $w \in D_{\emptyset,I}$. But $(\mathcal{W}, \rho)$ is cuspidal and irreducible. So $I = \emptyset$ (see (7.2.3)) and $(\mathcal{W}, \rho)$ is isomorphic to $(\mathbb{C}, w^{-1}(\chi))$ for some $w \in W$.

Now let us prove part (i) of (7.3.3). It follows from the above remark that for any non-zero subquotient $(\mathcal{V}, \pi)$ of $(I(\chi), i(\chi))$, we have $r_\Delta^\emptyset(\mathcal{V}, \pi) \neq (0)$. Therefore, the length of $(I(\chi), i(\chi))$ is bounded by the length of $r_\Delta^\emptyset(I(\chi), i(\chi))$ (see (7.1.3) (i)) and part (i) of (7.3.3) is a direct consequence of (7.3.4).

Part (ii) (a) of (7.3.3) is also a direct consequence of the above remark.

Let us prove part (iii) of (7.3.3). By adjunction (see (7.1.3) (i)), we have

$$\mathrm{Hom}_{\mathrm{Rep}_s(G(F))}((I(\chi), i(\chi)), (I(\chi'), i(\chi')))$$
$$\cong \mathrm{Hom}_{\mathrm{Rep}_s(T(F))}(r_\Delta^\emptyset(I(\chi), i(\chi)), (\mathbb{C}, \chi')).$$

Now, as $\chi' \in W(\chi)$, $(\mathbb{C}, \chi')$ is isomorphic to a subquotient of $r_\Delta^\emptyset(I(\chi), i(\chi))$ (see (7.3.4)) and $(\mathbb{C}, \chi')$ is isomorphic to a quotient of $r_\Delta^\emptyset(I(\chi), i(\chi))$ (see (D.4.11): $r_\Delta^\emptyset(I(\chi), i(\chi))$ has finite length thanks to (7.3.4). It follows that the above Hom is non-zero. Moreover, as the multiplicity of $(\mathbb{C}, \chi')$ (up to isomorphism) in $r_\Delta^\emptyset(I(\chi), i(\chi))$ is at most $|W(\chi, \chi')|$ (see (7.3.4)), the dimension of this Hom is at most $|W(\chi, \chi')|$.

To prove part (ii) (b) of (7.3.3), we can assume that $\chi' = s(\chi)$ for some simple reflection (induction on the length of $w \in W$ such that $\chi' = w(\chi)$). Let $\alpha \in \Delta$ be the corresponding simple root. As $i_{\{\alpha\}}^\Delta i_\emptyset^{\{\alpha\}} \cong i_\emptyset^\Delta$ (see (7.1.3) (iv)) and as $i_{\{\alpha\}}^\Delta$ is exact (see (7.1.3) (i)), to prove part (ii) (b) of (7.3.3), it suffices to prove that the semi-simplifications of $i_\emptyset^{\{\alpha\}}(\mathbb{C}, \chi)$ and $i_\emptyset^{\{\alpha\}}(\mathbb{C}, \chi')$ are isomorphic. Therefore, to prove part (ii) (b) of (7.3.3) we can (and we will) assume that $d = 2$ and that $\chi' = s(\chi)$ where s is the non-trivial element of W. Let us choose non-zero homomorphisms

$$\nu : (I(\chi), i(\chi)) \to (I(\chi'), i(\chi'))$$

and

$$\nu' : (I(\chi'), i(\chi')) \to (I(\chi), i(\chi))$$

(see part (iii) of (7.3.3) which is already proved). If ν or ν' is an isomorphism, we are done. Otherwise, we have exact sequences

$$
\begin{array}{ccccccc}
0 & \to & \mathrm{Ker}(\nu) & \to & (I(\chi), i(\chi)) & \to & \mathrm{Im}(\nu) & \to & 0, \\
0 & \to & \mathrm{Im}(\nu) & \to & (I(\chi'), i(\chi')) & \to & \mathrm{Coker}(\nu) & \to & 0, \\
0 & \to & \mathrm{Ker}(\nu') & \to & (I(\chi'), i(\chi')) & \to & \mathrm{Im}(\nu') & \to & 0, \\
0 & \to & \mathrm{Im}(\nu') & \to & (I(\chi), i(\chi)) & \to & \mathrm{Coker}(\nu') & \to & 0,
\end{array}
$$

where $\mathrm{Ker}(\nu)$, $\mathrm{Im}(\nu)$, $\mathrm{Coker}(\nu)$, $\mathrm{Ker}(\nu')$, $\mathrm{Im}(\nu')$ and $\mathrm{Coker}(\nu')$ are smooth irreducible representations of $G(F)$. Indeed, these representations are non-zero and the lengths of $(I(\chi), i(\chi))$ and $(I(\chi'), i(\chi'))$ are at most 2 (see part (i) of (7.3.3) which is already proved). Now either $\chi' = \chi$ and we are done, or $\chi' \neq \chi$ and $r_\Delta^0(\mathrm{Ker}(\nu))$, $r_\Delta^0(\mathrm{Coker}(\nu))$ and $r_\Delta^0(\mathrm{Im}(\nu'))$ (resp. $r_\Delta^0(\mathrm{Ker}(\nu'))$, $r_\Delta^0(\mathrm{Coker}(\nu'))$ and $r_\Delta^0(\mathrm{Im}(\nu))$) are isomorphic to $(\mathbb{C}, \chi)$ (resp. $(\mathbb{C}, \chi')$). Indeed, thanks to (7.1.3) (i), there exist non-trivial maps

$$
\begin{aligned}
r_\Delta^0(\mathrm{Ker}(\nu)) &\to (\mathbb{C}, \chi), \\
r_\Delta^0(\mathrm{Im}(\nu)) &\to (\mathbb{C}, \chi'), \\
r_\Delta^0(\mathrm{Ker}(\nu')) &\to (\mathbb{C}, \chi'), \\
r_\Delta^0(\mathrm{Im}(\nu')) &\to (\mathbb{C}, \chi)
\end{aligned}
$$

in $\mathrm{Rep}_s(T(F))$, thanks to (7.3.4), the semi-simplifications of $r_\Delta^0(I(\chi), i(\chi))$ and $r_\Delta^0(I(\chi'), i(\chi'))$ are isomorphic to

$$(\mathbb{C}, \chi) \oplus (\mathbb{C}, \chi')$$

and r_Δ^0 is exact (see (7.1.3) (i)). As the semi-simplification of $(I(\chi), i(\chi))$ (resp. $(I(\chi'), i(\chi'))$) is isomorphic to

$$\mathrm{Ker}(\nu) \oplus \mathrm{Im}(\nu)$$

and

$$\mathrm{Im}(\nu') \oplus \mathrm{Coker}(\nu')$$

(resp.

$$\mathrm{Im}(\nu) \oplus \mathrm{Coker}(\nu)$$

and

$$\mathrm{Ker}(\nu') \oplus \mathrm{Im}(\nu')),$$

we must have isomorphisms

$$\mathrm{Ker}(\nu) \cong \mathrm{Im}(\nu') \cong \mathrm{Coker}(\nu)$$

and

$$\mathrm{Coker}(\nu') \cong \mathrm{Im}(\nu) \cong \mathrm{Ker}(\nu').$$

This completes the proof of part (ii) (b) of (7.3.3). $\qquad\qquad\square$

Another consequence of (7.3.4) is the following useful proposition.

PROPOSITION (7.3.5). — *Let $(\mathcal{V}, \pi)$ be an irreducible subquotient of $(I(\chi), i(\chi))$ for some smooth character χ of $T(F)$. Then there exists $w \in W$ such that $(\mathcal{V}, \pi)$ is a subrepresentation (resp. a quotient) of $(I(w(\chi)), i(w(\chi)))$.*

Proof : The remark at the beginning of the proof of (7.3.3) tells us that $r_\Delta^0(\mathcal{V}, \pi) \neq (0)$. As r_Δ^0 is exact (see (7.1.3) (i)), $r_\Delta^0(\mathcal{V}, \pi)$ is a subquotient of $r_\Delta^0(I(\chi), i(\chi))$. Thanks to (7.3.4) it follows that, for some $w \in W$, $(\mathbb{C}, w(\chi))$ is isomorphic to a subquotient of $r_\Delta^0(\mathcal{V}, \pi)$. Applying (D.4.11) ($r_\Delta^0(\mathcal{V}, \pi)$ has finite length as a subquotient of $r_\Delta^0(I(\chi), i(\chi))$), we get that there exists an epimorphism

$$r_\Delta^0(\mathcal{V}, \pi) \longrightarrow (\mathbb{C}, w(\chi))$$

for some $w \in W$. By adjunction (see (7.1.3) (i)), it follows that there exists a non-zero map

$$(\mathcal{V}, \pi) \to (I(w(\chi)), i(w(\chi))).$$

As $(\mathcal{V}, \pi)$ is irreducible, this map is a monomorphism.

Now, if $(\mathcal{V}, \pi)$ is a subquotient of $(I(\chi), i(\chi))$, its contragredient representation $(\widetilde{\mathcal{V}}, \tilde{\pi})$ is a subquotient of $(\widetilde{I}(\chi), \widetilde{i}(\chi))$ (see (D.1)) which is canonically isomorphic to $(I(\chi^{-1}), i(\chi^{-1}))$ (see (7.1.3) (ii)). As before, we can find $w \in W$ and an embedding

$$(\widetilde{\mathcal{V}}, \tilde{\pi}) \hookrightarrow (I(w(\chi^{-1})), i(\chi^{-1})).$$

As the target of this embedding is admissible, by biduality (see (D.2.3)), we get an epimorphism

$$(I(w(\chi)), i(w(\chi))) \longrightarrow (\mathcal{V}, \pi)$$

(see (7.1.3) (ii)) and the proposition is completely proved. □

(7.4) Unramified principal series representations

An unramified character

$$\chi : T(F) \to \mathbb{C}$$

is a (smooth) character which is trivial on the maximal compact subgroup $T(\mathcal{O})$ of $T(F)$.

For each unramified character χ of $T(F)$, $(I(\chi), i(\chi))$ is the corresponding **unramified principal series representation** of $G(F)$.

Let $\mathcal{B}^\circ \subset G(F)$ be the standard Iwahori subgroup (see (5.1)). We have

$$(7.4.1) \qquad G(F) = B(F)G(\mathcal{O}) = \coprod_{w \in W} B(F) \, \dot{w} \, \mathcal{B}^\circ$$

(see (5.4.4)). Therefore, for any smooth character χ of $T(F)$, the map

$$(7.4.2) \qquad I(\chi)^{\mathcal{B}^\circ} \to \mathbb{C}^W, \quad v \mapsto (v(\dot{w}))_{w \in W}$$

is a $\mathbb{C}$-linear injective morphism.

LEMMA (7.4.3). — *If χ is an unramified character of $T(F)$, the map (7.4.2) is bijective.*

Proof : We have
$$B(F) \cap \dot{w}\, \mathcal{B}^{\circ}\, \dot{w}^{-1} \subset B(\mathcal{O}).$$

$\square$

THEOREM (7.4.4) (Casselman). — *Let $(\mathcal{V}, \pi)$ be a smooth irreducible representation of $G(F)$. Then the following conditions are equivalent :*

(i) $\mathcal{V}^{\mathcal{B}^{\circ}} \neq (0)$;

(ii) $(\mathcal{V}, \pi)$ *is isomorphic to a subrepresentation of $(I(\chi), i(\chi))$ for some unramified character χ of $T(F)$.*

In the proof of this theorem we will need the following property of the Hecke algebra

(7.4.5) $$\widetilde{\mathcal{H}} = \mathcal{C}_c^{\infty}(G(F)/\!\!/\mathcal{B}^{\circ})$$

of $\mathcal{B}^{\circ}$-bi-invariant functions $f : G(F) \rightarrow \mathbb{Q}$ with compact support (the product is the convolution with respect to the Haar measure dg on $G(F)$ which gives the volume 1 to $G(\mathcal{O})$):

PROPOSITION (7.4.6). — *For any $g \in G(F)$, the characteristic function*
$$1_{\mathcal{B}^{\circ}g\mathcal{B}^{\circ}} \in \widetilde{\mathcal{H}}$$
is invertible in $\mathbb{C} \otimes \widetilde{\mathcal{H}}$ (for the convolution product).

First of all, assuming this proposition, let us give the proof of the theorem (7.4.4).

Proof of (7.4.4) : Let us set $r_{\Delta}^{\emptyset}(\mathcal{V}, \pi) = (\mathcal{W}, \rho)$. As $(\mathcal{V}, \pi)$ is admissible (see (7.2.4) (ii)), the same is true for $(\mathcal{W}, \rho)$ (see (7.1.4) (i)). We have a canonical splitting
$$\mathcal{W} = \mathcal{W}^{T(\mathcal{O})} \oplus \mathrm{Ker}(\rho(e_{T(\mathcal{O})}))$$
of $\mathcal{W}$ as a representation of $T(\mathcal{O})$. But $T(F)$ is commutative, so it is also a splitting of $\mathcal{W}$ as a representation of $T(F)$ (see (D.1.3)). Therefore, there exists an unramified character χ of $T(F)$ such that
$$\mathrm{Hom}_{\mathrm{Rep}_s(T(F))}((\mathcal{W}, \rho), (\mathbb{C}, \chi)) \neq (0)$$
if and only if there exists an unramified character χ of $T(F)$ such that
$$\mathrm{Hom}_{\mathrm{Rep}_s(T(F))}((\mathcal{W}^{T(\mathcal{O})}, \rho|\mathcal{W}^{T(\mathcal{O})}), (\mathbb{C}, \chi)) \neq (0).$$

Obviously, this last condition is satisfied if and only if the finite dimensional $\mathbb{C}$-vector space $\mathcal{W}^{T(\mathcal{O})}$ is non-zero. By adjunction, it follows that condition (ii) of the theorem is equivalent to

$$\mathcal{W}^{T(\mathcal{O})} \neq (0).$$

Now $\mathcal{B}^\circ$ is in good position with respect to $(U(F), T(F), \widetilde{U}(F))$, i.e.

$$\mathcal{B}^\circ = (\mathcal{B}^\circ \cap U(F))(\mathcal{B}^\circ \cap T(F))(\mathcal{B}^\circ \cap \widetilde{U}(F))$$

with

$$\mathcal{B}^\circ \cap U(F) = U(\mathcal{O}),$$
$$\mathcal{B}^\circ \cap T(F) = T(\mathcal{O})$$

and

$$\mathcal{B}^\circ \cap \widetilde{U}(F) = \widetilde{U}(\mathcal{O}) \cap (1 + \varpi gl_d(\mathcal{O}))$$

and the hypotheses of Jacquet's theorem (D.3.8) are satisfied (see (7.1)). Therefore, we have

$$p_{U(F)}(\mathcal{V}^{\mathcal{B}^\circ}) = \mathcal{W}^{T(\mathcal{O})}$$

where

$$p_{U(F)} : \mathcal{V} \to \mathcal{V}/\mathcal{V}(U(F)) = \mathcal{W}$$

is the canonical projection. Moreover, if $\varepsilon \in]0,1]$ is small enough, the hypotheses of (D.3.9) are satisfied by any $t \in \mathcal{Z}_\emptyset^\emptyset(\varepsilon) \subset T(F)$ (see (7.1)) and, thanks to (7.4.6), we can apply the remark (D.3.10.2). It follows that $p_{U(F)}$ maps $\mathcal{V}^{\mathcal{B}^\circ}$ isomorphically onto $\mathcal{W}^{T(\mathcal{O})}$ and the theorem is proved. $\square$

Before we give the proof of (7.4.6), we need to recall the presentation of Iwahori and Matsumoto of $\widetilde{\mathcal{H}}$.

The Weyl group $W = \mathfrak{S}_d$ of (G, T) is generated by the set

$$S = \{s_1, \ldots, s_{d-1}\} \subset W$$

of simple reflexions $(s_i = (i, i+1), i = 1, \ldots, d-1)$ and the relations are

$$(s_i s_j)^{m_{ij}} = 1$$

where

$$m_{ij} = \begin{cases} 1 & \text{if} \quad i = j, \\ 2 & \text{if} \quad i+1 < j, \\ 3 & \text{if} \quad i+1 = j, \end{cases}$$

for $1 \leq i \leq j \leq d-1$ (see [Bou] Lie, Ch. VI).

The group W acts on $\mathbb{Z}^d$ by

$$w.\lambda = (\lambda_{w^{-1}(1)}, \ldots, \lambda_{w^{-1}(d)})$$

$(w \in W, \lambda = (\lambda_1, \ldots, \lambda_d) \in \mathbf{Z}^d)$ and the **affine Weyl group** is the semi-direct product

$$(7.4.7) \qquad \widetilde{W} = \mathbf{Z}^d \rtimes W .$$

We will use the notation $e^\lambda w$ for an element of $\widetilde{W}$ $(\lambda \in \mathbf{Z}^d, w \in W)$. We have

$$e^{\lambda'} w' e^{\lambda''} w'' = e^{\lambda' + w'(\lambda'')} w' w''$$

for all $\lambda', \lambda'' \in \mathbf{Z}^d$ and all $w', w'' \in W$. We set

$$\widetilde{S} = \{s_0\} \cup S$$

where

$$s_0 = e^{(-1,0,\ldots,0,1)} (1, d)$$

and

$$\Omega = \{\omega^i | i \in \mathbf{Z}\}$$

where

$$\omega = e^{(0,\ldots,0,1)} \begin{pmatrix} 1 & 2 & \cdots & d \\ d & 1 & \cdots & d-1 \end{pmatrix}.$$

We define the **length function**

$$\ell : \widetilde{W} \to \mathbf{N}$$

by

$$\ell(e^\lambda w) = \sum_{\substack{1 \le i < j \le d \\ w^{-1}(i) > w^{-1}(j)}} |\lambda_i - \lambda_j + 1| + \sum_{\substack{1 \le i < j \le d \\ w^{-1}(i) < w^{-1}(j)}} |\lambda_i - \lambda_j|.$$

Then it is easy to see that

$$\Omega = \{\tilde{w} \in \widetilde{W} | \ell(\tilde{w}' \tilde{w}) = \ell(\tilde{w} \tilde{w}') = \ell(\tilde{w}'), \forall \tilde{w}' \in \widetilde{W}\},$$

that the center of $\widetilde{W}$ is

$$\Omega^d = \{\omega^{di} | i \in \mathbf{Z}\} \subset \Omega$$

and that

$$\Omega \widetilde{S} = \{\tilde{w} \in \widetilde{W} | \ell(\tilde{w}) = 1\}.$$

The subset $\widetilde{S}$ of $\widetilde{W}$ generates the normal subgroup

$$W_a = \{\lambda \in \mathbf{Z}^d | \lambda_1 + \cdots + \lambda_d = 0\} \rtimes W$$

of $\widetilde{W}$ and $(W_a, \widetilde{S})$ is a Coxeter group with relations

$$(s_i s_j)^{m_{ij}} = 1$$

where

$$m_{ij} = \begin{cases} 0 & \text{if} \quad (i,j) = (0,1) \text{ and } d = 2, \\ 1 & \text{if} \quad i = j, \\ 2 & \text{if} \quad i+1 < j \text{ and } (i,j) \neq (0, d-1), \\ 3 & \text{if} \quad i+1 = j \text{ or if } (i,j) = (0, d-1) \text{ and } d > 2, \end{cases}$$

for $0 \leq i \leq j \leq d-1$ (see [Bou] Lie, Ch. VI). The length function of $(W_a, \widetilde{S})$ is the restriction of ℓ to W_a. The group $\widetilde{W}$ is the semi-direct product $W_a \rtimes \Omega$.

The braid group of $\widetilde{W}$ is the group $\mathcal{T}_{\widetilde{W}}$ with generators $T_{\tilde{w}}$ ($\tilde{w} \in \widetilde{W}$) and relations

$$T_{\tilde{w}'} T_{\tilde{w}''} = T_{\tilde{w}' \tilde{w}''}$$

for all $\tilde{w}', \tilde{w}'' \in \widetilde{W}$ such that

$$\ell(\tilde{w}') + \ell(\tilde{w}'') = \ell(\tilde{w}' \tilde{w}'').$$

In the group algebra (over $\mathbf{Q}$),

$$\mathbf{Q}[\mathcal{T}_{\widetilde{W}}],$$

of the braid group we consider the two-sided ideal

$$\mathcal{I}_{\widetilde{S}} \subset \mathbf{Q}[\mathcal{T}_{\widetilde{W}}]$$

generated by the elements

$$(T_{\tilde{s}} + 1)(T_{\tilde{s}} - q) \quad (\tilde{s} \in \widetilde{S}).$$

We can form the quotient $\mathbf{Q}$-algebra

(7.4.8) $$\mathbf{Q}[\mathcal{T}_{\widetilde{W}}]/\mathcal{I}_{\widetilde{S}} .$$

We can give another description of $\widetilde{W}$ which is more closely related to $G(F)$. Let $N_G(T)$ be the normalizer of T in G. It is a group scheme over $\mathcal{O}$ with connected component of the identity T and with group of connected component W. For each $w \in W$, let

$$\dot{w} \in N_G(T)(\mathcal{O}) \subset K$$

be the corresponding permutation matrix. The map

$$W \to N_G(T)(\mathcal{O}), \quad w \mapsto \dot{w} ,$$

is a group homomorphism and splits the exact sequence

$$1 \to T(F) \to N_G(T)(F) \to W \to 1 \ .$$

It follows that we have a group isomorphism

$$(T(F)/T(\mathcal{O}))\rtimes W \xrightarrow{\sim} N_G(T)(F)/T(\mathcal{O}).$$

For each $\lambda \in \mathbf{Z}^d$, let

$$\varpi^\lambda = \mathrm{diag}(\varpi^{\lambda_1},\ldots,\varpi^{\lambda_d}) \in T(F)$$

(recall that ϖ is a given uniformizer of $\mathcal{O}$). The map

$$\mathbf{Z}^d \to T(F)/T(\mathcal{O}), \ \lambda \mapsto \varpi^\lambda T(\mathcal{O}),$$

is a group isomorphism. For each $\tilde{w} = e^\lambda w \in \widetilde{W}$, let

$$\overset{\hat{}}{\tilde{w}} = \varpi^\lambda \ \hat{w}\in N_G(T)(F).$$

Then the map

$$\widetilde{W} \to N_G(T)(F), \ \tilde{w} \to \overset{\hat{}}{\tilde{w}} \ ,$$

is a group homomorphism and induces a group isomorphism

$$(7.4.9) \qquad\qquad \widetilde{W} \xrightarrow{\sim} N_G(T)(F)/T(\mathcal{O})$$

which is compatible with the semi-direct product decompositions and the above isomorphisms.

Now we have the **Bruhat–Tits decomposition**

$$(7.4.10) \qquad\qquad G(F) = \coprod_{\tilde{w}\in\widetilde{W}} \mathcal{B}^\circ \overset{\hat{}}{\tilde{w}}\mathcal{B}^\circ$$

with the following properties (see [Iw–Ma]§ 2) :

$$(7.4.11) \quad \mathcal{B}^\circ \overset{\hat{}}{\tilde{w}} \mathcal{B}^\circ \overset{\hat{}}{\tilde{s}} \ \mathcal{B}^\circ = \begin{cases} \mathcal{B}^\circ \ \overset{\hat{}}{\tilde{w}\tilde{s}} \ \mathcal{B}^\circ & \text{if } \ell(\tilde{w}\tilde{s}) = \ell(\tilde{w}) + 1, \\ (\mathcal{B}^\circ \ \overset{\hat{}}{\tilde{w}\tilde{s}} \ \mathcal{B}^\circ) \amalg (\mathcal{B}^\circ \ \overset{\hat{}}{\tilde{w}} \ \mathcal{B}^\circ) & \text{otherwise,} \end{cases}$$

for each $\tilde{w} \in \widetilde{W}$ and each $\tilde{s} \in \widetilde{S}$, and

$$(7.4.12) \qquad\qquad \omega\mathcal{B}^\circ\omega^{-1} = \mathcal{B}^\circ.$$

For each $\tilde{w} \in \widetilde{W}$, let us set

$$(7.4.13) \qquad\qquad t_{\tilde{w}} = \frac{1_{\mathcal{B}^\circ \overset{\hat{}}{\tilde{w}}\mathcal{B}^\circ}}{\mathrm{vol}(\mathcal{B}^\circ, dg)} \in \widetilde{\mathcal{H}} \ .$$

Then it follows from (7.4.10) that $(t_{\tilde{w}})_{\tilde{w}\in\widetilde{W}}$ is a basis of the $\mathbf{Q}$-vector space $\widetilde{\mathcal{H}}$.

LEMMA (7.4.14). — (i) *For any $\tilde{w} \in \widetilde{W}$, we have*

$$t_{\tilde{w}} * t_{\omega} = t_{\tilde{w}\omega}.$$

(ii) *For any $\tilde{w} \in \widetilde{W}$ and $\tilde{s} \in \widetilde{S}$, we have*

$$t_{\tilde{w}} * t_{\tilde{s}} = \begin{cases} t_{\tilde{w}\tilde{s}} & \text{if } \ell(\tilde{w}\tilde{s}) = \ell(\tilde{w}) + 1, \\ qt_{\tilde{w}\tilde{s}} + (q-1)t_{\tilde{w}} & \text{otherwise.} \end{cases}$$

Proof: Part (i) follows immediately from (7.4.12).
By the definition of the convolution product, we have

$$t_{\tilde{w}} * t_{\tilde{s}} = \sum_{\tilde{w}' \in \widetilde{W}} \alpha_{\tilde{w}'} t_{\tilde{w}'}$$

with

$$\alpha_{\tilde{w}'} = |\mathcal{B}^{\circ} \backslash ((\mathcal{B}^{\circ} \overset{\cdot}{\tilde{w}}{}^{-1} \mathcal{B}^{\circ} \overset{\cdot}{\tilde{w}'}) \cap (\mathcal{B}^{\circ} \overset{\cdot}{\tilde{s}} \mathcal{B}^{\circ}))|.$$

In particular, thanks to the property (7.4.11) of the Bruhat–Tits decomposition we have $\alpha_{\tilde{w}'} = 0$ unless $\tilde{w}' = \tilde{w}\tilde{s}$ if $\ell(\tilde{w}\tilde{s}) = \ell(\tilde{w}) + 1$ and $\alpha_{\tilde{w}'} = 0$ unless $\tilde{w}' = \tilde{w}\tilde{s}$ or $\tilde{w}' = \tilde{w}$ if $\ell(\tilde{w}\tilde{s}) = \ell(\tilde{w}) - 1$.

Let us compute the remaining $\alpha_{\tilde{w}'}$'s. The map $b \mapsto \overset{\cdot}{\tilde{s}} b^{-1}$ induces a bijection from $\mathcal{B}^{\circ}/(\mathcal{B}^{\circ} \cap (\overset{\cdot}{\tilde{s}} \mathcal{B}^{\circ} \overset{\cdot}{\tilde{s}}))$ onto $\mathcal{B}^{\circ} \backslash (\mathcal{B}^{\circ} \overset{\cdot}{\tilde{s}} \mathcal{B}^{\circ})$, so that

$$\alpha_{\tilde{w}'} = |(\mathcal{B}^{\circ} \cap (\overset{\cdot}{\tilde{w}'}{}^{-1} \mathcal{B}^{\circ} \overset{\cdot}{\tilde{w}} \mathcal{B}^{\circ} \overset{\cdot}{\tilde{s}}))/(\mathcal{B}^{\circ} \cap (\overset{\cdot}{\tilde{s}} \mathcal{B}^{\circ} \overset{\cdot}{\tilde{s}}))|.$$

If $\ell(\tilde{w}\tilde{s}) = \ell(\tilde{w}) + 1$ and $\tilde{w}' = \tilde{w}\tilde{s}$, we have

$$(\overset{\cdot}{\tilde{s}} \mathcal{B}^{\circ} \overset{\cdot}{\tilde{s}} \mathcal{B}^{\circ} \overset{\cdot}{\tilde{s}}) \cap (\overset{\cdot}{\tilde{w}'}{}^{-1} \mathcal{B}^{\circ} \overset{\cdot}{\tilde{w}} \mathcal{B}^{\circ} \overset{\cdot}{\tilde{s}}) = \overset{\cdot}{\tilde{s}}\overset{\cdot}{\tilde{w}}{}^{-1}((\overset{\cdot}{\tilde{w}} \mathcal{B}^{\circ} \overset{\cdot}{\tilde{s}} \mathcal{B}^{\circ}) \cap (\mathcal{B}^{\circ} \overset{\cdot}{\tilde{w}} \mathcal{B}^{\circ})) \overset{\cdot}{\tilde{s}} = \emptyset$$

$(\overset{\cdot}{\tilde{w}} \mathcal{B}^{\circ} \overset{\cdot}{\tilde{s}} \mathcal{B}^{\circ} \subset \mathcal{B}^{\circ} \overset{\cdot}{\tilde{w}\tilde{s}} \mathcal{B}^{\circ})$ and

$$\mathcal{B}^{\circ} \subset (\overset{\cdot}{\tilde{s}} \mathcal{B}^{\circ} \overset{\cdot}{\tilde{s}}) \amalg (\overset{\cdot}{\tilde{s}} \mathcal{B}^{\circ} \overset{\cdot}{\tilde{s}} \mathcal{B}^{\circ} \overset{\cdot}{\tilde{s}})$$

$(\overset{\cdot}{\tilde{s}} \mathcal{B}^{\circ} \overset{\cdot}{\tilde{s}} \subset \mathcal{B}^{\circ} \amalg (\mathcal{B}^{\circ} \overset{\cdot}{\tilde{s}} \mathcal{B}^{\circ}))$. Therefore, we have

$$\mathcal{B}^{\circ} \cap (\overset{\cdot}{\tilde{w}'}{}^{-1} \mathcal{B}^{\circ} \overset{\cdot}{\tilde{w}} \mathcal{B}^{\circ} \overset{\cdot}{\tilde{s}}) \subset \overset{\cdot}{\tilde{s}} \mathcal{B}^{\circ} \overset{\cdot}{\tilde{s}}$$

and

$$\alpha_{\tilde{w}'} = 1 \; .$$

If $\ell(\tilde{w}\tilde{s}) = \ell(\tilde{w}) - 1$ and $\tilde{w}' = \tilde{w}\tilde{s}$, we have

$$\dot{\tilde{w}}'\mathcal{B}^\circ\,\dot{\tilde{s}} \subset \mathcal{B}^\circ\,\dot{\tilde{w}}\,\mathcal{B}^\circ$$

($\tilde{w} = \tilde{w}'\tilde{s}$ and $\ell(\tilde{w}) = \ell(\tilde{w}') + 1$), i.e.

$$\mathcal{B}^\circ \subset \dot{\tilde{w}}'^{-1}\mathcal{B}^\circ\,\dot{\tilde{w}}\,\mathcal{B}^\circ\,\dot{\tilde{s}}\,.$$

Therefore, we have

$$\alpha_{\tilde{w}'} = |\mathcal{B}^\circ/(\mathcal{B}^\circ \cap (\dot{\tilde{s}}\,\mathcal{B}^\circ\,\dot{\tilde{s}}))|.$$

If $\ell(\tilde{w}\tilde{s}) = \ell(\tilde{w}) - 1$ and $\tilde{w}' = \tilde{w}$, we have

$$(\dot{\tilde{w}}'^{-1}\mathcal{B}^\circ\,\dot{\tilde{w}}\,\mathcal{B}^\circ\,\dot{\tilde{s}}) \cap (\dot{\tilde{s}}\,\mathcal{B}^\circ\,\dot{\tilde{s}}) = \emptyset$$

$((\mathcal{B}^\circ\,\dot{\tilde{w}}\,\mathcal{B}^\circ) \cap (\dot{\tilde{w}}\dot{\tilde{s}}\,\mathcal{B}^\circ) = \emptyset)$, so that

$$\mathcal{B}^\circ \cap (\dot{\tilde{w}}'^{-1}\mathcal{B}^\circ\,\dot{\tilde{w}}\,\mathcal{B}^\circ\,\dot{\tilde{s}}) \subset \mathcal{B}^\circ - (\mathcal{B}^\circ \cap (\dot{\tilde{s}}\,\mathcal{B}^\circ\,\dot{\tilde{s}})).$$

Conversely, we have

$$\mathcal{B}^\circ - (\mathcal{B}^\circ \cap (\dot{\tilde{s}}\,\mathcal{B}^\circ\,\dot{\tilde{s}})) \subset \dot{\tilde{s}}\,\mathcal{B}^\circ\,\dot{\tilde{s}}\,\mathcal{B}^\circ\,\dot{\tilde{s}}$$

($\dot{\tilde{s}}\,\mathcal{B}^\circ\,\dot{\tilde{s}} \subset \mathcal{B}^\circ \amalg \mathcal{B}^\circ\,\dot{\tilde{s}}\,\mathcal{B}^\circ$),

$$\dot{\tilde{s}}\,\mathcal{B}^\circ\,\dot{\tilde{s}}\,\mathcal{B}^\circ\,\dot{\tilde{s}} = \dot{\tilde{w}}'^{-1}\,\dot{\tilde{w}}\dot{\tilde{s}}\,\mathcal{B}^\circ\,\dot{\tilde{s}}\,\mathcal{B}^\circ\,\dot{\tilde{s}}$$

and

$$\dot{\tilde{w}}\dot{\tilde{s}}\,\mathcal{B}^\circ\,\dot{\tilde{s}}\,\mathcal{B}^\circ \subset \mathcal{B}^\circ\,\dot{\tilde{w}}\,\mathcal{B}^\circ$$

($\tilde{w} = \tilde{w}\tilde{s}\tilde{s}$ and $\ell(\tilde{w}) = \ell(\tilde{w}\tilde{s}) + 1$), so that

$$\mathcal{B}^\circ - (\mathcal{B}^\circ \cap (\dot{\tilde{s}}\,\mathcal{B}^\circ\,\dot{\tilde{s}})) \subset \mathcal{B}^\circ \cap (\dot{\tilde{w}}'^{-1}\mathcal{B}^\circ\,\dot{\tilde{w}}\,\mathcal{B}^\circ\,\dot{\tilde{s}}).$$

Therefore, we get

$$\alpha_{\tilde{w}'} = |\mathcal{B}^\circ/(\mathcal{B}^\circ \cap (\dot{\tilde{s}}\,\mathcal{B}^\circ\,\dot{\tilde{s}}))| - 1\,.$$

To finish the proof of part (ii) of the lemma, it suffices to check that

$$|\mathcal{B}^\circ/(\mathcal{B}^\circ \cap (\dot{\tilde{s}}\,\mathcal{B}^\circ\,\dot{\tilde{s}}))| = q$$

for all $\tilde{s} \in \tilde{S}$. Clearly, it suffices to prove this statement for $d = 2$ ($G = GL_2$). But, in this case,

$$\mathcal{B}^\circ = \begin{pmatrix} \mathcal{O}^\times & \mathcal{O} \\ \varpi\mathcal{O} & \mathcal{O}^\times \end{pmatrix},$$

$$\mathcal{B}^\circ \cap (\dot{s}_0\mathcal{B}^\circ\dot{s}_0) = \begin{pmatrix} \mathcal{O}^\times & \mathcal{O} \\ \varpi^2\mathcal{O} & \mathcal{O}^\times \end{pmatrix}$$

and

$$\mathcal{B}^\circ \cap (\dot{s}_1\mathcal{B}^\circ\dot{s}_1) = \begin{pmatrix} \mathcal{O}^\times & \varpi\mathcal{O} \\ \varpi\mathcal{O} & \mathcal{O}^\times \end{pmatrix}$$

and the statement is obvious. $\square$

As any element of $\widetilde{W}$ of length ℓ can be written as the product of an element of Ω and of ℓ elements of $\widetilde{S}$, it follows from (7.4.14) that

(7.4.15) $$t_{\tilde{w}'} * t_{\tilde{w}''} = t_{\tilde{w}'\tilde{w}''}$$

for all $\tilde{w}', \tilde{w}'' \in \widetilde{W}$ such that

$$\ell(\tilde{w}') + \ell(\tilde{w}'') = \ell(\tilde{w}'\tilde{w}'').$$

Thanks to (7.4.10), we have also

(7.4.16) $$(t_{\tilde{s}} + 1) * (t_{\tilde{s}} - q) = 0 .$$

Therefore, we have a well-defined homomorphism of $\mathbb{Q}$-algebras

(7.4.17) $$\mathbb{Q}[\mathcal{T}_{\widetilde{W}}]/\mathcal{I}_{\widetilde{S}} \to \mathcal{H}$$

which maps $T_{\tilde{w}}$ (modulo $\mathcal{I}_{\widetilde{S}}$) onto $t_{\tilde{w}}$ for each $\tilde{w} \in \widetilde{W}$.

THEOREM (7.4.18) (Iwahori and Matsumoto). — *The homomorphism of $\mathbb{Q}$-algebras (7.4.17) is an isomorphism.*

Proof : As $(t_{\tilde{w}})_{\tilde{w} \in \widetilde{W}}$ is a basis of the $\mathbb{Q}$-vector space $\widetilde{\mathcal{H}}$, it suffices to check that $(T_{\tilde{w}})_{\tilde{w} \in \widetilde{W}}$ induces a system of generators of the $\mathbb{Q}$-vector space $\mathbb{Q}[\mathcal{T}_{\widetilde{W}}]/\mathcal{I}_{\widetilde{S}}$. This is left to the reader. $\square$

Now we can prove (7.4.6).

Proof of (7.4.6) : Let $g \in G(F)$. We have

$$1_{\mathcal{B}^\circ g \mathcal{B}^\circ} = \mathrm{vol}(\mathcal{B}^\circ, dg) t_{\tilde{w}}$$

for some $\tilde{w} \in \widetilde{W}$ (see (7.4.10)).

Now $\tilde{w}$ can be written as

$$\tilde{w} = \omega^i s_{j_1} \cdots s_{j_\ell}$$

for some $i \in \mathbb{Z}$ and some $j_1, \ldots, j_\ell \in \{0, 1, \ldots, d-1\}$ where ℓ is the length of $\tilde{w}$. Then we have

$$t_{\tilde{w}} = t_{\omega^i} * t_{s_{j_1}} * \cdots * t_{s_{j_\ell}}.$$

Therefore to prove the proposition (7.4.6) it suffices to check that t_{ω^i} (for any $i \in \mathbb{Z}$) and $t_{\tilde{s}}$ (for any $\tilde{s} \in \widetilde{S}$) are invertible.

But this is the case: we have

$$(t_{\omega^i})^{-1} = t_{\omega^{-i}} \in \widetilde{H}$$

and

$$(t_{\tilde{s}})^{-1} = q^{-1}(t_{\tilde{s}} + 1 - q) \in \widetilde{H}$$

thanks to (7.4.18) (or (7.4.15) and (7.4.16)). $\square$

Let us finish this section with the following remark.

REMARK (7.4.19). — For any unramified character χ of $T(F)$ and for any subquotient $(\mathcal{V}, \pi)$ of $(I(\chi), i(\chi))$ in $\mathrm{Rep}_s(G(F))$, we have $\mathcal{V}^{\mathcal{B}^\circ} \neq (0)$. Indeed, we can assume that $(\mathcal{V}, \pi)$ is an irreducible representation of $G(F)$. Then, thanks to (7.3.5), there exists $w \in W$ such that $(\mathcal{V}, \pi)$ is a subrepresentation of $(I(w(\chi)), i(w(\chi)))$ and our assertion follows directly from (7.4.4).   □

(7.5) Spherical representations

A **spherical representation** of $G(F)$ is a smooth irreducible representation $(\mathcal{V}, \pi)$ of $G(F)$ such that
$$\mathcal{V}^K \neq (0).$$

If χ is an unramified character of $T(F)$, we have

(7.5.1) $\dim_{\mathbb{C}}(I(\chi)^K) = 1.$

Indeed, we have
$$G(F) = B(F)K$$
(Iwasawa's decomposition) and
$$T(F) \cap K = T(\mathcal{O})$$
so that the $\mathbb{C}$-linear map
$$I(\chi)^K \to \mathbb{C}, v \mapsto v(1)$$
is an isomorphism. As the functor of K-invariants is exact (see (D.1.5)), it follows that $(I(\chi), i(\chi))$ admits one and only one spherical subquotient in $\mathrm{Rep}_s(G(F))$. Moreover, if we denote by

(7.5.2) $(V(\chi), \pi(\chi)) \hookrightarrow\!\!\!\rightarrow (I(\chi), i(\chi))$

this spherical subquotient, we have

(7.5.3) $\dim_{\mathbb{C}}(V(\chi)^K) = 1.$

THEOREM (7.5.4) (Satake). — (i) *Any spherical representation $(\mathcal{V}, \pi)$ of $G(F)$ is isomorphic to $(V(\chi), \pi(\chi))$ for some unramified character χ of $T(F)$. In particular, we have*
$$\dim_{\mathbb{C}}(\mathcal{V}^K) = 1.$$

(ii) *Let χ and χ' be two unramified characters of $G(F)$. Then the spherical representations $(V(\chi), \pi(\chi))$ and $(V(\chi'), \pi(\chi'))$ of $G(F)$ are isomorphic if and only if there exists $w \in W$ such that $\chi' = w(\chi)$.*

Proof: If $(\mathcal{V}, \pi)$ is a spherical representation, we have

$$\mathcal{V}^{\mathcal{B}^\circ} \supset \mathcal{V}^K \neq (0).$$

Therefore, thanks to Casselman's theorem (7.4.4), $(\mathcal{V}, \pi)$ is isomorphic to a subrepresentation of $(I(\chi), i(\chi))$ for some unramified character χ of $T(F)$. This subrepresentation being spherical is automatically $(V(\chi), \pi(\chi))$ and part (i) is proved.

Part (ii) is a direct consequence of (7.3.3) (ii).   □

We can identify the set of unramified characters χ of $T(F)$ with $(\mathbb{C}^\times)^d$ in the following way: to $z = (z_1, \ldots, z_d) \in (\mathbb{C}^\times)^d$ we associate the unramified character

(7.5.5) $$\chi_z : T(F) \to \mathbb{C}^\times$$

which is defined by

$$\chi_z(\varpi^\lambda \tau) = z^\lambda \Big(\overset{\mathrm{dfn}}{=} \prod_{j=1}^d z_j^{\lambda_j}\Big)$$

for any $\lambda \in \mathbb{Z}^d$ and any $\tau \in T(\mathcal{O})$. If we let $W = \mathfrak{G}_d$ act on $(\mathbb{C}^\times)^d$ by

$$w(z) = (z_{w^{-1}(1)}, \ldots, z_{w^{-1}(d)})$$

for each $w \in W$ and each $z \in (\mathbb{C}^\times)^d$, we have

$$w(\chi_z) = \chi_{w(z)}.$$

Therefore, we can reformulate the theorem (7.5.4) in the following way: the map

$$z \mapsto (V(\chi_z), \pi(\chi_z))$$

induces a bijection from

$$\mathfrak{G}_d \backslash (\mathbb{C}^\times)^d$$

onto the set of isomorphism classes of spherical representations of $G(F)$.

If $(\mathcal{V}, \pi)$ is a spherical representation of $G(F)$, we will denote by

$$z(\pi) = (z_1(\pi), \ldots, z_d(\pi)) \in (\mathbb{C}^\times)^d$$

any $z \in (\mathbb{C}^\times)^d$ such that $\mathfrak{G}_d z$ maps onto the isomorphism class of $(\mathcal{V}, \pi)$ by the above bijection, i.e. such that $(\mathcal{V}, \pi)$ is isomorphic to $(V(\chi_z), \pi(\chi_z))$. The complex numbers $z_1(\pi), \ldots, z_d(\pi)$ are called the **Hecke eigenvalues of** $(\mathcal{V}, \pi)$ (they are well-defined up to a permutation).

THEOREM (7.5.6). — *Let $f \in \mathbb{C} \otimes C_c^\infty(G(F)/\!/K)$ and let $(\mathcal{V}, \pi)$ be an admissible irreducible representation of $G(F)$. Then:*

(i) $\operatorname{tr}\pi(f) = 0$ *(see (D.2.1)) unless $(\mathcal{V}, \pi)$ is spherical;*

(ii) *if $(\mathcal{V}, \pi)$ is spherical, we have*

$$\operatorname{tr}\pi(f) = f^\vee(z_1(\pi), \ldots, z_d(\pi))$$

where

$$f^\vee(z) \in \mathbb{C}[z_1, z_1^{-1}, \ldots, z_d, z_d^{-1}]^{\mathfrak{S}_d}$$

is the Satake transform of f (see (4.1.12) and (4.1.17)) and where

$$z(\pi) = (z_1(\pi), \ldots, z_d(\pi)) \in (\mathbb{C}^\times)^d$$

are the Hecke eigenvalues of $(\mathcal{V}, \pi)$.

In the proof of the theorem we will need the particular case $I = \emptyset$ of the following lemma.

LEMMA (7.5.7). — *Let $I \subset \Delta$, let $f \in \mathbb{C} \otimes C_c^\infty(G(F))$ and let $(\mathcal{W}, \rho)$ be an admissible representation of $M_I(F)$. Then, if we set $i_I^\Delta(\mathcal{W}, \rho) = (\mathcal{V}, \pi)$, we have*

$$\operatorname{tr}\pi(f) = \operatorname{tr}\rho(f^{P_I})$$

(recall that $(\mathcal{V}, \pi)$ is admissible thanks to (7.1.3) (iii), so that $\operatorname{tr}\pi(f)$ is well-defined) where $f^{P_I} \in \mathbb{C} \otimes C_c^\infty(M_I(F))$ is the K-invariant constant term of f along P_I (for each $m_I \in M_I(F)$, we have

$$f^{P_I}(m_I) = \delta_{P_I(F)}^{1/2}(m_I) \int_{N_I(F)} \int_K f(k^{-1}m_I n_I k) dk\, dn_I).$$

Proof: Let $\mathcal{V}^\circ$ be the $\mathbb{C}$-vector space of locally constant functions

$$v^\circ : K \to \mathcal{W}$$

such that

$$v^\circ(m_I n_I k) = \rho(m_I)(v^\circ(k))$$

for any $m_I \in M_I(\mathcal{O})$, $n_I \in N_I(\mathcal{O})$ and $k \in K$ and let π° be the left action of K on $\mathcal{V}^\circ$ which is induced by the right translation on K. The map

$$(\mathcal{V}, \pi|K) \to (\mathcal{V}^\circ, \pi^\circ), \quad v \mapsto v|K$$

is an isomorphism. Thanks to this isomorphism the operator $\pi(f)$ induces an operator on $\mathcal{V}^\circ$ that we will also denote by $\pi(f)$. For each $v^\circ \in \mathcal{V}^\circ$ and each $k \in K$, we have

$$(\pi(f)(v^\circ))(k) = \int_K \rho(\psi_{k,k'})(v^\circ(k')) dk'$$

where
$$\psi_{k,k'} \in \mathbb{C} \otimes \mathcal{C}_c^\infty(M_I(F))$$
is defined by
$$\psi_{k,k'}(m_I) = \delta_{P_I(F)}^{1/2}(m_I) \int_{N_I(F)} f(k^{-1}m_I n_I k')dn_I$$

for any $m_I \in M_I(F)$. Indeed, let $v \in \mathcal{V}$ be such that $v|K = v^\circ$, then we have

$$\pi(f)(v^\circ)(k) = \int_{G(F)} f(g)v(kg)dg$$

$$= \int_{G(F)} f(k^{-1}g)v(g)dg$$

$$= \int_{M_I(F)} \int_{N_I(F)} \int_K f(k^{-1}m_I n_I k')v(m_I n_I k')dk'dn_I dm_I$$

(see (4.1.7)) and

$$v(m_I n_I k') = \delta_{P_I(F)}^{1/2}(m_I)\rho(m_I)(v^\circ(k')).$$

There exists a non-negative integer n such that f is bi-invariant under the principal congruence subgroup of level n, $K(n)$. As $K(n)$ is a normal subgroup of K, $\psi_{k,k'}$ is bi-invariant under $K(n) \cap M_I(F)$ for each $k, k' \in K$ and the same is true for

$$f^{P_I} = \int_K \psi_{k,k}dk .$$

Moreover, any $v^\circ \in (\mathcal{V}^\circ)^{K(n)}$ takes its values in $\mathcal{W}^{M_I(F) \cap K(n)}$. It follows that

$$\pi(f)(\mathcal{V}^\circ) \subset (\mathcal{V}^\circ)^{K(n)},$$

$$\rho(\psi_{k,k'})(\mathcal{W}) \subset \mathcal{W}^{K(n) \cap M_I(F)}$$

and

$$\rho(f^{P_I})(\mathcal{W}) \subset \mathcal{W}^{K(n) \cap M_I(F)} ,$$

so that to compute the trace of $\pi(f)$ (resp. $\rho(\psi_{k,k'})$ and $\rho(f^{P_I})$), we can restrict this (resp. these) operator(s) to $(\mathcal{V}^\circ)^{K(n)}$ (resp. $\mathcal{W}^{K(n) \cap M_I(F)}$). Now we have an operator $\pi(f)$ on a subspace of the $\mathbb{C}$-vector space of all maps from the finite set $K/K(n)$ into the finite dimensional $\mathbb{C}$-vector space $\mathcal{W}^{K(n) \cap M_I(F)}$ and this operator is given by a kernel with values in $\mathrm{End}_{\mathbb{C}}(\mathcal{W}^{K(n) \cap M_I(F)})$. It is well known (and easy to prove) that the trace of this operator is the integral over the diagonal of the trace of the kernel.

Therefore, we get

$$\mathrm{tr}\,\pi(f) = \int_K \mathrm{tr}\rho(\psi_{k,k})dk = \mathrm{tr}\rho(f^{P_I})$$

as required. $\qquad\square$

Proof of (7.5.6) : If $f \in \mathbb{C} \otimes \mathcal{C}_c^\infty(G(F)/\!\!/K)$, we have $\pi(f)(\mathcal{V}) \subset \mathcal{V}^K$ and

$$\mathrm{tr}\,\pi(f) = \mathrm{tr}(\pi(f)|\mathcal{V}^K)$$

(see (D.2)), therefore part (i) is obvious.

If $(\mathcal{V}, \pi)$ is spherical, thanks to (7.5.4), we can assume that

$$(\mathcal{V}, \pi) = (V(\chi_z), \pi(\chi_z))$$

for $z = z(\pi) \in (\mathbb{C}^\times)^d$. Applying part (i) of (7.5.6) which is already proved we get that

$$\mathrm{tr}\,\pi(\chi_z)(f) = \mathrm{tr}\,i(\chi_z)(f)$$

$((V(\chi_z), \pi(\chi_z))$ is the unique spherical subquotient of $(I(\chi_z), i(\chi_z)))$.

Now, thanks to (7.5.7), we have

$$\mathrm{tr}\,i(\chi_z)(f) = \int_{T(F)} \chi_z(t) f^B(t)\,dt \ .$$

But this last integral is nothing else than $f^\vee(z)$ (see (4.1.13)) and the proof of (7.5.6) is completed. $\qquad\square$

EXAMPLES (7.5.8.1). — If $f = 1_K$, we have

$$\mathrm{tr}\,\pi(f) = \dim_{\mathbb{C}}(\mathcal{V}^K) = 1$$

for any spherical representation $(\mathcal{V}, \pi)$ of $G(F)$.

(7.5.8.2). — If $f = f_r$ is the Drinfeld function of level r (see (4.2.5)), we have

$$\mathrm{tr}\,\pi(f) = q^{r(d-1)/2}(z_1(\pi)^r + \cdots + z_d(\pi)^r)$$

for any spherical representation $(\mathcal{V}, \pi)$ of $G(F)$. $\qquad\square$

(7.6) Comments and references

There are two complete presentations of the material of this chapter, one by Bernstein and Zelevinsky ([Be–Ze 1], [Be–Ze 2]) and the other by Casselman ([Cas 1]). Other very useful references are Cartier's survey ([Car]) and Jacquet's lectures ([Ja]). Let me also mention the fundamental papers of Iwahori and Matsumoto ([Iw–Ma]) and Satake ([Sat]) for the sections (7.4) and (7.5).

A stronger version of the key lemma (7.3.4) is proved in [Be–Ze 2] (2.12) and [Cas 1] (6.3). For further reading about the admissible representations having vectors fixed under $\mathcal{B}^\circ$, we suggest Borel's paper ([Bo]).

8

Euler–Poincaré functions as pseudo-coefficients of the Steinberg representation

(8.0) Introduction

We will use the same notations as in chapters 4, 5 and 7.

The purpose of this chapter is to prove that the Euler–Poincaré function (5.1.2) is a pseudo-coefficient of the Steinberg representation. This result is due to Casselman and Kottwitz and we will follow [Bo–Wa] and [Ko 1].

(8.1) The Steinberg representation

For any $I \subset \Delta$, the induced representation

$$i_I^\Delta(\mathbb{C}, \delta_{P_I(F)}^{-1/2})$$

is nothing else than

$$(\mathbb{C} \otimes \mathcal{C}^\infty(P_I(F) \backslash G(F)), \rho_I)$$

where $\mathcal{C}^\infty(P_I(F) \backslash G(F))$ is the $\mathbb{Q}$-vector space of locally constant functions on $G(F)$ with values in $\mathbb{Q}$ which are left $P_I(F)$-invariant and where ρ_I is the left action of $G(F)$ on $\mathbb{C} \otimes \mathcal{C}^\infty(P_I(F) \backslash G(F))$ which is induced by the right translation on $P_I(F) \backslash G(F)$.

For any $I \subset J \subset \Delta$, we have a natural commutative diagram of monomorphisms

$$(\mathbb{C} \otimes \mathcal{C}^\infty(P_J(F) \backslash G(F)), \rho_J) \hookrightarrow (\mathbb{C} \otimes \mathcal{C}^\infty(B(F) \backslash G(F)), \rho_\emptyset)$$
$$\searrow \qquad \nearrow$$
$$(\mathbb{C} \otimes \mathcal{C}^\infty(P_I(F) \backslash G(F)), \rho_I)$$

in $\mathrm{Rep}_s(G(F))$ (a left $P_J(F)$-invariant (resp. $P_I(F)$-invariant) function is automatically left $P_I(F)$-invariant (resp. $B(F)$-invariant) as $B(F) \subset P_I(F) \subset P_J(F)$).

For any $I', I'' \subset \Delta$, we have

$$(\mathbb{C} \otimes \mathcal{C}^\infty(P_{I'}(F)\backslash G(F))) \cap (\mathbb{C} \otimes \mathcal{C}^\infty(P_{I''}(F)\backslash G(F)))$$
$$= \mathbb{C} \otimes \mathcal{C}^\infty(P_{I' \cup I''}(F)\backslash G(F))$$

in $\mathbb{C} \otimes \mathcal{C}^\infty(B(F)\backslash G(F))$ as $P_{I' \cup I''}(F)$ is the subgroup of $G(F)$ which is generated by $P_{I'}(F)$ and $P_{I''}(F)$ (for any $I \subset \Delta$, $P_I(F)$ is the subgroup of $G(F)$ which is generated by $B(F)$ and $\{\hat{s}_i | i \in I\}$).

For each $I \subset \Delta$, let us denote by

(8.1.1) $(V_I, \pi_I) \in \mathrm{ob}\,\mathrm{Rep}_s(G(F))$

the cokernel of the morphism

$$\bigoplus_{\substack{J \supset I \\ \neq}} (\mathbb{C} \otimes \mathcal{C}^\infty(P_J(F)\backslash G(F)), \rho_J) \to (\mathbb{C} \otimes \mathcal{C}^\infty(P_I(F)\backslash G(F)), \rho_I)$$

(sum of the natural monomorphisms) in $\mathrm{Rep}_s(G(F))$. It is clear that

$$(V_\Delta, \pi_\Delta) = (\mathbb{C} \otimes \mathcal{C}^\infty(G(F)\backslash G(F)), \rho_\Delta)$$

is isomorphic to the trivial representation $(\mathbb{C}, 1)$ of $G(F)$. The smooth representation $(V_\emptyset, \pi_\emptyset)$ is the so-called **Steinberg representation** of $G(F)$ and is also denoted by $(\mathrm{St}_{G(F)}, \mathrm{st}_{G(F)})$ or simply $(\mathrm{St}, \mathrm{st})$.

THEOREM (8.1.2) (Casselman). — (i) *For each $I \subset \Delta$, (V_I, π_I) is irreducible in $\mathrm{Rep}_s(G(F))$ and, for any $I', I'' \subset \Delta$, $(V_{I'}, \pi_{I'})$ and $(V_{I''}, \pi_{I''})$ are isomorphic if and only if $I' = I''$.*

(ii) *For each $I \subset \Delta$, the Jordan–Hölder subquotients of $(\mathbb{C} \otimes \mathcal{C}^\infty(P_I(F)\backslash G(F)), \rho_I)$ are exactly the (V_J, π_J)'s for $I \subset J \subset \Delta$, each of them occurring with multiplicity one.*

(iii) *For each $I \subset \Delta$, the smooth representation $r_\Delta^\emptyset(V_I, \pi_I)$ of $T(F)$ is isomorphic to*

$$\bigoplus_{\substack{w \in W \\ w(R^+) \cap \Delta = I}} (\mathbb{C}, w^{-1}(\delta_{B(F)}^{-1/2})).$$

In the proof of (8.1.2), we will need the following lemmas.

LEMMA (8.1.3). — *Let χ be a smooth character of $T(F)$. Let us assume that χ is regular, i.e. $w(\chi) \neq \chi$ for all $w \in W - \{1\}$. Then the principal series representation $(I(\chi), i(\chi))$ (see (7.3.1)) has a unique irreducible subrepresentation.*

For each $w \in W$, let us denote by $(Z(w(\chi)), z(w(\chi)))$ the unique irreducible subrepresentation of $(I(w(\chi)), i(w(\chi)))$ ($w(\chi)$ is also regular) and let Σ be a system of representatives of the classes in W for the equivalence relation

$$(w' \sim w'') \iff \begin{cases} (Z(w'(\chi)), z(w'(\chi))) \text{ and } (Z(w''(\chi)), z(w''(\chi))) \\ \text{are isomorphic.} \end{cases}$$

Then, up to isomorphism, the Jordan–Hölder subquotients of $(I(\chi), i(\chi))$ are exactly the $(Z(w(\chi)), z(w(\chi)))$'s for $w \in \Sigma$, each of them occurring with multiplicity one. In particular, the length of $(I(\chi), i(\chi))$ is equal to $|W/\sim|$.

Finally, for each $w \in W$, the smooth representation $r_\Delta^\emptyset(Z(w(\chi)), z(w(\chi)))$ of $T(F)$ is isomorphic to

$$\bigoplus_{\substack{w' \in W \\ w' \sim w}} (\mathbb{C}, w'(\chi))$$

in $\mathrm{Rep}_s(T(F))$.

Proof: If $(\mathcal{V}, \pi)$ is a subrepresentation of $(I(\chi), i(\chi))$, then $r_\Delta^\emptyset(\mathcal{V}, \pi)$ has a quotient isomorphic to $(\mathbb{C}, \chi)$ (see (7.1.3) (i)). If $(\mathcal{V}_1, \pi_1)$ and $(\mathcal{V}_2, \pi_2)$ were two distinct irreducible subrepresentations of $(I(\chi), i(\chi))$, then $(\mathcal{V}_1, \pi_1) \oplus (\mathcal{V}_2, \pi_2)$ would also be a subrepresentation of $(I(\chi), i(\chi))$ and $r_\Delta^\emptyset((\mathcal{V}_1, \pi_1) \oplus (\mathcal{V}_2, \pi_2))$ would have two distinct quotients isomorphic to $(\mathbb{C}, \chi)$. As $r_\Delta^\emptyset$ is exact (see (7.1.3) (i)), $(\mathbb{C}, \chi)$ would occur, up to isomorphism, at least twice as a Jordan–Hölder subquotient of $r_\Delta^\emptyset(I(\chi), i(\chi))$. But this would contradict (7.3.4) (χ is regular). This proves the first assertion of the lemma.

Now, if $(\mathcal{V}, \pi)$ is an irreducible subquotient of $(I(\chi), i(\chi))$, thanks to (7.3.5), there exists $w \in W$ such that $(\mathcal{V}, \pi)$ is isomorphic to an irreducible subrepresentation of $(I(w(\chi)), i(w(\chi)))$. Therefore, it follows from the first assertion which is already proved that $(\mathcal{V}, \pi)$ is isomorphic to $(Z(w(\chi)), z(w(\chi)))$ ($w(\chi)$ is regular). Moreover the multiplicity of $(\mathcal{V}, \pi)$ as a Jordan–Hölder subquotient of $(I(\chi), i(\chi))$ is equal to the multiplicity of $(Z(w(\chi)), z(w(\chi)))$ as a Jordan–Hölder subquotient of $(I(w(\chi)), i(w(\chi)))$ (see (7.3.3) (ii) (b)), i.e. to one. Recall that $r_\Delta^\emptyset$ is exact (see (7.1.3) (i)) and that the multiplicity of $(\mathbb{C}, w(\chi))$ in the semi-simplification of $r_\Delta^\emptyset(I(w(\chi)), i(w(\chi)))$ is one (see (7.3.4)). A last application of (7.3.5) shows that, for any $w \in W$, $(Z(w(\chi)), z(w(\chi)))$ is isomorphic to a subquotient of $(I(\chi), i(\chi))$ and the second assertion of the lemma is proved.

The last assertion of (8.1.3) follows from the previous ones and (7.3.4) ($r_\Delta^\emptyset$ is exact and, as χ is regular, $r_\Delta^\emptyset(Z(w(\chi)), z(w(\chi)))$ is semi-simple). $\qquad \square$

LEMMA (8.1.4). — *Let χ be a smooth character of $T(F)$. Let us assume that χ is regular (see (8.1.3)). Then, if $(\mathcal{V}_0, \pi_0), \ldots, (\mathcal{V}_L, \pi_L)$ are subrepresentations of $(I(\chi), i(\chi))$, we have*

$$\mathcal{V}_0 \cap (\mathcal{V}_1 + \cdots + \mathcal{V}_L) = (\mathcal{V}_0 \cap \mathcal{V}_1) + \cdots + (\mathcal{V}_0 \cap \mathcal{V}_L).$$

in $I(\chi)$.

Proof : Thanks to (7.3.4), the semi-simplification of $r_\Delta^\emptyset(I(\chi), i(\chi))$ is isomorphic to $\bigoplus_{w \in W} (\mathbb{C}, w^{-1}(\chi))$ in $\mathrm{Rep}_s(T(F))$. As χ is regular, it follows that $r_\Delta^\emptyset(I(\chi), i(\chi))$ is semi-simple and is isomorphic to $\bigoplus_{w \in W} (\mathbb{C}, w^{-1}(\chi))$ in $\mathrm{Rep}_s(T(F))$. Therefore, if we set

$$(\mathcal{W}_\ell, \rho_\ell) = r_\Delta^\emptyset(\mathcal{V}_\ell, \pi_\ell) \quad (\ell = 0, \ldots, L),$$

we have

$$\mathcal{W}_0 \cap (\mathcal{W}_1 + \cdots + \mathcal{W}_L) = (\mathcal{W}_0 \cap \mathcal{W}_1) + \cdots + (\mathcal{W}_0 \cap \mathcal{W}_L)$$

in $r_\Delta^\emptyset(I(\chi), i(\chi))$.

Now we have the obvious inclusion

$$(\mathcal{V}_0 \cap \mathcal{V}_1) + \cdots + (\mathcal{V}_0 \cap \mathcal{V}_L) \subset \mathcal{V}_0 \cap (\mathcal{V}_1 + \cdots + \mathcal{V}_L).$$

Let $\mathcal{V}$ be its cokernel and let π be the action of $G(F)$ on $\mathcal{V}$ induced by $i(\chi)$, so that $(\mathcal{V}, \pi)$ is a subquotient of $(I(\chi), i(\chi))$ in $\mathrm{Rep}_s(G(F))$. As $r_\Delta^\emptyset$ is exact (see (7.1.3) (i)), we have

$$r_\Delta^\emptyset(\mathcal{V}, \pi) = (0).$$

But this implies that $\mathcal{V} = (0)$ and the lemma. Otherwise, $(\mathcal{V}, \pi)$ would admit at least one irreducible subquotient $(\mathcal{V}', \pi')$ in $\mathrm{Rep}_s(G(F))$ and we would have $r_\Delta^\emptyset(\mathcal{V}', \pi') = (0)$ by exactness of $r_\Delta^\emptyset$; applying (7.3.5) and (7.1.3) (i), this would lead to a contradiction. □

Proof of (8.1.2) : For each $I \subset \Delta$, let us choose a numbering

$$I = J_0, J_1, \ldots, J_{N-1} = \Delta$$

of the set $\{J | I \subset J \subset \Delta\}$ such that

$$(J_n \subset J_m) \implies (n \leq m)$$

for all $n, m \in \{0, \ldots, N-1\}$ $(N = 2^{|\Delta - I|})$. For each $n = 0, \ldots, N$, let

$$F^n = F^n(\mathbb{C} \otimes \mathcal{C}_c^\infty(P_I(F) \backslash G(F)), \rho_I)$$

be the sum of the images of the monomorphisms

$$(\mathbb{C} \otimes \mathcal{C}^\infty(P_{J_m}(F)\backslash G(F)), \rho_{J_m}) \hookrightarrow (\mathbb{C} \otimes \mathcal{C}^\infty(P_I(F)\backslash G(F)), \rho_I)$$

for all $m = n, \ldots, N - 1$. Then we have a filtration

$$(0) = F^N \subset F^{N-1} \subset \cdots \subset F^0 = (\mathbb{C} \otimes \mathcal{C}^\infty(P_I(F)\backslash G(F)), \rho_I)$$

such that, for each $n = 0, \ldots, N - 1$,

$$\mathrm{gr}_F^n = F^n/F^{n+1}$$

is isomorphic to (V_{J_n}, π_{J_n}) in $\mathrm{Rep}_s(G(F))$ (see (8.1.4) and the remarks just before (8.1.1)).

Now the proof of the theorem is divided in three steps.

First step: for each $I \subset \Delta$, $V_I \neq (0)$, so that

$$\mathrm{length}(\mathbb{C} \otimes \mathcal{C}^\infty(P_I(F)\backslash G(F)), \rho_I) \geq N = 2^{|\Delta - I|}$$

with equality if and only if (V_J, π_J) is irreducible for each subset J of Δ containing I. In fact, we will prove that, for each $I \subset \Delta$, we have

$$\dim_{\mathbb{C}}((V_I)^{\mathcal{B}^\circ}) = |\{w \in W | w(R^+) \cap \Delta = I\}| > 0 .$$

For each $I \subset \Delta$, we have a bijection

$$D_{I,\emptyset} \xrightarrow{\sim} P_I(F)\backslash G(F)/\mathcal{B}^\circ, w \mapsto P_I(F)\dot{w}\mathcal{B}^\circ$$

where

$$D_{I,\emptyset} = \{w \in W | w(R^+) \cap \Delta \supset I\}$$

(see (5.4.4) and Iwasawa's decomposition). Therefore, for each $I \subset \Delta$, we have

$$\dim_{\mathbb{C}}((\mathbb{C} \otimes \mathcal{C}^\infty(P_I(F)\backslash G(F)))^{\mathcal{B}^\circ}) = |D_{I,\emptyset}|.$$

As the functor $(-)^{\mathcal{B}^\circ}$ on $\mathrm{Rep}_s(G(F))$ is exact (see (D.1.5)), it follows that, for each $I \subset \Delta$, we have

$$\sum_{I \subset J \subset \Delta} \dim_{\mathbb{C}}((V_J)^{\mathcal{B}^\circ}) = |D_{I,\emptyset}|$$

and the equality

$$\dim_{\mathbb{C}}((V_I)^{\mathcal{B}^\circ}) = |\{w \in W | w(R^+) \cap \Delta = I\}|$$

follows.

Now, for each $I \subset \Delta$, let $w_{\Delta-I}$ be the longest element of $W_{\Delta-I} \subset W$ (see (5.1)). Then we have

$$w_{\Delta-I}^{-1} = w_{\Delta-I},$$

$$w_{\Delta-I}(\Delta - I) \subset -(\Delta - I)$$

and

$$w_{\Delta-I}(I) \subset R^+,$$

so that

$$w_{\Delta-I}(R^+) \cap \Delta = I$$

and

$$\{w \in W | w(R^+) \cap \Delta = I\} \neq \emptyset .$$

Second step: *for any $w', w'' \in W$ with*

$$w'(R^+) \cap \Delta = w''(R^+) \cap \Delta ,$$

the two representations

$$(I(w'^{-1}(\delta_{B(F)}^{-1/2})), i(w'^{-1}(\delta_{B(F)}^{-1/2})))$$

and

$$(I(w''^{-1}(\delta_{B(F)}^{-1/2})), i(w''^{-1}(\delta_{B(F)}^{-1/2})))$$

of $G(F)$ are isomorphic, so that $w'^{-1} \sim w''^{-1}$ with the notations of (8.1.3) ($\delta_{B(F)}^{-1/2}$ is regular). First of all, let us assume that

$$w'' = w's$$

for some simple reflexion $s = s_i$ and let $\alpha = \alpha_i$ be the simple root corresponding to s ($i \in \{1, \ldots, d-1\}$). Then we have canonical isomorphisms

$$(I(w'^{-1}(\delta_{B(F)}^{-1/2})), i(w'^{-1}(\delta_{B(F)}^{-1/2})) \cong i_{\{\alpha\}}^{\Delta} i_{\emptyset}^{\{\alpha\}}(\mathbb{C}, w'^{-1}(\delta_{B(F)}^{-1/2}))$$

and

$$(I(w''^{-1}(\delta_{B(F)}^{-1/2})), i(w''^{-1}(\delta_{B(F)}^{-1/2}))) \cong i_{\{\alpha\}}^{\Delta} i_{\emptyset}^{\{\alpha\}}(\mathbb{C}, w''^{-1}(\delta_{B(F)}^{-1/2}))$$

(see (7.1.3) (iv)) and it suffices to prove that $i_{\emptyset}^{\{\alpha\}}(\mathbb{C}, w'^{-1}(\delta_{B(F)}^{-1/2}))$ and $i_{\emptyset}^{\{\alpha\}}(\mathbb{C}, w''^{-1}(\delta_{B(F)}^{-1/2}))$ are isomorphic in $\mathrm{Rep}_s(M_{\{\alpha\}}(F))$. Let us identify the pair $T \subset M_{\{\alpha\}}$ with the pair

$$(GL_1)^{d-2} \times (T_2 \subset GL_2),$$

where T_2 is the maximal torus of diagonal matrices in GL_2. Then we have

$$w'^{-1}(\delta_{B(F)}^{-1/2})(t_1, \ldots, t_d) = \mu(t_1, \ldots, t_{i-1}, t_{i+2}, \ldots, t_d)\chi'(t_i, t_{i+1})$$

and

$$w''^{-1}(\delta_{B(F)}^{-1/2})(t_1, \ldots, t_d) = \mu(t_1, \ldots, t_{i-1}, t_{i+2}, \ldots, t_d)\chi''(t_i, t_{i+1})$$

where we have set

$$\mu(t_1, \ldots, t_{i-1}, t_{i+2}, \ldots, t_d) = \prod_{\substack{j=1 \\ j \neq i, i+1}}^{d} |t_j|^{(w'(j)-d-1)/2,}$$

$$\chi' = \delta_{B_2(F)}^{(w'(i)-w'(i+1))/2}(t_i, t_{i+1})\nu(t_i t_{i+1}),$$
$$\chi'' = \delta_{B_2(F)}^{(w'(i+1)-w'(i))/2}(t_i, t_{i+1})\nu(t_i t_{i+1})$$

and

$$\nu(t) = |t|^{(w'(i)+w'(i+1)-d-1)/2}$$

for all $t_1, \ldots, t_d, t \in F^\times$ (here, B_2 is the Borel subgroup of upper triangular matrices in GL_2 and we have

$$w'(j) = w''(j)$$

for all $j \in \{1, \ldots, d\} - \{i, i+1\}$,

$$w''(i+1) - w''(i) = w'(i) - w'(i+1)$$

and

$$w''(i) + w''(i+1) = w'(i+1) + w'(i)).$$

Therefore, it suffices to prove that

$$i_{T_2(F),B_2(F)}^{GL_2(F)}(\mathbb{C}, \delta_{B_2(F)}^{(w'(i)-w'(i+1))/2})$$

and

$$i_{T_2(F),B_2(F)}^{GL_2(F)}(\mathbb{C}, \delta_{B_2(F)}^{(w'(i+1)-w'(i))/2})$$

are isomorphic in $\mathrm{Rep}_s(GL_2(F))$. But, thanks to the hypothesis

$$w'(R^+) \cap \Delta = w''(R^+) \cap \Delta,$$

we have

$$w'(i) - w''(i+1) \neq \pm 1 .$$

So, in order to finish the proof of the second step of the theorem in the case $w'' = w's$, it suffices to prove that

$$i_{T_2(F),B_2(F)}^{GL_2(F)}(\mathbb{C}, \delta_{B_2(F)}^x)$$

and

$$i_{T_2(F),B_2(F)}^{GL_2(F)}(\mathbb{C}, \delta_{B_2(F)}^{-x})$$

are isomorphic in $\text{Rep}_s(GL_2(F))$, for any $x \in \mathbb{C} - \{0, \pm 1\}$.

To prove this assertion, we can assume that $G = GL_2$, $T = T_2$ and $B = B_2$, that $\Delta = \{\alpha_1\}$ and that $s = s_1$ is the non-trivial element of the Weyl group $\mathfrak{S}_2$ of (GL_2, T_2). We will denote by χ a regular unramified character of $T_2(F)$. Then, thanks to (7.3.4), $r_\Delta^0(I(\chi), i(\chi))$ is isomorphic to $\chi \oplus s(\chi)$ and, thanks to (7.1.3) (i), the $\mathbb{C}$-vector space

$$\text{Hom}_{\text{Rep}_s(G(F))}((I(\chi), i(\chi)), (I(s(\chi)), i(s(\chi))))$$

is one dimensional. More precisely, following the proof of (7.3.4), we have a $B(F)$-invariant filtration

$$(0) = \mathcal{V}_0 \subset \mathcal{V}_1 \subset \mathcal{V}_2 = I(\chi)$$

where $\mathcal{V}_1$ is the $\mathbb{C}$-vector space of locally constant functions $v : G(F) \to \mathbb{C}$ such that

$$v(tug) = (\delta_{B(F)}^{1/2}\chi)(t)v(g)$$

for any $t \in T(F)$, any $u \in U(F)$ and any $g \in G(F)$ (i.e. $v \in I(\chi)$) and such that

$$\text{Supp}(v) \subset B(F)\dot{s}B(F)$$

and the $\mathbb{C}$-linear map

$$\lambda_1 : \mathcal{V}_1 \to \mathbb{C}, \; v \mapsto \int_{U(F)} v(\dot{s}u)du$$

(resp.

$$\lambda_2 : \mathcal{V}_2 \to \mathbb{C}, v \mapsto v(1))$$

is non-zero, vanishes on

$$\mathcal{V}_1(U(F)) \subset \mathcal{V}_1$$

(resp.

$$\mathcal{V}_2(U(F)) + \mathcal{V}_1 \subset \mathcal{V}_2)$$

and intertwines

$$\delta_{B(F)}^{-1/2} \otimes (i(\chi)|B(F))$$

and

$$s(\chi) \quad (\text{resp. } \chi)$$

(any character of $T(F)$ can be viewed as a character of $B(F)$ which is trivial on $U(F)$). If

$$\left| \chi \begin{pmatrix} \varpi & 0 \\ 0 & \varpi^{-1} \end{pmatrix} \right| < 1 \,,$$

the integral

$$\int_{U(F)} v(\dot{s}u)\,du$$

is absolutely convergent for each $v \in \mathcal{V}_2 = I(\chi)$. Indeed, if we set

$$v_K(tuk) = (\delta_{B(F)}^{1/2}\chi)(t)$$

for any $t \in T(F)$, any $u \in U(F)$ and any $k \in K$ (see Iwasawa's decomposition), then $v_K \in I(\chi)$ (it is a basis of $I(\chi)^K$, see (7.5.1)), we have the splitting

$$I(\chi) = \mathcal{V}_1 + \mathbf{C}v_K$$

(for any $v \in I(\chi)$, $v - v(1)v_K$ belongs to $\mathcal{V}_1$) and

$$\int_{U(F)} v_K(\dot{s}u)\,du = \int_{\mathcal{O}} v_K\left(\begin{pmatrix} 0 & 1 \\ 1 & x \end{pmatrix} \right) dx$$

$$+ \int_{F-\mathcal{O}} v_K\left(\begin{pmatrix} x^{-1} & 1 \\ 0 & x \end{pmatrix} \begin{pmatrix} -1 & 0 \\ x^{-1} & 1 \end{pmatrix} \right) dx$$

$$= 1 + \sum_{n=1}^{\infty} (\delta_{B(F)}^{1/2}\chi) \begin{pmatrix} \varpi^n & 0 \\ 0 & \varpi^{-n} \end{pmatrix} (q^n - q^{n-1})$$

$$= \frac{q - \chi \begin{pmatrix} \varpi & 0 \\ 0 & \varpi^{-1} \end{pmatrix}}{q\left(1 - \chi \begin{pmatrix} \varpi & 0 \\ 0 & \varpi^{-1} \end{pmatrix}\right)} < +\infty$$

(here dx is the Haar measure on F which is normalized by $\mathrm{vol}(\mathcal{O}, dx) = 1$ and we have $\mathrm{vol}(\varpi^{-n}\mathcal{O}^\times, dx) = q^n - q^{n-1}$ for each $n \in \mathbf{Z}$). Therefore, if

$$\left| \chi \begin{pmatrix} \varpi & 0 \\ 0 & \varpi^{-1} \end{pmatrix} \right| < 1 \,,$$

the $\mathbf{C}$-linear map

$$\lambda : I(\chi) \to \mathbf{C}, \ v \mapsto \int_{U(F)} v(\dot{s}u)\,du$$

is non-zero (it extends λ_1) and intertwines $\delta_{B(F)}^{-1/2} \otimes (i(\chi)|B(F))$ and $s(\chi)$. In general, let us define the $\mathbf{C}$-linear map

$$\lambda : I(\chi) \to \mathbf{C}$$

by

$$\lambda|\mathcal{V}_1 = \lambda_1$$

and

$$\lambda(v_K) = \frac{q - \chi\begin{pmatrix} \varpi & 0 \\ 0 & \varpi^{-1} \end{pmatrix}}{q(1 - \chi\begin{pmatrix} \varpi & 0 \\ 0 & \varpi^{-1} \end{pmatrix})}$$

(as χ is regular, we have

$$\chi\begin{pmatrix} \varpi & 0 \\ 0 & \varpi^{-1} \end{pmatrix} \neq 1).$$

By a direct computation or by "analytic continuation in χ", it is easy to see that this non-zero $\mathbb{C}$-linear map intertwines $\delta_{B(F)}^{-1/2} \otimes (i(\chi)|B(F))$ and $s(\chi)$. Then

$$J_s(\chi) : I(\chi) \to I(s(\chi)), \quad v \mapsto (g \mapsto \lambda(i(\chi)(g)(v)))$$

is a non-zero intertwining operator between $i(\chi)$ and $i(s(\chi))$. Moreover, if we define $v'_K \in I(s(\chi))$ by

$$v'_K(tuk) = (\delta_{B(F)}^{1/2} s(\chi))(t)$$

for any $t \in T(F)$, any $u \in U(F)$ and any $k \in K$, v'_K is a basis of $I(s(\chi))^K$ (see (7.5.1)) and

$$J_s(\chi)(v_K) = \frac{q - \chi\begin{pmatrix} \varpi & 0 \\ 0 & \varpi^{-1} \end{pmatrix}}{q(1 - \chi\begin{pmatrix} \varpi & 0 \\ 0 & \varpi^{-1} \end{pmatrix})} v'_K$$

($J_s(\chi)(v_K) \in I(s(\chi))^K$ as $J_s(\chi)$ intertwines $i(\chi)$ and $i(s(\chi))$, so that $J_s(\chi)(v_K) = \alpha v'_K$ with $\alpha = J_s(\chi)(v_K)(1) = \lambda(v_K)$).

Replacing χ by $s(\chi)$, we get similarly a non-zero intertwining operator

$$J_s(s(\chi)) : I(s(\chi)) \to I(\chi)$$

between $i(s(\chi))$ and $i(\chi)$ such that

$$J_s(s(\chi))(v'_K) = \frac{q - s(\chi)\begin{pmatrix} \varpi & 0 \\ 0 & \varpi^{-1} \end{pmatrix}}{q(1 - s(\chi)\begin{pmatrix} \varpi & 0 \\ 0 & \varpi^{-1} \end{pmatrix})} v_K.$$

Then $J_s(s(\chi)) \circ J_s(\chi)$ (resp. $J_s(\chi) \circ J_s(s(\chi))$) is an endomorphism of $(I(\chi), i(\chi))$ (resp. $(I(s(\chi)), i(s(\chi)))$) in $\mathrm{Rep}_s(G(F))$ such that

$$J_s(s(\chi)) \circ J_s(\chi)(v_K) = c(\chi)v_K$$

(resp.

$$J_s(\chi) \circ J_s(s(\chi))(v'_K) = c(s(\chi))v'_K)$$

with

$$c(\chi) = c(s(\chi)) = \frac{(q - s(\chi)\begin{pmatrix} \varpi & 0 \\ 0 & \varpi^{-1} \end{pmatrix})(q - \chi\begin{pmatrix} \varpi & 0 \\ 0 & \varpi^{-1} \end{pmatrix})}{q^2(1 - s(\chi)\begin{pmatrix} \varpi & 0 \\ 0 & \varpi^{-1} \end{pmatrix})(1 - \chi\begin{pmatrix} \varpi & 0 \\ 0 & \varpi^{-1} \end{pmatrix})}.$$

But, thanks to (7.1.3) (i), the $\mathbb{C}$-vector space

$$\mathrm{End}_{\mathrm{Rep}_s(G(F))}(I(\chi), i(\chi))$$

(resp.

$$\mathrm{End}_{\mathrm{Rep}_s(G(F))}(I(s(\chi)), i(s(\chi))))$$

is one dimensional, so that

$$J_s(s(\chi)) \circ J_s(\chi) = c(\chi)id$$

(resp.

$$J_s(s(\chi)) \circ J_s(\chi) = c(s(\chi))id).$$

In particular, if

$$c(\chi) = c(s(\chi)) \neq 0$$

for a regular unramified character χ of $T(F)$, then $(I(\chi), i(\chi))$ and $(I(s(\chi)), i(s(\chi)))$ are canonically isomorphic in $\mathrm{Rep}_s(G(F))$.
As we have

$$c(\delta^x_{B(F)}) = c(\delta^{-x}_{B(F)}) = \frac{(q - q^{2x})(q - q^{-2x})}{q^2(1 - q^{2x})(1 - q^{-2x})} \neq 0$$

for any $x \in \mathbb{C} - \{0, \pm 1\}$, this finishes the second step of the proof of the theorem in the case $w'' = w's$.
In general, if $w', w'' \in W$ satisfy

$$w'(R^+) \cap \Delta = w''(R^+) \cap \Delta,$$

let us denote by I this subset of Δ. Let V be the hyperplane

$$\{x = (x_1, \ldots, x_d) \in \mathbb{R}^d | x_1 + \cdots + x_d = 0\},$$

let us identify ε_i to the coordinate $x \mapsto x_i$ for $i = 1, \ldots, d$, so that each root $\alpha \in R$ (see (5.1)) induces an $\mathbb{R}$-linear form on V, let W act on V by

$$(w, x) \mapsto (x_{w^{-1}(1)}, \ldots, x_{w^{-1}(d)}),$$

so that

$$\alpha(w(x)) = w^{-1}(\alpha)(x)$$

for any $\alpha \in R$, any $w \in W$ and any $x \in V$, and let

$$C = \{x \in V | \alpha(x) > 0, \forall \alpha \in \Delta\}$$

be the standard Weyl chamber (recall that the Weyl chambers are the connected components of

$$V - \bigcup_{\alpha \in R} \{x \in V | \alpha(x) = 0\}$$

and that any Weyl chamber is equal to $w(C)$ for a unique $w \in W$). Then, for any $w \in W$, we have

$$w(R^+) \cap \Delta = I$$

if and only if $w(C)$ is contained in the set

$$C_I = \{x \in V | \alpha(x) > 0, \forall \alpha \in I, \text{ and } \alpha(x) < 0, \forall \alpha \in \Delta - I\}.$$

As this set C_I is convex, there exists at least one chain

$$w'(C) = w^{(1)}(C), \ldots, w^{(M+1)}(C) = w''(C)$$

of Weyl chambers, which are all contained in C_I, such that $w^{(m)}(C)$ and $w^{(m+1)}(C)$ are adjacent for all $m = 0, \ldots, M - 1$ (choose general enough points $x' \in w'(C)$ and $x'' \in w''(C)$ and take the chain of Weyl chambers which intersect the piece of straight line from x' to x''). Here, we recall that two Weyl chambers C_1 and C_2 are adjacent if $C_1 \neq C_2$ and there exists $\alpha \in R$ such that the interiors of the sets $\overline{C}_1 \cap \{x \in V | \alpha(x) = 0\}$ and $\overline{C}_2 \cap \{x \in V | \alpha(x) = 0\}$ are non-empty and equal ($\overline{C}_1$ and $\overline{C}_2$ are the closures of C_1 and C_2 in V). But, for each $m = 1, \ldots, M$, $w^{(m)}(C)$ and $w^{(m+1)}(C)$ are adjacent if and only if the Weyl chambers C and $(w^{(m)})^{-1}(w^{(m+1)}(C))$ are adjacent, i.e. if and only if $(w^{(m)})^{-1}w^{(m+1)}$ is a simple reflexion $s^{(m)}$. In other words, we have proved the existence of a sequence $(s^{(1)}, \ldots, s^{(M)})$ of simple reflexions such that

$$w'' = w's^{(1)} \cdots s^{(M)}$$

and such that

$$w's^{(1)} \cdots s^{(m)}(R^+) \cap \Delta = I$$

for any $m \in \{0, \dots, M\}$. Then the second step of the proof of the theorem follows from its particular case $(M = 1)$, which is already proved, by a trivial induction on M.

Third step : conclusion. Thanks to the second step of the proof of the theorem, for any $I \subset \Delta$, the subset

$$\{w \in W | w(R^+) \cap \Delta = I\}$$

is entirely contained in one equivalence class for the relation $w'^{-1} \sim w''^{-1}$. Therefore, we have at most 2^{d-1} equivalence classes for this relation and, therefore, also for the relation $w' \sim w''$. Thanks to (8.1.3), it follows that

$$\text{length}(\mathbb{C} \otimes \mathcal{C}^\infty(B(F)\backslash G(F)), \rho_\emptyset) = |W/\sim| \le 2^{d-1}.$$

Now the first step of the proof tells us that this length is exactly 2^{d-1}. It follows on the one hand that (V_I, π_I) is irreducible for each $I \subset \Delta$ and on the other hand that the equivalence classes for $w'^{-1} \sim w''^{-1}$ are exactly the (non-empty) subsets $\{w \in W | w(R^+) \cap \Delta = I\}$ of W, for $I \subset \Delta$.

Using again the first step of the proof, we deduce from the irreducibility of the (V_J, π_J)'s $(J \subset \Delta)$ that, for each $I \subset \Delta$, we have

$$\text{length}(\mathbb{C} \otimes \mathcal{C}^\infty(P_I(F)\backslash G(F)), \rho_I) = 2^{|\Delta - I|}$$

and the Jordan–Hölder subquotients of $(\mathbb{C} \otimes \mathcal{C}^\infty(P_I(F)\backslash G(F)), \rho_I)$ are exactly the (V_J, π_J)'s for $I \subset J \subset \Delta$.

For each $I \subset \Delta$, let $w_{\Delta - I} \in W_{\Delta - I} \subset W$ be the longest element as in the first step of the proof, so that

$$w_{\Delta - I}(R^+) \cap \Delta = I .$$

Then $(w_{\Delta - I})_{I \subset \Delta}$ is a system of representatives of the equivalence classes in W for the relation $w'^{-1} \sim w''^{-1}$. Thanks to (8.1.3), the Jordan–Hölder subquotients of $(\mathbb{C} \otimes \mathcal{C}^\infty(B(F)\backslash G(F)), \rho_\emptyset)$ are, up to isomorphism, the

$$(Z(w_{\Delta - I}^{-1}(\delta_{B(F)}^{-1/2})), z(w_{\Delta - I}^{-1}(\delta_{B(F)}^{-1/2})))$$

for $I \subset \Delta$ and they all occur with multiplicity one. This implies that there exists a bijection ι from $\{I \subset \Delta\}$ onto itself such that (V_I, π_I) is isomorphic to

$$(Z(w_{\Delta - \iota(I)}^{-1}(\delta_{B(F)}^{-1/2})), z(w_{\Delta - \iota(I)}^{-1}(\delta_{B(F)}^{-1/2})))$$

for any $I \subset \Delta$ and this implies that, for any $I', I'' \subset \Delta$, $(V_{I'}, \pi_{I'})$ and $(V_{I''}, \pi_{I''})$ are isomorphic if and only if $I' = I''$. In particular, that completes the proof of parts (i) and (ii) of the theorem.

Finally, let us prove that the bijection ι is the identity of $\{I \subset \Delta\}$. Thanks to (8.1.3), this will imply part (iii) of the theorem. In fact, we will prove that $\iota(I) = I$ by induction on $|\Delta - I|$.

For $I = \Delta$, (V_Δ, π_Δ) is isomorphic to $(\mathbb{C}, 1)$, so that $r_\Delta^0(V_\Delta, \pi_\Delta)$ is isomorphic to $(\mathbb{C}, \delta_{B(F)}^{-1/2})$. We have $w_\Delta = 1$ and, by construction, (V_Δ, π_Δ) is contained in $(I(\delta_{B(F)}^{-1/2}), i(\delta_{B(F)}^{-1/2}))$. Therefore, $\iota(\Delta) = \Delta$.

Let $I \subset \Delta$ with $|\Delta - I| > 0$ and let us assume that $\iota(J) = J$ (and therefore that part (iii) of the theorem holds) for any $J \subset \Delta$ with $|\Delta - J| < |\Delta - I|$. To prove that $\iota(I) = I$ it suffices to prove that $(\mathbb{C}, w_{\Delta-I}^{-1}(\delta_{B(F)}^{-1/2}))$ occurs, up to isomorphism, in the semi-simplification of $r_\Delta^0(\mathbb{C} \otimes \mathcal{C}^\infty(P_I(F)\backslash G(F)), \rho_I)$. Indeed, if this is the case, $(\mathbb{C}, w_{\Delta-I}^{-1}(\delta_{B(F)}^{-1/2}))$ occurs, up to isomorphism, in the semi-simplification of at least one of the (V_J, π_J)'s with $J \supset I$ (r_Δ^0 is exact (see (7.1.3) (i)) and we can apply part (ii) of the theorem which is already proved); but, thanks to the induction hypothesis, $(\mathbb{C}, w_{\Delta-I}^{-1}(\delta_{B(F)}^{-1/2}))$ cannot be isomorphic to a subquotient of $r_\Delta^0(V_J, \pi_J)$ if $J \underset{\neq}{\supset} I$, so it must be isomorphic to a subquotient of $r_\Delta^0(V_I, \pi_I)$ and we must have $w_{\Delta-\iota(I)}^{-1} \sim w_{\Delta-I}^{-1}$ (see (8.1.3)), i.e.

$$\iota(I) = w_{\Delta-\iota(I)}(R^+) \cap \Delta = w_{\Delta-I}(R^+) \cap \Delta = I$$

(see the beginning of the third step of the proof).

Now we have

$$B = N_{\Delta-I} P_\emptyset^{\Delta-I},$$

$$\dot{w}_{\Delta-I} N_{\Delta-I} \dot{w}_{\Delta-I}^{-1} = N_{\Delta-I},$$

$$P_I \supset N_{\Delta-I}$$

and

$$P_I \cap \dot{w}_{\Delta-I} P_\emptyset^{\Delta-I} \dot{w}_{\Delta-I} = T$$

(see [Ca] (2.8.7)). Therefore, we have

$$P_I \dot{w}_{\Delta-I} B = P_I \dot{w}_{\Delta-I} N_\emptyset^{\Delta-I}$$

and the factorization in the right hand side is unique. In other words, we have a locally closed embedding (for the Zariski topology)

$$N_\emptyset^{\Delta-I} \hookrightarrow P_I \backslash G, \quad n_\emptyset^{\Delta-I} \mapsto P_I \dot{w}_{\Delta-I} n_\emptyset^{\Delta-I}$$

and, if we let B act on $N_\emptyset^{\Delta-I}$ by

$$(n_\emptyset^{\Delta-I}, b) \mapsto n_\emptyset^{\Delta-I} u_\emptyset^{\Delta-I}$$

for any $n_\emptyset^{\Delta-I} \in N_\emptyset^{\Delta-I}$ and any $b \in B$, where we have factorized b into $u_{\Delta-I} t u_\emptyset^{\Delta-I}$ with $u_{\Delta-I} \in N_{\Delta-I}$, $t \in T$ and $u_\emptyset^{\Delta-I} \in N_\emptyset^{\Delta-I}$, and if we let B act on $P_I \backslash G$ by right translation, this embedding is B-equivariant. In particular, we get a $B(F)$-equivariant locally closed embedding (for the ϖ-adic topology)

$$N_\emptyset^{\Delta-I}(F) \hookrightarrow P_I(F) \backslash G(F).$$

Then any choice of a $B(F)$-invariant open subset of $P_I(F) \backslash G(F)$ which contains $N_\emptyset^{\Delta-I}(F)$ as a closed subset (for the ϖ-adic topology) defines an isomorphism from $\mathbb{C} \otimes \mathcal{C}_c^\infty(N_\emptyset^{\Delta-I}(F))$, together with the smooth action of $B(F)$ induced by the above action of $B(F)$ on $N_\emptyset^{\Delta-I}(F)$, onto a subquotient of $(\mathbb{C} \otimes \mathcal{C}^\infty(P_I(F) \backslash G(F)), \rho_I | B(F))$ (as in the proof of (7.3.4), consider the space of functions in $\mathbb{C} \otimes \mathcal{C}^\infty(P_I(F) \backslash G(F))$ which have their supports contained in this $B(F)$-invariant open subset).

Let us consider the $\mathbb{C}$-linear form

$$\mathbb{C} \otimes \mathcal{C}_c^\infty(N_\emptyset^{\Delta-I}(F)) \to \mathbb{C}, \; f \mapsto \int_{N_\emptyset^{\Delta-I}(F)} f(n_\emptyset^{\Delta-I}) dn_\emptyset^{\Delta-I},$$

where $dn_\emptyset^{\Delta-I}$ is the Haar measure on $N_\emptyset^{\Delta-I(F)}$ which is normalized by $\mathrm{vol}(N_\emptyset^{\Delta-I}(\mathcal{O}), dn_\emptyset^{\Delta-I}) = 1$. If we let $B(F)$ act on $\mathbb{C}$ through the character $\delta_{P_\emptyset^{\Delta-I}(F)}$ of its reductive quotient $T(F) = M_\emptyset^{\Delta-I}(F)$, this $\mathbb{C}$-linear form is $B(F)$-equivariant and, in particular, it vanishes on the subspace

$$(\mathbb{C} \otimes \mathcal{C}_c^\infty(N_\emptyset^{\Delta-I}(F)))(U(F)) \subset \mathbb{C} \otimes \mathcal{C}_c^\infty(N_\emptyset^{\Delta-I}(F)).$$

As this $\mathbb{C}$-linear form is not identically zero (it takes the value 1 on the characteristic function of $N_\emptyset^{\Delta-I}(\mathcal{O}) \subset N_\emptyset^{\Delta-I}(F)$), we have proved that $(\mathbb{C}, \delta_{B(F)}^{-1/2} \delta_{P_\emptyset^{\Delta-I}(F)})$ is isomorphic to a subquotient of $r_\Delta^\emptyset(\mathbb{C} \otimes \mathcal{C}_c^\infty(P_I(F) \backslash G(F)), \rho_I)$.

But we have

$$\delta_{B(F)} = \delta_{P_\emptyset^{\Delta-I}}(\delta_{P_{\Delta-I}} | T(F)),$$

$$w_{\Delta-I}^{-1}(\delta_{P_\emptyset^{\Delta-I}}) = \delta_{P_\emptyset^{\Delta-I}}^{-1}$$

and

$$w_{\Delta-I}^{-1}(\delta_{P_{\Delta-I}} | T(F)) = \delta_{P_{\Delta-I}} | T(F)$$

($\dot{w}_{\Delta-I}$ normalizes $N_{\Delta-I}(F)$). Therefore, we get

$$\delta_{B(F)}^{-1/2} \delta_{P_\emptyset^{\Delta-I}(F)} = w_{\Delta-I}^{-1}(\delta_{B(F)}^{-1/2})$$

and the proof of part (iii) of the theorem is completed. $\qquad\square$

REMARK (8.1.5). — A stronger form of (7.3.4) (see [Cas 1] (6.3) or [Be–Ze 2] (2.12)) implies directly that the smooth representation $r_\Delta^0(i_I^\Delta(\mathbb{C}, \delta_{P_I(F)}^{-1/2}))$ of $T(F)$ is isomorphic to

$$\bigoplus_{w \in D_{I,\emptyset}} (\mathbb{C}, w^{-1}(\delta_{B(F)}^{-1/2}))$$

and gives a simpler proof of part (iii) of the theorem (r_Δ^0 is exact, see (7.1.3) (i)) and consequently a simpler proof of $V_I \neq (0)$ for all $I \subset \Delta$. In fact, the argument at the end of the third step of the proof of (8.1.2) is a part of the argument which is used to prove that stronger form of (7.3.4).   □

(8.2) Main theorem

Let $I \subset \Delta$, let $(\mathcal{V}, \pi)$ be a smooth representation of $M_I(F)$ which admits a central character $\omega_\pi : Z_I(F) \to \mathbb{C}^\times$ and let $f : M_I(F) \to \mathbb{C}$ be a locally constant function such that

$$f(z_I m_I) = \omega_\pi^{-1}(z_I) f(m_I)$$

for any $z_I \in Z_I(F)$ and any $m_I \in Z_I(F)$ and such that

$$Z_I(F) \backslash \mathrm{Supp}(f) \subset Z_I(F) \backslash M_I(F)$$

is compact. Then we have a well-defined operator

$$\pi(f) : \mathcal{V} \to \mathcal{V}$$

with

$$\pi(f)(v) = \int_{Z_I(F) \backslash M_I(F)} f(m_I) \pi(m_I)(v) \frac{dm_I}{dz_I}$$

(dm_I, dz_I are the Haar measures on $M_I(F)$, $Z_I(F)$ respectively which are normalized by $\mathrm{vol}(M_I(\mathcal{O}), dm_I) = \mathrm{vol}(Z_I(\mathcal{O}), dz_I) = 1$).

If $(\mathcal{V}, \pi)$ is admissible, this operator has finite rank (f is right invariant by a small enough compact open subgroup Γ of $M_I(F)$ and $\pi(f)(\mathcal{V}) \subset \mathcal{V}^\Gamma$) and has a well-defined trace

$$\mathrm{tr}\ \pi(f) \in \mathbb{C}$$

(if f is right invariant by a compact open subgroup $\Gamma \subset M_I(F)$,

$$\mathrm{tr}\ \pi(f) = \mathrm{tr}(\pi(f)|\mathcal{V}^\Gamma).$$

The main purpose of this chapter is to prove the following theorem.

THEOREM (8.2.1) (Casselman, Kottwitz). — *Let*

$$f \in \mathcal{C}_c^\infty(F^\times \backslash G(F) /\!/ \mathcal{B}^\circ)$$

be our very cuspidal Euler–Poincaré function (see (5.1.2) and (5.1.3)) and let $(\mathcal{V}, \pi)$ *be an admissible irreducible representation of* $F^\times \backslash G(F)$, *i.e. of* $G(F)$ *with trivial central character (see (D.1.12)). Then:*

(i) $\operatorname{tr} \pi(f) = 0$ *unless there exists* $I \subset \Delta$ *such that* $(\mathcal{V}, \pi)$ *is isomorphic to* $(\mathcal{V}_I, \pi_I)$ *(see (8.1.2))* ;

(ii) *for each* $I \subset \Delta$, *we have*

$$\operatorname{tr} \pi_I(f) = (-1)^{|\Delta - I|}$$

(recall that $(\mathcal{V}_I, \pi_I)$ *is a subquotient of* $i_\emptyset^\Delta(\mathbb{C}, \delta_{B(F)}^{-1/2})$ *which admits a trivial central character (see (7.3.2)), so that* $(\mathcal{V}_I, \pi_I)$ *has a trivial central character).*

In fact, the conclusions of the theorem are still satisfied if we replace f by any Kottwitz function f_S (see (5.2.1)). But some parts of the proof of (8.2.1) are simpler under the assumption that f is very cuspidal.

The proof of (8.2.1) will be given in the next two sections.

(8.3) Some easy vanishing results

As f is $\mathcal{B}^\circ$-bi-invariant, we have

$$(8.3.1) \qquad\qquad \operatorname{tr} \pi(f) = \operatorname{tr}(\pi(f)|\mathcal{V}^{\mathcal{B}^\circ})$$

for any admissible representation $(\mathcal{V}, \pi)$ of $F^\times \backslash G(F)$. Therefore, thanks to (7.4.4), we get

LEMMA (8.3.2). — *Let* $(\mathcal{V}, \pi)$ *be an admissible irreducible representation of* $F^\times \backslash G(F)$. *Then* $\operatorname{tr} \pi(f) = 0$ *unless* $(\mathcal{V}, \pi)$ *is isomorphic to a subrepresentation of an unramified principal series representation* $(I(\chi), i(\chi))$ *of* $G(F)$. $\qquad \square$

As f is very cuspidal, we have

LEMMA (8.3.3). — *Let* $I \underset{\neq}{\subset} \Delta$ *and let* $(\mathcal{W}, \rho)$ *be any admissible representation of* $M_I(F)$ *such that*

$$\rho(z) = id_\mathcal{W}$$

for any $z \in F^\times = Z_\Delta(F) \subset M_I(F)$. *Then, if we set* $i_I^\Delta(\mathcal{W}, \rho) = (\mathcal{V}, \pi)$, $(\mathcal{V}, \pi)$ *admits a trivial central character and*

$$\operatorname{tr} \pi(f) = 0 .$$

Proof: Any $\varphi \in \mathbb{C} \otimes \mathcal{C}_c^\infty(F^\times \backslash M_I(F))$ induces an operator $\rho(\varphi)$ on $\mathcal{W}$ (if $w \in \mathcal{W}$, we set

$$\rho(\varphi)(w) = \int_{F^\times \backslash M_I(F)} \varphi(m_I)\rho(m_I)(w)\frac{dm_I}{dz}).$$

Now the same arguments as in the proof of (7.5.7) give that

$$\text{tr } \pi(f) = \text{tr}\rho(f^{P_I}) = 0$$

where

$$f^{P_I} \equiv 0$$

is the K-invariant constant term along P_I. $\square$

In particular, we have

(8.3.4) $\text{tr } i(\chi)(f) = 0$

for any smooth character χ of $F^\times \backslash T(F)$ and

(8.3.5) $\text{tr } \rho_I(f) = 0$

for any $I \underset{\neq}{\subset} \Delta$ (see (8.1)).

COROLLARY (8.3.6). — *For each $I \subset \Delta$, we have*

$$\text{tr } \pi_I(f) = (-1)^{|\Delta - I|}\text{tr } \pi_\Delta(f).$$

Proof: Thanks to (8.1.2) (ii) and (8.3.5), for each $I \subset \Delta$, we have

$$0 = \text{tr } \rho_I(f) = \sum_{J \supset I} \text{tr } \pi_J(f)$$

$((\mathcal{V}, \pi) \mapsto \text{tr } \pi(f)$ is additive on the short exact sequences in $\text{Rep}_a(F^\times \backslash G(F))$, see (D.2)). $\square$

Let us finish this section by a direct computation of $\text{tr } \pi_\Delta(f)$.

LEMMA (8.3.7). — *We have*

$$\text{tr } \pi_\Delta(f) = \sum_{\substack{I \subset \Delta \\ (s'-1)s'' \text{ even}}} \frac{(-1)^{|\Delta-I|}s'}{|\Delta - I| + 1} = 1$$

where s' and s'' are defined by

$$\begin{cases} |\Delta - I| + 1 = s's'', \\ d_j = d_{j+s''} = \cdots = d_{j+(s'-1)s''} \ (j = 1, \ldots, s'' - 1), \\ s' \text{ is maximal for these first two properties,} \end{cases}$$

if

$$d_I = (d_1, \ldots, d_s)$$

is the partition of d corresponding to I.

Proof: By definition (V_Δ, π_Δ) is isomorphic to $(\mathbb{C}, 1)$, so that

$$\operatorname{tr} \pi_\Delta(f) = \int_{F^\times \backslash G(F)} f(g) \frac{dg}{dz},$$

i.e.

$$\operatorname{tr} \pi_\Delta(f) = \sum_{I \subset \Delta} \frac{(-1)^{|\Delta - I|} \int_{F^\times \backslash G(F)} \chi_I(g) \frac{dg}{dz}}{(|\Delta + I| + 1) \operatorname{vol}(\mathcal{P}_I^\circ, dg)}$$

(see (5.1.2)). But, for each $I \subset \Delta$, we have

$$\int_{F^\times \backslash G(F)} \chi_I(g) \frac{dg}{dz} = \operatorname{vol}(\mathcal{P}_I^\circ, dg) \sum_{F^\times g \mathcal{P}_I^\circ \in F^\times \backslash \mathcal{P}_I / \mathcal{P}_I^\circ} \chi_I(F^\times g \mathcal{P}_I^\circ),$$

the set

$$\{ F^\times \varepsilon^{d''n} \mid n = 0, \ldots, s' - 1 \} \subset F^\times \backslash \mathcal{P}_I$$

is a system of representatives of the classes in $F^\times \backslash \mathcal{P}_I / \mathcal{P}_I^\circ$, where

$$d'' = d_1 + \cdots + d_{s''} = d/s'$$

and where ε is defined in (5.1), and we have

$$\chi_I(F^\times \varepsilon^{d''n}) = (-1)^{(s'-1)s''n}$$

for each $n = 0, \ldots, s' - 1$. Therefore, for each $I \subset \Delta$, we have

$$\frac{\int_{F^\times \backslash G(F)} \chi_I(g) \frac{dg}{dz}}{\operatorname{vol}(\mathcal{P}_I^\circ, dg)} = \sum_{n=0}^{s'-1} (-1)^{(s'-1)s''n}$$

$$= \begin{cases} s' & \text{if } (s'-1)s'' \text{ is even,} \\ 0 & \text{otherwise,} \end{cases}$$

and the first equality of the lemma is proved.

Let us prove the second equality. The number of subsets I of Δ such that $|\Delta - I| + 1 = s$ is equal to

$$P(d, s) = \binom{d-1}{d-s}$$

for each non-negative integer s. If s' divides the g.c.d. (d, s) of d and s, let $P_{s'}(d, s)$ be the number of subsets I of Δ such that $|\Delta - I| + 1 = s$ and such that $s'' = s/s'$ is the exact period of the partition d_I. Then we have

$$P(d, s) = \sum_{s' \mid (d,s)} P_{s'}(d, s)$$

and, by Möbius inversion (see [Co] ch. III, Compléments et exercices 16), it follows that

$$P_{s'}(d,s) = \sum_{t \mid \frac{(d,s)}{s'}} \mu(t) P\left(\frac{d}{s't}, \frac{s}{s't}\right)$$

for any s' dividing (d,s). Here, s', t are positive integers, μ is the Möbius function and we have

$$P_{s'}(d,s) = P_1\left(\frac{d}{s'}, \frac{s}{s'}\right)$$

and

$$\frac{(d,s)}{s'} = \left(\frac{d}{s'}, \frac{s}{s'}\right)$$

if s' divides (d,s). Therefore, we obtain the formula

$$(*) \qquad \sum_{\substack{I \subset \Delta \\ (s'-1)s'' \text{ even}}} \frac{(-1)^{|\Delta-I|} s'}{|\Delta - I| + 1} = \sum_{\substack{s',s'',t \\ s't \mid d \\ t \mid s'' \\ (s'-1)s'' \text{ even}}} \frac{(-1)^{s's''-1}\mu(t)}{s''}\left(\begin{array}{c}(d/s't)-1 \\ (d/s't)-(s''/t)\end{array}\right).$$

But we have

$$\left(\begin{array}{c}(d/s't)-1 \\ (d/s't)-(s''/t)\end{array}\right) = \frac{s's''}{d}\left(\begin{array}{c}d/s't \\ (d/s't)-(s''/t)\end{array}\right),$$

$$\frac{1+(-1)^{(s'-1)s''}}{2} = \begin{cases} 1 & \text{if } (s'-1)s'' \text{ is even,} \\ 0 & \text{otherwise,} \end{cases}$$

and

$$\sum_{\sigma''}(-1)^{x\sigma''-1}\left(\begin{array}{c}d/s't \\ (d/s't)-\sigma''\end{array}\right) = 1 - (1+(-1)^x)^{d/s't}$$

for any integer x (σ'' runs through the set of all positive integers), so that the right hand side of the formula $(*)$ is equal to

$$\sum_{\substack{s',t \\ s't \mid d}} \frac{s'\mu(t)}{2d}[2 - (1+(-1)^t)^{d/s't} - (1+(-1)^{s't})^{d/s't}].$$

In particular, if d is odd, s' and t are automatically odd, so that

$$1+(-1)^t = 1+(-1)^{s't} = 0$$

and the right hand side of the formula $(*)$ is equal to

$$\sum_{\substack{s',t \\ s't \mid d}} \frac{s'\mu(t)}{d} = 1 \,,$$

as

$$\sum_{t|\delta} \mu(t) = \begin{cases} 0 & \text{if } \delta > 1, \\ 1 & \text{if } \delta = 1, \end{cases}$$

for any positive integer δ.

Now let us assume that d is even and let us set $\bar{d} = d/2$. We have, by Möbius inversion again,

$$\sum_{t|n} \mu(t)\frac{n}{t} = \varphi(n)$$

for any positive integer n, where φ is the Euler function (see [Co] Ch. IV, §6), and it follows that

$$\sum_{\substack{s',t \\ s't|d}} s'\mu(t)(1 + (-1)^{s't})^{d/s't} = \sum_{\bar{n}|\bar{d}} \varphi(2\bar{n})2^{\bar{d}/\bar{n}}$$

(take $\bar{n} = s't/2$ if $s't$ is even). Similarly, for any positive integer $\bar{n}$, we have

$$\sum_{\bar{t}|\bar{n}} \mu(2\bar{t})\frac{\bar{n}}{\bar{t}} = - \sum_{\substack{\bar{t}|\bar{n} \\ \bar{t} \text{ odd}}} \mu(\bar{t})\frac{\bar{n}}{\bar{t}} = -\varphi(2\bar{n})$$

(if $\bar{n} = 2^\nu m$ with $(2,m) = 1$, we have

$$\varphi(2\bar{n}) = (2^{\nu+1} - 2^\nu)\varphi(m) = 2^\nu\varphi(m)$$

and

$$\sum_{\bar{t}|m} \mu(t)\frac{m}{\bar{t}} = \varphi(m))$$

and it follows that

$$\sum_{\substack{s',t \\ s't|d}} s'\mu(t)(1 + (-1)^t)^{d/s't} = - \sum_{\bar{n}|\bar{d}} \varphi(2\bar{n})2^{\bar{d}/\bar{n}}$$

(take $\bar{n} = s't/2$ if t is even). Therefore, we have proved that the right hand side of $(*)$ is again equal to

$$\sum_{\substack{s',t \\ s't|d}} \frac{s'\mu(t)}{d} = 1 .$$

The proof of the lemma is completed. $\square$

Proof of assertion (ii) *of* (8.2.1). Put together (8.3.5), (8.3.6) and (8.3.7).

$\square$

(8.4) Cohomological interpretation of $\mathrm{tr}\pi(f)$

Let $\mathcal{I}$ be the Bruhat–Tits building of $F^{\times}\backslash G(F)$ (see (5.2)). For each facet σ of $\mathcal{I}$ of positive dimension, let us choose once for all an orientation $\mathrm{or}(\sigma)$ of σ. Then, for each pair (σ, τ) of facets of $\mathcal{I}$ with

$$\begin{cases} \tau \subset \bar{\sigma}, \\ \dim(\tau) = \dim(\sigma) - 1, \end{cases}$$

we have a sign

$$\varepsilon(\sigma, \tau) \in \{\pm 1\}.$$

If $\dim(\tau) > 0$, this is defined by

$$\varepsilon(\sigma, \tau) = 1$$

if $\mathrm{or}(\tau)$ is equal to the orientation $\mathrm{or}(\sigma)|\tau$ of τ which is induced by $\mathrm{or}(\sigma)$ and by

$$\varepsilon(\sigma, \tau) = -1$$

otherwise; if σ is an edge, it is defined by

$$\varepsilon(\sigma, \tau_i) = (-1)^i \quad (i = 1, 2)$$

where $\{\tau_1, \tau_2\}$ is the numbering of the set of vertices of σ such that $\mathrm{or}(\sigma)$ is as shown on the picture

$$\underset{\tau_1}{\circ} \overset{\sigma}{\longrightarrow} \underset{\tau_2}{\circ} \ .$$

For each facet σ of $\mathcal{I}$ and each $g \in F^{\times}\backslash G(F)$ we also have a sign

$$\varepsilon(g, \sigma) \in \{\pm 1\}.$$

If $\dim(\sigma) > 0$, it is defined by

$$\varepsilon(g, \sigma) = 1$$

if $\mathrm{or}(g(\sigma))$ is the image $g(\mathrm{or}(\sigma))$ of $\mathrm{or}(\sigma)$ by g and by

$$\varepsilon(g, \sigma) = -1$$

otherwise; if σ is a vertex, it is defined by

$$\varepsilon(g, \sigma) = 1 \ .$$

For any facets σ, τ of $\mathcal{I}$ such that $\tau \subset \bar{\sigma}$ and $\dim(\tau) = \dim(\sigma) - 1$ and for any $g \in F^\times \backslash G(F)$, we have

$$\varepsilon(g(\sigma), g(\tau)) = \varepsilon(g, \sigma)\varepsilon(\sigma, \tau)\varepsilon(g, \tau)$$

and, for any facet σ of $\mathcal{I}$ and any $g, h \in F^\times \backslash G(F)$, we have

$$\varepsilon(gh, \sigma) = \varepsilon(g, h(\sigma))\varepsilon(h, \sigma).$$

Now we can define an augmented complex of $\mathbf{Q}$-vector spaces

(8.4.1) $$(C_\bullet(\mathcal{I}), \partial_\bullet) \xrightarrow{\varepsilon} \mathbf{Q}$$

in the following way. For each integer $n \geq 0$, we set

$$C_n(\mathcal{I}) = \mathbf{Q}^{\mathcal{F}_n}$$

where $\mathcal{F}_n$ is the set of n-dimensional facets of $\mathcal{I}$ (see (5.2)) (in particular, $C_n(\mathcal{I}) = (0)$ if $n > d - 1$) and we define

$$\partial_n : C_{n+1}(\mathcal{I}) \to C_n(\mathcal{I})$$

by

$$\partial_n(e_\sigma) = \sum_{\substack{\tau \in \mathcal{F}_n \\ \tau \subset \bar{\sigma}}} \varepsilon(\sigma, \tau)e_\tau$$

for any $\sigma \in \mathcal{F}_{n+1}$, where $(e_\sigma)_{\sigma \in \mathcal{F}_{n+1}}$ and $(e_\tau)_{\tau \in \mathcal{F}_n}$ are the canonical bases of $C_{n+1}(\mathcal{I})$ and $C_n(\mathcal{I})$ respectively; we define

$$\varepsilon : C_0(\mathcal{I}) \to \mathbf{Q}$$

by

$$\varepsilon(e_\sigma) = 1$$

for any vertex σ. It is easy to check that

$$\partial_n \circ \partial_{n+1} = 0$$

for any non-negative integer n and that

$$\varepsilon \circ \partial_0 = 0 .$$

LEMMA (8.4.2). — *The augmented complex $(C_\bullet(\mathcal{I}), \partial_\bullet) \xrightarrow{\varepsilon} \mathbf{Q}$ is acyclic.*

Proof: The complex $(C_\bullet(\mathcal{I}), \partial_\bullet)$ computes the singular homology of $\mathcal{I}$ and $\mathcal{I}$ is contractible. $\qquad\square$

For each non-negative integer n, let us set

$$(8.4.3) \qquad\qquad \rho_n(g)(e_\sigma) = \varepsilon(g,\sigma)e_{g(\sigma)}$$

for any $g \in F^\times\backslash G(F)$ and any $\sigma \in \mathcal{F}_n$. This defines an action of $F^\times\backslash G(F)$ on $C_n(\mathcal{I})$ and on $\mathbb{C} \otimes C_n(\mathcal{I})$ (see the above formula for $\varepsilon(gh,\sigma)$). Then

$$(\mathbb{C} \otimes C_n(\mathcal{I}), \rho_n)$$

is a smooth representation of $\mathrm{Rep}_s(F^\times\backslash G(F))$ (for each $\sigma \in \mathcal{F}$, $F^\times\backslash G(F)_\sigma$ is a (compact) open subgroup of $F^\times\backslash G(F)$) and

$$\mathbb{C} \otimes \partial_n : (\mathbb{C} \otimes C_{n+1}(\mathcal{I}), \rho_{n+1}) \to (\mathbb{C} \otimes C(\mathcal{I}), \rho_n)$$

is a morphism in $\mathrm{Rep}_s(F^\times\backslash G(F))$ (see the above formula for $\varepsilon(g(\sigma), g(\tau))$); moreover,

$$\mathbb{C} \otimes \varepsilon : (\mathbb{C} \otimes C_0(\mathcal{I}), \rho_0) \to (\mathbb{C}, 1)$$

is also a morphism in $\mathrm{Rep}_s(F^\times\backslash G(F))$. Therefore, thanks to (8.4.2),

$$(8.4.4) \qquad\qquad ((\mathbb{C} \otimes C_\bullet(\mathcal{I}), \rho_\bullet), \mathbb{C} \otimes \partial_\bullet) \xrightarrow{\mathbb{C} \otimes \varepsilon} (\mathbb{C}, 1)$$

is a resolution of $(\mathbb{C}, 1)$ in $\mathrm{Rep}_s(F^\times\backslash G(F))$.

PROPOSITION (8.4.5). — *For each non-negative integer n, $(\mathbb{C} \otimes C_n(\mathcal{I}), \rho_n)$ is a projective object in $\mathrm{Rep}_s(F^\times\backslash G(F))$. In other words, (8.4.4) is a projective resolution of $(\mathbb{C}, 1)$ in $\mathrm{Rep}_s(F^\times\backslash G(F))$.*

Proof : Let $n \in \{0, \dots, d-1\}$ and let $\mathcal{S}_n \subset \mathcal{F}_n$ be a system of representatives of the $(F^\times\backslash G(F))$-orbits.

Then we have a canonical isomorphism in $\mathrm{Rep}_s(F^\times\backslash G(F))$ (see (D.5.8)),

$$(8.4.6) \qquad \bigoplus_{\sigma \in \mathcal{S}_n} c\text{-} \mathrm{Ind}_{F^\times\backslash G(F)_\sigma}^{F^\times\backslash G(F)} (\mathbb{C}, \mathrm{sgn}_\sigma) \xrightarrow{\sim} (\mathbb{C} \otimes C_n(\mathcal{I}), \rho_n),$$

where $\mathrm{sgn}_\sigma : F^\times\backslash G(F)_\sigma \to \{\pm 1\}$ is defined in (5.2) for each $\sigma \in \mathcal{F}$ (it is a smooth character of $F^\times\backslash G(F)_\sigma$ as it is trivial on $G(F)_\sigma^\circ$). Indeed, we have an obvious isomorphism

$$\bigoplus_{\sigma \in \mathcal{S}_n} (\mathbb{C}^{(F^\times\backslash G(F))\cdot\sigma}, \rho_n|\mathbb{C}^{(F^\times\backslash G(F))\cdot\sigma}) \xrightarrow{\sim} (\mathbb{C} \otimes C_n(\mathcal{I}), \rho_n)$$

and, for each $\sigma \in \mathcal{S}_n$, we have a canonical isomorphism

$$c\text{-} \mathrm{Ind}_{F^\times\backslash G(F)_\sigma}^{F^\times\backslash G(F)} (\mathbb{C}, \mathrm{sgn}_\sigma) \xrightarrow{\sim} (\mathbb{C}^{(F^\times\backslash G(F))\cdot\sigma}, \rho_n|\mathbb{C}^{(F^\times\backslash G(F))\cdot\sigma})$$

which maps a locally constant function with compact support

$$v : F^\times \backslash G(F) \to \mathbb{C}$$

such that

$$v(hg) = \operatorname{sgn}_\sigma(h) v(g)$$

for any $h \in F^\times \backslash G(F)_\sigma$ and any $g \in F^\times \backslash G(F)$ onto

$$\bigoplus_g v(g^{-1}) \varepsilon(g, \sigma) e_{g(\sigma)}$$

where g runs through a system of representatives in $F^\times \backslash G(F)$ of the classes in $F^\times \backslash G(F) / G(F)_\sigma$ (we have

$$\operatorname{sgn}_\sigma(h) = \varepsilon(h, \sigma)$$

and

$$v(h^{-1} g^{-1}) \varepsilon(gh, \sigma) e_{gh(\sigma)} = v(g^{-1}) \operatorname{sgn}_\sigma(h^{-1}) \varepsilon(g, h(\sigma)) \varepsilon(h, \sigma) e_{g(\sigma)}$$
$$= v(g^{-1}) \varepsilon(g, \sigma) e_{g(\sigma)}$$

for any $h \in G(F)_\sigma$ and any $h \in F^\times \backslash G(F)$). Thanks to (D.5.9), this completes the proof of the proposition. $\square$

For each non-negative integer n, let

(8.4.7) $$H^n(F^\times \backslash G(F), (\mathcal{V}, \pi))$$

be the n-th **Yoneda extension $\mathbb{C}$-vector space** of $(\mathbb{C}, 1)$ by $(\mathcal{V}, \pi)$ in the abelian and $\mathbb{C}$-linear category $\operatorname{Rep}_s(F^\times \backslash G(F))$. As $\operatorname{Rep}_s(F^\times \backslash G(F))$ has enough injective objects (see (D.5.1)), the functor

$$(\mathcal{V}, \pi) \mapsto H^n(F^\times \backslash G(F), (\mathcal{V}, \pi))$$

is also the n-th derived functor of the left exact functor

$$(\mathcal{V}, \pi) \mapsto H^0(F^\times \backslash G(F), (\mathcal{V}, \pi)) = \mathcal{V}^{F^\times \backslash G(F)}.$$

For any projective resolution

$$((\mathcal{W}_\bullet, \rho_\bullet), \partial_\bullet) \to (\mathbb{C}, 1)$$

of $(\mathbb{C}, 1)$ in $\operatorname{Rep}_s(F^\times \backslash G(F))$, $H^n(F^\times \backslash G(F), (\mathcal{V}, \pi))$ is canonically isomorphic to the n-th cohomology $\mathbb{C}$-vector space of the complex

$$\operatorname{Hom}_{\operatorname{Rep}_s(F^\times \backslash G(F))}(((\mathcal{W}_\bullet, \rho_\bullet), \partial_\bullet), (\mathcal{V}, \pi)).$$

COROLLARY (8.4.8). — (i) *Let* $(\mathcal{V}, \pi) \in \mathrm{ob}\,\mathrm{Rep}_s(F^\times \backslash G(F))$. *Then* $H^\bullet(F^\times \backslash G(F), (\mathcal{V}, \pi))$ *is canonically isomorphic to the cohomology of the complex*

$$(C^\bullet(\mathcal{I}, \mathcal{V})^{F^\times \backslash G(F)}, \partial^\bullet | C^\bullet(\mathcal{I}, \mathcal{V})^{F^\times \backslash G(F)})$$

where

$$C^\bullet(\mathcal{I}, \mathcal{V}) = \mathrm{Hom}_{\mathbb{Q}}(C_\bullet(\mathcal{I}), \mathcal{V})$$

and

$$\partial^\bullet = \mathrm{Hom}_{\mathbb{Q}}(\partial_\bullet, \mathcal{V})$$

and where $F^\times \backslash G(F)$ *acts on* $C^\bullet(\mathcal{I}, \mathcal{V})$ *by*

$$(g, c^\bullet) \mapsto \pi(g) \circ c^\bullet \circ \rho_\bullet(g^{-1})$$

for any $g \in F^\times \backslash G(F)$ *and any* $c^\bullet \in C^\bullet(\mathcal{I}, \mathcal{V})$. *In particular, we have*

$$H^n(F^\times \backslash G(F), (\mathcal{V}, \pi)) = (0)$$

for all integers $n > d - 1$.

(ii) *Let us assume moreover that* $(\mathcal{V}, \pi)$ *is admissible. Then, for all non-negative integers* n, *we have*

$$\dim_{\mathbb{C}} H^n(F^\times \backslash G(F), (\mathcal{V}, \pi)) < +\infty$$

and, for any system $\mathcal{S}$ *of representatives of the* $(F^\times \backslash G(F))$-*orbits in* $\mathcal{F}$, *the Euler–Poincaré characteristic*

$$\mathrm{EP}(\mathcal{V}, \pi) = \sum_{n=0}^{d-1} (-1)^n \dim_{\mathbb{C}} H^n(F^\times \backslash G(F), (\mathcal{V}, \pi))$$

is equal to

$$\sum_{\sigma \in \mathcal{S}} (-1)^{\dim(\sigma)} \dim_{\mathbb{C}}(\mathcal{V}_\sigma)$$

where

$$\mathcal{V}_\sigma = \{v \in \mathcal{V} | \pi(h)(v) = \mathrm{sgn}_\sigma(h)v, \ \forall h \in F^\times \backslash G(F)_\sigma\}$$

for each $\sigma \in \mathcal{F}$.

Proof : Part (i) follows directly from (8.4.5). Now, thanks to (8.4.6), for each non-negative integer n, the $\mathbb{C}$-vector space

$$C^n(\mathcal{I}, \mathcal{V})^{F^\times \backslash G(F)}$$

is canonically isomorphic to

$$\bigoplus_{\sigma \in \mathcal{S}_n} \mathrm{Hom}_{\mathrm{Rep}_s(F^\times \backslash G(F))}(c\text{-}\mathrm{Ind}_{F^\times \backslash G(F)_\sigma}^{F^\times \backslash G(F)}(\mathbb{C}, \mathrm{sgn}_\sigma), (\mathcal{V}, \pi))$$

where $\mathcal{S}_n = \mathcal{S} \cap \mathcal{F}_n$, if $\mathcal{S}$ is any system of representatives of the $(F^\times \backslash G(F))$-orbits in $\mathcal{F}$. But, by adjunction (see (D.5)), this last direct sum is canonically isomorphic to

$$\bigoplus_{\sigma \in \mathcal{S}_n} \operatorname{Hom}_{\operatorname{Rep}_s(F^\times \backslash G(F)_\sigma)}((\mathbb{C}, \operatorname{sgn}_\sigma), (\mathcal{V}, \pi | (F^\times \backslash G(F)_\sigma))),$$

i.e. is canonically isomorphic to

$$\bigoplus_{\sigma \in \mathcal{S}_n} \mathcal{V}_\sigma \ ,$$

and part (ii) of the corollary follows (we have

$$\mathcal{V}_\sigma \subset \mathcal{V}^{F^\times \backslash G(F)^\circ_\sigma}$$

for each $\sigma \in \mathcal{F}$). $\qquad\square$

An immediate consequence of this corollary is the following cohomological interpretation of $\operatorname{tr} \pi(f)$ for our very cuspidal Euler–Poincaré function.

PROPOSITION (8.4.9). — *For any admissible representation $(\mathcal{V}, \pi)$ of $F^\times \backslash G(F)$ and for any system $\mathcal{S}$ of representatives of the $(F^\times \backslash G(F))$-orbits in $\mathcal{F}$, we have*

$$\operatorname{tr} \pi(f_\mathcal{S}) = \operatorname{EP}(\mathcal{V}, \pi)$$

where $f_\mathcal{S}$ is the corresponding Kottwitz function (see (5.2.1)). In particular,

$$\operatorname{tr} \pi(f) = \operatorname{EP}(\mathcal{V}, \pi)$$

where f is our very cuspidal Euler–Poincaré function (see (5.2.2)). $\qquad\square$

If χ is a smooth character of $F^\times \backslash T(F)$, we have proved that

$$\operatorname{tr} i(\chi)(f) = 0$$

(see (8.3.4)), so that

$$\operatorname{EP}(I(\chi), i(\chi)) = 0 \ ,$$

thanks to the above proposition. In fact, if

$$\chi \neq \delta_{B(F)}^{-1/2},$$

this vanishing of the Euler–Poincaré characteristic is a consequence of the vanishing of the cohomology itself.

LEMMA (8.4.10). — *For any smooth character χ of $F^\times\backslash T(F)$ such that*

$$\chi \neq \delta_{B(F)}^{-1/2}$$

we have

$$H^\bullet(F^\times\backslash G(F),(I(\chi),i(\chi))) = (0).$$

Proof: Thanks to (D.5.6), it suffices to prove that

$$H^n(F^\times\backslash T(F),(\mathbb{C},\delta_{B(F)}^{-1/2}\chi)) = (0)$$

for any non-negative integer, if $\chi \neq \delta_{B(F)}^{-1/2}$. But this follows immediately from (D.5.7). $\square$

Now this lemma can be reinforced.

PROPOSITION (8.4.11) (Casselman). — *For any smooth character χ of $F^\times\backslash T(F)$ such that*

$$\chi \notin W(\delta_{B(F)}^{-1/2})$$

and for any subquotient $(\mathcal{V},\pi)$ of $(I(\chi),i(\chi))$ in $\mathrm{Rep}_s(F^\times\backslash G(F))$, we have

$$H^\bullet(F^\times\backslash G(F),(\mathcal{V},\pi)) = (0)$$

and, in particular (see (8.4.9)),

$$\mathrm{tr}\ \pi(f) = \mathrm{EP}(\mathcal{V},\pi) = 0.$$

Proof : By induction on the non-negative integer n, let us prove the following assertion:

$$H^n(F^\times\backslash G(F),(\mathcal{V},\pi)) = (0)$$

for any smooth character χ of $F^\times\backslash T(F)$ such that $\chi \notin W(\delta_{B(F)}^{-1/2})$ and for any subquotient $(\mathcal{V},\pi)$ of $(I(\chi),i(\chi))$ in $\mathrm{Rep}_s(F^\times\backslash G(F))$.

Let us assume that $n = 0$ or that $n > 0$ and the assertion is proved for $n - 1$. We want to prove the assertion for n. As $(I(\chi),i(\chi))$ has finite length (see (7.3.3) (i)), we can assume that $(\mathcal{V},\pi)$ is irreducible. Then, thanks to (7.3.5), we can assume that $(\mathcal{V},\pi)$ is a subrepresentation of $(I(\chi'),i(\chi'))$ for some smooth character $\chi' \neq \delta_{B(F)}^{-1/2}$ of $F^\times\backslash T(F)$. Let us denote by $(\mathcal{W},\rho)$ the quotient of $(I(\chi'),i(\chi'))$ by $(\mathcal{V},\pi)$ in $\mathrm{Rep}_s(F^\times\backslash G(F))$. We have an exact sequence

$$H^{n-1}(F^\times\backslash G(F),(\mathcal{W},\rho)) \quad \rightarrow \quad H^n(F^\times\backslash G(F),(\mathcal{V},\pi))$$
$$\rightarrow \quad H^n(F^\times\backslash G(F),(I(\chi'),i(\chi'))).$$

But $H^{n-1}(F^\times\backslash G(F),(\mathcal{W},\rho)) = (0)$ (by definition if $n = 0$ and by induction hypothesis if $n > 0$) and $H^n(F^\times\backslash G(F),(I(\chi'),i(\chi'))) = (0)$, thanks to (8.4.10). Therefore, $H^n(F^\times\backslash G(F),(\mathcal{V},\pi))$ vanishes and the proposition is proved. $\square$

Summary of the proof of (8.2.1). Assertion (i) follows from (8.3.2), and (8.4.11). Assertion (ii) follows from (8.3.6) and (8.3.7). $\qquad\square$

(8.5) Unitarizable representations

A smooth representation $(\mathcal{V}, \pi)$ of $F^{\times}\backslash G(F)$ is **unitarizable** if there exists a $(F^{\times}\backslash G(F))$-invariant, positive definite, Hermitian scalar product $(\,,\,)$ on $\mathcal{V}$.

If $(\mathcal{V}, \pi)$ is irreducible, this Hermitian scalar product (if it exists) is unique up to a multiple by a positive real number (see (7.2.4) (ii) and (D.6.3) (ii)).

Obviously, the trivial representation $(\mathbb{C}, 1)$ of $F^{\times}\backslash G(F)$ is unitarizable. The purpose of this section is to prove that, among the representations (V_I, π_I)'s of $F^{\times}\backslash G(F)$, only the trivial representation $(\mathbb{C}, 1) = (V_{\Delta}, \pi_{\Delta})$ and the Steinberg representation $(\mathrm{St}, \mathrm{st}) = (V_{\emptyset}, \pi_{\emptyset})$ are unitarizable.

For the Steinberg representation of $F^{\times}\backslash G(F)$ we will even prove a stronger statement. A **square-integrable representation** of $F^{\times}\backslash G(F)$ is an admissible representation $(\mathcal{V}, \pi)$ of $F^{\times}\backslash G(F)$ such that, for each $v \in \mathcal{V}$ and each $\tilde{v} \in \tilde{\mathcal{V}}$, the matrix coefficient

$$g \longmapsto <\tilde{v}, \pi(g)(v)>$$

is a square-integrable (locally constant) function on $F^{\times}\backslash G(F)$, i.e.

$$\int_{F^{\times}\backslash G(F)} |<\tilde{v}, \pi(g)(v)>|^2 \frac{dg}{dz} < +\infty .$$

Here, $(\tilde{\mathcal{V}}, \tilde{\pi})$ is the contragredient representation of $(\mathcal{V}, \pi)$.

PROPOSITION (8.5.1) (Casselman). — *The Steinberg representation of* $F^{\times}\backslash G(F)$ *is square-integrable and therefore unitarizable.*

Proof : Any square-integrable irreducible representation of $F^{\times}\backslash G(F)$ is unitarizable (see (D.6.4)) and $(V_{\emptyset}, \pi_{\emptyset})$ is irreducible (see (8.1.2) (i)). So it suffices to prove that $(V_{\emptyset}, \pi_{\emptyset})$ is square-integrable.

To do this, let us use the Cartan decomposition (see (4.1))

$$F^{\times}\backslash G(F) = \coprod_{\lambda+\mathbf{Z}} F^{\times}\backslash(K\varpi^{\lambda+\mathbf{Z}}K),$$

where $\lambda + \mathbf{Z}$ runs through the set

$$\Lambda = \{\lambda + \mathbf{Z} \in \mathbf{Z}^d/\mathbf{Z} | \lambda_1 \geq \lambda_2 \geq \cdots \geq \lambda_d\}$$

($\mathbf{Z}$ is diagonally embedded in $\mathbf{Z}^d$). For any $v_0 \in V_{\emptyset}$ and any $\tilde{v}_0 \in \tilde{V}_{\emptyset}$, we get that

$$\int_{F^{\times}\backslash G(F)} |<\tilde{v}_0, \pi_{\emptyset}(g)(v_0)>|^2 \frac{dg}{dz}$$

is equal to

$$\sum_{\lambda+\mathbb{Z}} \int_{F^\times\backslash(K\varpi^{\lambda+\mathbb{Z}}K)} | < \tilde{v}_0, \pi_\emptyset(g)(v_0) > |^2 \frac{dg}{dz}$$

and we want to prove that this series is convergent. As $\pi_\emptyset(K)(v_0)$ and $\tilde{\pi}_\emptyset(K)(\tilde{v}_0)$ are finite sets, we have

$$\int_{F^\times\backslash(K\varpi^{\lambda+\mathbb{Z}}K)} | < \tilde{v}_0, \pi_\emptyset(g)(v_0) > |^2 \frac{dg}{dz}$$

$$\leq \mathrm{vol}(\lambda) \sum_{\substack{v\in\pi_\emptyset(K)(v_0) \\ \tilde{v}\in\tilde{\pi}_\emptyset(K)(\tilde{v}_0)}} | < \tilde{v}, \pi_\emptyset(\varpi^\lambda)(v) > |^2$$

for each $\lambda + \mathbb{Z}$, where we have set

$$\mathrm{vol}(\lambda) = \mathrm{vol}(F^\times\backslash(K\varpi^{\lambda+\mathbb{Z}}K), \frac{dg}{dz}).$$

Therefore, in order to prove the convergence of the above series, it suffices to prove the convergence of the series

$$\sum_{\lambda+\mathbb{Z}} \mathrm{vol}(\lambda)| < \tilde{v}, \pi_\emptyset(\varpi^\lambda)(v) > |^2$$

for all $v \in \pi_\emptyset(K)(v_0)$ and all $\tilde{v} \in \tilde{\pi}_\emptyset(K)(\tilde{v}_0)$. Therefore, in order to prove the proposition, it suffices to prove the convergence of the series

$$\sum_{\lambda+\mathbb{Z}} \mathrm{vol}(\lambda)| < \tilde{v}, \pi_\emptyset(\varpi^\lambda)(v) > |^2$$

for all $v \in V_\emptyset$ and all $\tilde{v} \in \tilde{V}_\emptyset$.

Let us fix $v \in V_\emptyset$ and $\tilde{v} \in \tilde{V}_\emptyset$. Thanks to (7.1.5), there exists $\varepsilon' \in]0,1]$ such that, for any $J \subset \Delta$, we have

$$< \tilde{v}, \pi_\emptyset(z_\emptyset')(v) > = \delta_{P_J(F)}^{1/2}(z_\emptyset') < \iota_J(p_J^\vee(\tilde{v})), \sigma_J(z_\emptyset')(p_J(v)) >$$

for all $z_\emptyset' \in \mathcal{Z}_{\emptyset,J}^\Delta(\varepsilon')$. Here,

$$p_J : V_\emptyset \longrightarrow V_\emptyset/V_\emptyset(N_J(F))$$

and

$$p_J^\vee : \tilde{V}_\emptyset \longrightarrow \tilde{V}_\emptyset/\tilde{V}_\emptyset(\tilde{N}_J(F))$$

are the canonical projections, we have set

$$r_\Delta^J(V_\emptyset, \pi_\emptyset) = (V_\emptyset/V_\emptyset(N_J(F)), \sigma_J)$$

and ι_J is the canonical isomorphism from $\tilde{r}_\Delta^J(\tilde{V}_\emptyset, \tilde{\pi}_\emptyset)$ onto the contra-gredient representation of $r_\Delta^J(V_\emptyset, \pi_\emptyset)$. We will use this result to bound $|<\tilde{v}, \pi_\emptyset(\varpi^\lambda)(v)>|^2$ on

$$\Lambda_J = \{\lambda + \mathbb{Z} \mid \varepsilon' < q^{\lambda_{j+1} - \lambda_j} \leq 1 \text{ if } j \in J \text{ and } q^{\lambda_{j+1} - \lambda_j} \leq \varepsilon' \text{ if } j \in \Delta - J\}$$

for all $J \subset \Delta$.

First of all, let us show that, for any $J \subset \Delta$, $r_\Delta^J(V_\emptyset, \pi_\emptyset)$ is irre-ducible and that its central character (see (D.1.12)) is equal to $\delta_{B(F)}^{1/2}|Z_J(F)$. If $J = \emptyset$, this follows from (8.1.2) (ii). Indeed, the longest element $w_0 \in W$ is the unique element of W such that $w_0(R^+) \cap \Delta = \emptyset$ and it satisfies $w_0(R^+) = R^-$, so that $w_0(\delta_{B(F)}^{-1/2}) = \delta_{B(F)}^{1/2}$. In general, thanks to (7.1.3) (i) and the definition of $(V_\emptyset, \pi_\emptyset)$, $r_\Delta^J(V_\emptyset, \pi_\emptyset)$ is a quotient of $r_\Delta^J(i_\emptyset^\Delta(\mathbb{C}, \delta_{B(F)}^{-1/2}))$. Therefore, for any irreducible subquotient $(\mathcal{W}, \rho)$ of $r_\Delta^J(V_\emptyset, \pi_\emptyset)$ in $\text{Rep}_s(M_J(F))$, there exists at least one $w \in D_{\emptyset, J}$ such that $(\mathcal{W}, \rho)$ is isomorphic to a subquotient of $i_\emptyset^J(\mathbb{C}, w^{-1}(\delta_{B(F)}^{-1/2}))$ (see (7.3.4)). Then an obvious variant of (7.3.5) (replace G by $M_J = GL_{d_1} \times \cdots \times GL_{d_s}$ if $d_J = (d_1, \ldots, d_s)$) implies that $(\mathcal{W}, \rho)$ is isomorphic to a subrepresen-tation of $i_\emptyset^J(\mathbb{C}, w_J w^{-1}(\delta_{B(F)}^{-1/2}))$ for at least one $w_J \in W_J$ and it follows that $(\mathbb{C}, w_J w^{-1}(\delta_{B(F)}^{-1/2}))$ is isomorphic to a quotient of $r_J^\emptyset(\mathcal{W}, \rho)$ (see (7.1.3) (i)). In particular, $r_J^\emptyset(\mathcal{W}, \rho) \neq (0)$. But $r_J^\emptyset$ is exact and $r_J^\emptyset(r_\Delta^J(V_\emptyset, \pi_\emptyset))$ is isomorphic to $r_\Delta^\emptyset(V_\emptyset, \pi_\emptyset)$ (see (7.1.3) (i) and (iv)). Hence, the irreducibility of $r_\Delta^\emptyset(V_\emptyset, \pi_\emptyset)$ implies the irreducibility of $r_\Delta^J(V_\emptyset, \pi_\emptyset)$. Now, if ω is the central character of $r_\Delta^J(V_\emptyset, \pi_\emptyset)$, $Z_J(F)$ acts through ω on $r_\Delta^\emptyset(V_\emptyset, \pi_\emptyset)$, i.e. on $r_\Delta^J(V_\emptyset, \pi_\emptyset)$ and $\delta_{B(F)}^{1/2}|Z_J(F) = \omega$.

Then let us fix $J \subset \Delta$ and let Γ_J be the additive semi-group

$$\{\lambda + \mathbb{Z} \mid \lambda_j = \lambda_{j+1} \text{ if } j \in J \text{ and } \lambda_j \leq \lambda_{j+1} \text{ if } j \in \Delta - J\}.$$

We have

$$\Gamma_J + \Lambda_J \subset \Lambda_J$$

and there exists a finite subset Σ_J of Λ_J such that

$$\Lambda_J = \Gamma_J + \Sigma_J.$$

If $\lambda + \mathbb{Z} \in \Lambda_J$ (resp. Γ_J), $\varpi^\lambda \in \mathcal{Z}_{\emptyset, J}^\Delta(\varepsilon')$ (resp. $\mathcal{Z}_J^\Delta(1) \subset Z_J(F)$) (see (7.1)). Therefore, we have

$$<\tilde{v}, \pi_\emptyset(\varpi^{\lambda + \mu})(v)> = (\delta_{P_J(F)} \delta_{B(F)})^{1/2}(\varpi^\lambda) <\tilde{v}, \pi_\emptyset(\varpi^\mu)(v)>$$

for any $\lambda + \mathbb{Z} \in \Gamma_J$ and any $\mu + \mathbb{Z} \in \Lambda_J$ and the series

$$\sum_{\lambda + \mathbb{Z} \in \Lambda_J} \text{vol}(\lambda)|<\tilde{v}, \pi_\emptyset(\varpi^\lambda)(v)>|^2$$

converges if and only if the series

$$\sum_{\lambda + \mathbf{Z} \in \Gamma_J} \text{vol}(\lambda)\delta^2_{P_J(F)}(\varpi^\lambda)$$

converges $(\delta_{B(F)}(\varpi^\lambda) = \delta_{P_J(F)}(\varpi^\lambda)$ if $\lambda + \mathbf{Z} \in \Gamma_J)$.

Finally, as the Haar measures dg and dz are normalized by $\text{vol}(K, dg) = \text{vol}(\mathcal{O}^\times, dz) = 1$, we have

$$\text{vol}(\lambda) = |K/(K \cap \varpi^\lambda K \varpi^{-\lambda})|,$$

i.e.

$$\text{vol}(\lambda) = \prod_{n=-1}^{\infty} |K(n+1)\backslash K(n)/(K(n) \cap (\varpi^\lambda K \varpi^{-\lambda}))|$$

$(K(-1) = K)$, for all $\lambda + \mathbf{Z} \in \Gamma_J$. But, the map $g \mapsto g(\text{mod } \varpi gl_d(\mathcal{O}))$ (resp. $1 + \varpi^{n+1}\xi \mapsto \xi$) induces a bijection from $K(1)\backslash K/(K(1) \cap (\varpi^\lambda K \varpi^{-\lambda}))$ (resp. $K(n+1)\backslash K(n)/(K(n) \cap (\varpi^\lambda K \varpi^{-\lambda}))$ for $n \geq 0$) onto $G(k)/P_I(k)$ (resp.

$$gl_d(\mathcal{O})/(\varpi gl_d(\mathcal{O}) + gl_d(\mathcal{O}) \cap \varpi^{-n-1}(\varpi^\lambda gl_d(\mathcal{O})\varpi^{-\lambda}))),$$

so that

$$\text{vol}(\lambda) = |G(k)/P_I(k)|\delta^{-1}_{P_J(F)}(\varpi^\lambda)$$

for all $\lambda + \mathbf{Z} \in \Gamma_J$. It follows that

$$\sum_{\lambda + \mathbf{Z} \in \Gamma_J} \text{vol}(\lambda)\delta^2_{P_J(F)}(\varpi^\lambda) = |G(k)/P_I(k)| \sum_{\lambda + \mathbf{Z} \in \Gamma_J} \delta_{P_J(F)}(\varpi^\lambda) < +\infty.$$

As $(\Lambda_J)_{J \subset \Delta}$ is a partition of Λ, that completes the proof of the proposition.
□

THEOREM (8.5.2) (Casselman). — *Let $I \subset \Delta$. Then the admissible irreducible representation (V_I, π_I) of $F^\times\backslash G(F)$ (see (8.1)) is unitarizable if and only if $I = \Delta$ (the trivial representation) or $I = \emptyset$ (the Steinberg representation).*

Putting together (8.2.1) and (8.5.2), we get

COROLLARY (8.5.3). — *Let $f \in \mathcal{C}^\infty_c(F^\times\backslash G(F)/\!/\mathcal{B}^\circ)$ be our very cuspidal Euler-Poincaré function (see (5.1.2) and (5.1.3)) and let $(\mathcal{V}, \pi)$ be an unitarizable admissible irreducible representation of $F^\times\backslash G(F)$. Then $\text{tr } \pi(f) = 0$ unless $(\mathcal{V}, \pi)$ is isomorphic to the trivial representation $(\mathbb{C}, 1)$ or to the Steinberg representation (St, st) of $F^\times\backslash G(F)$. Moreover*

$$\text{tr}(1(f)) = 1$$

and

$$\text{tr}(\text{st}(f)) = (-1)^{d-1}.$$

□

A basic ingredient of the proof of (8.5.2) is the following criterion of non-unitarizability.

THEOREM (8.5.4) (Howe and Moore). — *Let* $(\mathcal{V}, \pi)$ *be a unitarizable admissible irreducible representation of* $F^\times \backslash G(F)$. *Let us assume that the restriction of* $(\mathcal{V}, \pi)$ *to* $SL_d(F) \subset GL_d(F) = G(F)$ *is non-trivial. Then, for any* $v \in \mathcal{V}$ *and any* $\tilde{v} \in \tilde{\mathcal{V}}$ *(* $(\tilde{\mathcal{V}}, \tilde{\pi})$ *is the contragredient representation of* $(\mathcal{V}, \pi)$ *), the restriction to* $F^\times \backslash T(F)$ *of the matrix coefficient* $g \mapsto < \tilde{v}, \pi(g)(v) >$ *is square-integrable, i.e.*

$$\int_{F^\times \backslash T(F)} | < \tilde{v}, \pi(t)(v) > |^2 \frac{dt}{dz} < +\infty .$$

REMARK (8.5.5). — The square-integrability of the matrix coefficients on $F^\times \backslash T(F)$ does not imply the square-integrability of the matrix coefficients on $F^\times \backslash G(F)$. The first condition is equivalent to the convergence of the series

$$\sum_{\lambda + \mathbf{Z} \in \mathbf{Z}^d / \mathbf{Z}} | < \tilde{v}, \pi(\varpi^\lambda)(v) > |^2 ,$$

for all $v \in \mathcal{V}$ and $\tilde{v} \in \tilde{\mathcal{V}}$, and the second one is equivalent to the convergence of the series

$$\sum_{\lambda + \mathbf{Z} \in \Lambda} \mathrm{vol}(\lambda) | < \tilde{v}, \pi(\varpi^\lambda)(v) > |^2 ,$$

for all $v \in \mathcal{V}$ and $\tilde{v} \in \tilde{\mathcal{V}}$ (see the proof of (8.5.1)). $\qquad \square$

At first, let us assume the theorem (8.5.4) and let us prove (8.5.2).

Proof of (8.5.2) : The "if" part has been already proved (see (8.5.1)).

Let I be a non-empty subset of Δ. Let us denote by

$$p : V_I \longrightarrow V_I / V_I(U(F))$$

and

$$\check{p} : \tilde{V}_I \longrightarrow \tilde{V}_I / \tilde{V}_I(\tilde{U}(F))$$

the canonical projections, let us denote by ι the canonical isomorphism from $\tilde{r}_\Delta^\emptyset (\tilde{V}_I, \tilde{\pi}_I)$ onto the contragredient representation of $r_\Delta^\emptyset(V_I, \pi_I)$ (see (7.1.4) (ii)) and let us set

$$r_\Delta^\emptyset(V_I, \pi_I) = (V_I / V_I(U(F)), \sigma).$$

Thanks to (8.1.2) (iii), the smooth representation $r_\Delta^\emptyset(V_I, \pi_I)$ of $T(F)$ is isomorphic to

$$\bigoplus_{\substack{w \in W \\ w(R^+) \cap \Delta = I}} (\mathbf{C}, w^{-1}(\delta_{B(F)}^{-1/2})).$$

Let $w_{\Delta-I}$ be the longest element of $W_{\Delta-I}$. In the proof of (8.1.2), we have seen that

$$w_{\Delta-I}(R^+) \cap \Delta = I$$

and that

$$w_{\Delta-I}^{-1}(\delta_{B(F)}^{-1/2}) = \delta_{B(F)}^{-1/2} \delta_{P_\emptyset^{\Delta-I}(F)}.$$

Therefore, there exists $v \in V_I$, $v \neq 0$, such that

$$\sigma(t)(p(v)) = (\delta_{B(F)}^{-1/2} \delta_{P_\emptyset^{\Delta-I}(F)})(t)p(v)$$

for all $t \in T(F)$. Let us fix such a v in V_I and let $\tilde{v} \in \tilde{V}_I$ be an arbitrary fixed element such that $< \iota(\check{p}(\tilde{v})), p(v) > \neq 0$. Thanks to (7.1.4) (ii), there exists $\varepsilon \in]0,1]$ such that

$$< \tilde{v}, \pi_I(t)(v) > = \delta_{B(F)}^{1/2}(t) < \iota(\check{p}(\tilde{v})), \sigma(t)(p(v)) >,$$

i.e.

$$< \tilde{v}, \pi_I(t)(v) > = \delta_{P_\emptyset^{\Delta-I}(F)}(t) < \iota(\check{p}(v)), p(v) >,$$

for all $t \in T(F)$ with $|\alpha(t)| \leq \varepsilon$ for each $\alpha \in \Delta$.

Now, if we arbitrarily fix $t_0 \in T(F)$ with $|\alpha(t_0)| \leq \varepsilon$ for each $\alpha \in \Delta$ and if we denote by A the set of $t \in T(F)$ such that

$$|\alpha(t)| \leq 1$$

for all $\alpha \in I$ and

$$|\alpha(t)| = 1$$

for all $\alpha \in \Delta - I$, we have

$$< \tilde{v}, \pi_I(tt_0)(v) > = \delta_{P_\emptyset^{\Delta-I}(F)}(t_0) < \iota(\check{p}(v)), p(v) >$$

for all $t \in A$, so that the matrix coefficient $g \mapsto < \tilde{v}, \pi_I(g)(v) >$ is constant over the subset

$$At_0 \subset T(F) \subset G(F).$$

As

$$\mathrm{vol}(F^\times \backslash At_0, \frac{dt}{dz}) = +\infty$$

(here we use the hypothesis $I \neq \emptyset$), that implies that the restriction of this matrix coefficient to $F^\times \backslash T(F)$ cannot be square-integrable. Therefore, thanks to (8.5.4), (V_I, π_I) is unitarizable only if π_I is trivial on $SL_d(F) \subset G(F)$.

But π_I is trivial on $SL_d(F)$ if and only if $I = \emptyset$. Indeed, if π_I is trivial on $SL_d(F)$, it factors through the quotient

$$\det : F^\times \backslash G(F) \to (F^\times)^d \backslash F^\times \;;$$

so V_I must be one dimensional ((V_I, π_I) is irreducible), i.e. (V_I, π_I) must be isomorphic to $(\mathbb{C}, \chi \circ \det)$ for some smooth character χ of $(F^\times)^d \backslash F^\times$; if $I = \Delta$, it is the case; if $I \neq \Delta$, it is impossible as

$$r_\Delta^\emptyset(\mathbb{C}, \chi \circ \det) = (\mathbb{C}, \delta_{B(F)}^{-1/2} \chi \circ \det)|(F^\times \backslash T(F)))$$

cannot be equal to $(\mathbb{C}, w^{-1}(\delta_{B(F)}^{-1/2}))$ for any $w \in W$ with $w(R^+) \cap \Delta = I$. This completes the proof of (8.5.2). $\qquad\square$

(8.6) Proof of Howe and Moore's criterion of non-unitarizability
For each root $\alpha \in R$, $\alpha = \varepsilon_i - \varepsilon_j$ $(1 \leq i, j \leq d, i \neq j)$, let us denote by

$$x_\alpha : \mathbb{G}_a \to SL_d \subset GL_d$$

the one parameter subgroup which maps t onto

$$s_\alpha(t) = 1 + tE_{i,j}$$

($E_{i,j}$ is the elementary matrix with all entries 0 except the entry on the i-th row and the j-th column which is equal to 1).

Let us fix a unitarizable smooth representation $(\mathcal{V}, \pi)$ of $F^\times \backslash G(F)$ and a $(F^\times \backslash G(F))$-invariant, positive definite, Hermitian scalar product $(\,,\,)$ on $\mathcal{V}$. We denote by $\widehat{\mathcal{V}}$ the completion of $\mathcal{V}$ with respect to the norm $\|-\| = (-,-)^{1/2}$ and, for any $g \in F^\times \backslash G(F)$ (or $G(F)$), by $\hat{\pi}(g)$ the natural extension of $\pi(g)$ to $\widehat{\mathcal{V}}$. Then $\widehat{\mathcal{V}}$, together with the positive definite, Hermitian scalar product and the norm which are induced by $(-,-)$ and $\|-\|$ and that we denote again by $(-,-)$ and $\|-\|$, is a Hilbert space; for each $g \in F^\times \backslash G(F)$, $\hat{\pi}(g)$ is a unitary operator on this Hilbert space and the map

$$(F^\times \backslash G(F)) \times \widehat{\mathcal{V}} \to \widehat{\mathcal{V}}, \; (g, \hat{v}) \mapsto \hat{\pi}(g)(\hat{v})$$

is continuous.

In the proof of (8.5.4), we will need the following lemma and its corollary.

LEMMA (8.6.1). — *Let $\hat{v} \in \widehat{\mathcal{V}}$ with $\|\hat{v}\| = 1$. If there exists $\alpha \in R$ such that*

$$\hat{\pi}(x_\alpha(t))(\hat{v}) = \hat{v}$$

for all $t \in F$, then we have

$$\hat{\pi}(g)(\hat{v}) = \hat{v}$$

for all $g \in SL_d(F) \subset G(F)$.

Proof : If $\alpha = \varepsilon_i - \varepsilon_j$ there exists $w \in \mathfrak{G}_d = W$ such that $w(i) = 1$ and $w(j) = 2$. For such a $w \in W$, we have

$$\dot{w} E_{i,j} \dot{w}^{-1} = E_{w(i),w(j)} = E_{1,2},$$

so that

$$\dot{w} x_\alpha(t) \dot{w}^{-1} = x_{\varepsilon_1 - \varepsilon_2}(t)$$

and

$$\hat{\pi}(x_{\varepsilon_1 - \varepsilon_2}(t))(\hat{\pi}(\dot{w})(v)) = \hat{\pi}(\dot{w})(v)$$

for all $t \in F$. Therefore, as $SL_d(F)$ is a normal subgroup of $G(F)$, we can (and we will) assume that $\alpha = \varepsilon_1 - \varepsilon_2$.

Let us consider the embedding

$$\iota : SL_2 \to GL_d$$

defined by

$$\iota(A) = \begin{pmatrix} A & 0 \\ 0 & 1_{d-2} \end{pmatrix}$$

for any $A \in SL_2$ (1_{d-2} is the identity matrix of GL_{d-2}). We have

$$\pi(\iota(\begin{pmatrix} 1 & t \\ 0 & 1 \end{pmatrix})))(\hat{v}) = \hat{v}$$

for any $t \in F$. Let us consider the continuous function

$$\varphi : SL_2(F) \to \mathbb{R}_{>0}$$

defined by

$$\varphi(A) = (\hat{v}, \hat{\pi}(\iota(A))(\hat{v}))$$

for any $A \in SL_2(F)$. We have

$$\varphi(\begin{pmatrix} 1 & t' \\ 0 & 1 \end{pmatrix} A \begin{pmatrix} 1 & t'' \\ 0 & 1 \end{pmatrix}) = \varphi(A)$$

for any $t', t'' \in F$ and any $A \in SL_2(F)$. Therefore, we have

$$\varphi(A) = \varphi(\begin{pmatrix} 1 & 0 \\ a_{21} & 1 \end{pmatrix}))$$

for any

$$A = \begin{pmatrix} a_{11} & a_{12} \\ a_{21} & a_{22} \end{pmatrix} \in SL_2(F)$$

with $a_{21} \neq 0$. Indeed, it is easy to check that

$$\begin{pmatrix} 1 & a_{21}^{-1}(1 - a_{11}) \\ 0 & 1 \end{pmatrix} A \begin{pmatrix} 1 & a_{21}^{-1}(1 - a_{22}) \\ 0 & 1 \end{pmatrix} = \begin{pmatrix} 1 & 0 \\ a_{21} & 1 \end{pmatrix}.$$

In particular

$$\varphi(\begin{pmatrix} \varpi & 0 \\ 0 & \varpi^{-1} \end{pmatrix}) = \lim_{\substack{u \to 0 \\ u \neq 0}} \varphi(\begin{pmatrix} \varpi & 0 \\ u & \varpi^{-1} \end{pmatrix})$$

is equal to

$$\lim_{\substack{u \to 0 \\ u \neq 0}} \varphi(\begin{pmatrix} 1 & 0 \\ u & 1 \end{pmatrix}) = \varphi(\begin{pmatrix} 1 & 0 \\ 0 & 1 \end{pmatrix}) = 1$$

$(u \in F)$. It follows that

$$\hat{\pi}(\iota(\begin{pmatrix} \varpi & 0 \\ 0 & \varpi^{-1} \end{pmatrix})))(\hat{v}) = \hat{v} .$$

Indeed, if we set

$$\hat{v}_1 = \hat{\pi}(g)(\hat{v}), \text{ with } g = \iota(\begin{pmatrix} \varpi & 0 \\ 0 & \varpi^{-1} \end{pmatrix}),$$

we have

$$\begin{aligned} (\hat{v}_1 - \hat{v}, \hat{v}_1 - \hat{v}) &= (\hat{v}_1, \hat{v}_1) - (\hat{v}, \hat{v}_1) - (\hat{v}_1, \hat{v}) + (\hat{v}, \hat{v}) \\ &= 2(\hat{v}, \hat{v}) - 2Re(\hat{v}, \hat{v}_1) \\ &= 2 - 2Re\varphi(g) = 0 . \end{aligned}$$

Let us consider the subgroups $N_{\Delta - \{\alpha\}}(F)$ and $\tilde{N}_{\Delta - \{\alpha\}}(F)$ of $G(F)$. If $n \in N_{\Delta - \{\alpha\}}(F)$ (resp. $\tilde{n} \in \tilde{N}_{\Delta - \{\alpha\}}(F)$), the sequence

$$g_k = \iota(\begin{pmatrix} \varpi^k & 0 \\ 0 & \varpi^{-k} \end{pmatrix})n\iota(\begin{pmatrix} \varpi^{-k} & 0 \\ 0 & \varpi^k \end{pmatrix}))$$

(resp.

$$\tilde{g}_k = \iota(\begin{pmatrix} \varpi^{-k} & 0 \\ 0 & \varpi^k \end{pmatrix})\tilde{n}\iota(\begin{pmatrix} \varpi^k & 0 \\ 0 & \varpi^{-k} \end{pmatrix})))$$

$(k \in \mathbb{Z}_{>0})$ of elements of $G(F)$ converges to 1 as $k \to +\infty$. But, for each $k \in \mathbb{Z}_{>0}$, we have

$$(\hat{\pi}(g_k)(\hat{v}) - \hat{v}, \hat{\pi}(g_k)(\hat{v}) - \hat{v}) = (\hat{\pi}(n)(\hat{v}) - \hat{v}, \hat{\pi}(n)(\hat{v}) - \hat{v})$$

(resp.

$$(\hat{\pi}(\tilde{g}_k)(\hat{v}) - \hat{v}, \hat{\pi}(\tilde{g}_k)(\hat{v}) - \hat{v}) = (\hat{\pi}(\tilde{n})(\hat{v}) - \hat{v}, \hat{\pi}(\tilde{n})(\hat{v}) - \hat{v}))$$

as

$$\hat{\pi}(\iota(\begin{pmatrix} \varpi & 0 \\ 0 & \varpi^{-1} \end{pmatrix})))(\hat{v}) = \hat{v} .$$

Taking the limit as $k \to +\infty$, we obtain

$$(\hat{\pi}(n)(\hat{v}) - \hat{v}, \hat{\pi}(n)(\hat{v}) - \hat{v}) = 0$$

(resp.

$$(\hat{\pi}(\tilde{n})(\hat{v}) - \hat{v}, \hat{\pi}(\tilde{n})(\hat{v}) - \hat{v}) = 0) ,$$

i.e.

$$\hat{\pi}(n)(\hat{v}) = \hat{v}$$

(resp.

$$\hat{\pi}(\tilde{n})(\hat{v}) = \hat{v}).$$

This is the so-called Mautner phenomenon.

We have proved that

$$\hat{\pi}(g)(\hat{v}) = \hat{v}$$

for any $g \in N_{\Delta-\{\alpha\}}(F) \cup \widetilde{N}_{\Delta-\{\alpha\}}(F)$ and, therefore, for any g in the subgroup

$$< N_{\Delta-\{\alpha\}}(F), \widetilde{N}_{\Delta-\{\alpha\}}(F) > \subset SL_d(F)$$

generated by $N_{\Delta-\{\alpha\}}(F)$ and $\widetilde{N}_{\Delta-\{\alpha\}}(F)$. In fact, we have

$$< N_{\Delta-\{\alpha\}}(F), \widetilde{N}_{\Delta-\{\alpha\}}(F) > = SL_d(F).$$

This is well known and can be checked in the following way. We have seen that

$$\begin{pmatrix} a_{11} & a_{12} \\ a_{21} & a_{22} \end{pmatrix} = \begin{pmatrix} 1 & a_{21}^{-1}(a_{11} - 1) \\ 0 & 1 \end{pmatrix} \begin{pmatrix} 1 & 0 \\ a_{21} & 1 \end{pmatrix} \begin{pmatrix} 1 & a_{21}^{-1}(a_{22} - 1) \\ 0 & 1 \end{pmatrix}$$

if $a_{21} \neq 0$, so that

$$\begin{pmatrix} a_{11} & a_{12} \\ a_{21} & a_{22} \end{pmatrix} \in < U(F), \widetilde{U}(F) > \subset SL_2(F)$$

as long as $a_{21} \neq 0$ or $a_{12} \neq 0$ (the subgroup $< U(F), \widetilde{U}(F) >$ of $SL_2(F)$ generated by $U(F)$ and $\widetilde{U}(F)$ is stable by transposition); moreover, we have

$$\begin{pmatrix} a & 0 \\ 0 & a^{-1} \end{pmatrix} = \begin{pmatrix} a & a \\ 0 & a^{-1} \end{pmatrix} \begin{pmatrix} 1 & -1 \\ 0 & 1 \end{pmatrix}$$

for any $a \in F^\times$. Therefore, we have

$$< U(F), \widetilde{U}(F) > = SL_2(F).$$

It follows that, for $d \geq 2$, we have

$$\iota(SL_2(F)) \subset\, <N_{\Delta-\{\alpha\}}(F), \widetilde{N}_{\Delta-\{\alpha\}}(F)> \subset SL_d(F).$$

But it is easy to check that

$$N_{\Delta-\{\alpha,\alpha'\}}(F) \subset N_{\Delta-\{\alpha\}}(F)\iota(A)N_{\Delta-\{\alpha\}}(F)\iota(A^{-1})$$

and

$$\widetilde{N}_{\Delta-\{\alpha,\alpha'\}}(F) \subset \widetilde{N}_{\Delta-\{\alpha\}}(F)\iota(A)\widetilde{N}_{\Delta-\{\alpha\}}(F)\iota(A^{-1})$$

where $\alpha' = \varepsilon_2 - \varepsilon_3$ and $A = \begin{pmatrix} 0 & 1 \\ -1 & 0 \end{pmatrix} \in SL_2(F)$, so that

$$N_{\Delta-\{\alpha,\alpha'\}}(F) \subset\, <N_{\Delta-\{\alpha\}}(F), \widetilde{N}_{\Delta-\{\alpha\}}(F)>$$

and

$$\widetilde{N}_{\Delta-\{\alpha,\alpha'\}}(F) \subset\, <N_{\Delta-\{\alpha\}}(F), \widetilde{N}_{\Delta-\{\alpha\}}(F)>.$$

By induction on d, let us assume that our assertion holds for $SL_{d-1}(F)$. Then it follows that

$$\iota'(SL_{d-1}(F)) \subset\, <N_{\Delta-\{\alpha\}}(F), \widetilde{N}_{\Delta-\{\alpha\}}(F)>,$$

where

$$\iota' : SL_{d-1}(F) \hookrightarrow SL_d(F)$$

is the embedding defined by

$$\iota'(A') = \begin{pmatrix} 1 & 0 \\ 0 & A' \end{pmatrix}$$

for any $A' \in SL_{d-1}(F)$, so that

$$P_{\Delta-\{\alpha\}}(F) \cap SL_d(F) \subset\, <N_{\Delta-\{\alpha\}}(F), \widetilde{N}_{\Delta-\{\alpha\}}(F)>$$

and

$$\widetilde{P}_{\Delta-\{\alpha\}}(F) \cap SL_d(F) \subset\, <N_{\Delta-\{\alpha\}}(F), \widetilde{N}_{\Delta-\{\alpha\}}(F)>.$$

In particular, if

$$\begin{pmatrix} a & u \\ \tilde{u} & A' \end{pmatrix} \in SL_d(F)$$

with $a \in F$, $a \neq 0$, $A' \in gl_{d-1}(F)$, $u \in \mathrm{Mat}_{1\times(d-1)}(F)$ and $\tilde{u} \in \mathrm{Mat}_{(d-1)\times 1}(F)$, we have

$$\begin{pmatrix} a & u \\ \tilde{u} & A' \end{pmatrix} \in\, <N_{\Delta-\{\alpha\}}(F), \widetilde{N}_{\Delta-\{\alpha\}}(F)>$$

as

$$\begin{pmatrix} 1 & 0 \\ -a^{-1}\tilde{u} & 1_{d-1} \end{pmatrix} \begin{pmatrix} a & u \\ \tilde{u} & A' \end{pmatrix} \in P_{\Delta-\{\alpha\}}(F) \cap SL_d(F).$$

The equality

$$< N_{\Delta-\{\alpha\}}(F), \widetilde{N}_{\Delta-\{\alpha\}}(F) > = SL_d(F)$$

follows (for any

$$\begin{pmatrix} 0 & u \\ \tilde{u} & A' \end{pmatrix} \in SL_d(F)$$

with $u \in \mathrm{Mat}_{1\times(d-1)}(F)$, $\tilde{u} \in \mathrm{Mat}_{(d-1)\times 1}(F)$ and $A' \in gl_d(F)$, there exists $u' \in \mathrm{Mat}_{1\times(d-1)}(F)$ such that $u'\tilde{u} \neq 0$ and we have

$$\begin{pmatrix} 1 & u' \\ 0 & 1_{d-1} \end{pmatrix} \begin{pmatrix} 0 & u \\ \tilde{u} & A' \end{pmatrix} = \begin{pmatrix} u'\tilde{u} & u'A' + u \\ \tilde{u} & A' \end{pmatrix})$$

and the proof of the lemma is completed.   □

COROLLARY (8.6.2). — *Let $(\mathcal{V}, \pi)$ be a unitarizable admissible irreducible representation of $F^\times \backslash G(F)$. If there exist $\hat{v} \in \widehat{\mathcal{V}}$, $\hat{v} \neq 0$, and $\alpha \in R$ such that*

$$\hat{\pi}(x_\alpha(t))(\hat{v}) = \hat{v}$$

for all $t \in F$, then the restriction of $(\mathcal{V}, \pi)$ to $SL_d(F)$ is isomorphic to the trivial representation $(\mathbb{C}, 1)$ of $SL_d(F)$.

Proof: It suffices to prove that

$$\hat{\pi}(g) = id_{\widehat{\mathcal{V}}}$$

for any $g \in SL_d(F)$. Indeed, if this is the case, π factors through the commutative quotient

$$\det : PGL_d(F) \longrightarrow (F^\times)^d \backslash F^\times,$$

so that $(\mathcal{V}, \pi)$ can be viewed as an admissible irreducible representation of $(F^\times)^d \backslash F^\times$; but any admissible irreducible representation of $(F^\times)^d \backslash F^\times$ is isomorphic to $(\mathbb{C}, \chi)$ for some smooth character χ of $(F^\times)^d \backslash F^\times$ (thanks to (D.1.12), it admits a central character) and $(\mathcal{V}, \pi)$ is isomorphic to $(\mathbb{C}, \chi \circ \det)$ for some smooth character χ of $(F^\times)^d \backslash F^\times$.

Now let $\hat{v} \in \widehat{\mathcal{V}}$, $\hat{v} \neq 0$, and $\alpha \in R$ be such that

$$\hat{\pi}(x_\alpha(t))(\hat{v}) = \hat{v}$$

for all $t \in F$. Replacing $\hat{v}$ by $\hat{v}/\|\hat{v}\|$, we can assume that $\|\hat{v}\| = 1$. Then, thanks to (8.6.1), we get that

$$\hat{\pi}(g)(\hat{v}) = \hat{v}$$

for all $g \in SL_d(F) \subset G(F)$ and it follows that

$$\hat{\pi}(g)(\hat{w}) = \hat{w}$$

for any $g \in SL_d(F) \subset G(F)$ and any $\hat{w}$ in the closure in $\hat{\mathcal{V}}$ of the $\mathbb{C}$-linear span of $\hat{\pi}(G(F))(\hat{v})$ ($SL_d(F)$ is a normal subgroup of $G(F)$ and $\hat{\pi}(g)$ is continuous for any $g \in SL_d(F)$).

To complete the proof of the corollary, let us check that $(\hat{\mathcal{V}}, \hat{\pi})$ is topologically irreducible, so that the closure in $\hat{\mathcal{V}}$ of the $\mathbb{C}$-linear span of $\hat{\pi}(G(F))(\hat{v})$ is equal to $\hat{\mathcal{V}}$. If $\hat{\mathcal{V}}_1$ is an $(F^\times \backslash G(F))$-invariant, non-zero, closed $\mathbb{C}$-vector subspace of $\hat{\mathcal{V}}$, we want to prove that $\hat{\mathcal{V}}_1$ is equal to $\hat{\mathcal{V}}$. Let

$$\mathcal{W}_1 = \hat{\mathcal{V}}_1 \cap \mathcal{V} \; ;$$

it is an $(F^\times \backslash G(F))$-invariant $\mathbb{C}$-vector subspace of $\mathcal{V}$, so that $\mathcal{W}_1 = (0)$ or $\mathcal{W}_1 = \mathcal{V}$ ($(\mathcal{V}, \pi)$ is irreducible). If $\mathcal{W}_1 = \mathcal{V}$, i.e. if $\mathcal{V} \subset \hat{\mathcal{V}}_1$, then $\hat{\mathcal{V}}_1 = \hat{\mathcal{V}}$ as $\hat{\mathcal{V}}_1$ is closed in $\hat{\mathcal{V}}$. Therefore, it suffices to prove that $\mathcal{W}_1 \neq (0)$. But, for any compact open subgroup Γ of $G(F)$, we have

$$(\hat{\mathcal{V}})^\Gamma = \mathcal{V}^\Gamma.$$

Indeed, $\mathcal{V}^\Gamma$ is dense in $(\hat{\mathcal{V}})^\Gamma$ (if $\hat{v} \in (\hat{\mathcal{V}})^\Gamma$ and if $\varepsilon \in \mathbb{R}_{>0}$, there exists $v \in \mathcal{V}$ such that $\|\hat{v} - v\| < \varepsilon$ and we have

$$\|\hat{v} - \pi(e_\Gamma)(v)\| = \|\hat{\pi}(e_\Gamma)(\hat{v} - v)\| \leq \|\hat{v} - v\| < \varepsilon$$

with $\pi(e_\Gamma)(v) \in \mathcal{V}^\Gamma$) and $\dim_\mathbb{C}(\mathcal{V}^\Gamma) < +\infty$.

This implies that

$$\mathcal{W}_1 = \bigcup_\Gamma (\hat{\mathcal{V}}_1)^\Gamma$$

where Γ runs through the set of all compact open subgroups of $G(F)$. Let $\hat{v}_1 \in \hat{\mathcal{V}}_1$ with $\|\hat{v}_1\| = 1$. There exists $v \in \mathcal{V}$ such that $\|\hat{v}_1 - v\| < \frac{1}{2}$ and there exists a compact open subgroup Γ of $G(F)$ such that $v \in \mathcal{V}^\Gamma$. Then we have

$$\|\hat{\pi}(e_\Gamma)(\hat{v}_1) - v\| = \|\hat{\pi}(e_\Gamma)(\hat{v}_1 - v)\| \leq \|\hat{v}_1 - v\| < \frac{1}{2},$$

so that

$$\|\hat{\pi}(e_\Gamma)(\hat{v}_1) - \hat{v}_1\| < 1$$

and

$$\hat{\pi}(e_\Gamma)(\hat{v}_1) \neq 0 .$$

As we have

$$\hat{\pi}(e_\Gamma)(\hat{v}_1) \in (\hat{\mathcal{V}}_1)^\Gamma \subset \mathcal{W}_1,$$

we have proved that $\mathcal{W}_1 \neq (0)$ and the corollary. $\qquad\square$

In the proof of (8.5.4), we will also need the following rudiments of spectral decomposition of unitary representations (see [Na] §17 and 31).

Let us fix a non-trivial unitary additive character

$$\psi : F \to \mathbb{C}^\times$$

such that

$$\psi|\mathcal{O} \equiv 1$$

but

$$\psi|\varpi^{-1}\mathcal{O} \not\equiv 1$$

(see [We 2] Ch. II, §5). Then the topological group of all continuous unitary additive characters of F^{d-1} is isomorphic to F^{d-1} with its ϖ-adic topology by

$$x' \mapsto (x \mapsto \psi(< x', x >))$$

where

$$< x', x > = \sum_{i=1}^{d-1} x_i' x_i$$

(see [We 2] Ch. II, §5). Let dx (or dx') be the Haar measure on F^{d-1} which gives the volume 1 to $\mathcal{O}^{d-1}$. The Fourier transformation

$$\mathcal{F}_\psi : \mathbb{C} \otimes \mathcal{C}_c^\infty(F^{d-1}) \to \mathbb{C} \otimes \mathcal{C}_c^\infty(F^{d-1})$$

is the $\mathbb{C}$-linear isomorphism defined by

$$\mathcal{F}_\psi(f)(x') = \int_{F^{d-1}} \psi(< x', x >) f(x) dx,$$

for any $f \in \mathbb{C} \otimes \mathcal{C}_c^\infty(F^{d-1})$ and any $x' \in F^{d-1}$, and its inverse $\mathcal{F}_\psi'$ is given by

$$\mathcal{F}_\psi'(f')(x) = \int_{F^{d-1}} \psi(- < x', x >) f'(x') dx',$$

for any $f' \in \mathbb{C} \otimes \mathcal{C}_c^\infty(F^{d-1})$ and any $x \in F^{d-1}$ (see [We 2] Ch. VII, §2). For any $f_1', f_2' \in \mathbb{C} \otimes \mathcal{C}_c^\infty(F^{d-1})$ we have

$$\mathcal{F}_\psi'(f_1' f_2') = \mathcal{F}_\psi'(f_1') * \mathcal{F}_\psi'(f_2'),$$

where

$$(f_1 * f_2)(x) = \int_{F^{d-1}} f_1(x - y) f_2(y) dy$$

for any $f_1, f_2 \in \mathbb{C} \otimes \mathcal{C}_c^\infty(F^{d-1})$ and any $x \in F^{d-1}$. For any $f' \in \mathbb{C} \otimes \mathcal{C}_c^\infty(F^{d-1})$, we have

$$\mathcal{F}_\psi'(\bar{f}') = (\mathcal{F}_\psi'(f'))^*,$$

where
$$f^*(x) = \overline{f(-x)}$$
for any $f \in \mathbb{C} \otimes \mathcal{C}_c^\infty(F^{d-1})$ and any $x \in F^{d-1}$.

Now let $(\mathcal{V}, \pi)$ be a **unitary continuous** representation of F^{d-1}: $\mathcal{V}$ is a separable Hilbert space with positive definite, Hermitian scalar product $(\,,\,)$ and norm $\| - \| = (-,-)^{1/2}$, π is a representation of F^{d-1} on $\mathcal{V}$, for each $x \in F^{d-1}$, $\pi(x)$ is a unitary operator on $\mathcal{V}$ and the map

$$F^{d-1} \times \mathcal{V} \to \mathcal{V}, \ (x,v) \mapsto \pi(x)(v),$$

is continuous. The convolution algebra $\mathbb{C} \otimes \mathcal{C}_c^\infty(F^{d-1})$ acts on $\mathcal{V}$ by

$$(f,v) \mapsto \pi(f)(v) = \int_{F^{d-1}} f(x)\pi(x)(v)dx \ .$$

More precisely, $\pi(f)(v)$ is the unique element of $\mathcal{V}$ such that

$$(v', \pi(f)(v)) = \int_{F^{d-1}} f(x)(v', \pi(x)(v))dx$$

for all $v' \in \mathcal{V}$. Moreover, for any $f \in \mathbb{C} \otimes \mathcal{C}_c^\infty(F^{d-1})$, $\pi(f)$ is a bounded operator on the Hilbert space $(\mathcal{V}, \| - \|)$ with

$$\|\pi(f)\| \leq \int_{F^{d-1}} |f(x)|dx \stackrel{\mathrm{dfn}}{=} \|f\|_1$$

and its adjoint operator $\pi(f)^*$ is equal to $\pi(f^*)$.

Firstly, let us assume that $(\mathcal{V}, \pi)$ is **cyclic**, i.e. that there exists $v \in \mathcal{V} - \{0\}$ such that

$$\pi(\mathbb{C} \otimes \mathcal{C}_c^\infty(F^{d-1}))(v) \subset \mathcal{V}$$

is dense. Then let us fix a **cyclic vector** v, i.e. a vector v as above, and let us consider the $\mathbb{C}$-linear form

$$\lambda : \mathbb{C} \otimes \mathcal{C}_c^\infty(F^{d-1}) \to \mathbb{C}$$

defined by

$$\lambda(f') = (v, \pi(\mathcal{F}'_\psi(f'))(v))$$

(the Hermitian scalar product is anti-linear in the first variable and linear in the second one). This functional is positive, i.e.

$$\lambda(f') \geq 0$$

for any $f' \in \mathbf{C} \in C_c^\infty(F^{d-1})$ which takes only non-negative real values. Indeed, if $g' \in \mathbf{C} \otimes C_c^\infty(F^{d-1})$, we have

$$\lambda(|g'|^2) = (v, \pi(\mathcal{F}'_\psi(|g'|^2))(v))$$
$$= (v, \pi((\mathcal{F}'_\psi(g'))^*)(\pi(\mathcal{F}'_\psi(g'))(v)))$$
$$= (\pi(\mathcal{F}'_\psi(g'))(v), \pi(\mathcal{F}'_\psi(g'))(v)) \geq 0$$

and any $f' \in \mathbf{C} \otimes C_c^\infty(F^{d-1})$ which takes only non-negative real values is equal to $g'^2 = |g'|^2$ for some $g' \in \mathbf{C} \otimes C_c^\infty(F^{d-1})$ which also takes only non-negative real values. Therefore, thanks to the Riesz representation theorem (see [Ru] (2.14) and (2.17)), there exists a unique regular positive Borel measure μ on F^{d-1}, giving a finite volume to any compact subset of F^{d-1}, such that

$$\lambda(f') = \int_{F^{d-1}} f' d\mu$$

for any $f' \in \mathbf{C} \otimes C_c^\infty(F^{d-1})$.

Now, let us consider the $\mathbf{C}$-linear map

$$\iota : \mathbf{C} \otimes C_c^\infty(F^{d-1}) \to \mathcal{V}, \ f' \mapsto \pi(\mathcal{F}'_\psi(f'))(v).$$

Obviously, for any $f' \in \mathbf{C} \otimes C_c^\infty(F^{d-1})$, we have

$$\|\iota(f')\| = (\pi(\mathcal{F}'_\psi(f'))(v), \pi(\mathcal{F}'_\psi(f'))(v))^{1/2}$$
$$= (v, \pi(\mathcal{F}'_\psi(|f'|^2))(v))^{1/2}$$
$$= \lambda(|f'|^2)^{1/2}$$
$$= \left(\int_{F^{d-1}} |f'|^2 d\mu\right)^{1/2} \overset{\text{dfn}}{=} \|f'\|_{\mu,2}$$

and the completion of $\mathbf{C} \otimes C_c^\infty(F^{d-1})$ with respect to $\|-\|_{\mu,2}$ is nothing else than $L^2(F^{d-1}, \mu)$ (see [Ru] (3.14)). Therefore, ι is the restriction of a unique isometry, which we will also denote by ι, from $L^2(F^{d-1}, \mu)$ into $\mathcal{V}$. Moreover, as

$$\iota(\mathbf{C} \otimes C_c^\infty(F^{d-1})) = \pi(\mathbf{C} \otimes C_c^\infty(F^{d-1}))(v)$$

is dense in $\mathcal{V}$ by hypothesis, this isometry is surjective and

$$\iota : (L^2(F^{d-1}, \mu), \|-\|_{\mu,2}) \overset{\sim}{\to} (\mathcal{V}, \|-\|)$$

is an isomorphism of Hilbert spaces.

Let σ be the action of F^{d-1} on $\mathbf{C} \otimes C_c^\infty(F^{d-1})$ given by

$$(\sigma(x)(f'))(x') = \psi(<x', x>)f'(x')$$

for any $x, x' \in F^{d-1}$ and any $f' \in \mathbb{C} \otimes C_c^\infty(F^{d-1})$. For any $x \in F^{d-1}$ and any $f' \in \mathbb{C} \otimes C_c^\infty(F^{d-1})$, we have

$$\|\sigma(x)(f')\|_{\mu,2} = \|f'\|_{\mu,2}$$

and, for any $x \in F^{d-1}$, we have

$$\iota \circ \sigma(x) = \pi(x) \circ \iota \, .$$

Therefore, for each $x \in F^{d-1}$, $\sigma(x)$ extends to a unique unitary operator on $L^2(F^{d-1}, \mu)$, which we will also denote by $\sigma(x)$, $(L^2(F^{d-1}, \mu), \sigma)$ is a unitary continuous representation of F^{d-1} and

$$\iota : (L^2(F^{d-1}, \mu), \sigma) \xrightarrow{\sim} (\mathcal{V}, \pi)$$

is an isomorphism of unitary continuous representations of F^{d-1}.

Secondly, let us consider the general case. Let $\mathcal{S}$ be the set of all subsets $S = \{v_s | s \in S\}$ of $\mathcal{V} - \{0\}$ such that, for any $s_1, s_2 \in S$, $s_1 \neq s_2$, and any $f_1, f_2 \in \mathbb{C} \otimes C_c^\infty(F^{d-1})$, we have

$$(\pi(f_1)(v_{s_1}), \pi(f_2)(v_{s_2})) = 0 \, .$$

This set $\mathcal{S}$ is partially ordered by inclusion and, for any totally ordered subset $\mathcal{T}$ of $\mathcal{S}$,

$$T = \bigcup_{S \in \mathcal{T}} S$$

belongs to $\mathcal{S}$. Therefore, by Zorn's lemma, we can find a maximal element $S = \{v_s | s \in S\}$ in $\mathcal{S}$. For each $s \in S$, let us denote by $\mathcal{V}_s$ the closure in $\mathcal{V}$ of

$$\pi(\mathbb{C} \otimes C_c^\infty(F^{d-1}))(v_s) \subset \mathcal{V} \, .$$

Then, for each $s \in S$, $\mathcal{V}_s$ is an F^{d-1}-invariant, closed, $\mathbb{C}$-vector subspace of $\mathcal{V}$ and the unitary continuous representation

$$(\mathcal{V}_s, \pi_s) = (\mathcal{V}_s, \pi | \mathcal{V}_s)$$

is cyclic, with cyclic vector v_s. For any $s_1, s_2 \in S$, $s_1 \neq s_2$, v_{s_1} is orthogonal to v_{s_2}, so that the closure $\mathcal{V}'$ in $\mathcal{V}$ of the sum

$$\sum_{s \in S} \mathcal{V}_s \subset \mathcal{V}$$

is isomorphic to the Hilbertian orthogonal sum

$$\widehat{\bigoplus_{s \in S}} \mathcal{V}_s \, .$$

By maximality of S, we have $\mathcal{V}' = \mathcal{V}$. Otherwise, we could add to S any non-zero vector in $\mathcal{V}$ which is orthogonal to $\mathcal{V}'$. Finally, as $\mathcal{V}$ is separable, S is countable. We have proved

LEMMA (8.6.3). — *Any unitary continuous representation* $(\mathcal{V}, \pi)$ *of* F^{d-1} *is isomorphic to the Hilbertian orthogonal sum*

$$(\bigoplus_{n=1}^{N})^{\wedge}(L^2(F^{d-1}, \mu_n), \sigma_n)$$

for some $N \in \mathbb{Z}_{\geq 0} \cup \{+\infty\}$ *and some family* $(\mu_n)_{n=1,\ldots,N}$ *of non-zero, regular, positive, Borel measures on* F^{d-1} *giving a finite volume to any compact subset of* F^{d-1}. *Here, for each* $n = 1, \ldots, N, \sigma_n$ *is the unitary continuous representation of* F^{d-1} *on* $L^2(F^{d-1}, \mu_n)$ *defined by*

$$\sigma_n(x)(f'_n)(x') = \psi(< x', x >)f'_n(x')$$

for every $x \in F^{d-1}$, *every* $f'_n \in L^2(F^{d-1}, \mu_n)$ *and for almost all* $x' \in F^{d-1}$ *(with respect to* μ_n). □

The family of measures $(\mu_n)_{n=1,\ldots,N}$ in (8.6.3) is not unique. To get a uniqueness statement, let us replace it by the following operator valued measure P. Let us fix an isomorphism

$$\iota : (\bigoplus_{n=1}^{N})^{\wedge}(L^2(F^{d-1}, \mu_n), \sigma_n) \overset{\sim}{\to} (\mathcal{V}, \pi)$$

as in (8.6.3). For each Borel subset B in F^{d-1}, let $P(B)$ be the operator on $\mathcal{V}$ which is defined by

$$P(B) \circ \iota = (\bigoplus_{n=1}^{N})^{\wedge} Q_n(B)$$

where

$$Q_n(B)(f'_n) = 1_B f'_n$$

for any $f'_n \in L^2(F^{d-1}, \mu_n)$ $(n = 1, \ldots, N)$. Then P is a **spectral measure** on F^{d-1} for $\mathcal{V}$, i.e. it has the following properties:

(i) for each Borel subset B of F^{d-1}, $P(B)$ is a bounded (in fact $\|P(B)\| = 0$ or 1), self-adjoint, projection operator $(P(B) \circ P(B) = P(B))$ on $\mathcal{V}$;

(ii) $P(F^{d-1}) = id_{\mathcal{V}}$;

(iii) for each pair (B_1, B_2) of Borel subsets of F^{d-1} such that $B_1 \cap B_2 = \emptyset$, we have $P(B_1 \cup B_2) = P(B_1) + P(B_2)$ (in particular, $P(\emptyset) = 0$);

(iv) for each pair (B_1, B_2) of Borel subsets of F^{d-1}, we have $P(B_1 \cap B_2) = P(B_1) \circ P(B_2)$ (in particular, $P(B_1) \circ P(B_2) = P(B_2) \circ P(B_1)$);

(v) for each pair (v_1, v_2) of vectors in $\mathcal{V}$, the additive complex function

$$B \mapsto \nu_{v_1, v_2}(B) = (v_1, P(B)(v_2))$$

on the σ-algebra of Borel subsets B in F^{d-1} is countably additive, i.e. is a complex, Borel measure on F^{d-1} (it is automatically regular, see [Ru] (2.18) and (6.4)).

Moreover, for each $x \in F^{d-1}$, we have the **spectral decomposition**

$$\pi(x) = \int_{F^{d-1}} \psi(<-, x>) dP \; ,$$

i.e. by definition,

$$(v_1, \pi(x)(v_2)) = \int_{F^{d-1}} \psi(<-, x>) d\nu_{v_1, v_2}$$

for all $v_1, v_2 \in \mathcal{V}$ (if $v_i = \iota(f_i')$ for some

$$f_i' = (f_{i,n}')_{n=1,\dots,N} \in (\bigoplus_{n=1}^{N})^{\wedge} L^2(F^{d-1}, \mu_n)$$

$(i = 1, 2)$, both sides of this last formula are equal to

$$\sum_{n=1}^{N} \int_{F^{d-1}} \psi(<-, x>) \bar{f}_{1,n}' f_{2,n}' d\mu_n).$$

Therefore, we have proved the existence part of the following theorem.

THEOREM (8.6.4). — *For any unitary continuous representation $(\mathcal{V}, \pi)$ of F^{d-1}, there exists a unique spectral measure P on F^{d-1} for $\mathcal{V}$ such that the spectral decomposition*

$$\pi(x) = \int_{F^{d-1}} \psi(<-, x>) dP$$

holds for all $x \in F^{d-1}$.

Proof: The existence being already proved, let us check the uniqueness. If P' and P'' are two spectral measures on F^{d-1} for $\mathcal{V}$ such that

$$\int_{F^{d-1}} \psi(<-, x>) d\nu_{v_1, v_2}' = \int_{F^{d-1}} \psi(<-, x>) d\nu_{v_1, v_2}''$$

for some $v_1, v_2 \in \mathcal{V}$ ($\nu_{v_1,v_2}^{(i)}(B) = (v_1, P^{(i)}(B)(v_2))$) for $i = 1, 2$ and all Borel subsets B in F^{d-1}), we have

$$\int_{F^{d-1}} \mathcal{F}_\psi(f) d\nu'_{v_1,v_2} = \int_{F^{d-1}} \mathcal{F}_\psi(f) d\nu''_{v_1,v_2}$$

for all $f \in \mathbb{C} \otimes \mathcal{C}_c^\infty(F^{d-1})$ and, therefore,

$$\int_{F^{d-1}} f' d\nu'_{v_1,v_2} = \int_{F^{d-1}} f' d\nu''_{v_1,v_2}$$

for all $f' \in \mathbb{C} \otimes \mathcal{C}_c^\infty(F^{d-1})$. It follows that

$$\nu'_{v_1,v_2} = \nu''_{v_1,v_2}$$

(see [Ru] (6.18) and (6.19)). But, if, for some Borel subset B in F^{d-1}, we have

$$(v_1, P'(B)(v_2)) = (v_1, P''(B)(v_2))$$

for all $v_1, v_2 \in \mathcal{V}$, we have

$$P'(B) = P''(B)$$

and the uniqueness is proved.   □

Finally, in the proof of (8.5.4), we will need the following particular case of Mackey's theory of unitary continuous representations of semidirect products.

Let $(F^\times)^{d-1}$ act on F^{d-1} by

$$(z, x) \mapsto z(x) = (z_1 x_1, \ldots, z_{d-1} x_{d-1})$$

and let us consider the corresponding semidirect product

$$F^{d-1} \rtimes (F^\times)^{d-1},$$

where

$$z \cdot x \cdot z^{-1} = z(x)$$

for any $z \in (F^\times)^{d-1}$ and any $x \in F^{d-1}$.

If $(\mathcal{W}, \rho)$ is a unitary continuous representation of F^{d-1}, following Mackey, we can unitarily induce it from F^{d-1} to $F^{d-1} \rtimes (F^\times)^{d-1}$. We get a unitary continuous representation

$$\mathrm{Ind}(\mathcal{W}, \rho) = (\mathcal{V}, \pi)$$

which can be described in the following way. Its space

$$\mathcal{V} = L^2((F^\times)^{d-1}, \mathcal{W})$$

is the space of all functions

$$v : (F^\times)^{d-1} \to \mathcal{W}$$

such that, for each $w \in \mathcal{W}$, the function

$$(F^\times)^{d-1} \to \mathbb{C}, \ t \mapsto (w, v(t))_{\mathcal{W}}$$

is measurable and such that

$$\int_{(F^\times)^{d-1}} \|v(t)\|^2_{\mathcal{W}} d^\times t \ .$$

Here $d^\times t$ is the Haar measure on $(F^\times)^{d-1}$ normalized by $\mathrm{vol}((\mathcal{O}^\times)^{d-1}, d^\times t) = 1$. The positive definite, Hermitian scalar product $(-, -)_{\mathcal{V}}$ on the Hilbert space $\mathcal{V}$ is given by

$$(v_1, v_2)_{\mathcal{V}} = \int_{(F^\times)^{d-1}} (v_1(t), v_2(t))_{\mathcal{W}} d^\times t$$

for any $v_1, v_2 \in \mathcal{V}$. The representation π on $\mathcal{V}$ is defined by

$$(\pi(y \cdot u)(v))(t) = \rho(t(y))(v(tu))$$

for any $y \in F^{d-1}$ and any $u, t \in (F^\times)^{d-1}$ ($\pi(y \cdot u)$ is unitary and the map

$$(F^{d-1} \rtimes (F^\times)^{d-1}) \times \mathcal{V} \to \mathcal{V}, \ (y \cdot u, v) \mapsto \pi(y \cdot u)(v),$$

is continuous).

THEOREM (8.6.5) (Mackey). — *Let $(\mathcal{V}, \pi)$ be a unitary continuous representation of $F^{d-1} \rtimes (F^\times)^{d-1}$ and let P be the spectral measure on F^{d-1} for $\mathcal{V}$ which gives the spectral decomposition of $(\mathcal{V}, \pi|F^{d-1})$ (see (8.6.4)). Let us set*

$$O' = \{x' \in F^{d-1} | x_1 \cdots x_{d-1} \neq 0\} \subset F^{d-1}$$

and let us assume that P is concentrated on O', i.e.

$$P(F^{d-1} - O') = 0 \ .$$

Then, up to isomorphism, $(\mathcal{V}, \pi)$ is unitarily induced from some unitary continuous representation $(\mathcal{W}, \rho)$ of F^{d-1}.

Proof: Using the same argument as the one preceding (8.6.3), we see that $(\mathcal{V}, \pi)$ is isomorphic to the Hilbertian orthogonal sum of a countable family of cyclic unitary continuous representations of $F^{d-1} \rtimes (F^\times)^{d-1}$. But the unitary induction commutes with countable Hilbertian orthogonal sums. Therefore, we can assume that $(\mathcal{V}, \pi)$ is cyclic. Let us fix a cyclic vector v for $(\mathcal{V}, \pi)$, i.e. a non-zero vector v in $\mathcal{V}$ such that

$$\pi(\mathbb{C} \otimes \mathcal{C}_c^\infty(F^{d-1} \rtimes (F^\times)^{d-1}))(v) \subset \mathcal{V}$$

is dense. For simplicity, we assume that $\|v\| = 1$.

For each $f \in \mathbb{C} \otimes \mathcal{C}_c^\infty((F^\times)^{d-1})$, let us set

$$P(f) = \int_{(F^\times)^{d-1}} f(z) dP(z(1))$$

(O' is the $(F^\times)^{d-1}$-orbit of $1 = (1, \ldots, 1) \in F^{d-1}$). We have

$$P(f_1 f_2) = P(f_1) \circ P(f_2),$$

$$P(f)^* = P(\bar{f})$$

and

$$\pi(z) \circ P(f) \circ \pi(z^{-1}) = P(z_1 \mapsto f(z_1 z))$$

for any $f_1, f_2, f \in \mathbb{C} \otimes \mathcal{C}_c^\infty((F^\times)^{d-1})$ and any $z \in (F^\times)^{d-1}$. Indeed, if $f_1 = 1_{\Gamma_1}$, $f_2 = 1_{\Gamma_2}$ and $f = 1_\Gamma$ where Γ_1, Γ_2 and Γ are compact open subsets of $(F^\times)^{d-1}$, the first two formulas are obvious and the third one follows from the uniqueness of P (see (8.6.4)). Moreover, we have

$$\|P(f)\| = \|f\|_\infty \overset{\text{dfn}}{=\!=} \sup\{|f(z)| \,|\, z \in (F^\times)^{d-1}\}$$

for any $f \in \mathbb{C} \otimes \mathcal{C}_c^\infty((F^\times)^{d-1})$. Indeed,

$$\|P(f)(v)\| = \sum_{\lambda \in \mathbb{C}} |\lambda| \, \|P(f^{-1}(\lambda))(v)\|$$

and

$$\|v\| = \sum_{\lambda \in \mathbb{C}} \|P(f^{-1}(\lambda))(v)\|$$

for any $v \in \mathcal{V}$, so that

$$\|P(f)\| = \sup\{|\lambda| \,|\, \lambda \in \mathbb{C}, P(f^{-1}(\lambda)) \neq 0\}$$

and

$$\|P(f)\| \leq \sup\{|\lambda| \,|\, \lambda \in \mathbb{C}, f^{-1}(\lambda) \neq \emptyset\} = \|f\|_\infty$$

with equality if $P(f^{-1}(\lambda)) \neq 0$ for any $\lambda \in \mathbb{C}-\{0\}$ such that $f^{-1}(\lambda) \neq \emptyset$; but, for any compact open subset Γ of $(F^\times)^{d-1}$, $P(\Gamma) \neq 0$: otherwise, $P((F^\times)^{d-1})$ would also be zero ($(F^\times)^{d-1}$ is a countable union of subsets of the form $z^{-1}\Gamma$ ($z \in (F^\times)^{d-1}$) and

$$P(z^{-1}\Gamma) = \pi(z) \circ P(\Gamma) \circ \pi(z^{-1}) = 0).$$

For each $\varphi \in \mathbb{C} \otimes C_c^\infty(F^{d-1} \rtimes (F^\times)^{d-1})$, we set

$$\pi(\varphi) = \int_{F^{d-1}} \left(\int_{(F^\times)^{d-1}} \varphi(x \cdot z)\pi(x \cdot z)d^\times z \right) dx$$

(as usual), so that

$$\|\pi(\varphi)\| \le v(\varphi)\|\varphi\|_\infty .$$

Here $v(\varphi)$ is the volume of $\mathrm{Supp}(\varphi)$ for $dx \, d^\times z$ and

$$\|\varphi\|_\infty \overset{\mathrm{dfn}}{=\!=} \sup\{|\varphi(x \cdot z)| \,|\, x \in F^{d-1}, z \in (F^\times)^{d-1}\} .$$

Then, by the Cauchy–Schwarz inequality, we get

$$|(v, P(f)(\pi(\varphi_2)(v)))| \le v(\varphi_2)\|f\|_\infty\|\varphi_2\|_\infty$$

and there exists a unique Radon measure λ on $(F^\times)^{d-1} \times (F^{d-1} \rtimes (F^\times)^{d-1})$ such that

$$(v, P(f)(\pi(\varphi_2)(v))) = \int_{(F^\times)^{d-1} \times (F^{d-1} \rtimes (F^\times)^{d-1})} f(z)\varphi_2(x_2 \cdot z_2)d\lambda(z, x_2 \cdot z_2)$$

(the image of $\mathbb{C} \otimes C_c^\infty((F^\times)^{d-1}) \otimes C_c^\infty(F^{d-1} \rtimes (F^\times)^{d-1})$ by

$$f \otimes \varphi_2 \mapsto ((z, x_2 \cdot z_2) \mapsto f(z)\varphi_2(x_2 \cdot z_2))$$

is dense in the $\mathbb{C}$-vector space of all compactly supported continuous complex functions on $(F^\times)^{d-1} \times (F^{d-1} \rtimes (F^\times)^{d-1})$ for the topology of uniform convergence on each compact subset). It follows that

$$(\pi(\varphi_1)(v), P(f)(\pi(\varphi_2)(v)))$$

is equal to

$$\int_{F^{d-1}} \left(\int_{(F^\times)^{d-1}} |z_1|\overline{\varphi_1(x_1 - z_1)} \left(\int_{(F^\times)^{d-1} \times (F^{d-1} \rtimes (F^\times)^{d-1})} f(zz_1^{-1}) \right. \right.$$
$$\left. \left. \times \varphi_2((x_1 + z_1(x_2))z_1z_2)d\lambda(z, x_2 \cdot z_2) \right) d^\times z_1 \right) dx_1$$

(we have

$$\pi(x_1 \cdot z_1)^{-1} \circ P(f) = P(z_1 \mapsto f(zz_1^{-1})) \circ \pi(x_1 \cdot z_1)^{-1}$$

and

$$\pi(x_1 \cdot z_1)^{-1} \circ \pi(\varphi_2) = |z_1|\pi(x_2 \cdot z_2 \mapsto \varphi_2((z_1(x_2) + x_1) \cdot z_1 z_2)),$$

where

$$|z_1| = \prod_{i=1}^{d-1} |z_{1,i}| = \frac{d(z_1(x_2))}{dx_2})$$

for any $\varphi_1, \varphi_2 \in \mathbb{C} \otimes C_c^\infty(F^{d-1} \rtimes (F^\times)^{d-1})$ and any $f \in \mathbb{C} \otimes C_c^\infty((F^\times)^{d-1})$.
Let us apply the Fubini theorem for Radon measures (see [Bou] Int. III, § 4,
Thm. 2) to permute the order of integrations and let us make the changes of
variables $z_1 \mapsto t = zz_1^{-1}$, $x_1 \mapsto x_1 := t(x_1)$. Then we obtain the formula

$$(\pi(\varphi_1)(v), P(f)(\pi(\varphi_2)(v))) = \int_{(F^\times)^{d-1}} f(t)(\varphi_{1,t}, \varphi_{2,t})_W d^\times t$$

where we have set

$$\varphi_{i,t}(x \cdot z) = |t^{-1}|\varphi_i(t^{-1} \cdot (x \cdot z)) = |t^{-1}|\varphi_i(t^{-1}(x) \cdot t^{-1}z)$$

for any $x \cdot z \in F^{d-1} \rtimes (F^\times)^{d-1}$ $(i = 1, 2)$ and where

$$(\psi_1, \psi_2)_W = \int_{F^{d-1}} \left(\int_{(F^\times)^{d-1} \times (F^{d-1} \rtimes (F^\times)^{d-1})} |z|\psi_1(x_1 \cdot z)\psi_2((x_1 + z(x_2) \cdot zz_2))d\lambda(z, x_2 \cdot z_2) \right) dx_1$$

for any $\psi_1, \psi_2 \in \mathbb{C} \otimes C_c^\infty(F^{d-1} \rtimes (F^\times)^{d-1})$. Using the fact that

$$\overline{(\pi(\varphi_1)(v), P(f)(\pi(\varphi_2)(v)))} = (\pi(\varphi_2)(v), P(\bar{f})(\pi(\varphi_1)(v)))$$

for any $f \in \mathbb{C} \otimes C_c^\infty((F^\times)^{d-1})$, we get that

$$(\overline{\psi_1, \psi_2})_W = (\psi_2, \psi_1)_W$$

for any $\psi_1, \psi_2 \in \mathbb{C} \otimes C_c^\infty(F^{d-1} \rtimes (F^\times)^{d-1})$. Using the fact that

$$(\pi(\varphi)(v), P(f)(\pi(\varphi)(v))) \geq 0$$

for any $f \in \mathbb{C} \otimes C_c^\infty((F^\times)^{d-1})$ which takes only non-negative real values
(write $f = f'^2$ for some $f' \in \mathbb{C} \otimes C_c^\infty((F^\times)^{d-1})$), we get that

$$(\psi, \psi)_W \geq 0$$

for any $\psi \in \mathbb{C} \otimes \mathcal{C}_c^\infty(F^{d-1} \rtimes (F^\times)^{d-1})$.

Now let $\mathcal{W}$ be the Hilbert completion of $\mathbb{C} \otimes \mathcal{C}_c^\infty(F^{d-1} \rtimes (F^\times)^{d-1})$ for the positive, Hermitian scalar product $(-,-)_\mathcal{W}$. We will also denote by $(-,-)_\mathcal{W}$ the positive definite, Hermitian scalar product of $\mathcal{W}$. If $\varphi \in \mathbb{C} \otimes \mathcal{C}_c^\infty(F^{d-1} \rtimes (F^\times)^{d-1})$ and $y \in F^{d-1}$, let $\rho(y)(\varphi) \in \mathbb{C} \otimes \mathcal{C}_c^\infty(F^{d-1} \rtimes (F^\times)^{d-1})$ be defined by

$$(\rho(y)(\varphi))(x \cdot z) = \varphi((x-y) \cdot z) \quad (\forall x \cdot z \in F^{d-1} \rtimes (F^\times)^{d-1}).$$

Then it is clear that

$$(\rho(y)(\psi_1), \rho(y)(\psi_2))_\mathcal{W} = (\psi_1, \psi_2)_\mathcal{W}$$

for any $\psi_1, \psi_2 \in \mathbb{C} \otimes \mathcal{C}_c^\infty(F^{d-1} \rtimes (F^\times)^{d-1})$. Hence ρ induces a unitary continuous representation of F^{d-1} on $\mathcal{W}$. We are going to prove that $(\mathcal{V}, \pi)$ is isomorphic to $\mathrm{Ind}(\mathcal{W}, \rho)$.

Let us consider the $\mathbb{C}$-linear map

$$\iota : \mathbb{C} \otimes \mathcal{C}_c^\infty((F^\times)^{d-1} \times (F^{d-1} \rtimes (F^\times)^{d-1})) \to \mathcal{V}$$

defined by

$$\iota(\Phi) = \int_{(F^\times)^{d-1}} \left(\int_{F^{d-1}} \left(\int_{(F^\times)^{d-1}} |t| \Phi(t, t(x) \cdot tz) dP(t(1)) (\pi(x \cdot z)(v)) \right) dx \right) d^\times z.$$

In particular, if Φ has the form

$$\Phi(t, x \cdot z) = f(t)|t^{-1}|\varphi(t^{-1}(x) \cdot t^{-1}(z)) = f(t)\varphi_t(x \cdot z)$$

$(t, z \in (F^\times)^{d-1}, x \in F^{d-1})$ for some $f \in \mathbb{C} \otimes \mathcal{C}_c^\infty((F^\times)^{d-1})$ and $\varphi \in \mathbb{C} \otimes \mathcal{C}_c^\infty(F^{d-1} \rtimes (F^\times)^{d-1})$, we have

$$\iota(\Phi) = P(f)(\pi(\varphi)(v)).$$

It follows that

$$(\iota(\Phi_1), \iota(\Phi_2)) = \int_{(F^\times)^{d-1}} (\Phi_1(t,-), \Phi_2(t,-))_\mathcal{W} d^\times t$$

for any $\Phi_1, \Phi_2 \in \mathbb{C} \otimes \mathcal{C}_c^\infty((F^\times)^{d-1} \times (F^{d-1} \rtimes (F^\times)^{d-1}))$. Indeed, we can assume that

$$\Phi_i(t, x \cdot z) = f_i(t)|t^{-1}|\varphi_i(t^{-1}(x) \cdot t^{-1}(z))$$

for some $f_i \in \mathbb{C} \otimes \mathcal{C}_c^\infty((F^\times)^{d-1})$ and $\varphi_i \in \mathbb{C} \otimes \mathcal{C}_c^\infty(F^{d-1} \rtimes (F^\times)^{d-1})$ $(i = 1, 2)$ and we have

$$(P(f_1)(\pi(\varphi_1)(v)), P(f_2)(\pi(\varphi_2)(v))) = (\pi(\varphi_1)(v), P(\bar{f}_1 f_2)(\pi(\varphi_2)(v)))$$
$$= \int_{(F^\times)^{d-1}} \overline{f_1(t)} f_2(t)(\varphi_{1,t}, \varphi_{2,t}) w \, d^\times t \ .$$

Therefore, ι induces an isometry, which we will also denote ι, from $L^2((F^\times)^{d-1}, \mathcal{W})$ into $\mathcal{V}$. Moreover, if $\Phi \in \mathbb{C} \otimes \mathcal{C}_c^\infty((F^\times)^{d-1} \times (F^{d-1} \rtimes (F^\times)^{d-1}))$, if $y \cdot u \in F^{d-1} \rtimes (F^\times)^{d-1}$ and if Φ' is defined by

$$\Phi'(t, x \cdot z) = \Phi(tu, (x - t(y)) \cdot z),$$

for any $x \cdot z \in F^{d-1} \rtimes (F^\times)^{d-1}$, we have

$$\iota(\Phi') = \pi(y \cdot u)(\iota(\Phi))$$

(we have

$$\pi(y \cdot u) \circ dP(t(1)) \circ \pi(x \cdot z) = dP(u^{-1}t(1)) \circ \pi((y + u(x)) \cdot uz)).$$

Therefore, the isometry ι is in fact an embedding

$$\iota : \mathrm{Ind}(\mathcal{W}, \rho) \hookrightarrow (\mathcal{V}, \pi)$$

of unitary continuous representations of $F^{d-1} \rtimes (F^\times)^{d-1}$.

At this point of the proof we have not yet used the main hypothesis $P(F^{d-1} - O') = 0$. Let us use it to prove that ι is surjective. By hypothesis, $\pi(\mathbb{C} \otimes \mathcal{C}_c^\infty(F^{d-1} \rtimes (F^\times)^{d-1}))(v)$ is dense in $\mathcal{V}$. Therefore, if $(P_n)_{n \in \mathbb{Z}_{>0}}$ is a sequence of bounded operators on the Hilbert space $\mathcal{V}$ such that

$$\|v - P_n(v)\| \to 0$$

for any $v \in \mathcal{V}$ as $n \to +\infty$, the set

$$\bigcup_{n \in \mathbb{Z}_{>0}} P_n(\pi(\mathbb{C} \otimes \mathcal{C}_c^\infty(F^{d-1} \rtimes (F^\times)^{d-1}))(v))$$

is also dense in $\mathcal{V}$. But we can take

$$P_n = P(1_{\Gamma_n}) = P(\Gamma_n(1))$$

where

$$\Gamma_n = \{z \in (F^\times)^{d-1} | \frac{1}{n} \leq |z_i| \leq n, \forall i = 1, \dots, d\}$$

is compact open in $(F^\times)^{d-1}$ for each $n \in \mathbb{Z}_{>0}$. Indeed, we have

$$\Gamma_1(1) \subset \Gamma_2(1) \subset \Gamma_3(1) \subset \cdots \subset \bigcup_{n \in \mathbb{Z}_{>0}} \Gamma_n(1) = O',$$

so that

$$\nu_{v,v}(\Gamma_n(1)) \to \nu_{v,v}(O')$$

for any $v \in \mathcal{V}$, as $n \to +\infty$, we have

$$\nu_{v,v}(O') = \nu_{v,v}(F^{d-1}) = (v, v)$$

for any $v \in \mathcal{V}$ $(P(F^{d-1} - O') = 0)$, and we have

$$\|v - P(\Gamma_n(1))(v)\| = (v, v - P(\Gamma_n(1))(v))$$
$$= (v, v) - \nu_{v,v}(\Gamma_n(1))$$

for any $v \in \mathcal{V}$ and any $n \in \mathbb{Z}_{>0}$ $(P(\Gamma_n(1))^* = P(\Gamma_n(1))$ and $P(\Gamma_n(1))^2 = P(\Gamma_n(1)))$. Therefore,

$$P(\mathbb{C} \otimes \mathcal{C}_c^\infty((F^\times)^{d-1}))(\pi(\mathbb{C} \otimes \mathcal{C}_c^\infty(F^{d-1} \rtimes (F^\times)^{d-1}))(v))$$

is dense in $\mathcal{V}$ and the theorem is proved.   □

Proof of (8.5.4) : Let us identify the subgroup $F^\times \backslash N_{\Delta - \{\alpha_1\}}(F)T(F)$ of $F^\times \backslash G(F)$, where $\alpha_1 = \varepsilon_1 - \varepsilon_2$, with the semidirect product $F^{d-1} \rtimes (F^\times)^{d-1}$ of (8.6.5) by

$$n \mapsto (n_{12}, \ldots, n_{1d}) \in F^{d-1}$$

and

$$t \mapsto (t_{11}/t_{22}, \ldots, t_{11}/t_{dd}) \in (F^\times)^{d-1}$$

for any $n \in N_{\Delta - \{\alpha_1\}}(F)$ and any $t \in T(F)$.

Let us fix an $(F^\times \backslash G(F))$-invariant, positive definite, Hermitian scalar product $(,)$ on the unitarizable admissible irreducible representation $(\mathcal{V}, \pi)$ of $F^\times \backslash G(F)$ and let us denote by $(\widehat{\mathcal{V}}, \hat{\pi})$ its completion with respect to the norm $\| - \| = (-, -)^{1/2}$.

If P is the spectral measure on F^{d-1} for $\mathcal{V}$ which gives the spectral decomposition of $(\widehat{\mathcal{V}}, \hat{\pi}|N_{\Delta - \{\alpha_1\}}(F))$ (see (8.6.4)), P is concentrated on

$$O' = \{x' \in F^{d-1} | x_1' \cdots x_d' \neq 0\} .$$

Otherwise, there would exist $i \in \{1, \ldots, d-1\}$ such that

$$P(H_i') \neq 0 ,$$

where
$$H'_i = \{x' \in F^{d-1} | x'_i = 0\}$$

(we have
$$P(F^{d-1} - O') = \sum_{\substack{I \subset \{1,\dots,d-1\} \\ I \neq \emptyset}} (\prod_{i \in I} P(H_i)) P(U'_{\{1,\dots,d-1\}-I}),$$

where
$$U'_J = \{x' \in F^{d-1} | \prod_{j \in J} x'_j \neq 0\}$$

for any $J \subset \{1, \dots, d-1\})$ and any non-zero $\hat{v} \in P(H'_i)(\widehat{\mathcal{V}})$ would be fixed by
$$\hat{\pi}(x_{\varepsilon_1 - \varepsilon_{i+1}}(t))$$

for any $t \in F$; but this would contradict (8.6.2) as $\pi | SL_d(F)$ is non-trivial by hypothesis.

Then, applying Mackey's theorem (8.6.5), we obtain that $(\widehat{\mathcal{V}}, \hat{\pi} | (F^\times \backslash N_{\Delta - \{\alpha_1\}}(F)T(F)))$ is (up to isomorphism) unitarily induced by a unitary continuous representation $(\mathcal{W}, \rho)$ of $F^\times \backslash T(F)$.

But, if we identify $(\widehat{\mathcal{V}}, \hat{\pi} | (F^\times \backslash N_{\Delta - \{\alpha_1\}}(F)T(F)))$ with $\mathrm{Ind}(\mathcal{W}, \rho)$, we have

$$(\hat{v}', \hat{\pi}(t)(\hat{v})) = \int_{F^\times \backslash T(F)} (\hat{v}'(t'), \hat{v}(t't))_\mathcal{W} \frac{dt'}{dz'}$$

for any $\hat{v}, \hat{v}' \in \widehat{\mathcal{V}} = L^2(F^\times \backslash T(F), \mathcal{W})$ and we get

$$\int_{F^\times \backslash T(F)} |(\hat{v}', \hat{\pi}(t)(\hat{v}))|^2 \frac{dt}{dz} = \int_{(F^\times \backslash T(F)) \times (F^\times \backslash T(F))} |(\hat{v}'(t'), \hat{v}(t))_\mathcal{W}|^2 \frac{dt'}{dz'} \times \frac{dt}{dz}$$

by the Fubini theorem and an obvious change of variable. By the Cauchy–Schwarz inequality it follows that

$$\int_{F^\times \backslash T(F)} |(\hat{v}', \hat{\pi}(t)(\tilde{v}))|^2 \frac{dt}{dz} \leq (\int_{F^\times \backslash T(F)} \|\hat{v}'(t')\|_\mathcal{W}^2 \frac{dt'}{dz'})(\int_{F^\times \backslash T(F)} \|\hat{v}(t)\|_\mathcal{W}^2 \frac{dt}{dz}),$$

i.e. that
$$\int_{F^\times \backslash T(F)} |(\hat{v}', \hat{\pi}(t)(\hat{v}))|^2 \frac{dt}{dz} \leq \|\hat{v}'\|^2 \|\hat{v}\|^2,$$

for any $\hat{v}, \hat{v}' \in \widehat{\mathcal{V}}$. Therefore, we have proved that, for any $\hat{v}, \hat{v}' \in \widehat{\mathcal{V}}$, the restriction to $F^\times \backslash T(F)$ of the function

$$F^\times \backslash G(F) \to \mathbb{C}, \quad g \mapsto (\hat{v}', \hat{\pi}(g)(\hat{v}))$$

is square-integrable.

To finish the proof of Howe's theorem, it suffices to remark that any matrix coefficient $g \mapsto \; <\tilde{v}, \pi(g)(v)>$ of $(\mathcal{V}, \pi)$ is equal to $g \mapsto (v', \hat{\pi}(g)(v))$ for some $v' \in \mathcal{V}$ (see (D.6.3) (i)). $\qquad\square$

(8.7) Comments and references

The basic references for sections 1 to 5 of this chapter are [Bo–Wa] (Ch. X, §2, 3, 4, Ch. XI, §4) and [Ko 1] §2.

Some simplifications occur in the proofs of (8.3.3) and (8.3.7) due to the explicit form of f and to the fact that f is very cuspidal.

Let $\mathcal{Z}$ be the center of the Hecke algebra

$$\mathbb{C} \otimes \widetilde{\mathcal{H}}^{ad} = \mathbb{C} \otimes \mathcal{C}_c^\infty(F^\times \backslash G(F) /\!\!/ \mathcal{B}^\circ).$$

Following Bernstein, $\mathcal{Z}$ can be identified with the $\mathbb{C}$-subalgebra

$$\mathbb{C}[F^\times \backslash T(F)/T(\mathcal{O})]^W$$

of the group $\mathbb{C}$-algebra
$$\mathbb{C}[F^\times \backslash T(F)/T(\mathcal{O})].$$

In particular, any unramified character χ of $F^\times \backslash T(F)$ which can be viewed as a character of the $\mathbb{C}$-algebra $\mathbb{C}[F^\times \backslash T(F)/T(\mathcal{O})]$ induces a character $\chi_{|\mathcal{Z}}$ of the $\mathbb{C}$-algebra $\mathcal{Z}$. Obviously, we have

$$w(\chi)_{|\mathcal{Z}} = \chi_{|\mathcal{Z}}$$

for each $w \in W$. Then the proof of the second assertion of (8.4.11) would be much simpler if one could prove by a direct computation that

$$z * f - \chi_{|\mathcal{Z}}(z)f \in [\mathbb{C} \otimes \widetilde{\mathcal{H}}^{ad}, \mathbb{C} \otimes \widetilde{\mathcal{H}}^{ad}]$$

for any $z \in \mathcal{Z}$ and some (any) $\chi \in W(\delta_{B(F)}^{-1/2})$ (in particular, one would not need group cohomology to prove (8.2)).

There are two other proofs of (8.5.2) (see [Bo–Wa] (Ch. XI, 3.7) and [Cas 2]).

The proof of (8.5.4) which is given in section 6 of this chapter is the one given in [Ho–Mo] (5.4). In fact, our form of Howe and Moore's theorem is slightly weaker than the original one but is sufficient to prove (8.5.2). Other useful references for this section are chapter 2 of [Mac], [Mar] (Ch. II, §3), [Na] and [Ru].

Appendices

A

Central simple algebras

(A.0) Central simple algebras

In this appendix, we will review some well known facts about central simple algebras (see [Bou] Alg. Ch. 8, [Re] and [We 2]).

Let K be a commutative field. A K-**algebra** A is a ring (associative and with a unit element) endowed with a injective ring homomorphism of K into the center of A which maps 1 to 1. The K-algebra A is said to be **central** if the image of K in A is exactly the center of A. The K-algebra A is said to be **simple** if any right A-module is semi-simple and if, up to isomorphisms, there is only one simple right A-module.

We will consider only simple K-algebras which are of finite dimension as K-vector spaces. If A is such a simple K-algebra, one denotes by $[A : K]$ its dimension over K. If we assume moreover that A is central then one has $[A : K] = d^2$ for some positive integer d.

If V is "the" unique simple right A-module for some central simple K-algebra A as before, $D = \mathrm{Hom}_A(V, V)$ is a **central division algebra over** K (i.e. a skew field with center K) and A is canonically isomorphic to $\mathrm{Hom}_D(V, V)$. In particular, if we choose a basis of the finite dimensional right D-vector space V, we get an isomorphism of A with the matrix algebra $gl_r(D)$ $(r = \dim_D(V))$.

Two central simple K-algebras A_1 and A_2 (which are finite dimensional as K-vector spaces) are said to be **equivalent** if there exist a central division algebra D over K and two positive integers r_1, r_2 such that A_i is isomorphic to $gl_{r_i}(D)$ $(i = 1, 2)$. The set of equivalence classes is denoted by $\mathrm{Br}(K)$ and the tensor product over K induces a commutative group law on $\mathrm{Br}(K)$. The commutative group $\mathrm{Br}(K)$ is the **Brauer group** of K.

(A.1) Bicommutant theorem

Let A be a central simple K-algebra (of finite dimension as a K-vector space) and let $B \subset A$ be a simple K-subalgebra. Its **commutant** is the K-subalgebra

$$(A.1.1) \qquad B' = \{a \in A | ab = ba, \forall b \in B\}$$

of A.

THEOREM (A.1.2). — (i) B' is again a simple K-subalgebra of A.

(ii) The commutant B'' of B' in A coincides with B.

(iii) We have
$$[A : K] = [B : K][B' : K].$$

(iv) If V is "the" simple right A-module, the K-algebra homomorphism

$$D \otimes_K B' \to Hom_B(V, V),$$

where $D = Hom_A(V, V)$, which maps $d \otimes b'$ onto $(v \mapsto d(b'v) = b'd(v))$, is an isomorphism.

Proof: See [Bou] Alg. Ch. 8, § 10, Thm. 2 and Prop. 2 or [Re] (7.11) and (7.13). □

THEOREM (A.1.3) (Skolem–Noether). — Let B be a simple K-algebra (of finite dimension as a K-vector space). Then, if u, v are K-isomorphisms of B onto K-subalgebras of A, there exists $a \in A$ such that

$$v(b) = a^{-1}u(b)a$$

for all $b \in B$ (i.e. u and v are conjugate in A).

Proof: See [Bou] Alg. Ch. 8, § 10, Thm. 1 or [Re] (7.21). □

PROPOSITION (A.1.4). — Let L be a finite extension of degree n of K (L is a commutative field) and let D be a central division algebra over K with $[D : K] = d^2$ ($d \in \mathbb{Z}, d \geq 1$). Then there exists a K-isomorphism of L onto a K-subalgebra of D if and only if n divides d and the central simple L-algebra $D \otimes_K L$ is L-isomorphic to $gl_n(\Delta)$ for some central division algebra Δ over L with

$$[\Delta : L] = (d/n)^2.$$

Moreover, any two such K-isomorphisms are conjugate in D.

Proof : The last assertion is a particular case of (A.1.3).

If we have a K-isomorphism of L onto a K-subalgebra of D, the commutant Δ of L in D is a central division algebra over L with

$$[\Delta : L] = (d/n)^2$$

and the map

$$D \otimes_K L \to \mathrm{Hom}_\Delta(D, D) \ (\cong gl_n(\Delta)),$$

$$d \otimes \ell \mapsto (x \mapsto dx\ell)$$

(D is viewed as a right Δ-vector space by right multiplication) is an isomorphism of L-algebras (apply (A.1.2) (i), (ii), (iii) to $B = L$ and then apply (A.1.2) (iv) to $B = \Delta$).

Conversely, if n divides d and $D \otimes_K L$ is L-isomorphic to $gl_n(\Delta)$, then Δ^n is a simple right $gl_n(\Delta)$-module and therefore a right D-module such that

$$L \subset \mathrm{Hom}_D(\Delta^n, \Delta^n).$$

But we have

$$\dim_K(\Delta^n) = d^2 = [D : K]$$

so that Δ^n is isomorphic to D as a right D-module. Choosing an isomorphism of right D-modules between Δ^n and D, we get an embedding of K-algebras

$$L \subset \mathrm{Hom}_D(D, D) = D.$$

$\square$

(A.2) Central simple algebras over local fields

Let us assume in this section that K is a non-archimedean local field with finite residue field k. For each positive integer d let K_d be "the" unramified extension of degree d of K : its residue field k_d is "the" extension of degree d of k. The Galois group $\mathrm{Gal}(k_d/k)$ is generated by the arithmetic Frobenius element, i.e. the q-th power if k has q elements. Let $\sigma \in \mathrm{Gal}(K_d/K)$ be the unique lifting of this arithmetic Frobenius.

Let us fix a uniformizer ϖ of K. Then for each integer e with $(e, d) = 1$ we can form the cyclic K-algebra

(A.2.1) $$D_{d,e} = K_d[f]/(f^d - \varpi^e)$$

where $K_d[f]$ is the non-commutative polynomial ring in one variable f over K_d with commutation rule

$$f \cdot \alpha = \sigma(\alpha) \cdot f$$

for each $\alpha \in K_d$. It is a central division algebra over K of dimension d^2. Obviously, it depends only on the class of $e/d \in \mathbb{Q}$ in $\mathbb{Q}/\mathbb{Z}$.

THEOREM (A.2.2). — *There exists a unique group isomorphism*

$$\mathrm{inv}_K = \mathrm{inv} : \mathrm{Br}(K) \xrightarrow{\sim} \mathbb{Q}/\mathbb{Z}$$

such that

$$\mathrm{inv}_K(D_{d,e}) = \frac{e}{d} + \mathbb{Z}$$

for each pair of integers (d, e) with $d \geq 1$ and $(d, e) = 1$.

Moreover, this isomorphism is independent of the choice of the uniformizer ϖ.

Proof: See [We 2] Ch. XII, §2, Thm. 1 or [Re] (31.8). □

PROPOSITION (A.2.3). — *Let L be a finite extension of degree n of K (L is also a non-archimedean local field). Let*

$$\mathrm{res}_{K/L} : \mathrm{Br}(K) \to \mathrm{Br}(L)$$

be the unique group homomorphism such that

$$\mathrm{res}_{K/L}(A) = A \otimes_K L$$

for each central simple K-algebra A. Then the following diagram commutes:

$$
\begin{array}{ccc}
\mathrm{Br}(K) & \xrightarrow{\mathrm{inv}_K} & \mathbb{Q}/\mathbb{Z} \\
{\scriptstyle \mathrm{res}_{K/L}} \downarrow & & \downarrow {\scriptstyle n} \\
\mathrm{Br}(L) & \xrightarrow{\mathrm{inv}_L} & \mathbb{Q}/\mathbb{Z}
\end{array}
$$

Proof: See [We 2] Ch. XII, §2, Cor. 2 of Thm. 2 or [Re] (31.9). □

COROLLARY (A.2.4). — *Let D be a central division algebra over K, let L be a commutative subfield of D containing K and let Δ be the commutant of L in D. Let us recall that $[L : K]$ divides $[D : K]^{1/2}$ and that Δ is a central division algebra over L of dimension $[D : K]/[L : K]^2$ (see (A.1.4) and its proof). Then we have*

$$\mathrm{inv}_L(\Delta) = [L : K]\,\mathrm{inv}_K(D).$$

Proof : Thanks to (A.1.4) and its proof, we have an isomorphism of L-algebras

$$D \otimes_K L \cong gl_{[L:K]}(\Delta).$$

Therefore, the corollary follows from (A.2.3).   □

Let (d, e) be a pair of integers with $d \geq 1$ and $(d, e) = 1$. The central division algebra

$$D_{d,e} = K_d[f]/(f^d - \varpi^e)$$

has another presentation. Let d', e' be integers such that

$$dd' + ee' = 1$$

and let us set

$$\tau = \varpi^{d'} f^{e'}.$$

We have

$$\tau^d = \varpi,$$
$$f = \tau^e$$

and the commutation relation

$$\tau \cdot \alpha = \sigma^{e'}(\alpha) \cdot \tau$$

for each $\alpha \in K_d$. Therefore, if $K_d[\tau]$ is the non-commutative polynomial ring in τ over K_d with the above commutation rule, we have

(A.2.5) $$D_{d,e} = K_d[\tau]/(\tau^d - \varpi).$$

If $\mathcal{O}$ (resp. $\mathcal{O}_d$) is the ring of integers of K (resp. K_d), then

(A.2.6) $$\mathcal{D}_{d,e} = \mathcal{O}_d[\tau]/(\tau^d - \varpi) \subset K_d[\tau]/(\tau^d - \varpi) = D_{d,e}$$

is the maximal $\mathcal{O}$-order of $D_{d,e}$ and $\tau \mathcal{D}_{d,e} = \mathcal{D}_{d,e}\tau$ is the maximal ideal of $\mathcal{D}_{d,e}$ (see [Re] §14).

(A.3) Central simple algebras over function fields

Let now K be a function field with finite field of constants. We denote by $|K|$ the set of places of K.

For each $v \in |K|$, let

$$\mathrm{res}_v : \mathrm{Br}(K) \to \mathrm{Br}(K_v)$$

be the group homomorphism induced by

$$A \mapsto A \otimes_K K_v.$$

(A is a central simple K-algebra), where K_v is the completion of K at v (a non-archimedean local field).

If A is a central simple K-algebra, we have

$$\mathrm{inv}_{K_v}(\mathrm{res}_v(A)) = 0$$

for almost all $v \in |K|$ (see [We 2] Ch. XI, §1, Thm. 1). In particular, there is a well-defined group homomorphism (the **invariant**)

$$\mathrm{inv}_K = \mathrm{inv} : \mathrm{Br}(K) \to (\mathbb{Q}/\mathbb{Z})^{(|K|)}$$

with v-component (the **invariant at** v)

$$\mathrm{inv}_{K,v} = \mathrm{inv}_v = \mathrm{inv}_{K_v} \circ \mathrm{res}_v : \mathrm{Br}(K) \to \mathbb{Q}/\mathbb{Z}.$$

Let us consider the sequence of group homomorphisms

$$(\mathrm{A.3.1}) \qquad 0 \to \mathrm{Br}(K) \xrightarrow{\mathrm{inv}_K} (\mathbb{Q}/\mathbb{Z})^{(|K|)} \xrightarrow{\Sigma_K} \mathbb{Q}/\mathbb{Z} \to 0$$

where Σ_K is the sum of the v-components ($v \in |K|$).

LEMMA (A.3.2). — *Let L be a finite extension of K (L is also a function field). Then the following diagram commutes:*

$$
\begin{array}{ccccccccc}
0 & \to & \mathrm{Br}(K) & \xrightarrow{\mathrm{inv}_K} & (\mathbb{Q}/\mathbb{Z})^{(|K|)} & \xrightarrow{\Sigma_K} & \mathbb{Q}/\mathbb{Z} & \to & 0 \\
& & {\scriptstyle \mathrm{res}_{K/L}}\downarrow & & \downarrow & & {\scriptstyle [L:K]}\downarrow & & \\
0 & \to & \mathrm{Br}(K) & \xrightarrow{\mathrm{inv}_L} & (\mathbb{Q}/\mathbb{Z})^{(|L|)} & \xrightarrow{\Sigma_L} & \mathbb{Q}/\mathbb{Z} & \to & 0
\end{array}
$$

where

$$\mathrm{res}_{K/L}(A) = A \otimes_K L$$

for each central simple K-algebra A and where the middle vertical arrow maps $(\lambda_v)_{v \in |K|}$ onto $(\mu_w)_{w \in |L|}$ with

$$\mu_w = [L_w : K_v]\lambda_v$$

if $w \in |L|$ divides $v \in |K|$.

Proof : This is a direct consequence of (A.2.3). $\qquad\qquad \square$

THEOREM (A.3.3). — *The sequence (A.3.1) is exact.*
Moreover, if D is a central division algebra over K with $[D : K] = d^2$ for some positive integer d, then d is the smallest positive integer such that $d \operatorname{inv}_{K,v}(D) \equiv 0$ (modulo $\mathbb{Z}$) for each $v \in |K|$.

Proof: See [We 2] Ch. XI, § 2, Thm. 2, Ch. XIII, § 3, Thm. 2 and Ch. XIII, Thm. 4 or [Re] (32.13) for the first assertion.
See [Re] (32.19) for the second assertion.   □

COROLLARY (A.3.4). — *Let D be a central division algebra over K with $[D : K] = d^2$ for some positive integer d and let L be a finite extension of K of degree n (L is also a function field). Then there exists a K-isomorphism of L onto a K-subalgebra of D if and only if n divides d and*

$$\frac{d}{n}[L_w : K_v] \operatorname{inv}_{K,v}(D) \equiv 0 \ (\text{modulo } \mathbb{Z})$$

for each $w \in |L|$ dividing $v \in |L|$.
Moreover, any two such K-isomorphisms are conjugate in D.

Proof: Thanks to (A.1.4), we can assume that n divides d and it suffices to check that $D \otimes_K L$ is L-isomorphic to $gl_n(\Delta)$ for some central division algebra Δ over L if and only if

$$\frac{d}{n}[L_w : K_v] \operatorname{inv}_{K,v}(D) \equiv 0 \ (\text{modulo } \mathbb{Z})$$

for each $w \in |L|$ dividing $v \in |K|$.
But, thanks to (A.3.2),

$$\operatorname{inv}_{L,w}(D \otimes_K L) = [L_w : K_v] \operatorname{inv}_{K,v}(D)$$

for each $w \in |L|$ dividing $v \in |K|$. Moreover, thanks to (A.3.3), d is the smallest positive integer such that

$$d \operatorname{inv}_{K,v}(D) \equiv 0 \ (\text{modulo } \mathbb{Z})$$

and $D \otimes_K L$ is L-isomorphic to $gl_n(\Delta)$ for some central division algebra Δ over L with $[\Delta : L] = (d/n)^2$ if and only if $\dfrac{d}{n}$ is the smallest positive integer such that

$$\frac{d}{n} \operatorname{inv}_{L,w}(D \otimes_K L) \equiv 0 \ (\text{modulo } \mathbb{Z})$$

for each $w \in |L|$.
Now the corollary is obvious.   □

(A.4) Comments and references

The material of this appendix is classical. The basic references are [Bou] Alg. Ch. 8, [Re] and [We 2].

B

Dieudonné's theory : some proofs

(B.1) Proof of (2.4.5)

We will denote simply by $\mathcal{O}_{\bar{o}}$ the completion of the maximal unramified extension of $\mathcal{O}_o$ with residue field $\overline{\kappa(o)}$ and by $F_{\bar{o}}$ its field of fractions. We denote again by o the natural extension of the valuation o to $\mathcal{O}_{\bar{o}}$.

We can form the non-commutative polynomial ring

(B.1.1)
$$F_{\bar{o}}[f]$$

with commutation rule

$$f \cdot a = \sigma_o(a) \cdot f$$

for each $a \in F_{\bar{o}}$.

Obviously, the category of Dieudonné F_o-modules over $\overline{\kappa(o)}$ is a full subcategory of the category of left $F_{\bar{o}}[f]$-modules: we identify (N, f) with N endowed with the left $F_{\bar{o}}[f]$-module structure given by the $F_{\bar{o}}$-vector space structure of N and by the σ_o-linear endomorphism f. In particular, for each pair of integers (r, s) with $r \geq 1$ and $(r, s) = 1$, we have

(B.1.2)
$$N_{r,s} = F_{\bar{o}}[f]/F_{\bar{o}}[f](f^r - \varpi_o^s).$$

LEMMA (B.1.3). — *Let N be a left $F_{\bar{o}}[f]$-module of finite dimension over $F_{\bar{o}}$ (i.e. as a $F_{\bar{o}}$-vector space). Then N is a finitely generated torsion left $F_{\bar{o}}[f]$-module.*
$\square$

For each $P(f) \in F_{\bar{o}}[f]$, let $\deg(P(f))$ be its degree in f $(\deg(0) = -\infty)$. Then we have

$$\begin{cases} \deg(P_1(f) + P_2(f)) \leq \sup(\deg(P_1(f)), \deg(P_2(f))), \\ \deg(P_1(f)P_2(f)) = \deg(P_1(f)) + \deg(P_2(f)), \end{cases}$$

for each $P_1(f)$, $P_2(f) \in F_{\bar{o}}[f]$. Moreover, we have a left Euclidean algorithm on $F_{\bar{o}}[f]$: $F_{\bar{o}}[f]$ doesn't have zero divisors and if $P_1(f)$, $P_2(f)$ are in $F_{\bar{o}}[f]$ and, if $P_2(f) \neq 0$, there exist unique $Q(f)$ and $R(f)$ in $F_{\bar{o}}[f]$ such that

$$\begin{cases} P_1(f) = Q(f)P_2(f) + R(f), \\ \deg(R(f)) < \deg(P_2(f)). \end{cases}$$

Therefore we get

LEMMA (B.1.4). — (i) *Every left ideal of $F_{\bar{o}}[f]$ is principal. In particular, $F_{\bar{o}}[f]$ is left noetherian.*

(ii) *If $P(f) \in F_{\bar{o}}[f]$ and $P(f) \neq 0$, the cyclic left $F_{\bar{o}}[f]$-module*

$$F_{\bar{o}}[f]/F_{\bar{o}}[f]P(f)$$

is of dimension $\deg(P(f))$ over $F_{\bar{o}}$. In particular, it is torsion.

(iii) *Every finitely generated torsion left $F_{\bar{o}}[f]$-module is isomorphic to a direct sum of cyclic $F_{\bar{o}}[f]$-modules $F_{\bar{o}}[f]/F_{\bar{o}}[f]P(f)$ where $P(f) \neq 0$. In particular, every finitely generated torsion left $F_{\bar{o}}[f]$-module is of finite dimension over $F_{\bar{o}}$.* $\square$

Our next goal is to study more closely the torsion cyclic left $F_{\bar{o}}[f]$-modules.

LEMMA (B.1.5). — *Let $P(f) \in F_{\bar{o}}[f]$ be unitary (i.e. $P(f) \neq 0$ and the coefficient of the leading term is 1). Then there exists an integer m, with $1 \leq m \leq d!$, such that, over $F_{\bar{o}}[f][\varpi_o^{1/m}] = F_{\bar{o}}[\varpi_o^{1/m}][f]$ $(f \cdot \varpi_o^{1/m} = \varpi_o^{1/m} \cdot f)$, we can decompose $P(f)$ into*

$$P(f) = \prod_{i=1}^{d}(f - x_i)$$

where $d = \deg(P(f))$ and $x_i \in F_{\bar{o}}[\varpi_o^{1/m}]$ $(i = 1, \ldots, d)$.

REMARK (B.1.6). — $F_{\bar{o}}[\varpi_o^{1/m}]$ is nothing else than $\overline{\kappa(o)}((\varpi_o^{1/m}))$; in particular, it has an obvious discrete valuation. The corresponding discrete valuation ring is $\mathcal{O}_{\bar{o}}[\varpi_o^{1/m}] = \overline{\kappa(o)}[[\varpi_o^{1/m}]]$ and $\varpi_o^{1/m}$ is a uniformizer. $\square$

Proof of (B.1.5) : By induction on d, it is enough to find an integer m, with $1 \leq m \leq d$, and $x \in F_{\bar{o}}[\varpi_o^{1/m}]$ such that

$$P(f) = Q(f)(f - x)$$

for some $Q(f) \in F_{\bar{o}}[\varpi_o^{1/m}][f]$. Let us write

$$P(f) = f^d + \sum_{i=0}^{d-1} a_i f^i.$$

Then we see by equating the coefficients that it is the same to find an integer $m > 0$ and $x \in F_{\bar{o}}[\varpi_o^{1/m}]$ such that

$$a_0 + a_1 x + a_2 x \sigma_o(x) + \cdots + a_{d-1} x \sigma_o(x) \cdots \sigma_o^{d-2}(x) + x \sigma_o(x) \cdots \sigma_o^{d-1}(x) = 0.$$

If $a_0 = 0$, we can take $x = 0$. So let us assume that $a_0 \neq 0$ and let us set

$$\frac{n}{m} = \mathrm{Inf}\{\frac{o(a_i)}{d-i} \mid i = 0, \ldots, d-1\}$$

$(o(0) = +\infty)$ where $m, n \in \mathbb{Z}$, $m \geq 1$ and $(m, n) = 1$. Replacing F_o by $F_o[\varpi_o^{1/m}]$ and $o(-)$ by $m \cdot o(-)$, we are reduced to the case $m = 1$. In that case, we can write

$$a_i = \varpi_o^{n(d-i)} b_i$$

with $b_i \in \mathcal{O}_{\bar{o}}$ $(i = 0, \ldots, d-1)$ and at least one of the b_i's is a unit in $\mathcal{O}_{\bar{o}}$ (i.e. $o(b_i) = 0$). Putting

$$x = \varpi_o^n y,$$

it suffices to find $y \in \mathcal{O}_{\bar{o}}$ such that

$$(*) \quad b_0 + b_1 y + b_2 y \sigma_o(y) + \cdots + b_{d-1} y \sigma_o(y) \cdots \sigma_o^{d-2}(y) + y \sigma_o(y) \cdots \sigma_o^{d-1}(y) = 0.$$

As

$$\sigma_o(y) \equiv y^{p^{deg(o)}} \pmod{\varpi_o}$$

and $\overline{\kappa(o)}$ is algebraically closed, the equation $(*)$ has a non-trivial solution modulo ϖ_o (at least one of the b_i's is non-zero modulo ϖ_o). Now let us assume that $y_n \in \mathcal{O}_{\bar{o}}$ is a solution of $(*)$ modulo ϖ_o^n for some integer $n \geq 1$ such that y_n is non-zero modulo ϖ_o. Then

$$y_{n+1} = y_n + \varpi_o^n z$$

with $z \in \mathcal{O}_{\bar{o}}$ will be a solution of $(*)$ modulo ϖ_o^{n+1} if and only if z is a solution of the congruence

$$(**) \quad c + \sum_{i=1}^{d} b_i \sum_{j=0}^{i-1} y_n \sigma_o(y_n) \cdots \sigma_o^{j-1}(y_n) \sigma_o^j(z) \sigma_o^{j+1}(y_n) \cdots \sigma_o^{i-1}(y_n)$$

$$\equiv 0 \pmod{\varpi_o}$$

where $b_d = 1$ and $c \in \mathcal{O}_{\bar{o}}$ satisfies the congruence

$$b_0 + b_1 y_n + b_2 y_n \sigma_o(y_n) + \cdots + b_{d-1} y_n \sigma_o(y_n) \cdots \sigma_o^{d-2}(y_n)$$
$$+ y_n \sigma_o(y_n) + \cdots + \sigma_o^{d-1}(y_n) \equiv c \varpi_o^n \pmod{\varpi_o^{n+1}}.$$

But, as before, $(**)$ has a solution in $\mathcal{O}_{\bar{o}}$. By induction on the integer n we get a true solution of $(*)$ in $\mathcal{O}_{\bar{o}}$ ($\mathcal{O}_{\bar{o}}$ is complete) and the lemma is proved.
□

EXAMPLE (B.1.7). — Following the proof of (B.1.5) it is easy to see that for each pair of integers (r, s) with $r \geq 1$ and $(r, s) = 1$, the unitary polynomial $f^r - \varpi_o^s$ splits over $F_{\bar{o}}[\varpi_o^{1/r}]$ into

$$f^r - \varpi_o^s = \prod_{i=1}^{r} (f - u_i \varpi_o^{s/r})$$

where $u_1, \ldots, u_r$ are units in $\mathcal{O}_{\bar{o}}[\varpi_o^{1/r}]$.
□

LEMMA (B.1.8). — *For each $a \in F_{\bar{o}}$, the left $F_{\bar{o}}[f]$-module*

$$N_a = F_{\bar{o}}[f]/F_{\bar{o}}[f](f - a)$$

is simple. If $a, b \in F_{\bar{o}}$, then N_a is isomorphic to N_b if and only if $o(a) = o(b)$. In particular, if $a \in F_{\bar{o}}$, $a \neq 0$, then N_a is isomorphic to $N_{1,o(a)}$ and, if $s, s' \in \mathbb{Z}$, $N_{1,s}$ is isomorphic to $N_{1,s'}$ if and only if $s = s'$.

Moreover, if $a', a'' \in F_{\bar{o}}$ and if $a' \neq 0$ or $a'' \neq 0$, any extension of left $F_{\bar{o}}[f]$-modules of $N_{a''}$ by $N_{a'}$ splits.

CAUTION. — There are non-trivial extensions of N_0 by itself (for example $F_{\bar{o}}[f]/F_{\bar{o}}[f]f^2$).

Proof of (B.1.8) : The simplicity of N_a follows from $\dim_{F_{\bar{o}}}(N_a) = 1$.

If $N_a \xrightarrow{u} N_b$ is an isomorphism of left $F_{\bar{o}}[f]$-modules and if $u(1) = c \cdot 1$ with $c \in F_{\bar{o}}^{\times}$ (1 is the canonical basis of N_a or N_b over $F_{\bar{o}}$), we have

$$a = \frac{\sigma(c)}{c} b$$

Therefore, u exists if and only if $o(a) = o(b)$.

Now let

$$0 \longrightarrow N_{a'} \xrightarrow{u} N \xrightarrow{v} N_{a''} \longrightarrow 0$$

be an extension of $N_{a''}$ by $N_{a'}$. Then we have $\dim_{F_{\bar{o}}}(N) = 2$ and N admits a basis (e_1, e_2) such that

$$\begin{cases} f(e_1) = a' e_1, \\ f(e_2) = b e_1 + a'' e_2, \end{cases}$$

for some $b \in F_{\bar{o}}$ ($e_1 = u(1)$ and $v(e_2) = 1$). To finish the proof of the lemma it is enough to find $x \in F_{\bar{o}}$ such that

$$f(xe_1 + e_2) = a''(xe_1 + e_2),$$

i.e.

$$b + a'\sigma_o(x) = a''x.$$

If $b = 0$ we can just take $x = 0$. If $b \neq 0$, by multiplying a' and a'' by $\varpi_o^{-o(b)}$ we can assume that $o(b) = 0$. Then let

$$r = \operatorname{Inf}(o(a'), o(a'')).$$

Replacing a', a'' and x by $\varpi_o^{-r}a'$, $\varpi_o^{-r}a''$ and $\varpi_o^r x$ respectively, we can assume that

$$\operatorname{Inf}(o(a'), o(a'')) = 0.$$

Finally, we get a solution $x \in \mathcal{O}_{\bar{o}}^{\times}$ of our equation with $(o(b) = \operatorname{Inf}(o(a'), o(a'')) = 0)$ by successive approximation as in the proof of (B.1.5). $\quad\square$

COROLLARY (B.1.9). — *If*

$$P(f) = \prod_{i=1}^{d}(f - x_i) \in F_{\bar{o}}[f]$$

with $x_i \in F_{\bar{o}}$ ($i = 1, \ldots, d$), $x_i \neq 0$ ($i = 1, \ldots, t$) and $x_i = 0$ ($i = t+1, \ldots, d$) ($0 \leq t \leq d$), then the left $F_{\bar{o}}[f]$-module

$$N = F_{\bar{o}}[f]/F_{\bar{o}}[f]P(f)$$

has a canonical decomposition into

$$N = N_{\mathrm{ss}} \oplus N_{\mathrm{nil}}$$

where N_{ss} is semi-simple and isomorphic to

$$\bigoplus_{i=1}^{t} N_{1,s_i}$$

with $s_i = o(x_i)$ ($i = 1, \ldots, r$) and N_{nil} is indecomposable of length $\ell = d - r$ and is isomorphic to

$$F_{\bar{o}}[f]/F_{\bar{o}}[f]f^{\ell}.$$

Moreover, the length ℓ of N_{nil} and the sequence of integers $(s_1, \ldots, s_t)$ (up to a permutation of the indices) are uniquely determined by the isomorphism class of the left $F_{\bar{o}}[f]$-module N.

Proof : Let

$$P_{ss}(f) = \prod_{i=1}^{t}(f - x_i)$$

and

$$P_{nil}(f) = f^{\ell}$$

so that

$$P(f) = P_{ss}(f)P_{nil}(f).$$

We have a canonical exact sequence

$$0 \longrightarrow N_{ss} \xrightarrow{u} N \xrightarrow{v} N_{nil} \longrightarrow 0$$

where

$$\begin{cases} N_{ss} = F_{\bar{o}}[f]/F_{\bar{o}}[f]P_{ss}(f), \\ N_{nil} = F_{\bar{o}}[f]/F_{\bar{o}}[f]P_{nil}(f) \end{cases}$$

and

$$\begin{cases} u(1) \equiv P_{nil}(f) \ (\text{modulo } F_{\bar{o}}[f]P(f)), \\ v(1) \equiv 1 \ (\text{modulo } F_{\bar{o}}[f]P_{nil}(f)). \end{cases}$$

Obviously N_{nil} is killed by f^{ℓ} and not killed by $f^{\ell-1}$ (if $\ell \geq 1$). It follows that any simple subquotient of N_{nil} is isomorphic to

$$N_0 = F_{\bar{o}}[f]/F_{\bar{o}}[f]f$$

and that N_{nil} is indecomposable. As there is no non-trivial extension of N_0 by $N_{1,s}$ $(s \in \mathbf{Z})$, it suffices to prove the corollary when $d = t$.

Let us set

$$\widetilde{P}(f) = \prod_{i=1}^{d-1}(f - x_i).$$

Then we have an exact sequence

$$0 \to \widetilde{N} \to N \to N_{x_d} \to 0$$

of left $F_{\bar{o}}[f]$-modules where

$$\widetilde{N} = F_{\bar{o}}[f]/F_{\bar{o}}[f]\widetilde{P}(f)$$

and

$$N_{x_d} = F_{\bar{o}}[f]/F_{\bar{o}}[f](f - x_d).$$

By induction on d, we can assume that $\widetilde{N}$ is isomorphic to

$$\bigoplus_{i=1}^{d-1} N_{1,s_i}.$$

But N_{x_d} is isomorphic to N_{1,s_d} and the corollary follows from (B.1.8). $\qquad\square$

LEMMA (B.1.10). — *For any pair of integers (r, s) with $r \geq 1$ and $(r, s) = 1$, the cyclic left $F_{\bar{o}}[f]$-module $N_{r,s}$ is simple. If (r', s') is another pair of integers with $r' \geq 1$ and $(r', s') = 1$, the cyclic left $F_{\bar{o}}[f]$-modules $N_{r,s}$ and $N_{r',s'}$ are isomorphic if and only if $r = r'$ and $s = s'$.*

Proof : Thanks to (B.1.7) and (B.1.9) we have a decomposition over $F_{\bar{o}}[\varpi_o^{1/r}]$

$$F_{\bar{o}}[\varpi_o^{1/r}] \otimes_{F_{\bar{o}}} N_{r,s} \cong (N_{1,s})^r$$

where

$$N_{1,s} = F_{\bar{o}}[\varpi_o^{1/r}](f) / F_{\bar{o}}[\varpi_o^{1/r}][f](f - \varpi_o^{s/r}).$$

If m is any positive integer, then

$$F_{\bar{o}}[\varpi_o^{1/mr}] \otimes_{F_{\bar{o}}[\varpi_o^{1/r}]} N_{1,s}$$

is a simple left $F_{\bar{o}}[\varpi_o^{1/mr}][f]$-module which is isomorphic to

$$N_{1,ms} = F_{\bar{o}}[\varpi_o^{1/mr}][f] / F_{\bar{o}}[\varpi_o^{1/mr}][f](f - \varpi_o^{ms/mr})$$

and we have a decomposition over $F_{\bar{o}}[\varpi_o^{1/mr}]$

$$F_{\bar{o}}[\varpi_o^{1/mr}] \otimes_{F_{\bar{o}}} N_{r,s} \cong (N_{1,ms})^r.$$

In particular, it follows from (B.1.9) that the pair of integers (r, s) is uniquely determined by the isomorphism class of the left $F_{\bar{o}}[f]$-module $N_{r,s}$. If $N_{r,s}$ is not simple, there exists a proper cyclic left $F_{\bar{o}}[f]$-module

$$N \subsetneq N_{r,s},$$

$$N \cong F_{\bar{o}}[f] / F_{\bar{o}}[f]P(f)$$

with $P(f)$ unitary and

$$\deg(P(f)) = \dim_{F_{\bar{o}}}(N) < \dim_{F_{\bar{o}}}(N_{r,s}) = r.$$

Following the proof of (B.1.5), there exists an integer m with

$$1 \leq m \leq \deg(P(f)) < r$$

such that

$$P(f) = Q(f)(f - x)$$

for some $x \in F_{\bar{o}}[\varpi_o^{1/m}]$ and $Q(f) \in F_{\bar{o}}[\varpi_o^{1/m}][f]$. Therefore the left $F_{\bar{o}}[\varpi_o^{1/m}]$-module

$$N_x = F_{\bar{o}}[\varpi_o^{1/m}][f] / F_{\bar{o}}[\varpi_o^{1/m}][f](f - x)$$

is a quotient of $F_{\bar{o}}[\varpi_o^{1/m}] \otimes_{F_{\bar{o}}} N$ and the left $F_{\bar{o}}[\varpi_o^{1/mr}][f]$-module

$$F_{\bar{o}}[\varpi_o^{1/mr}] \otimes_{F_{\bar{o}}[\varpi_o^{1/m}]} N_x$$

is a subquotient of $F_{\bar{o}}[\varpi_o^{1/mr}] \otimes_{F_{\bar{o}}} N_{r,s}$. From our previous discussion, it follows that we have an isomorphism of left $F_{\bar{o}}[\varpi_o^{1/mr}][f]$-modules

$$F_{\bar{o}}[\varpi_o^{1/mr}] \otimes_{F_{\bar{o}}[\varpi_o^{1/m}]} N_x \cong N_{1,ms}$$

$(\dim_{F_{\bar{o}}[\varpi_o^{1/m}]}(N_x) = 1)$. Clearly, this implies $x \neq 0$ and

$$x = u\varpi_o^{ms/mr} = u\varpi_o^{s/r}$$

for some unit u in $\mathcal{O}_{\bar{o}}[\varpi_o^{1/mr}]$. But this is impossible as $m < r$ and $x \in F_{\bar{o}}[\varpi_o^{1/m}]$ (recall that $(r,s) = 1$). Hence $N_{r,s}$ is simple and the lemma is proved. $\qquad\square$

LEMMA (B.1.11). — *Let* $P'(f)$, $P''(f) \in F_{\bar{o}}[f]$ *be unitary and let us consider the exact sequence of left* $F_{\bar{o}}[f]$-*modules*

$$0 \longrightarrow N' \xrightarrow{u} N \xrightarrow{v} N'' \longrightarrow 0$$

where

$$\begin{cases} N' &= F_{\bar{o}}[f]/F_{\bar{o}}[f]P'(f), \\ N &= F_{\bar{o}}[f]/F_{\bar{o}}[f]P'(f)P''(f), \\ N'' &= F_{\bar{o}}[f]/F_{\bar{o}}[f]P''(f) \end{cases}$$

and

$$\begin{cases} u(1) \equiv P''(f) \ (modulo \ F_{\bar{o}}[f]P'(f)P''(f)), \\ v(1) \equiv 1 \ (modulo \ F_{\bar{o}}[f]P''(f)). \end{cases}$$

Then this exact sequence splits if and only if there exist $Q'(f)$, $Q''(f) \in F_{\bar{o}}[f]$ *such that*

$$Q'(f)P'(f) + P''(f)Q''(f) = 1$$

and in that case

$$w(1) \equiv 1 - Q''(f)P''(f) \ (modulo \ F_{\bar{o}}[f]P'(f)P''(f))$$

defines a section w *of* v.

Proof: If there exist $Q'(f)$, $Q''(f) \in F_{\bar{o}}[f]$ such that

$$Q'(f)P'(f) + P''(f)Q''(f) = 1,$$

then

$$P''(f)(1 - Q''(f)P''(f)) = (1 - P''(f)Q''(f))P''(f) = Q'(f)P'(f)P''(f)$$

and the map

$$w : F_{\bar{o}}[f] \to F_{\bar{o}}[f]/F_{\bar{o}}[f]P'(f)P''(f),$$
$$1 \mapsto 1 - Q''(f)P''(f)$$

factors through the quotient $F_{\bar{o}}[f]/F_{\bar{o}}[f]P''(f)$ of $F_{\bar{o}}[f]$ and gives a section of v.

Conversely, if w is a section of v, there exists $Q''(f) \in F_{\bar{o}}[f]$ such that

$$w(1) \equiv 1 - Q''(f)P''(f) \pmod{F_{\bar{o}}[f]P'(f)P''(f)}$$

and as

$$w(P''(f)) = 0$$

there exists $Q'(f)$ such that

$$P''(f)(1 - Q''(f)P''(f)) = Q'(f)P'(f)P''(f),$$

i.e.

$$(1 - Q'(f)P'(f) - P''(f)Q''(f))P''(f) = 0.$$

But $F_{\bar{o}}[f]$ doesn't have zero divisors and we get

$$Q'(f)P'(f) + P''(f)Q''(f) = 1$$

as we want. $\qquad\square$

COROLLARY (B.1.12). — *Let*

$$0 \to N' \to N \to N'' \to 0$$

be an exact sequence of torsion cyclic left $F_{\bar{o}}[f]$-modules as in (B.1.11) and let m be a positive integer. Then this exact sequence splits over $F_{\bar{o}}[f]$ if and only if it splits over $F_{\bar{o}}[\varpi_o^{1/m}]$.

Proof: For each $Q(f) \in F_{\bar{o}}[\varpi_o^{1/m}][f]$ there is a unique decomposition

$$Q(f) = Q_0(f) + Q_1(f)\varpi_o^{1/m} + \cdots + Q_{m-1}(f)\varpi_o^{(m-1)/m}$$

with $Q_i(f) \in F_{\bar{o}}[f]$ $(i = 0, \cdots, m-1)$. Therefore if $P'(f)$, $P''(f) \in F_{\bar{o}}[f]$ are unitary and if $Q'(f)$, $Q''(f) \in F_{\bar{o}}[\varpi_o^{1/m}][f]$ satisfies

$$Q'(f)P'(f) + P''(f)Q''(f) = 1$$

we also have

$$Q'_0(f)P'(f) + P''(f)Q''_0(f) = 1.$$

Now the corollary immediately follows from (B.1.11). $\qquad\square$

LEMMA (B.1.13). — *Every torsion cyclic left $F_{\bar{o}}[f]$-module*

$$N = F_{\bar{o}}[f]/F_{\bar{o}}[f]P(f)$$

($P(f) \in F_{\bar{o}}[f]$, $P(f)$ unitary) admits a canonical splitting

$$N = N_{\mathrm{ss}} \oplus N_{\mathrm{nil}}$$

where

(i) N_{ss} *is semi-simple and is isomorphic to a direct sum*

$$N_{\mathrm{ss}} = \bigoplus_{i=1}^{t} N_{r_i, s_i}$$

for some sequence of pairs of integers

$$((r_1, s_1), \ldots, (r_t, s_t))$$

with $r_i \geq 1$ and $(r_i, s_i) = 1$ $(i = 1, \ldots, t)$,

(ii) *N is indecomposable and isomorphic to*

$$F_{\bar{o}}[f]/F_{\bar{o}}[f]f^\ell$$

for some integer $\ell \geq 0$.

Moreover, the sequence $((r_1, s_1), \ldots, (r_t, s_t))$ (up to a permutation of the indices) and the integer ℓ are uniquely determined by the isomorphism class of N.

Proof: We can decompose $P(f)$ into

$$P(f) = P_{\mathrm{ss}}(f)P_{\mathrm{nil}}(f)$$

where the constant term of $P_{\mathrm{ss}}(f)$ is non-zero and $P_{\mathrm{nil}}(f) = f^\ell$ for some integer $\ell \geq 0$. Accordingly we have an exact sequence of left $F_{\bar{o}}[f]$-modules

$$0 \to N_{\mathrm{ss}} \to N \to N_{\mathrm{nil}} \to 0$$

as in the proof of (B.1.9). Thanks to (B.1.5) and (B.1.9), there exists an integer $m \geq 1$ such that this exact sequence splits over $F_{\bar{o}}[\varpi_o^{1/m}]$ and such that $F_{\bar{o}}[\varpi_o^{1/m}] \otimes_{F_{\bar{o}}} N_{\mathrm{ss}}$ is isomorphic to a direct sum of a finite number of $N_{1,n}$'s $(n \in \mathbb{Z})$.

If $N_{1,n}$ is a direct factor of $F_{\bar{o}}[\varpi_o^{1/m}] \otimes_{F_{\bar{o}}} N_{\mathrm{ss}}$ for some n, let (r,s) be the unique pair of integers with $r \geq 1$ and $(r,s) = 1$ such that

$$\frac{s}{r} = \frac{n}{m}.$$

Then, if $x \in F_{\bar{o}}[\varpi_o^{1/m}] \otimes_{F_{\bar{o}}} N_{\mathrm{ss}}$ is the image of canonical generator $1 \in N_{1,n}$, we have

$$(f - \varpi_o^{s/r})x = 0$$

and

$$x = \sum_{j=0}^{m-1} \varpi_o^{j/m} x_j$$

where $x_j \in N_{\mathrm{ss}}$ $(j = 0, \ldots, m-1)$. Therefore, we have

$$(f^r - \varpi_o^s)x_j = 0$$

for all $j = 0, \ldots, m-1$. Taking $j \in \{0, \ldots, m-1\}$ such that $x_j \neq 0$ $(x \neq 0)$, we get a non-trivial map

$$N_{r,s} \to N_{\mathrm{ss}} \ , \ 1 \mapsto x_j$$

where 1 is now the canonical generator of $N_{r,s}$. As $N_{r,s}$ is simple (see (B.1.10)) this map is injective.

Finally, if we apply (B.1.12), we see that the exact sequences

$$0 \to N_{\mathrm{ss}} \to N \to N_{\mathrm{nil}} \to 0$$

and

$$0 \to N_{r,s} \to N \to N/N_{r,s} \to 0$$

split (they split over $F_{\bar{o}}[\varpi_o^{1/m}]$) and the lemma easily follows. $\qquad \square$

Putting all the preceeding results together, we get

PROPOSITION (B.1.14). — *The abelian category of finitely generated torsion left $F_{\bar{o}}[f]$-modules has the following properties*

(i) *every object is of finite length and, in particular, is a direct sum of indecomposable objects;*

(ii) *the isomorphism classes of indecomposable objects are represented by the*

$$N_{r,s} = F_{\bar{o}}[f]/F_{\bar{o}}[f](f^r - \varpi_o^s)$$

for all pairs of integers (r, s) with $r \geq 1$ and $(r, s) = 1$ and the

$$F_{\bar{o}}[f]/F_{\bar{o}}[f]f^\ell$$

for all integers $\ell \geq 0$;

(iii) *the isomorphism classes of simple objects are represented by the $N_{r,s}$ for all pairs of integers (r, s) with $r \geq 1$ and $(r, s) = 1$ and*

$$N_0 = F_{\bar{o}}[f]/F_{\bar{o}}[f]f.$$

□

Proof of (2.4.5) : It is now easy to see that the Dieudonné F_o-modules over $\kappa(o)$ are exactly the finitely generated torsion left $F_{\bar{o}}[f]$-modules without f-torsion, i.e. without direct factors $F_{\bar{o}}[f]/F_{\bar{o}}[f]f^\ell$ ($\ell \in \mathbb{Z}$, $\ell > 0$) and part (i) is a direct consequence of (B.1.14).

As $N_{r,s}$ is a simple left $F_{\bar{o}}[f]$-module, $\mathrm{End}_{F_{\bar{o}}[f]}(N_{r,s}) = \Delta_{r,s}$ is a skew field. Obviously $\Delta_{r,s}$ contains F_o in its center. To finish the proof of (2.4.5) let us check that $\Delta_{r,s}$ is a central division algebra over F_o with invariant $-\frac{s}{r} + \mathbb{Z} \in \mathbb{Q}/\mathbb{Z}$. By definition, the opposite F_o-algebra $\Delta_{r,s}^{opp}$ to $\Delta_{r,s}$ is the quotient of the F_o-subalgebra

$$\{P(f) \in F_{\bar{o}}[f] \,|\, (f^r - \varpi_o^s)P(f) \in F_{\bar{o}}[f](f^r - \varpi_o^s)\}$$

of $F_{\bar{o}}[f]$ by its two-sided ideal

$$F_{\bar{o}}[f](f^r - \varpi_o^s).$$

But, thanks to the left Euclidean algorithm, we can write any $P(f) \in F_{\bar{o}}[f]$ as

$$P(f) = Q(f)(f^r - \varpi_o^s) + R(f)$$

with $\deg(R(f)) < r$ and we have

$$(f^r - \varpi_o^s)P(f) \in F_{\bar{o}}[f](f^r - \varpi_o^s)$$

if and only if $R(f) \in F_{o,r}[f]$ where $F_o \subset F_{o,r} \subset F_{\bar{o}}$ is "the" maximal unramified extension of degree r of F_o. Therefore, we have

$$\Delta_{r,s}^{opp} = F_{o,r}[f]/(f^r - \varpi_o^s).$$

Hence, the F_o-algebra $\Delta_{r,s}^{opp}$ is F_o-isomorphic to the central division algebra $D_{r,s}$ over F_o and its invariant is $\frac{s}{r} + \mathbb{Z} \in \mathbb{Q}/\mathbb{Z}$ (see (A.2.2)). Theorem (2.4.5) is now completely proved.   □

(B.2) Proof of (2.4.6)

Let (M, f) be a Dieudonné $\mathcal{O}_o$-module over k $(\kappa(o) \subset k \subset \overline{\kappa(o)})$, let

$$(\overline{M}, \overline{f}) = \overline{\kappa(o)} \otimes_k (M, f)$$

be its base change to $\overline{\kappa(o)}$ and let

$$(\overline{N}, \overline{f}) = (F_{\bar{o}} \otimes_{\mathcal{O}_{\bar{o}}} \overline{M}, \ F_{\bar{o}} \otimes_{\mathcal{O}_{\bar{o}}} \overline{f})$$

be the corresponding Dieudonné F_o-module over $\overline{\kappa(o)}$.

We denote by

$$\lambda_1 \leq \lambda_2 \leq \cdots \leq \lambda_r$$

the non-decreasing sequence of slopes of (M, f), i.e. of $(\overline{N}, \overline{f})$, where the slope $\lambda \in \mathbb{Q}$ appears as many times as its multiplicity in (M, f).

LEMMA (B.2.1). — *We have* $\lambda_i \geq 0$ *for each* $i = 1, \dots, t$. *Moreover, if* μ *is a positive multiple of* $r!$, *then* $\mu\lambda_i \in \mathbb{Z}$ *for* $i = 1, \dots, r$, *there exists a basis* $(\overline{m}_1, \dots, \overline{m}_r)$ *of the* $\mathcal{O}_{\bar{o}}[\varpi_o^{1/\mu}]$-*module* $\mathcal{O}_{\bar{o}}[\varpi_o^{1/\mu}] \otimes_{\mathcal{O}_o} \overline{M}$ *such that*

$$\overline{f}(\overline{m}_i) \equiv \varpi_o^{\lambda_i} \overline{m}_i \ (modulo \ \sum_{j=1}^{i-1} \mathcal{O}_{\bar{o}}[\varpi_o^{1/\mu}]\overline{m}_j),$$

i.e. such that the matrix of $\overline{f}$ *in this basis is upper triangular with diagonal entries*

$$\varpi_o^{\lambda_1}, \dots, \varpi_o^{\lambda_r},$$

and the sequence $(\lambda_1, \dots, \lambda_r)$ *is uniquely determined by these properties.*

Proof : Obviously $r!$ is a positive multiple of the common denominator of $\lambda_1, \dots, \lambda_t$ and, if μ is any positive multiple of this common denominator, there exists a basis $(\overline{n}_1, \dots, \overline{n}_r)$ of the $F_{\bar{o}}[\varpi_o^{1/\mu}]$-vector space $F_{\bar{o}}[\varpi_o^{1/\mu}] \otimes_{F_{\bar{o}}} \overline{N}$ such that

$$\overline{f}(\overline{n}_i) = \varpi_o^{\lambda_i} \overline{n}_i$$

for $i = 1, \dots, r$.

Now, by induction on r, we can find a basis $(\overline{m}_1, \dots, \overline{m}_r)$ of the $\mathcal{O}_{\bar{o}}[\varpi_o^{1/\mu}]$-module $\mathcal{O}_{\bar{o}}[\varpi_o^{1/\mu}] \otimes_{F_{\bar{o}}} \overline{M}$ such that

$$\overline{m}_i \in \sum_{j=1}^{i-1} F_{\bar{o}}[\varpi_o^{1/\mu}]\overline{n}_j$$

for $i = 1, \dots, r$. Then

$$\overline{f}(\overline{m}_i) \equiv \varpi_o^{\lambda_i} \overline{m}_i \ (modulo \ \sum_{j=1}^{i-1} F_{\bar{o}}[\varpi_o^{1/\mu}]\overline{m}_j)$$

for $i = 1, \dots, r$. But $\overline{f}(\overline{m}_i) \in \mathcal{O}_{\bar{o}}[\varpi_o^{1/\mu}] \otimes_{F_{\bar{o}}} \overline{M}$ for $i = 1, \dots, r$ and the lemma follows. $\square$

COROLLARY (B.2.2). — *All the slopes λ_i ($i = 1, \ldots, r$) are zero (resp. positive) if and only if $f \colon M \to M$ is bijective (resp. f is topologically nilpotent, i.e. there exists a positive integer n such that $f^n(M) \subset \varpi_o M$).*

$\square$

Proof of (2.4.6) : Let

$$M^{et} = \bigcap_{i \geq 0} f^i(M) \subset M$$

and

$$M^c = \{ m \in M | \forall j > 0, \exists i \geq 0, f^i(m) \in \varpi_o^j M \}.$$

We have

$$f(M^{et}) \subset M^{et}$$

(resp.

$$f(M^c) \subset M^c)$$

and, if we denote by

$$f^{et} \colon M^{et} \to M^{et}$$

(resp.

$$f^c \colon M^c \to M^c)$$

the restriction of f to M^{et} (resp. M^c), then (M^{et}, f^{et}) (resp. (M^c, f^c)) is a Dieudonné $\mathcal{O}_o$-module over k with f^{et} bijective (resp. f^c topologically nilpotent), i.e. such that all its slopes are equal to zero (resp. positive) thanks to (B.2.2). To finish the proof of (2.4.6), it suffices to check that

$$M = M^{et} \oplus M^c.$$

For each positive integer, let

$$M_j^{et} = \bigcap_{i \geq 0} (f^i(M) + \varpi_o^j M)$$

and

$$M_j^c = \bigcup_{i \geq 0} f^{-i}(\varpi_o^j M).$$

We have

$$M_j^{et} \supset \varpi_o^j M,$$

$$M_j^c \supset \varpi_o^j M$$

and a commutative diagram

$$
\begin{array}{ccc}
M^{et} & \xrightarrow{\ u^{et}\ } & \varprojlim_{j}(M_j^{et}/\varpi_o^j M) \\[2ex]
\downarrow & & \downarrow \\[2ex]
M & \xrightarrow[u]{\sim} & \varprojlim_{j}(M/\varpi_o^j M) \\[2ex]
\uparrow & & \uparrow \\[2ex]
M^c & \xrightarrow{\ u^c\ } & \varprojlim_{j}(M_j^c/\varpi_o^j M).
\end{array}
$$

As the $\mathcal{O}_k$-modules $M/\varpi_o^j M$, $M^{et}/\varpi_o^j M$ and $M_j^c/\varpi_o^j M$ are artinian for all positive integers j, the above inverse systems satisfy the Mittag–Leffler condition and the inverse limit is exact. Therefore to finish the proof of (2.4.6) it suffices to check that u^{et} and u^c are bijective and that

$$
M/\varpi_o^j M = (M_j^{et}/\varpi_o^j M) \oplus (M_j^c/\varpi_o^j M)
$$

for each positive integer j.

For any positive integer j, there exists a non-negative integer $i(j)$ such that

$$
f^i(M) + \varpi_o^j M = M_j^{et}
$$

and

$$
f^{-i}(\varpi_o^j M) = M_j^c
$$

for all $i \geq i(j)$ ($M/\varpi_o^j M$ is artinian and M is noetherian). So, for each $m \in M$, there exists $m' \in M$ such that

$$
f^{i(j)}(m) \in f^{2i(j)}(m') + \varpi_o^j M
$$

and we have

$$
m - f^{i(j)}(m') \in f^{-i(j)}(\varpi_o^j M) = M_j^c
$$

with

$$
f^{i(j)}(m') \in f^{i(j)}(M) + \varpi_o^j M = M_j^{et}.
$$

In other words, we have

$$
M = M_j^{et} + M_j^c.
$$

Similarly, let us check that

$$
M_j^{et} \cap M_j^c = \varpi_o^j M.
$$

If m belongs to this intersection, there exists $m' \in M$ such that

$$m = f^{i(j)}(m')$$

and in fact

$$m' \in f^{-2i(j)}(\varpi_o^j M).$$

But we have

$$f^{-2i(j)}(\varpi_o^j M) = M_j^c = f^{-i(j)}(\varpi_o^j M)$$

and

$$m \in \varpi_o^j M$$

as desired. Therefore, we have proved that

$$M/\varpi_o^j M = (M_j^{et}/\varpi_o^j M) \oplus (M_j^c/\varpi_o^j M).$$

The surjectivity (and hence the bijectivity) of u^{et} (resp. u^c) is equivalent to the inclusion

$$\bigcap_{j>0} M_j^{et} \subset M^{et}$$

(resp.

$$\bigcap_{j>0} M_j^c \subset M^c).$$

The inclusion

$$\bigcap_{j>0} M_j^c \subset M^c$$

is obvious by definition of M^c. Let $m \in \bigcap_{j>0} M_j^{et}$, we want to prove that $m \in f^i(M)$ for each non-negative integer i. Let us choose $m_j' \in M$ such that

$$m \in f^i(m_j') + \varpi_o^j M$$

for each positive integer j. For each pair of integers (j, j') with $j' \geq j > 0$ we have

$$m_{j'}' - m_j' \in f^{-i}(\varpi_o^j M).$$

But there exists a non-negative integer n such that

$$\varpi_o^n M \subset f^i(M)$$

(the cokernel of f has finite length over $\mathcal{O}_k$). Therefore

$$f^{-i}(\varpi_o^j M) \subset \varpi_o^{j-n} M$$

for all integers $j \geq n$ (f is injective) and the sequence $(m_j')_{j>0}$ is a Cauchy sequence for the ϖ_o-adic topology in M. As M is complete for this topology, $(m_j')_{j>0}$ has a limit m' in M and obviously

$$m = f^i(m').$$

This completes the proof of (2.4.6). $\square$

(B.3) Proof of (2.4.11)

Let $\mathcal{M}_k$ be the category of pairs (M, f) where M is a finite dimensional k-vector space and f is a $p^{\deg(o)}$-linear endomorphism of M and let $\mathcal{G}_k$ be the category of finite k-schemes in $\kappa(o)$-vector spaces which can be embedded into $\mathbf{G}_{a,k}^N$ for some integer $N \geq 0$.

For each $(M, f) \in \mathrm{ob}\,\mathcal{M}_k$ we consider the symmetric algebra of M over k

$$\mathrm{Sym}_k(M) = \bigoplus_{n \geq 0} \mathrm{Sym}_k^n(M)$$

and its ideal $I_{(M,f)}$ generated by the elements

$$m^{p^{\deg(o)}} - f(m) \quad (m \in M)$$

($m \in M = \mathrm{Sym}_k^1(M)$, $m^n \in \mathrm{Sym}_k^n(M)$ for each integer $n \geq 0$, $f(m) \in M = \mathrm{Sym}_k^1(M)$) and we set

(B.3.1) $$R_{G_k(M,f)} = \mathrm{Sym}_k(M)/I_{(M,f)}.$$

Let

$$\mu_{G_k(M,f)}^* : R_{G_k(M,f)} \to R_{G_k(M,f)} \otimes_k R_{G_k(M,f)}$$

and

$$[\alpha]_{G_k(M,f)}^* : R_{G_k(M,f)} \to R_{G_k(M,f)} \quad (\alpha \in \kappa(o))$$

be the homomorphisms of k-algebras defined by

$$\mu_{G_k(M,f)}^*(\overline{m}) = 1 \otimes \overline{m} + \overline{m} \otimes 1$$

and

$$[\alpha]_{G_k(M,f)}^*(\overline{m}) = \alpha\overline{m}$$

for each $m \in M = \mathrm{Sym}_k^1(M)$ with class $\overline{m}$ modulo $I_{(M,f)}$. Then

(B.3.2) $$G_k(M, f) = \mathrm{Spec}(R_{G_k(M,f)})$$

is a k-scheme in $\kappa(o)$-vector spaces ($\mu_{G_k(M,f)}$ is the group law and $[\alpha]_{G_k(M,f)}$ is the multiplication by $\alpha \in \kappa(o)$). Moreover, $G_k(M, f)$ is canonically embedded as a closed k-subscheme in $\kappa(o)$-vector spaces into

$$\mathbf{V}(M) = \mathrm{Spec}(\mathrm{Sym}_k(M)).$$

LEMMA (B.3.3). — *As a k-vector space, $R_{G_k(M,f)}$ is finite dimensional of dimension*

$$\dim_k(R_{G_k(M,f)}) = p^{\deg(o)\dim_k(M)},$$

so that $G_k(M, f) \in \mathrm{ob}\,\mathcal{G}_k$, and we have

$$G_k(M, f)(R) = \{g \in \mathrm{Hom}_k(M, R) | g(f(m)) = g(m)^{p^{\deg(o)}}, \forall m \in M\}$$

for each k-algebra R. $\qquad\qquad\square$

For each $G \in \mathrm{ob}\,\mathcal{G}_k$, we consider its k-bialgebra

$$(R_G, \mu_G^*, [\alpha]_G^* \quad (\alpha \in \kappa(o)))$$

(R_G is a k-algebra which is finite dimensional as a k-vector space, μ_G is the group law of G and $[\alpha]$ is the multiplication by $\alpha \in \kappa(o)$) and we set

(B.3.4) $M_k(G) = \{m \in R_G | \mu_G^*(m) = 1 \otimes m + m \otimes 1,\ [\alpha]_G^*(m) = \alpha m,$
$$\forall \alpha \in \kappa(o)\}$$

and

(B.3.5) $$f_k(G) : M_k(G) \to M_k(G),\ m \mapsto m^{p^{\deg(o)}}.$$

Then $M_k(G)$ is a finite dimensional k-vector space and $f_k(G)$ is $p^{\deg(o)}$-linear.

LEMMA (B.3.6). — $M_k(G)$ is nothing else than the k-vector space of homomorphisms of k-schemes in $\kappa(o)$-vector spaces

$$G \to \mathbb{G}_{a,k}$$

and $f_k(G)$ is induced by the Frobenius endomorphism of $\mathbb{G}_{a,k}$ relative to $\kappa(o)$. $\square$

The above constructions are clearly functorial and we get two functors

(B.3.7) $$G_k : \mathcal{M}_k^{opp} \to \mathcal{G}_k,$$

(B.3.8) $$(M_k, f_k) : \mathcal{G}_k \to \mathcal{M}_k^{opp}.$$

LEMMA (B.3.9). — (i) The functors G_k and (M_k, f_k) are additive.

(ii) G_k (resp. (M_k, f_k)) is left (resp. right) exact.

(iii) The functor (M_k, f_k) is a left adjoint to the functor G_k, i.e. we have a bi-functorial bijection

$$\mathrm{Hom}_{\mathcal{G}_k}(G, G_k(M, f)) \cong \mathrm{Hom}_{\mathcal{M}_k}((M, f), (M_k(G), f_k(G)))$$

for $G \in \mathrm{ob}\,\mathcal{G}_k$ and $(M, f) \in \mathrm{ob}\,\mathcal{M}_k$.

Proof : Part (i) is obvious.

The right exactitude of (M_k, f_k) (i.e. its left exactitude as a contravariant functor) follows from (B.3.6). Moreover, if we prove part (iii), the left exactitude of G_k will follow from the right exactitude of (M_k, f_k).

Finally, part (iii) is a consequence of the following remark. Both sets

$$\mathrm{Hom}_{\mathcal{G}_k}(G, G_k(M, f))$$

and

$$\mathrm{Hom}_{\mathcal{M}_k}((M, f), (M_k(G), f_k(G)))$$

are in canonical bijection with

$$\{g \in \mathrm{Hom}_k(M, R_G) | \forall m \in M, g(f(m)) = g(m)^{p^{\deg(o)}},$$
$$\mu_G^*(g(m)) = 1 \otimes g(m) + g(m) \otimes 1,$$
$$[\alpha]_G^*(g(m)) = \alpha g(m), \forall \alpha \in \kappa(o)\}.$$

$\square$

LEMMA (B.3.10). — (i) *For any* $(M, f) \in \mathrm{ob}\mathcal{M}_k$, *we have a canonical splitting*

$$(M, f) = (M_{\mathrm{ss}}, f_{\mathrm{ss}}) \oplus (M_{\mathrm{nil}}, f_{\mathrm{nil}})$$

in $\mathcal{M}_k$ *where* f_{ss} *is bijective and* f_{nil} *is nilpotent.*

(ii) *If* $(M, f) \in \mathrm{ob}\,\mathcal{M}_k$ *and* f *is bijective (resp. nilpotent),* $G_k(M, f)$ *is étale (resp. connected).*

Proof : If $(M, f) \in \mathrm{ob}\,\mathcal{M}_k$, we set

$$M_{\mathrm{ss}} = \bigcap_{n \geq 1} \mathrm{Im}(f^n)$$

and

$$M_{\mathrm{nil}} = \bigcup_{n \geq 1} \mathrm{Ker}(f^n)$$

$(n \in \mathbb{Z})$. There exists $N \in \mathbb{Z}$, $N \geq 1$, such that

$$M_{\mathrm{ss}} = \mathrm{Im}(f^N)$$

and

$$M_{\mathrm{nil}} = \mathrm{Ker}(f^N).$$

In particular

$$\dim_k(M_{\mathrm{ss}}) + \dim_k(M_{\mathrm{nil}}) = \dim_k(M).$$

But, if $m \in M_{\mathrm{ss}} \cap M_{\mathrm{nil}}$, there exists $m' \in M$ with $f^N(m') = m$ and we have $f^{2N}(m') = 0$. As $\mathrm{Ker}(f^{2N}) = \mathrm{Ker}(f^N)$, we also have $f^N(m') = 0$ and we have proved that

$$M_{\mathrm{ss}} \cap M_{\mathrm{nil}} = \{0\}.$$

Part (i) follows (see also (B.1)).

If $\Phi = (\varphi_{ij})_{1 \leq i,j \leq d} \in GL_d(k)$, the morphism of k-schemes

$$\mathbb{A}_k^d \to \mathbb{A}_k^d, \ (x_i)_{1 \leq i \leq d} \mapsto (x_i^{p^{\deg(o)}} - \sum_{j=1}^d \varphi_{ij} x_j)_{1 \leq i \leq d}$$

is étale: its Jacobian matrix is $-\Phi$. If $\Phi = (\varphi_{ij})_{1 \leq i,j \leq d} \in gl_d(k)$ and

$$\Phi^{(N-1)} \cdots \Phi^{(1)} \Phi^{(0)} = 0,$$

for some positive integer N, where we have set

$$\Phi^{(n)} = (\varphi_{ij}^{p^{n \deg(o)}})_{1 \leq i,j \leq d}$$

for each $n \in \mathbb{Z}$, the system of equations

$$x_i^{p^{\deg(o)}} - \sum_{j=1}^d \varphi_{ij} x_j = 0 \quad (1 \leq i \leq d)$$

has no other solutions than $(0, \dots, 0)$ in any reduced k-algebra : it implies the system of equations

$$x_i^{p^{n \deg(o)}} = 0 \quad (1 \leq i \leq d).$$

Part (ii) follows. $\square$

Let

(B.3.11) $\qquad\qquad G \xrightarrow{u_G} G_k(M_k(G), f_k(G))$

and

(B.3.12) $\qquad\qquad (M, f) \xrightarrow{v_{(M,f)}} (M_k(G_k(M, f)), f_k(G_k(M, f)))$

be the adjunction maps for $G \in \mathrm{ob}\,\mathcal{G}_k$ and $(M, f) \in \mathrm{ob}\,\mathcal{M}_k$. To finish the proof of (2.4.11) it suffices to prove

PROPOSITION (B.3.13). — *For each $G \in \mathrm{ob}\,\mathcal{G}_k$ and each $(M, f) \in \mathrm{ob}\,\mathcal{M}_k$, the adjunction maps u_G and $v_{(M,f)}$ are isomorphisms.*

In the proof of this proposition we will need the following lemmas.

LEMMA (B.3.14). — *Let G be a finite commutative k-group scheme, let R_G be its affine k-algebra and let*

$$M_G = \{m \in R_G | \mu_G^*(m) = 1 \otimes m + m \otimes 1\}$$

be the k-vector space of the primitive elements in the k-coalgebra (R_G, μ_G^*). Then we have the inequality

$$\dim_k(R_G) \geq p^{\dim_k(M_G)}.$$

Proof: Let G^* be the Cartier dual of G. Let us recall that

$$R_{G^*} = \mathrm{Hom}_k(R_G, k),$$

that the product on R_{G^*} is the transpose of μ_G^* and that the coproduct $\mu_{G^*}^*$ on R_{G^*} is the transpose of the product $R_G \otimes_k R_G \to R_G$ on R_G. The finite commutative k-group scheme G^* represents the functor

$$R \mapsto \mathrm{Hom}_{k\text{-}gr}(G, \mathbb{G}_{m,k})(R) = \mathrm{Hom}_{gr}(G(R), R^\times)$$

on the category of commutative k-algebras R. Therefore, we can identify the Lie algebra, $\mathrm{Lie}(G^*)$, of G^* with the k-vector space

$$\mathrm{Hom}_{k\text{-}gr}(G, \mathbb{G}_{a,k}).$$

But we can also identify this last k-vector space with M_G. So we get an identification of $\mathrm{Lie}(G^*)$ with M_G.

In particular, our inequality is equivalent to the inequality

$$\dim_k(R_{G^*}) \geq p^{\dim_k(\mathrm{Lie}(G^*))}$$

$(\dim_k(R_{G^*}) = \dim_k(R_G))$. So the problem is reduced to proving that, for any finite commutative k-group scheme G, we have the inequality

$$\dim_k(R_G) \geq p^{\dim_k(\mathrm{Lie}(G))}.$$

But we can split G into its connected part G^c and its étale part G^{et} (k is perfect) and we have

$$R_G = R_{G^c} \otimes_k R_{G^{et}}$$

and

$$\mathrm{Lie}(G) = \mathrm{Lie}(G^c)$$

$(\mathrm{Lie}(G^{et}) = \{0\})$. So we can assume G connected. Now it is known that the k-algebra R_G is non-canonically isomorphic to the k-algebra

$$k[x_1, \ldots, x_d]/(x_1^{p^{r_1}}, \ldots, x_d^{p^{r_d}}),$$

for some non-negative integer d and some positive integers $r_1, \ldots, r_d$ (see [Fon] (9.5.2) for example), so that

$$\dim_k(R_G) = p^{r_1 + \cdots + r_d}$$

and

$$\dim_k(\mathrm{Lie}(G)) = d.$$

The lemma follows. $\qquad\square$

LEMMA (B.3.15). — *Let N be a non-negative integer and let*

$$\mathcal{P} \subset k[x_1, \ldots, x_N]$$

be the k-vector subspace of additive polynomials. Then, for any finite k-group subscheme

$$i : G \hookrightarrow \mathbb{G}_{a,k}^N = \mathrm{Spec}(k[x_1, \ldots, x_N]),$$

the surjective k-algebra homomorphism

$$i^* : k[x_1, \ldots, x_N] \longrightarrow\!\!\!\rightarrow R_G$$

maps $\mathcal{P}$ onto M_G (same notations as in (B.3.14)).

Proof: It is obvious that $i^*(\mathcal{P}) \subset M_G$ and we want to prove the equality. If $\overline{k}$ is an algebraic closure of k, $\overline{k} \otimes_k \mathcal{P}$ is the set of additive polynomials in $\overline{k}[x_1, \ldots, x_N]$ and

$$\overline{k} \otimes_k M_G \subset M_{\overline{k} \otimes_k G}.$$

So, if we know that

$$(\overline{k} \otimes_k i)^*(\overline{k} \otimes_k \mathcal{P}) = M_{\overline{k} \otimes_k G},$$

we know that

$$\overline{k} \otimes_k i^*(\mathcal{P}) = (\overline{k} \otimes_k i)^*(\overline{k} \otimes_k \mathcal{P}) = \overline{k} \otimes_k M_G$$

and we know that

$$i^*(\mathcal{P}) = M_G.$$

Therefore we can assume that k is algebraically closed.

As in the proof of (B.3.14), we can identify M_G with

$$\mathrm{Hom}_{k\text{-}gr}(G, \mathbb{G}_{a,k})$$

and $\mathcal{P}$ with

$$\mathrm{Hom}_{k\text{-}gr}(\mathbb{G}_{a,k}^N, \mathbb{G}_{a,k}).$$

Then the map

$$i^* : \mathcal{P} \to M_G$$

is nothing else than

$$\mathrm{Hom}_{k\text{-}gr}(i, \mathbb{G}_{a,k}).$$

Let H be the quotient of $\mathbb{G}_{a,k}^N$ by $i(G)$ in the abelian category of affine k-group schemes (see [Fon] (6.6) for example). Then we have an exact sequence of k-vector spaces

$$\mathrm{Hom}_{k\text{-}gr}(\mathbb{G}_{a,k}^N, \mathbb{G}_{a,k}) \xrightarrow{i^*} \mathrm{Hom}(G, \mathbb{G}_{a,k}) \to \mathrm{Ext}^1_{k\text{-}gr}(H, \mathbb{G}_{a,k}).$$

Now it is clear that H is connected, unipotent, killed by p and of dimension N. So, thanks to [Se 3] VII, Prop. 11, H is non-canonically isomorphic to $\mathbb{G}_{a,k}^N$. Therefore, any extension

$$0 \to \mathbb{G}_{a,k} \to H' \to H \to 0$$

of H by $\mathbb{G}_{a,k}$ in the category of affine k-group schemes, such that H' is killed by p, splits. Indeed, thanks to loc. cit., H' is non-canonically isomorphic to $\mathbb{G}_{a,k}^{N+1}$ and $(e_1, \ldots, e_{N+1}) \in \mathrm{End}_{k\text{-}gr}(\mathbb{G}_{a,k})^{N+1}$ defines an injective homomorphism of k-group schemes

$$\mathbb{G}_{a,k} \hookrightarrow \mathbb{G}_{a,k}^{N+1}$$

if and only if the left ideal of the ring $\mathrm{End}_{k\text{-}gr}(\mathbb{G}_{a,k})$ generated by $e_1, \ldots, e_{N+1}$ is $\mathrm{End}_{k\text{-}gr}(\mathbb{G}_{a,k})$ itself (any left ideal in $\mathrm{End}_{k\text{-}gr}(\mathbb{G}_{a,k})$ is principal; see [Or]). In particular, the boundary map

$$\mathrm{Hom}_{k\text{-}gr}(G, \mathbb{G}_{a,k}) \to \mathrm{Ext}^1_{k\text{-}gr}(H, \mathbb{G}_{a,k})$$

is identically zero and the lemma is proved.   □

LEMMA (B.3.16). — *Let* $G \in \mathrm{ob}\,\mathcal{G}_k$. *Then we have the inequality*

$$\dim_k(R_G) \geq p^{\deg(o)\dim_k(M_k(G))}.$$

Proof: If we forget the structure of $\kappa(o)$-vector space, we can define

$$M_G = \{m \in R_G \mid \mu_G^*(m) = 1 \otimes m + m \otimes 1\}$$

and we have the inequality

$$\dim_k(R_G) \geq p^{\dim_k(M_G)}$$

thanks to (B.3.14). So it suffices to prove that

$$\dim_k(M_G) = \deg(o)\dim_k(M_k(G)).$$

Let

$$M_{G,j} = \{m \in M_G \mid [\alpha]_G^*(m) = \alpha^{p^j} m, \forall \alpha \in \kappa(o)\}$$

for $i \in \mathbb{Z}/\deg(o)\mathbb{Z}$. In particular, we have

$$M_{G,0} = M_k(G).$$

It is clear that

$$M_G = \bigoplus_{j \in \mathbf{Z}/\deg(o)\mathbf{Z}} M_{G,j}.$$

Let $f_G : M_G \to M_G$ be the restriction of the p-th power on R_G. It is clear that

$$f_G(M_{G,j}) \subset M_{G,j+1}$$

for any $j \in \mathbf{Z}/\deg(o)\mathbf{Z}$ and that

$$f_G^{\deg(o)} : M_{G,0} \to M_{G,0}$$

is $f_k(G)$.

To finish the proof of the lemma it suffices to check that

$$f_G(M_{G,j}) = M_{G,j+1}$$

for any $j \in \mathbf{Z}/\deg(o)\mathbf{Z}$, $j \neq 0$. But let us fix an embedding of k-schemes in $\kappa(o)$-vector spaces

$$G \overset{i}{\hookrightarrow} \mathbf{G}_{a,k}^N$$

and let $\mathcal{P}$ be the k-vector space of additive polynomials in $k[x_1, \ldots, x_N]$. We have

$$\mathcal{P} = \bigoplus_{j \in \mathbf{Z}/\deg(o)\mathbf{Z}} \mathcal{P}_j$$

where we have set

$$\mathcal{P}_j = \{P(x) \in \mathcal{P} \mid P(\alpha x) = \alpha^{p^j} P(x), \forall \alpha \in \kappa(o)\}$$

for any j. Thanks to (B.3.15), we have

$$i^*(\mathcal{P}) = M_G$$

and it follows that we also have

$$i^*(\mathcal{P}_j) = M_{G,j}$$

for any $j \in \mathbf{Z}/\deg(o)\mathbf{Z}$. Therefore $M_{G,j+1}/f_G(M_{G,j})$ is a quotient of $\mathcal{P}_{j+1}/(\mathcal{P}_j)^p$. As we have

$$(\mathcal{P}_j)^p = \mathcal{P}_{j+1}$$

for all $j \in \mathbf{Z}/\deg(o)\mathbf{Z}$, $j \neq 0$, the lemma is proved. $\qquad\square$

Proof of (B.3.13) : Let us begin with u_G. We fix an embedding of k-schemes in $\kappa(o)$-vector spaces

$$G \stackrel{i}{\hookrightarrow} \mathbf{G}_{a,k}^N.$$

Then we have the surjective k-algebra homomorphism

$$i^* : k[x_1, \dots, x_N] \longrightarrow R_G$$

and $i^*(x_1), \dots, i^*(x_N) \in M_k(G)$. So the homomorphism of k-algebras

$$\mathrm{Sym}_k(M_k(G)) \to R_G$$

induced by the inclusion $M_k(G) \subset R_G$ is surjective and the homomorphism of k-algebras

$$u_G^* : R_{G_k(M_k(G),f_k(G))} \to R_G$$

is surjective too. But thanks to (B.3.3) we have

$$\dim_k(R_{G_k(M_k(G),f_k(G))}) = p^{\deg(o)\dim_k(M_k(G))}$$

and thanks to (B.3.16) we have

$$\dim_k(R_G) \geq p^{\deg(o)\dim_k(M_k(G))}.$$

Therefore, u_G^* is an isomorphism of k-algebras and u_G is an isomorphism of k-group schemes.

Let us prove that $v_{(M,f)}$ is also an isomorphism. It is clear that

$$M \subset M_k(G_k(M,f)) \subset \mathrm{Sym}_k(M)/I_{(M,f)}$$

$(M \cap I_{(M,f)} = \{0\})$ and that $v_{(M,f)}$ is this inclusion. But, thanks to (B.3.3), we have

$$\dim_k(R_{G_k(M,f)}) = p^{\deg(o)\dim_k(M)}$$

and thanks to (B.3.16), we have

$$\dim_k(R_{G_k(M,f)}) \geq p^{\deg(o)\dim_k(M_k(G_k(M,f)))}.$$

Therefore, we get

$$\dim_k(M) \geq \dim_k(M_k(G_k(M,f)))$$

and $v_{(M,f)}$ is an isomorphism. $\square$

(B.4) Comments and references

The theory of Dieudonné $\mathcal{O}_o$-modules (resp. F_o-modules) over an algebraic closure of $\kappa(o)$ is completely analogous to the classical theory of Dieudonné $\mathbf{Z}_p$-modules (resp. $\mathbf{Q}_p$-modules) over an algebraically closed field of characteristic p.

Section (B.1) is essentially a copy of [Haz] (28.4). In sections (B.2) and (B.3) I have closely followed [Ka] (1.3) and [Fon] Ch. III. Other useful references for this appendix are [Ma] and [Zi].

C

Combinatorial formulas

(C.0) Introduction

In the proof of (4.6.1), we have made use of combinatorial formulas which involve the q-binomial coefficients. They are well known (see [MM] (Ch. V)) but for the convenience of the reader we will recall their proofs.

(C.1) q-binomial coefficients

Let x and q be two independent indeterminates over $\mathbb{Z}$. For any integers d, r with $d \geq 0$, the q-binomial coefficient $\binom{d}{r}_q$ is defined by

$$\binom{d}{r}_q = \begin{cases} \dfrac{(1-q)\cdots(1-q^d)}{(1-q)\cdots(1-q^r)(1-q)\cdots(1-q^{d-r})} & \text{if } 0 \leq r \leq d, \\ 0 & \text{otherwise.} \end{cases}$$

If $d \geq 1$, we have

$$\binom{d}{r}_q = \binom{d-1}{r-1}_q + q^r \binom{d-1}{r}_q$$

and

$$\binom{d}{r}_q = \binom{d-1}{r}_q + q^{d-r} \binom{d-1}{r-1}_q$$

and it follows by induction on d that

$$\binom{d}{r}_q \in \mathbb{Z}[q]$$

for any integers $d \geq 0$ and r.

LEMMA (C.1.1). — *For any positive integer d, we have*

$$\sum_{r=1}^{d}(x-q)\cdots(x-q^{r-1})\binom{d}{r}_q = \frac{1-x^d}{1-x}$$

in $\mathbf{Z}[x,q]$.

Proof : The formula is trivial if $d=1$. Let us prove it in general by induction on d. If $d \geq 2$, we have

$$\sum_{r=1}^{d}(x-q)\cdots(x-q^{r-1})\binom{d}{r}_q = \sum_{r=1}^{d}(x-q)\cdots(x-q^{r-1})\left(\binom{d-1}{r-1}_q + q^r\binom{d-1}{r}_q\right)$$

$$= 1 + x\sum_{r=1}^{d-1}(x-q)\cdots(x-q^{r-1})\binom{d-1}{r}_q$$

$$= 1 + x\left(\frac{1-x^{d-1}}{1-x}\right)$$

by the induction hypothesis and the lemma follows.   □

Let us set

$$F_d(x,q) = (x-q)\cdots(x-q^d) \in \mathbf{Z}[x,q]$$

for any non-negative integer d and let us denote by $\varphi_r^d(q)$ the coefficient of x^r in the expansion of $F_d(x,q)$ as a polynomial in x with coefficients in $\mathbf{Z}[q]$ ($r \in \mathbf{Z}$, $r \geq 0$; if $r \in \mathbf{Z}$, $r < 0$, we set $\varphi_r^d(q) = 0$).

LEMMA (C.1.2) (Cayley). — *For any non-negative integer d, we have*

$$F_d(x,q) = \sum_{r=0}^{d}(-1)^{d-r}x^r q^{(d-r)(d-r+1)/2}\binom{d}{r}_q \ ,$$

i.e.

$$\varphi_r^d(q) = (-1)^{d-r}q^{(d-r)(d-r+1)/2}\binom{d}{r}_q$$

for all $r \in \mathbf{Z}$.

Proof : We have

$$F_d(qx,q) = q^d(x-1)F_{d-1}(x,q).$$

Therefore, we have the induction relation

$$\varphi_r^d(q) = q^{d-r}(\varphi_{r-1}^{d-1}(q) - \varphi_r^{d-1}(q))$$

with

$$\varphi_0^0(q) = 1 \ .$$

But, using the relation

$$\binom{d}{r}_q = q^{d-r}\binom{d-1}{r-1}_q + \binom{d-1}{r}_q,$$

it is easy to check that

$$(-1)^{d-r}q^{(d-r)(d-r+1)/2}\binom{d}{r}_q$$

satisfies the same induction relation with the same initial value as $\varphi_r^d(q)$ and the lemma is proved. $\square$

COROLLARY (C.1.3). — *For any positive integer d, we have*

$$(1-q)\cdots(1-q^{d-1}) = \sum_{r=1}^d (-1)^{d-r}rq^{(d-r)(d-r-1)/2}\binom{d}{r}_q \ .$$

Proof : We have

$$\frac{\partial F_d}{\partial x}(q,q) = q^{d-1}(1-q)\cdots(1-q^{d-1})$$

by definition of $F_d(x,q)$ and it suffices to evaluate in q the x-derivative of the right hand side of the formula for $F_d(x,q)$ in (C.1.2). $\square$

COROLLARY (C.1.4). — *For any pair of integers (c,d) with $0 \le c \le d$, we have*

$$\sum_{r=c}^d (-1)^{d-r}q^{(d-r)(d-r-1)/2}\binom{d-c}{r-c}_q = \delta_{c,d} \ .$$

Proof : We have

$$F_{d-c}(q,q) = \delta_{c,d}$$

and the corollary follows directly from (C.1.2). $\square$

D

Representations of unimodular, locally compact, totally discontinuous, separated, topological groups

(D.0) Introduction

Let H be a unimodular, locally compact, totally discontinuous, separated, topological group. In particular, any neighborhood of the unit element 1 in H contains a compact open subgroup of H.

In this appendix, I will review some basic results of the theory of smooth representations of H (see [Be–Ze1]).

(D.1) Smooth representations of H

A **representation** of H is a pair $(\mathcal{V}, \pi)$ where $\mathcal{V}$ is a $\mathbb{C}$-vector space and π is a left action of H on $\mathcal{V}$ (for each $h \in H$, $\pi(h)$ is a $\mathbb{C}$-linear automorphism of $\mathcal{V}$; for any h_1, $h_2 \in H$, $\pi(h_1 h_2) = \pi(h_1) \circ \pi(h_2)$).

Let $(\mathcal{V}, \pi)$ be a representation of H. A vector $v \in \mathcal{V}$ is said to be **smooth** if its **stabilizer**

$$H_v = \{h \in H \mid \pi(h)(v) = v\}$$

is an open subgroup of H. One denotes by

$$\mathcal{V}^\infty \subset \mathcal{V}$$

the subset of all smooth vectors in $\mathcal{V}$. Obviously, $\mathcal{V}^\infty$ is a $\mathbb{C}$-vector subspace of $\mathcal{V}$ and $\pi(h)(\mathcal{V}^\infty) \subset \mathcal{V}^\infty$ for any $h \in H$. One denotes by π^∞ the restriction to $\mathcal{V}^\infty$ of the action π of H; $(\mathcal{V}^\infty, \pi^\infty)$ is again a representation of H.

A **smooth representation** of H is a representation $(\mathcal{V}, \pi)$ of H such that any $v \in \mathcal{V}$ is smooth, i.e. $\mathcal{V}^\infty = \mathcal{V}$.

The representations of H form an abelian category (a morphism between two representations of H is a $\mathbb{C}$-linear map between the underlying $\mathbb{C}$-vector spaces which commutes with the actions of H). Let us denote by $\text{Rep}(H)$ this category. Let $\text{Rep}_s(H)$ be the full subcategory of $\text{Rep}(H)$ with objects the smooth representations of H. This subcategory is stable by subquotients and by finite direct and inverse limits and is therefore abelian.

If dh is a Haar measure on H, for any two compact open subgroups I_1 and I_2 of H, the ratio

$$\text{vol}(I_1, dh)/\text{vol}(I_2, dh)$$

belongs to $\mathbb{Q}^\times$. So, if $\text{vol}(I, dh) \in \mathbb{Q}^\times$ for some compact open subgroup I of H, the same is true for any compact open subgroup I of H. A Haar measure dh on H such that $\text{vol}(I, dh) \in \mathbb{Q}^\times$ for any compact open subgroup I of H is said to be **rational**. Let us fix a rational Haar measure dh on H.

Let $\mathcal{C}_c^\infty(H)$ be the $\mathbb{Q}$-vector space of locally constant functions on H, with values in $\mathbb{Q}$ and with compact support in H. The convolution product with respect to dh,

$$(f_1 * f_2)(h') = \int_H f_1(h'h^{-1})f_2(h)dh$$

($f_1, f_2 \in \mathcal{C}_c^\infty(H)$, $h' \in H$), endows $\mathcal{C}_c^\infty(H)$ a $\mathbb{Q}$-algebra structure.

In general this algebra does not have a unit element (unless H is compact) but it has a lot of idempotent elements. Indeed, for any compact open subgroup I of H, the element

(D.1.1)
$$e_I = \frac{1_I}{\text{vol}(I, dh)} \in \mathcal{C}_c^\infty(H)$$

is idempotent (as usual 1_I is the characteristic function of I in H). In particular, the $\mathbb{Q}$-subspace

$$\mathcal{C}_c^\infty(H//I) = e_I * \mathcal{C}_c^\infty(H) * e_I \subset \mathcal{C}_c^\infty(H)$$

of I-bi-invariant functions is a $\mathbb{Q}$-subalgebra (with unit element e_I) and the right (resp. left) ideal

$$\mathcal{C}_c^\infty(I\backslash H) = e_I * \mathcal{C}_c^\infty(H) \subset \mathcal{C}_c^\infty(H)$$

(resp.

$$\mathcal{C}_c^\infty(H/I) = \mathcal{C}_c^\infty(H) * e_I \subset \mathcal{C}_c^\infty(H))$$

of left (resp. right) I-invariant functions is a left (resp. right) $\mathcal{C}_c^\infty(H//I)$-submodule of $\mathcal{C}_c^\infty(H)$.

For any $\mathbb{Q}$-algebra $\mathcal{A}$, we will simply denote by $\text{Mod}(\mathcal{A})$ the abelian category of left ($\mathbb{C}\otimes\mathcal{A}$)-modules. A left ($\mathbb{C}\otimes\mathcal{A}$)-module $\mathcal{V}$ is said to be **non-degenerate** if, for any $v \in \mathcal{V}$, there exists $a \in \mathbb{C}\otimes\mathcal{A}$ such that $v = a \cdot v$ (this condition is automatic if $\mathcal{A}$ has a unit element). Let us denote by $\text{Mod}_{nd}(\mathcal{A})$

the full subcategory of $\mathrm{Mod}(\mathcal{A})$ with objects the non-degenerate left $(\mathbb{C} \otimes \mathcal{A})$-modules. It is stable by subquotients. It is stable by finite direct and inverse limits and is therefore abelian if, for any finite family $(a_\alpha)_\alpha$ of elements of $\mathcal{A}$, there exists $a \in \mathcal{A}$ such that $a a_\alpha = a_\alpha$ for all α.

If $(\mathcal{V}, \pi) \in \mathrm{ob} \, \mathrm{Rep}_s(H)$, then, for each $v \in \mathcal{V}$ and each $f \in \mathcal{C}_c^\infty(H)$, the integral

$$\pi(f)(v) = f \cdot v = \int_H f(h)\pi(h)(v)dh$$

makes sense (it can be reduced to a finite sum of elements in $\mathcal{V}$) and defines a left $(\mathbb{C} \otimes \mathcal{C}_c^\infty(H))$-module structure on $\mathcal{V}$. If $\alpha : (\mathcal{V}_1, \pi_1) \longrightarrow (\mathcal{V}_2, \pi_2)$ is any morphism in $\mathrm{Rep}_s(H)$, the map $\alpha : \mathcal{V}_1 \longrightarrow \mathcal{V}_2$ is obviously $(\mathbb{C} \otimes \mathcal{C}_c^\infty(H))$-linear. So we have defined a functor from $\mathrm{Rep}_s(H)$ to $\mathrm{Mod}(\mathcal{C}_c^\infty(H))$.

LEMMA (D.1.2). — *For any $(\mathcal{V}, \pi) \in \mathrm{ob} \, \mathrm{Rep}_s(H)$, the left $(\mathbb{C} \otimes \mathcal{C}_c^\infty(H))$-module $\mathcal{V}$ is non-degenerate. The above functor induces an equivalence of categories from $\mathrm{Rep}_s(H)$ onto $\mathrm{Mod}_{nd}(\mathcal{C}_c^\infty(H))$.*

Proof : If $v \in \mathcal{V}$ and if $I \subset H_v$ is a compact open subgroup, we have $e_I \cdot v = v$ and the first assertion is proved.

If $\mathcal{V} \in \mathrm{Mod}_{nd}(\mathcal{C}_c^\infty(H))$ and if $v \in \mathcal{V}$, there exists a compact open subgroup I of H such that $e_I \cdot v = v$ (write $v = f \cdot v$ for some $f \in \mathbb{C} \otimes \mathcal{C}_c^\infty(H)$ and let us choose I such that f is left I-invariant). Now, if $h \in H$, we can set

$$\pi(h)(v) = (\delta_h * e_I) \cdot v$$

where $\delta_h * e_I \in \mathcal{C}_c^\infty(H)$ is defined by

$$(\delta_h * e_I)(h') = e_I(h^{-1}h')$$

for all $h' \in H$. This is independent of the choice of I and $(\mathcal{V}, \pi) \in \mathrm{ob} \, \mathrm{Rep}_s(H)$. The second assertion of the lemma is now easy.   □

From now on we will identify $\mathrm{Rep}_s(H)$ with $\mathrm{Mod}_{nd}(\mathcal{C}_c^\infty(H))$ using the above equivalence of categories.

Let $(\mathcal{V}, \pi) \in \mathrm{ob} \, \mathrm{Rep}_s(H)$ and let I be a compact open subgroup of H. The subspace

$$\mathcal{V}^I = \{v \in \mathcal{V} \mid I \subset H_v\} \subset \mathcal{V}$$

is equal to

$$\mathrm{Im}(\pi(e_I)) = \mathrm{Ker}(\pi(e_I) - id_\mathcal{V})$$

and we have a canonical splitting

(D.1.3) $\mathcal{V} = \mathcal{V}^I \oplus \mathrm{Ker}(\pi(e_I))$

So we have

$$\pi(\mathcal{C}_c^\infty(H//I))(\mathcal{V}^I) \subset \mathcal{V}^I,$$

a functor

(D.1.4) $\operatorname{Rep}_s(H) \longrightarrow \operatorname{Mod}(\mathcal{C}_c^\infty(H//I)),\ (\mathcal{V}, \pi) \longmapsto \mathcal{V}^I$,

and the lemma

LEMMA (D.1.5). — *This functor is exact.*   □

In fact, (D.1.4) has a left adjoint

(D.1.6) $\operatorname{Mod}(\mathcal{C}_c^\infty(H//I)) \longrightarrow \operatorname{Rep}_s(H),$

$$V \longmapsto \mathcal{C}_c^\infty(H/I) \otimes_{\mathcal{C}_c^\infty(H//I)} V,$$

and a right adjoint

(D.1.7) $\operatorname{Mod}(\mathcal{C}_c^\infty(H//I)) \longrightarrow \operatorname{Rep}_s(H),$

$$V \longmapsto \operatorname{Hom}_{\mathcal{C}_c^\infty(H//I)}(\mathcal{C}_c^\infty(I\backslash H), V)^\infty.$$

The adjunction maps

$$V \longrightarrow (\mathcal{C}_c^\infty(H/I) \otimes_{\mathcal{C}_c^\infty(H//I)} V)^I,\ v \longmapsto e_I \otimes v$$

and

$$(\operatorname{Hom}_{\mathcal{C}_c^\infty(H//I)}(\mathcal{C}_c^\infty(I\backslash H), V)^\infty)^I \longrightarrow V,\ \nu \longmapsto \nu(e_I)$$

are isomorphisms. The kernel and the cokernel of the adjunction maps

$$\mathcal{C}_c^\infty(H/I) \otimes_{\mathcal{C}_c^\infty(H//I)} \mathcal{V}^I \longrightarrow \mathcal{V},\ f \otimes v \longmapsto \pi(f)(v),$$

$$\mathcal{V} \longrightarrow \operatorname{Hom}_{\mathcal{C}_c^\infty(H//I)}(\mathcal{C}_c^\infty(I\backslash H), \mathcal{V}^I)^\infty,\ v \longmapsto (f \longmapsto (f * e_I) \cdot v),$$

have no non-zero fixed vector under I.

PROPOSITION (D.1.8). — *The functor* (D.1.4) *induces a bijection between the set of isomorphism classes of irreducible smooth representations* $(\mathcal{V}, \pi)$ *of* H *such that* $\mathcal{V}^I \neq (0)$ *and the set of isomorphism classes of irreducible left* $(\mathbb{C} \otimes \mathcal{C}_c^\infty(H//I))$*-modules.*

Proof : If $(\mathcal{V}, \pi) \in \operatorname{ob} \operatorname{Rep}_s(H)$ is irreducible and if V is a non-zero $(\mathbb{C} \otimes \mathcal{C}_c^\infty(H//I))$-submodule of $\mathcal{V}^I$, we consider the map

$$\mathcal{C}_c^\infty(H/I) \otimes_{\mathcal{C}_c^\infty(H//I)} V \longrightarrow \mathcal{V},\ f \otimes v \longmapsto \pi(f)(v) .$$

Taking the I-invariants, it gives back the inclusion $V \hookrightarrow \mathcal{V}^I$. As $V \neq \{0\}$, it is non-zero. As $(\mathcal{V}, \pi)$ is irreducible, it is therefore surjective. But this implies that the inclusion $V \hookrightarrow \mathcal{V}^I$ is also surjective (see (D.1.5)) and we have proved that $\mathcal{V}^I$ is an irreducible left $(\mathbb{C} \otimes \mathcal{C}_c^\infty(H//I))$-module as long as it is non-zero.

If $(\mathcal{V}, \pi) \in \mathrm{ob}\, \mathrm{Rep}_s(H)$, there exists a largest (non-degenerate) $(\mathbb{C} \otimes \mathcal{C}_c^\infty(H))$-submodule $\mathcal{V}_1$ of $\mathcal{V}$ such that $\mathcal{V}_1^I = (0)$: just take

$$\mathcal{V}_1 = \{v \in \mathcal{V} \mid \mathcal{C}_c^\infty(I \backslash H) \cdot v = \{0\}\} \ .$$

Then $\mathcal{V}_2 = \mathcal{V}/\mathcal{V}_1$ is the largest (non-degenerate) quotient of the $(\mathbb{C} \otimes \mathcal{C}_c^\infty(H))$-module $\mathcal{V}$ such that, for any non-degenerate $(\mathbb{C} \otimes \mathcal{C}_c^\infty(H))$-module $\mathcal{W}$ with $\mathcal{W}^I = \{0\}$, we have $\mathrm{Hom}_{\mathbb{C} \otimes \mathcal{C}_c^\infty(H)}(\mathcal{W}, \mathcal{V}_2) = \{0\}$. The quotient map $\mathcal{V} \longrightarrow \mathcal{V}_2$ induces an isomorphism $\mathcal{V}^I \xrightarrow{\sim} \mathcal{V}_2^I$ (see (D.1.5)).

Now, if V is an irreducible left $(\mathbb{C} \otimes \mathcal{C}_c^\infty(H//I))$-module, we can apply the above constructions to $\mathcal{C}_c^\infty(H/I) \otimes_{\mathcal{C}_c^\infty(H//I)} V$. Then the quotient module

$$(\mathcal{C}_c^\infty(H/I) \otimes_{\mathcal{C}_c^\infty(H//I)} V)_2$$

admits V as space of I-invariants and is irreducible. Indeed, if $\mathcal{V}$ is a non-zero (non-degenerate) $(\mathbb{C} \otimes \mathcal{C}_c^\infty(H))$-submodule of this quotient, $\mathcal{V}^I$ is a non-zero $(\mathbb{C} \otimes \mathcal{C}_c^\infty(H//I))$-submodule of V, so that $\mathcal{V}^I = V$; but V generates the $(\mathbb{C} \otimes \mathcal{C}_c^\infty(H))$-module $(\mathcal{C}_c^\infty(H/I) \otimes_{\mathcal{C}_c^\infty(H//I)} V)_2$ as required.

Finally, it is clear that the adjunction map induces an isomorphism

$$(\mathcal{C}_c^\infty(H/I) \otimes_{\mathcal{C}_c^\infty(H//I)} \mathcal{V}^I)_2 \xrightarrow{\sim} \mathcal{V}$$

for any irreducible $(\mathcal{V}, \pi) \in \mathrm{ob}\, \mathrm{Rep}_s(H)$.   □

The **contragredient representation** of $(\mathcal{V}, \pi) \in \mathrm{ob}\, \mathrm{Rep}_s(H)$ is the smooth representation

(D.1.9) $(\widetilde{\mathcal{V}}, \widetilde{\pi}) = (\mathcal{V}'^\infty, \pi'^\infty)$

of H, where $\mathcal{V}' = \mathrm{Hom}_{\mathbb{C}}(\mathcal{V}, \mathbb{C})$ and where π' is the representation of H on $\mathcal{V}'$ given by

$$\pi'(h)(v') = v' \circ \pi(h^{-1})$$

for any $h \in H$ and any $v' \in \mathcal{V}'$.

For any compact open subgroup I of H, the subspace

(D.1.10) $\widetilde{\mathcal{V}}^I = \mathcal{V}'^I \subset \mathcal{V}'$

is the orthogonal subspace of $\mathrm{Ker}(\pi(e_I)) \subset \mathcal{V}$ in $\mathcal{V}'$. Therefore, the splitting

$$\mathcal{V} = \mathcal{V}^I \oplus \mathrm{Ker}(\pi(e_I))$$

induces an isomorphism

$$\widetilde{\mathcal{V}}^I \xrightarrow{\sim} (\mathcal{V}^I)' \overset{\mathrm{dfn}}{=\!=} \mathrm{Hom}_{\mathbb{C}}(\mathcal{V}^I, \mathbb{C})$$

in $\mathrm{Mod}(\mathcal{C}_c^\infty(H/\!/I))$. In particular, the contravariant functor $(\mathcal{V}, \pi) \longmapsto (\widetilde{\mathcal{V}}, \widetilde{\pi})$ from $\mathrm{Rep}_s(H)$ into itself is exact (for each I, $(\mathcal{V}, \pi) \longmapsto \widetilde{\mathcal{V}}^I$ is exact).

For any compact open subgroup I of H and any $V \in \mathrm{ob}\, \mathrm{Mod}(\mathcal{C}_c^\infty(H/\!/I))$, the map

$$\mathrm{Hom}_{\mathcal{C}_c^\infty(H/\!/I)}(\mathcal{C}_c^\infty(I\backslash H), V') \longrightarrow (\mathcal{C}_c^\infty(H/I) \otimes_{\mathcal{C}_c^\infty(H/\!/I)} V)',$$

$$\nu' \longmapsto (f \otimes v \mapsto \nu'(f^\vee)(v)),$$

where $f^\vee(h) = f(h^{-1})$ for any $f \in \mathcal{C}_c^\infty(H)$ and any $h \in H$ ($f^\vee \in \mathcal{C}_c^\infty(H)$), is an isomorphism of representations of H. It induces an isomorphism

$$(D.1.11) \qquad \mathrm{Hom}_{\mathcal{C}_c^\infty(H/\!/I)}(\mathcal{C}_c^\infty(I\backslash H), V')^\infty \xrightarrow{\sim} (\mathcal{C}_c^\infty(H/I) \otimes_{\mathcal{C}_c^\infty(H/\!/I)} V)^\sim$$

in $\mathrm{Rep}_s(H)$.

Let Z be the center of H. One says that a smooth representation $(\mathcal{V}, \pi)$ of H **admits a central character** if, for any $z \in Z$, $\pi(z)$ is equal to

$$\omega_\pi(z) \mathrm{id}_\mathcal{V}$$

for some $\omega_\pi(z) \in \mathbb{C}^\times$. If this is the case,

$$\omega_\pi : Z \longrightarrow \mathbb{C}^\times$$

is a character, which is trivial on a small enough neighborhood of 1 in Z (for any $v \in V - \{0\}$, ω_π is trivial on $Z \cap H_v$). In other words, $(\mathbb{C}, \omega_\pi)$ is a one dimensional smooth representation of the unimodular, locally compact, totally discontinuous, separated, topological group Z. The character ω_π is called the **central character of** $(\mathcal{V}, \pi)$.

LEMMA (D.1.12) (Schur, Jacquet). — *Let us assume that H is countable at infinity and let $(\mathcal{V}, \pi)$ be a smooth irreducible representation of H. Then any endomorphism of $(\mathcal{V}, \pi)$ in $\mathrm{Rep}_s(H)$ is equal to $\lambda\, \mathrm{id}_\mathcal{V}$ for some $\lambda \in \mathbb{C}$. In particular, $(\mathcal{V}, \pi)$ admits a central character.*

Proof: Let ν be an endomorphism of $(\mathcal{V}, \pi)$ in $\mathrm{Rep}_s(H)$ and let us assume that $\nu \neq \lambda\, \mathrm{id}_\mathcal{V}$ for all $\lambda \in \mathbb{C}$. Then, for each $\lambda \in \mathbb{C}$, $\nu - \lambda\, \mathrm{id}_\mathcal{V}$ is an automorphism of $(\mathcal{V}, \pi)$ in $\mathrm{Rep}_s(H)$ (its kernel and its cokernel are proper subquotients of $(\mathcal{V}, \pi)$ and therefore are trivial). Let

$$R_\lambda = (\nu - \lambda\, \mathrm{id}_\mathcal{V})^{-1}$$

be the inverse of this automorphism and let us choose $v \in \mathcal{V} - \{0\}$. Then the vectors $R_\lambda(v)$ ($\lambda \in \mathbb{C}$) are linearly independent over $\mathbb{C}$. Indeed, if

$\lambda_1, \ldots, \lambda_m$ are distinct complex numbers and if $a_1, \ldots, a_m$ are non-zero complex numbers, the operator

$$\sum_{i=1}^{m} a_i R_{\lambda_i}$$

can be written as

$$\left(\prod_{i=1}^{m} R_{\lambda_i} \right) P(\nu)$$

where

$$P(T) = \sum_{i=1}^{m} a_i \prod_{\substack{j=1 \\ j \neq i}}^{m} (T - \lambda_j)$$

is a non-zero polynomial in $\mathbb{C}[T]$. Factorizing $P(T)$ into prime factors

$$P(T) = a \prod_{j=1}^{n} (T - \mu_j)$$

$(a \in \mathbb{C}^\times, \mu_1, \ldots, \mu_n \in \mathbb{C}, n \geq 0)$, we see that the endomorphism $P(\nu)$ of $(\mathcal{V}, \pi)$ in $\mathrm{Rep}_s(H)$ is invertible. Therefore, the endomorphism

$$\sum_{i=1}^{m} a_i R_{\lambda_i}$$

of $(\mathcal{V}, \pi)$ in $\mathrm{Rep}_s(H)$ is also invertible and our assertion follows.

Now $\mathcal{V}$ is generated as a representation of H by v and, therefore, as a $\mathbb{C}$-vector space by the family $(\pi(h)(v))_{h \in S}$, where $S \subset H$ is a system of representatives of the cosets in H/H_v. But, as H is countable at infinity, H/H_v is countable and any linearly independent family of elements in $\mathcal{V}$ must be countable. So we get a contradiction and the lemma is proved. $\square$

If $(\mathcal{V}, \pi)$ admits a central character, the same holds for $(\widetilde{\mathcal{V}}, \widetilde{\pi})$ and

(D.1.13) $$\omega_{\widetilde{\pi}} = \omega_{\pi}^{-1}.$$

(D.2) Admissible representations of H

An **admissible** representation of H is a smooth representation $(\mathcal{V}, \pi)$ of H such that, for any compact open subgroup I of H, $\mathcal{V}^I$ is finite dimensional over $\mathbb{C}$. Let $\mathrm{Rep}_a(H)$ be the full subcategory of $\mathrm{Rep}_s(H)$ with objects the admissible representations of H. This subcategory is stable by subquotients and finite sum and is therefore abelian. It is also stable by extensions in

$\text{Rep}_s(H)$. Let us recall that, for each compact open subgroup I of H, the functor $(\mathcal{V}, \pi) \longmapsto \mathcal{V}^I$ on $\text{Rep}_s(H)$ is exact.

If $(\mathcal{V}, \pi) \in \text{ob } \text{Rep}_a(H)$ and if $f \in \mathbb{C} \otimes \mathcal{C}_c^\infty(H)$, the endomorphism $\pi(f)$ of the $\mathbb{C}$-vector space $\mathcal{V}$ is of finite rank. Indeed, there exists a compact open subgroup I of H such that $f \in \mathbb{C} \otimes \mathcal{C}_c^\infty(I \backslash H)$ and, for any such I, we have

$$\pi(f)(\mathcal{V}) \subset \mathcal{V}^I.$$

Therefore, we can define the trace of $\pi(f)$ as the trace of the endomorphism

$$\pi(f) \mid \pi(f)(\mathcal{V}) : \pi(f)(\mathcal{V}) \longrightarrow \pi(f)(\mathcal{V}).$$

One denotes by $\text{tr } \pi(f)$ or $\Theta_\pi(f)$ this trace and the $\mathbb{C}$-linear form

(D.2.1) $$\text{tr } \pi(-) = \Theta_\pi : \mathbb{C} \otimes \mathcal{C}_c^\infty(H) \longrightarrow \mathbb{C}$$

is called the **character** of $(\mathcal{V}, \pi)$ (like $\pi(f)$, $\text{tr } \pi(f)$ depends on the choice of the Haar measure dh on H). If $f \in \mathbb{C} \otimes \mathcal{C}_c^\infty(I \backslash H)$ for some compact open subgroup I of H, $\text{tr } \pi(f)$ is also the trace of the endomorphism $\pi(f) \mid \mathcal{V}^I : \mathcal{V}^I \longrightarrow \mathcal{V}^I$. In particular, $\text{tr } \pi(f) = 0$ if $\mathcal{V}^I = (0)$ and, as $(\mathcal{V}, \pi) \longmapsto \mathcal{V}^I$ is exact on $\text{Rep}_a(H)$, the map $(\mathcal{V}, \pi) \longmapsto \text{tr } \pi(f)$ is additive on the short exact sequences in $\text{Rep}_a(H)$.

PROPOSITION (D.2.2). — *Let $(\mathcal{V}_\lambda, \pi_\lambda)_{\lambda \in L}$ be a finite family of admissible and irreducible representations of H. Assume that $(\mathcal{V}_{\lambda'}, \pi_{\lambda'})$ is non-isomorphic to $(\mathcal{V}_{\lambda''}, \pi_{\lambda''})$ for all $\lambda' \neq \lambda''$ in L. Then the characters $\Theta_{\pi_\lambda} (\lambda \in L)$ are linearly independent $\mathbb{C}$-linear forms, hence are mutually distinct.*

Proof : Let us fix a compact open subgroup I of H such that $\mathcal{V}_\lambda^I \neq (0)$ for each $\lambda \in L$. Then $(\mathcal{V}_\lambda^I)_{\lambda \in L}$ is a finite family of finite dimensional (over $\mathbb{C}$) and irreducible left $(\mathbb{C} \otimes \mathcal{C}_c^\infty(H // I))$-modules such that $\mathcal{V}_{\lambda'}^I$ and $\mathcal{V}_{\lambda''}^I$ are non-isomorphic for all $\lambda' \neq \lambda''$ in L (see (D.1.8)). Moreover, for any $f \in \mathbb{C} \otimes \mathcal{C}_c^\infty(H // I)$ and any $\lambda \in L$, $\Theta_{\pi_\lambda}(f)$ is the trace of the endomorphism

$$f \cdot (-) : \mathcal{V}_\lambda^I \longrightarrow \mathcal{V}_\lambda^I .$$

Let A be the image of $\mathbb{C} \otimes \mathcal{C}_c^\infty(H // I)$ in the product of the finite dimensional $\mathbb{C}$-algebras $\text{End}_\mathbb{C}(\mathcal{V}_\lambda^I)$ $(\lambda \in L)$. Then, A is a finite dimensional $\mathbb{C}$-algebra, $(\mathcal{V}_\lambda^I)_{\lambda \in L}$ is a finite family of finite dimensional and irreducible left A-modules such that $\mathcal{V}_{\lambda'}^I$ and $\mathcal{V}_{\lambda''}^I$ are non-isomorphic for all $\lambda' \neq \lambda''$ in L and it suffices to prove that the characters of those A-modules are linearly independent. Now the proposition follows from the Jacobson density theorem ([Bou] Alg. Ch. 8, § 13, Prop. 2). □

If $(\mathcal{V}, \pi)$ is an admissible representation of H, so is $(\widetilde{\mathcal{V}}, \widetilde{\pi})$ and the map

(D.2.3) $$(\mathcal{V}, \pi) \longrightarrow (\widetilde{\widetilde{\mathcal{V}}}, \widetilde{\widetilde{\pi}}), \; v \longmapsto (\widetilde{v} \longmapsto \widetilde{v}(v))$$

is an isomorphism in $\mathrm{Rep}_a(H)$ (for any compact open subgroup I of H, $\widetilde{\mathcal{V}}^I = (\mathcal{V}^I)'$ and $\widetilde{\widetilde{\mathcal{V}}}^I = (\mathcal{V}^I)''$). In particular, the contravariant functor $(\mathcal{V}, \pi) \longmapsto (\widetilde{\mathcal{V}}, \widetilde{\pi})$ is an equivalence of categories from $\mathrm{Rep}_a(H)^{opp}$ onto $\mathrm{Rep}_a(H)$ and $(\mathcal{V}, \pi) \in \mathrm{ob}\,\mathrm{Rep}_a(H)$ is irreducible if and only if $(\widetilde{\mathcal{V}}, \widetilde{\pi})$ is irreducible.

(D.3) Induction and restriction

Let $R \subset Q \subset H$ be closed subgroups of H and let $H^\circ \subset H$ be a compact open subgroup. We assume that

(A) R is normal in Q,

(B) any compact subset of R is contained in a compact open subgroup of R (in particular, R is the union of its compact open subgroups),

(C) $H = Q \cdot H^\circ$,

(D) for any Haar measure dr on R and any $q \in Q$, we have

$$d(qrq^{-1}) = \delta_Q(q)dr$$

where $\delta_Q : Q \longrightarrow \mathbb{R}_{>0}$ is the modulus character of Q (thanks to (B), R is unimodular).

Then the locally compact, totally discontinuous, separated, topological group

$$\overline{H} = R\backslash Q$$

is unimodular and any Haar measures $d\overline{h}$ on $\overline{H}$ and dr on R define a left (resp. right) Haar measure $d_\ell q$ (resp. $d_r q$) on Q such that

$$\int_Q f(q)d_\ell q = \int_{\overline{H}} \left(\int_R f(qr)dr \right) d(qR)$$

(resp.

$$\int_Q f(q)d_r q = \int_{\overline{H}} \left(\int_R f(rq)dr \right) d(Rq))$$

for any locally constant function with compact support f on Q. Moreover, we have

$$d_r q = \delta_Q(q)d_\ell q \,,$$

δ_Q is trivial on R and δ_Q factors through a character of $\overline{H}$ that we will also denote by δ_Q. This character $\delta_Q : \overline{H} \longrightarrow \mathbb{R}_{>0}$ is trivial on the compact open subgroup

$$\overline{H}^\circ = R^\circ\backslash Q^\circ$$

of H, where we have set

$$Q^\circ = Q \cap H^\circ$$

and

$$R^\circ = R \cap H^\circ .$$

Let us normalize the Haar measures dh, $d\overline{h}$ and dr by $\mathrm{vol}(H^\circ, dh) = 1$, $\mathrm{vol}(\overline{H}^\circ, d\overline{h}) = 1$ and $\mathrm{vol}(R^\circ, dr) = 1$, so that

$$dh = d_\ell q \; dh^\circ$$

where dh° is the Haar measure on H° which is normalized by $\mathrm{vol}(H^\circ, dh^\circ) = 1$.

Under the above assumptions, one can define two functors: the **induction**

$$\text{(D.3.1)} \qquad i^H_{\overline{H},Q} : \mathrm{Rep}_s(\overline{H}) \longrightarrow \mathrm{Rep}_s(H)$$

and the **restriction** (also called the **modified Jacquet functor**)

$$\text{(D.3.2)} \qquad r^{\overline{H},Q}_H : \mathrm{Rep}_s(H) \longrightarrow \mathrm{Rep}_s(\overline{H}) .$$

If $(\overline{V}, \overline{\pi}) \in \mathrm{ob} \, \mathrm{Rep}_s(\overline{H})$, then $i^H_{\overline{H},Q}(\overline{V}, \overline{\pi}) = (V, \pi)$ where V is the $\mathbb{C}$-vector space of locally constant functions

$$v : H \longrightarrow \overline{V}$$

such that

$$v(qh) = \delta^{1/2}_Q(Rq)\overline{\pi}(Rq)(v(h))$$

for any $h \in H$ and any $q \in Q$ and where π is the left action of H on such functions which is induced by the right translation on H.

As $H = QH^\circ$, any v in V is uniquely determined by its restriction to H° and, as a representation of $H^\circ \subset H$, $(V, \pi|H^\circ)$ is canonically isomorphic to (V°, π°) where V° is the $\mathbb{C}$-vector space of locally constant functions

$$v^\circ : H^\circ \longrightarrow \overline{V}$$

such that

$$v^\circ(q^\circ h^\circ) = \overline{\pi}(R^\circ q^\circ)(v^\circ(h^\circ))$$

for any $h^\circ \in H^\circ$ and any $q^\circ \in Q^\circ$ and where π° is the left action of H° on such functions which is induced by the right translation on H°.

If $(V, \pi) \in \mathrm{ob} \, \mathrm{Rep}_s(H)$, then $r^{\overline{H},Q}_H(V, \pi) = (\overline{V}, \overline{\pi})$ where

$$\overline{V} = V/V(R),$$

if we denote by $\mathcal{V}(R)$ the $\mathbb{C}$-vector subspace of $\mathcal{V}$ generated by the elements

$$\pi(r)(v) - v \quad (r \in R, \ v \in \mathcal{V}),$$

and where $\overline{\pi}$ is induced by

$$\delta_Q^{-1/2}(-) \otimes (\pi|_Q) : Q \longrightarrow \mathrm{Aut}_{\mathbb{C}}(\mathcal{V})$$

(recall that Q normalizes R).

Thanks to the hypothesis (B) on R, we also have

$$\mathcal{V}(R) = \bigcup_J \mathrm{Ker}(\pi_J)$$

where π_J is the $\mathbb{C}$-linear endomorphism of $\mathcal{V}$ which is given by

$$\pi_J(v) = \int_J \frac{\pi(r)(v)}{\mathrm{vol}(J, dr)} dr \quad (v \in \mathcal{V})$$

and where J runs through the set of all compact open subgroups of R. Indeed, if $v \in \mathcal{V}$ and $\pi_J(v) = 0$ for some compact open subgroup J of R, then $J \cap H_v$ is an open subgroup of J. So, if $(r_1, \ldots, r_n)$ is a system of representatives of the classes in $J/(J \cap H_v)$, we have

$$\sum_{i=1}^n \pi(r_i)(v) = [J : J \cap H_v] \pi_J(v) = 0$$

and

$$v = -\frac{1}{n} \sum_{i=1}^n (\pi(r_i)(v) - v) \in \mathcal{V}(R) .$$

Conversely, if $v_1, \ldots, v_n \in \mathcal{V}$ and $r_1, \ldots, r_n \in R$ there exists a compact open subgroup J of R such that $r_1, \ldots, r_n \in J$ and we have

$$\pi_J(\pi(r_i)(v_i) - v_i) = 0$$

for $i = 1, \ldots, n$.

LEMMA (D.3.3). — (i) *The functor* $r_H^{\overline{H},Q}$ *is left adjoint to the functor* $i_{\overline{H},Q}^H$, *i.e. we have a bifunctorial isomorphism*

$$\mathrm{Hom}_{\overline{H}}(r_H^{\overline{H},Q}(\mathcal{V}_1, \pi_1), (\overline{\mathcal{V}}_2, \overline{\pi}_2)) \cong \mathrm{Hom}_H((\mathcal{V}_1, \pi_1), \ i_{\overline{H},Q}^H(\overline{\mathcal{V}}_2, \overline{\pi}_2))$$

for $(\mathcal{V}_1, \pi_1) \in \mathrm{ob}\,\mathrm{Rep}_s(H)$ *and* $(\overline{\mathcal{V}}_2, \overline{\pi}_2) \in \mathrm{ob}\,\mathrm{Rep}_s(\overline{H})$ *(Frobenius reciprocity).*

(ii) *The functors* $i_{\overline{H},Q}^H$ *and* $r_H^{\overline{H},Q}$ *are exact.*

Proof : (i) The adjunction is given by

$$(\overline{\alpha} : r_H^{\overline{H},Q}(\mathcal{V}_1, \pi_1) \longrightarrow (\overline{\mathcal{V}}_2, \overline{\pi}_2)) \longleftrightarrow (\alpha : (\mathcal{V}_1, \pi_1) \longrightarrow i_{H,Q}^H(\overline{\mathcal{V}}_2, \overline{\pi}_2))$$

where

$$\alpha(v_1)(h) = \overline{\alpha}(\pi(h)(v_1) + \mathcal{V}(R))$$

and

$$\overline{\alpha}(v_1 + \mathcal{V}(R)) = \alpha(v_1)(1)$$

for any $v_1 \in \mathcal{V}_1$ and any $h \in H$.

(ii) It follows from (i) that $i_{H,Q}^H$ is left exact and that $r_H^{\overline{H},Q}$ is right exact.

If $(\overline{\mathcal{V}}, \overline{\pi}) \xrightarrow{\overline{\beta}} (\overline{\mathcal{V}}'', \overline{\pi}'')$ is an epimorphism in $\mathrm{Rep}_s(\overline{H})$, we can find a section $\overline{\gamma}$ of the epimorphism $\overline{\mathcal{V}} \xrightarrow{\overline{\beta}} \overline{\mathcal{V}}''$ of $\mathbb{C}$-vector spaces. Replacing it by

$$\int_{\overline{H}^\circ} \overline{\pi}(\overline{h}^{-1}) \circ \overline{\gamma} \circ \overline{\pi}''(\overline{h}) d\overline{h}$$

if it is necessary we can assume that

$$\overline{\gamma} \circ \overline{\pi}''(\overline{h}^\circ) = \overline{\pi}(\overline{h}^\circ) \circ \overline{\gamma}$$

for all $\overline{h}^\circ \in \overline{H}^\circ$. Then the map

$$v''^\circ \longmapsto \overline{\gamma} \circ v''^\circ$$

is a section of $i_{H,Q}^H(\overline{\beta})^\circ$ ($v''^\circ : H^\circ \longrightarrow \overline{\mathcal{V}}''$ is a locally constant function such that

$$v''^\circ(q^\circ h^\circ) = \overline{\pi}''(R^\circ q^\circ)(v^\circ(h^\circ))$$

for any $h^\circ \in H^\circ$ and any $q^\circ \in Q^\circ$) and $i_{H,Q}^H(\overline{\beta})$ is an epimorphism.

If $(\mathcal{V}', \pi') \xrightarrow{\alpha} (\mathcal{V}, \pi)$ is a monomorphism in $\mathrm{Rep}_s(H)$, we have

$$\mathcal{V}'(R) = \alpha^{-1}(\mathcal{V}(R)) \ .$$

Indeed, we have

$$\mathcal{V}(R) = \bigcup_J \mathrm{Ker}(\pi_J),$$

$$\mathcal{V}'(R) = \bigcup_J \mathrm{Ker}(\pi'_J)$$

and

$$\pi_J \circ \alpha = \alpha \circ \pi'_J$$

where J runs through the set of compact open subgroups of R. Therefore, the map

$$\mathcal{V}'/\mathcal{V}'(R) \longrightarrow \mathcal{V}/\mathcal{V}(R)$$

which is induced by α is injective and $r_H^{\overline{H},Q}(\alpha)$ is a monomorphism. $\square$

Let $(\overline{\mathcal{V}}, \overline{\pi}) \in \mathrm{ob} \, \mathrm{Rep}_s(\overline{H})$ and let $(\mathcal{V}, \pi) = i^H_{\overline{H}, Q}(\overline{\mathcal{V}}, \overline{\pi})$. Let us denote by

$$\mathcal{C}^\infty_c(H, \overline{\mathcal{V}}) = \mathcal{C}^\infty_c(H) \otimes_{\mathbb{Q}} \overline{\mathcal{V}}$$

the $\mathbb{C}$-vector space of locally constant functions with compact support on H and with values in $\overline{\mathcal{V}}$ and let us denote by λ the left action of H on $\mathcal{C}^\infty_c(H, \overline{\mathcal{V}})$ by right translations. Then $(\mathcal{C}^\infty_c(H, \overline{\mathcal{V}}), \lambda)$ is a smooth representation of H and

$$\mathcal{P}(\varphi)(h) = \int_Q \delta_Q^{-1/2}(q) \overline{\pi}(Rq^{-1})(\varphi(qh)) d_r q$$

$(\varphi \in \mathcal{C}^\infty_c(H, \overline{\mathcal{V}}), \, h \in H)$ defines a map

(D.3.4) $\mathcal{P} : (\mathcal{C}^\infty_c(H, \overline{\mathcal{V}}), \lambda) \longrightarrow (\mathcal{V}, \pi)$

in $\mathrm{Rep}_s(H)$. This map is an epimorphism. Indeed, if I is a compact open subgroup of H, if $h_0 \in H$ and if $f \in \mathcal{V}$ has its support which is contained in Qh_0I, then we have

$$\varphi \stackrel{\mathrm{dfn}}{=\!=} 1_{h_0 I} f \in \mathcal{C}^\infty_c(H, \overline{\mathcal{V}})$$

and

$$\mathcal{P}(\varphi) = \mathrm{vol}((h_0 I h_0^{-1}) \cap Q, d_r q) f.$$

Let $\overline{\mathcal{V}}' = \mathrm{Hom}_{\mathbb{C}}(\overline{\mathcal{V}}, \mathbb{C})$ and let λ' be the left action of H on $\mathcal{C}^\infty(H, \overline{\mathcal{V}}')$ by right translations ($\mathcal{C}^\infty(H, \overline{\mathcal{V}}')$ is the space of locally constant functions on H with values in $\overline{\mathcal{V}}'$). In general, the representation $(\mathcal{C}^\infty(H, \overline{\mathcal{V}}'), \lambda')$ is not smooth. Let $(\mathcal{C}^\infty(H, \overline{\mathcal{V}}')^\infty, \lambda'^\infty)$ be its maximal smooth subrepresentation. By definition, we have

$$\mathcal{C}^\infty(H, \overline{\mathcal{V}}')^\infty = \bigcup_I \mathcal{C}^\infty(H/I, \overline{\mathcal{V}}')$$

and $\mathcal{C}^\infty(H/I, \overline{\mathcal{V}}')$ is canonically isomorphic to $\mathrm{Hom}_{\mathbb{C}}(\mathcal{C}^\infty_c(H/I, \overline{\mathcal{V}}), \mathbb{C})$. Here I runs through the set of all compact open subgroups of H. It follows that $(\mathcal{C}^\infty(H, \overline{\mathcal{V}}')^\infty, \lambda'^\infty)$ is the contragredient representation of $(\mathcal{C}^\infty_c(H, \overline{\mathcal{V}}), \lambda)$, with canonical pairing given by

$$< \varphi', \varphi > = \int_H \varphi'(h)(\varphi(h)) dh$$

$(\varphi' \in \mathcal{C}^\infty(H, \overline{\mathcal{V}}')^\infty, \, \varphi \in \mathcal{C}^\infty_c(H, \overline{\mathcal{V}}))$.

Therefore, the contragredient representation of $(\mathcal{V}, \pi)$ can be described in the following way : $\widetilde{\mathcal{V}}$ is the orthogonal subspace of $\mathrm{Ker}(\mathcal{P})$ in $\mathcal{C}^\infty(H, \overline{\mathcal{V}}')^\infty$ and $\widetilde{\pi}$ is the restriction of λ'^∞ to $\widetilde{\mathcal{V}}$. But, by definition, the representation $i^H_{\overline{H}, Q}(\widetilde{\overline{\mathcal{V}}}, \widetilde{\overline{\pi}})$ is also a subrepresentation of $(\mathcal{C}^\infty(H, \overline{\mathcal{V}}')^\infty, \lambda'^\infty)$. So it makes sense to compare $(\widetilde{\mathcal{V}}, \widetilde{\pi})$ and $i^H_{\overline{H}, Q}(\widetilde{\overline{\mathcal{V}}}, \widetilde{\overline{\pi}})$.

LEMMA (D.3.5). — *The two subrepresentations* $(i_{H,Q}^H(\overline{\mathcal{V}}, \pi))^{\sim}$ *and* $i_{H,Q}^H(\widetilde{\overline{\mathcal{V}}}, \widetilde{\pi})$ *of* $(\mathcal{C}^\infty(H, \overline{\mathcal{V}'})^\infty, \lambda'^\infty)$ *coincide.*

Proof: Let us set $i_{H,Q}^H(\widetilde{\overline{\mathcal{V}}}, \widetilde{\pi}) = (\widetilde{\mathcal{V}}_1, \widetilde{\pi}_1)$. We want to prove that $\widetilde{\mathcal{V}}_1 = \widetilde{\mathcal{V}}$ as subspaces of $\mathcal{C}^\infty(H, \overline{\mathcal{V}'})^\infty$.

Let μ (resp. μ') be the right action of H on $\mathcal{C}_c^\infty(H, \overline{\mathcal{V}})$ (resp. $\mathcal{C}^\infty(H, \overline{\mathcal{V}'})$) by left translations. Then, for any $\varphi \in \mathcal{C}_c^\infty(H, \overline{\mathcal{V}})$ and any $q \in Q$, an easy computation shows that

$$\varphi - \delta_Q^{-1/2}(Rq)\overline{\pi}(Rq) \circ \mu(q^{-1})(\varphi) \in \mathrm{Ker}(\mathcal{P}) \ .$$

Therefore, if $\varphi' \in \widetilde{\mathcal{V}}$, we have

$$< \varphi', \varphi - \delta_Q^{-1/2}(Rq)\overline{\pi}(Rq) \circ \mu(q^{-1})(\varphi) > = 0 \ ,$$

i.e.

$$< \varphi' - \delta_Q^{-1/2}(Rq)\mu'(q)(\overline{\pi}'(Rq^{-1}) \circ \varphi'), \varphi > = 0 \ ,$$

for any $\varphi \in \mathcal{C}_c^\infty(H, \overline{\mathcal{V}})$ and any $q \in Q$, and it follows that

$$\varphi'(qh) = \delta_Q^{1/2}(Rq)\overline{\pi}'(Rq)(\varphi'(h))$$

for any $q \in Q$ and any $h \in H$. Now, if I is a compact open subgroup of H such that $\varphi' \in \widetilde{\mathcal{V}}^I$ and if $h \in H$, we have

$$\varphi'(qh) = \varphi'(h(h^{-1}qh)) = \varphi'(h)$$

for any $q \in (hIh^{-1}) \cap Q$. But

$$\varphi'(qh) = \overline{\pi}'(Rq)(\varphi'(h))$$

for any $q \in (hIh^{-1}) \cap Q$ (δ_Q is trivial on any compact subgroup) and it follows that $\varphi'(h) \in \widetilde{\overline{\mathcal{V}}} \subset \overline{\mathcal{V}'}$. So we have proved that $\varphi' \in \widetilde{\mathcal{V}}_1$ and $\widetilde{\mathcal{V}}$ is contained in $\widetilde{\mathcal{V}}_1$.

Conversely, let $\varphi' \in \widetilde{\mathcal{V}}_1$ and let $\varphi \in \mathrm{Ker}(\mathcal{P})$. To finish the proof of the lemma (i.e. to prove the inclusion $\widetilde{\mathcal{V}}_1 \subset \widetilde{\mathcal{V}}$) it suffices to check that $< \varphi', \varphi > = 0$. Let $\chi : H \longrightarrow \mathbb{R}_{>0}$ be the locally constant function which is defined by

$$\chi(qh^\circ) = \delta_Q(q)$$

for any $q \in Q$ and any $h^\circ \in H^\circ$ ($H = QH^\circ$ and δ_Q is trivial on Q°). Then, thanks to [We] ch. II, § 9, there exists a unique Haar measure $d(Qh)$ on $Q \backslash H$ such that

$$\int_H \psi(h)dh = \int_{Q \backslash H} \left(\int_Q \psi(qh)\chi(qh)d_r q \right) d(Qh)$$

for any $\psi \in \mathbb{C} \otimes C_c^\infty(H)$. In particular, we have

$$< \varphi', \varphi > = \int_{Q \backslash H} (\int_Q \varphi'(qh)(\varphi(qh))\chi(qh)d_r q)d(Qh) \ .$$

But we have

$$\varphi'(qh)(\varphi(qh))\chi(qh) = \varphi'(h)(\delta_Q^{-1/2}(Rq)\overline{\pi}(Rq^{-1})(\varphi(qh)))\chi(h)$$

and

$$\int_Q \delta_Q^{-1/2}(Rq)\overline{\pi}(Rq^{-1})(\varphi(qh))d_r q = \mathcal{P}(\varphi) = 0$$

by hypothesis. So $< \varphi', \varphi > = 0$ as required. $\square$

LEMMA (D.3.6). — *If $(\overline{\mathcal{V}}, \overline{\pi})$ is an admissible representation of $\overline{H}$, $i_{\overline{H},Q}^H(\overline{\mathcal{V}}, \overline{\pi})$ is an admissible representation of H.*

Proof: The space $\mathcal{V}$ of $i_{\overline{H},Q}^H(\overline{\mathcal{V}}, \overline{\pi})$ is canonically isomorphic to the space $\mathcal{V}^\circ$ of locally constant functions

$$v^\circ : H^\circ \longrightarrow \overline{\mathcal{V}}$$

such that

$$v^\circ(q^\circ h^\circ) = \overline{\pi}(R^\circ q^\circ)(v^\circ(h^\circ))$$

for any $q^\circ \in Q^\circ$ and any $h^\circ \in H^\circ$. Therefore, for any compact open subgroup I of H, we can identify $\mathcal{V}^I$ with a subspace of $(\mathcal{V}^\circ)^{I^\circ}$ where we have set $I^\circ = I \cap H^\circ$. To finish the proof of the lemma it suffices to check that $(\mathcal{V}^\circ)^{I^\circ}$ is finite dimensional over $\mathbb{C}$. But $(\mathcal{V}^\circ)^{I^\circ}$ is the space of locally constant functions

$$v^\circ : H^\circ/I^\circ \longrightarrow \overline{\mathcal{V}}$$

such that

$$v^\circ(q^\circ h^\circ I^\circ) = \overline{\pi}(R^\circ q^\circ)(f(h^\circ I^\circ))$$

for any $q^\circ \in Q^\circ$ and any $h^\circ \in H^\circ$. In particular,

$$v^\circ(h^\circ I^\circ) \in \overline{\mathcal{V}}^{\overline{I}(h^\circ)}$$

for any $h^\circ \in H^\circ$, where $\overline{I}(h^\circ)$ is the image of the compact open subgroup $Q^\circ \cap (h^\circ I^\circ (h^\circ)^{-1})$ of Q° by the canonical projection $Q^\circ \longrightarrow \overline{H}^\circ$. As H°/I° is finite and $\overline{\mathcal{V}}^{\overline{I}(h^\circ)}$ is finite dimensional over $\mathbb{C}$ for each $h^\circ \in H^\circ$ by hypothesis, the finite dimensionality of $(\mathcal{V}^\circ)^{I^\circ}$ follows. $\square$

Now let $R' \subset Q' \subset H$ be another pair of closed subgroups of H satisfying the same hypotheses (A) to (D) as $R \subset Q$. Moreover, let us assume that

(E) $R \cap R' = \{1\}$ and the canonical projections $Q \longrightarrow\!\!\!\!\rightarrow R \backslash Q = \overline{H}$ and $Q' \longrightarrow\!\!\!\!\rightarrow R' \backslash Q' = \overline{H}'$ induce isomorphisms from

$$L \stackrel{\mathrm{dfn}}{=\!=} Q \cap Q'$$

onto $\overline{H}$ and $\overline{H}'$ respectively,

(F) $\delta_{Q'}(\ell) = \delta_Q(\ell^{-1})$ for any $\ell \in L$.

Thanks to (E), we can and we will identify $\overline{H}$ and $\overline{H}'$ with L, so that Q and Q' are the semidirect products of L by R and R' respectively.

A compact open subgroup I of H is said **in good position with respect to** the triple (R, L, R') if it admits the so-called Iwahori decomposition

$$I = (I \cap R)(I \cap L)(I \cap R') \, .$$

LEMMA (D.3.7). — *Let I be a compact open subgroup of H in good position with respect to (R, L, R') and let $(\mathcal{V}, \pi) \in \mathrm{ob}\ \mathrm{Rep}_s(H)$. Then for any $v \in \mathcal{V}^{I \cap Q'}$, we have*

$$\pi(e_I)(v) = \pi_{I \cap R}(v) \ \left(= \int_{I \cap R} \frac{\pi(r)(v)}{\mathrm{vol}(I \cap R, dr)} dr \right).$$

Proof: Clearly, the measure

$$\frac{dr}{\mathrm{vol}(I \cap R, dr)} \times \frac{d\ell}{\mathrm{vol}(I \cap L, d\ell)} \times \frac{dr'}{\mathrm{vol}(I \cap R', dr')}$$

on $I = (I \cap R)(I \cap L)(I \cap R')$ is left invariant under $I \cap Q$, right invariant under $I \cap Q'$ and gives the volume 1 to I. The same is true for the measure

$$\frac{dh}{\mathrm{vol}(I, dh)}$$

on I. As the action of $(I \cap Q) \times (I \cap Q')$ on I given by $(q, q') \cdot h = qhq'^{-1}$ ($q \in I \cap Q$, $q' \in I \cap Q'$ and $h \in I$) is transitive, these two measures coincide (see [We3] (Ch. II, § 9)). □

THEOREM (D.3.8) (Jacquet). — *Let I be a compact open subgroup of H in good position with respect to (R, L, R') and let $\mathcal{V} \in$ ob $\mathrm{Rep}_s(H)$ be such that $\mathcal{V}^I$ is finite dimensional over $\mathbb{C}$. Let us assume that for every compact open subgroup J' of R' there exists $\ell \in L$ such that*

$$\ell J' \ell^{-1} \supset I \cap R'$$

and

$$\ell(I \cap L)\ell^{-1} = I \cap L .$$

Then, if we set

$$r_H^{L,Q}(\mathcal{V}, \pi) = (\mathcal{W}, \rho)$$

and if we denote by

$$p_R : \mathcal{V} \longrightarrow \mathcal{V}/\mathcal{V}(R) = \mathcal{W}$$

the canonical projection, $\mathcal{W}^{I \cap L}$ is also finite dimensional over $\mathbb{C}$ and p_R maps $\mathcal{V}^I$ onto $\mathcal{W}^{I \cap L}$.

Proof: If $\ell \in L$ and $v \in \mathcal{V}$, we have

$$\rho(\ell)(p_R(v)) = \delta_Q^{-1/2}(\ell)p_R(\pi(\ell)(v)) .$$

Therefore, we have

$$p_R(\mathcal{V}^I) \subset p_R(\mathcal{V}^{I \cap L}) \subset \mathcal{W}^{I \cap L} .$$

In fact, if $v \in \mathcal{V}$ and if $p_R(v) \in \mathcal{W}^{I \cap L}$, we have

$$p_R\left(\int_{I \cap L} \frac{\pi(\ell)(v)}{\mathrm{vol}(I \cap L, d\ell)} d\ell\right) = p_R(v)$$

and it follows that

$$p_R(\mathcal{V}^{I \cap L}) = \mathcal{W}^{I \cap L} .$$

Now let $w_1, \ldots, w_n$ be a finite family of elements of $\mathcal{W}^{I \cap L}$. Let us choose $v_1, \ldots, v_n$ in $\mathcal{V}^{I \cap L}$ such that

$$p_R(v_i) = w_i \quad (i = 1, \ldots, n) .$$

There exists a compact open subgroup J' of R' such that

$$v_i \in \mathcal{V}^{J'} \quad (i = 1, \ldots, n) .$$

Let us choose $\ell \in L$ such that

$$\ell J' \ell^{-1} \supset I \cap R'$$

and
$$\ell(I \cap L)\ell^{-1} = I \cap L .$$

Then we have
$$\pi(\ell)(v_i) \in \mathcal{V}^{I \cap Q'}$$

and
$$\rho(\ell)(w_i) = \delta_Q^{-1/2}(\ell) p_R(\pi(\ell)(v_i))$$

$(i = 1, \ldots, n)$. Let us set

$$v_i' = \pi(e_I)(\pi(\ell)(v_i)) \in \mathcal{V}^I .$$

Thanks to (D.3.7), we have

$$v_i' = \pi_{I \cap R}(\pi(\ell)(v_i))$$

and it follows that
$$p_R(v_i') = p_R(\pi(\ell)(v_i))$$

$(i = 1, \ldots, n)$. Therefore, we have proved that

$$\rho(\ell)(w_i) \in p_R(\mathcal{V}^I) \quad (i = 1, \ldots, n) .$$

In other words we have proved that for any finite dimensional $\mathbb{C}$-vector subspace W of $\mathcal{W}^{I \cap L}$ there exists $\ell \in L$ such that

$$\rho(\ell)(W) \subset p_R(\mathcal{V}^I)$$

and
$$\ell(I \cap L)\ell^{-1} = I \cap L .$$

It follows that
$$\dim_{\mathbb{C}}(W) = \dim_{\mathbb{C}}(\rho(\ell)(W))$$

is bounded by
$$\dim_{\mathbb{C}}(p_R(\mathcal{V}^I)) \leq \dim_{\mathbb{C}}(\mathcal{V}^I) < +\infty .$$

But this means that $\mathcal{W}^{I \cap L}$ is finite dimensional over $\mathbb{C}$. Taking $W = \mathcal{W}^{I \cap L}$ and using the fact that

$$\rho(\ell)(\mathcal{W}^{I \cap L}) = \mathcal{W}^{I \cap L}$$

$(\ell(I \cap L)\ell^{-1} = I \cap L)$, we get that

$$\mathcal{W}^{I \cap L} \subset p_R(\mathcal{V}^I) .$$

This completes the proof of Jacquet's theorem. $\qquad\qquad\square$

LEMMA (D.3.9). — *Let I be a compact open subgroup of H in good position with respect to (R, L, R') and let $(\mathcal{V}, \pi) \in \text{ob } \text{Rep}_s(H)$. Let us set*

$$r_H^{L,Q}(\mathcal{V}, \pi) = (\mathcal{W}, \rho)$$

and let us denote by

$$p_R : \mathcal{V} \longrightarrow \mathcal{V}/\mathcal{V}(R) = \mathcal{W}$$

the canonical projection. We assume that

$$p_R(\mathcal{V}^I) = \mathcal{W}^{I \cap L} .$$

Then, for any $\ell \in L$ such that

(i) $\ell(I \cap L)\ell^{-1} = I \cap L$ *and* $\ell(I \cap R')\ell^{-1} \supset I \cap R'$,

(ii) $\mathcal{V}^I \cap \mathcal{V}(R) \subset \text{Ker}(\pi_{\ell^{-1}(I \cap R)\ell})$,

we have

$$\mathcal{V}^I = \pi_{I \cap R}(\pi(\ell)(\mathcal{V}^I)) \oplus (\mathcal{V}^I \cap \mathcal{V}(R))$$

(in other words, $\pi_{I \cap R}(\pi(\ell)(\mathcal{V}^I))$ is a cross-section of $p_R : \mathcal{V}^I \longrightarrow \mathcal{W}^{I \cap L}$).

Proof: It follows from (D.3.7) that

$$\pi_{I \cap R}(\pi(\ell)(\mathcal{V}^I)) \subset \mathcal{V}^I$$

$(\ell(I \cap Q')\ell^{-1} \supset I \cap Q'$, so that $\pi(\ell)(\mathcal{V}^I) \subset \mathcal{V}^{I \cap Q'})$ and it suffices to prove that

$$p_R(\pi_{I \cap R}(\pi(\ell)(\mathcal{V}^I))) = \mathcal{W}^{I \cap L}$$

and

$$\pi_{I \cap R}(\pi(\ell)(\mathcal{V}^I)) \cap \mathcal{V}(R) = (0) .$$

The first assertion is obvious ($p_R \circ \pi_{I \cap R} = p_R$, $\rho(\ell) \circ p_R = \delta_Q^{-1/2}(\ell) p_R \circ \pi(\ell)$ and $\ell(I \cap L)\ell^{-1} = I \cap L$ so that $\rho(\ell)(\mathcal{W}^{I \cap L}) = \mathcal{W}^{I \cap L}$).

For the second assertion, let us remark that, if $v \in \mathcal{V}^I$ and $\pi_{I \cap R}(\pi(\ell)(v)) \in \mathcal{V}(R)$, we have also $v \in \mathcal{V}(R)$. Indeed, we have

$$p_R(\pi_{I \cap R}(\pi(\ell)(v))) = \delta_Q^{1/2}(\ell)\rho(\ell)(p_R(v)) .$$

But this means that, if $v \in \mathcal{V}^I$ and $\pi_{I \cap R}(\pi(\ell)(v)) \in \mathcal{V}(R)$, we have also

$$v \in \mathcal{V}^I \cap \mathcal{V}(R) \subset \text{Ker}(\pi_{\ell^{-1}(I \cap R)\ell}) ,$$

so that

$$\pi_{I \cap R}(\pi(\ell)(v)) = \pi(\ell)(\pi_{\ell^{-1}(I \cap R)\ell}(v)) = 0 .$$

$\square$

REMARKS (D.3.10.1). — If $\mathcal{V}^I$ is finite dimensional over $\mathbb{C}$, the same is true for $\mathcal{V}^I \cap \mathcal{V}(R)$ and there exists a compact open subgroup J of R such that

$$\mathcal{V}^I \cap \mathcal{V}(R) \subset \mathrm{Ker}(\pi_J)$$

(recall that $\mathcal{V}(R)$ is the union of the $\mathrm{Ker}(\pi_J)$'s). Therefore, a practical way to find $\ell \in L$ such that

$$\mathcal{V}^I \cap \mathcal{V}(R) \subset \mathrm{Ker}(\pi_{\ell^{-1}(I \cap R)\ell})$$

is to find $\ell \in L$ such that

$$\ell J \ell^{-1} \subset I \cap R \, .$$

(D.3.10.2) Let I be a compact open subgroup of H in good position with respect to (R, L, R'), let $\ell \in L$ be such that $\ell(I \cap R)\ell^{-1} \subset I \cap R$, $\ell(I \cap L)\ell^{-1} = I \cap L$ and $\ell(I \cap R')\ell^{-1} \supset I \cap R'$. Then, for any $(\mathcal{V}, \pi) \in \mathrm{ob}\, \mathrm{Rep}_s(H)$, we have

$$\pi_{I \cap R} \circ \pi(\ell) \mid \mathcal{V}^I = \delta_Q(\ell)\pi\left(\frac{1_{I\ell I}}{\mathrm{vol}(I, dh)}\right) \, .$$

Indeed, let S be a system of representatives of the classes in $(I \cap R)/$ $\ell(I \cap R)\ell^{-1}$; we have

$$I\ell I = \coprod_{r \in S} r\ell I \, ,$$

so that

$$\pi\left(\frac{1_{I\ell I}}{\mathrm{vol}(I, dh)}\right) = \sum_{r \in S} \pi(r) \circ \pi(\ell) \mid \mathcal{V}^I \, ,$$

and we have

$$\pi_{I \cap R} \circ \pi(\ell) \mid \mathcal{V}^I = \sum_{r \in S} \frac{\mathrm{vol}(\ell(I \cap R)\ell^{-1}, dr)\pi(r) \circ \pi(\ell) \mid \mathcal{V}^I}{\mathrm{vol}(I \cap R, dr)} \, .$$

Now let us assume moreover that the hypotheses of (D.3.9) are satisfied and that $\pi(1_{I\ell I})$ is invertible in the convolution $\mathbb{C}$-algebra $\mathbb{C} \otimes \mathcal{C}_c^\infty(H//I)$. Then it follows that $\mathcal{V}^I \cap \mathcal{V}(R) = (0)$, i.e. that p_R induces an isomorphism from $\mathcal{V}^I$ onto $\mathcal{W}^{I \cap L}$.   □

THEOREM (D.3.11) (Casselman). — *Let I be a compact open subgroup of H in good position with respect to (R, L, R'). Let $\Lambda \subset L$ be a subset such that $\ell(I \cap R)\ell^{-1} \subset I \cap R$, $\ell(I \cap L)\ell^{-1} = I \cap L$ and $\ell(I \cap R')\ell^{-1} \supset I \cap R'$ for each $\ell \in \Lambda$. Let $(\mathcal{V}, \pi)$ be a smooth representation of H and let $(\widetilde{V}, \widetilde{\pi})$ be its contragredient representation. We assume that*

(i) *for each compact open subgroup J of R and each compact open subgroup J' of R' there exists $\ell \in \Lambda$ such that*

$$\ell J \ell^{-1} \subset I \cap R$$

and

$$\ell J' \ell^{-1} \supset I \cap R' ,$$

(ii) $\mathcal{V}^I$ *is finite dimensional over* $\mathbb{C}$.

Then, if we identify $\widetilde{\mathcal{V}}^I$ *with the* $\mathbb{C}$*-linear dual of* $\mathcal{V}^I$ *(see (D.1.10)), we have the direct sum decompositions*

$$\mathcal{V}^I = (\widetilde{\mathcal{V}}^I \cap \widetilde{\mathcal{V}}(R'))^{\perp} \oplus (\mathcal{V}^I \cap \mathcal{V}(R))$$

and

$$\widetilde{\mathcal{V}}^I = (\mathcal{V}^I \cap \mathcal{V}(R))^{\perp} \oplus (\widetilde{\mathcal{V}}^I \cap \widetilde{\mathcal{V}}(R'))$$

Moreover, for each $\ell \in \Lambda$ *such that*

$$\mathcal{V}^I \cap \mathcal{V}(R) \subset \mathrm{Ker}(\pi_{\ell^{-1}(I \cap R)\ell})$$

and

$$\widetilde{\mathcal{V}}^I \cap \widetilde{\mathcal{V}}(R') \subset \mathrm{Ker}(\widetilde{\pi}_{\ell(I \cap R')\ell^{-1}}),$$

we have

$$\pi_{I \cap R}(\pi(\ell)(\mathcal{V}^I)) = (\widetilde{\mathcal{V}}^I \cap \widetilde{\mathcal{V}}(R'))^{\perp}$$

and

$$\widetilde{\pi}_{I \cap R'}(\widetilde{\pi}(\ell^{-1})(\widetilde{\mathcal{V}}^I)) = (\mathcal{V}^I \cap \mathcal{V}(R))^{\perp} .$$

COROLLARY (D.3.12) (Casselman). — *Let us keep the notations and the hypotheses of* (D.3.11). *Let us set*

$$r_H^{L,Q}(\mathcal{V}, \pi) = (\mathcal{W}, \rho)$$

and

$$r_H^{L,Q'}(\widetilde{\mathcal{V}}, \widetilde{\pi}) = (\widecheck{\mathcal{W}}, \widecheck{\rho}) .$$

Then there exists a unique $\mathbb{C}$*-linear isomorphism*

$$\iota : \widecheck{\mathcal{W}}^{I \cap L} \xrightarrow{\sim} (\mathcal{W}^{I \cap L})' = \mathrm{Hom}_{\mathbb{C}}(\mathcal{W}^{I \cap L}, \mathbb{C})$$

such that, for each $v \in \mathcal{V}^I$, *each* $\widetilde{v} \in \widetilde{\mathcal{V}}^I$ *and each* $\ell \in \Lambda$ *with*

$$\mathcal{V}^I \cap \mathcal{V}(R) \subset \mathrm{Ker}(\pi_{\ell^{-1}(I \cap R)\ell})$$

and

$$\widetilde{\mathcal{V}}^I \cap \widetilde{\mathcal{V}}(R') \subset \mathrm{Ker}(\widetilde{\pi}_{\ell(I \cap R')\ell^{-1}}) ,$$

we have

$$< \iota(p_{R'}(\widetilde{v})), \rho(\ell)(p_R(v)) > = \delta_Q^{-1/2}(\ell) < \widetilde{v}, \pi(\ell)(v) > ,$$

where

$$p_R : \mathcal{V} \longrightarrow \mathcal{V}/\mathcal{V}(R) = \mathcal{W}$$

and

$$p_{R'} : \widetilde{\mathcal{V}} \longrightarrow \widetilde{\mathcal{V}}/\widetilde{\mathcal{V}}(R') = \widecheck{\mathcal{W}}$$

are the canonical projections.

Proof of the corollary assuming the theorem : Thanks to (D.3.11), p_R and $p_{R'}$ induce isomorphisms from $(\widetilde{\mathcal{V}}^I \cap \widetilde{\mathcal{V}}(R'))^\perp$ and $(\mathcal{V}^I \cap \mathcal{V}(R))^\perp$ onto $\mathcal{W}^{I \cap L}$ and $\widecheck{\mathcal{W}}^{I \cap L}$ respectively. The identification of $\widetilde{\mathcal{V}}^I$ with the dual of $\mathcal{V}^I$ induces an isomorphism from $(\widetilde{\mathcal{V}}^I \cap \widetilde{\mathcal{V}}(R'))^\perp$ onto the dual of $(\mathcal{V}^I \cap \mathcal{V}(R))^\perp$ and therefore an isomorphism from $\widecheck{\mathcal{W}}^{I \cap L}$ onto the dual of $\mathcal{W}^{I \cap L}$. Let ι be this isomorphism.

If ℓ is as in the corollary, it follows from the theorem that

$$(\widetilde{\mathcal{V}}^I \cap \widetilde{\mathcal{V}}(R'))^\perp = \pi_{I \cap R}(\pi(\ell)(\mathcal{V}^I)) .$$

Now, if $\widetilde{v} \in \widetilde{\mathcal{V}}^I$ and $v \in \mathcal{V}^I$, the definition of ι implies that

$$< \iota(p_{R'}(\widetilde{v})), p_R(\pi_{I \cap R}(\pi(\ell)(v))) >$$

is equal to

$$< \widetilde{v}, \pi_{I \cap R}(\pi(\ell)(v)) > .$$

But we have

$$p_R(\pi_{I \cap R}(\pi(\ell)(v))) = \delta_Q^{1/2}(\ell)\rho(\ell)(p_R(v))$$

and

$$< \widetilde{v}, \pi_{I \cap R}(\pi(\ell)(v)) > = < \widetilde{\pi}_{I \cap R}(\widetilde{v}), \pi(\ell)(v) >$$

$$= < \widetilde{v}, \pi(\ell)(v) >$$

($\widetilde{\pi}_{I \cap R}(\widetilde{v}) = \widetilde{v}$ as $\widetilde{v} \in \widetilde{\mathcal{V}}^I \subset \widetilde{\mathcal{V}}^{I \cap R}$). Therefore, we get that

$$< \iota(p_{R'}(\widetilde{v})), \rho(\ell)(p_R(v)) > = \delta_Q^{-1/2}(\ell) < \widetilde{v}, \pi(\ell)(v) >$$

as required.

As $\rho(\ell)(\mathcal{W}^{I \cap L}) = \mathcal{W}^{I \cap L}$ ($\ell(I \cap L)\ell^{-1} = I \cap L$), the uniqueness of ι is obvious. $\qquad \square$

Proof of theorem (D.3.11) : The hypotheses of Jacquet's theorem are satisfied by $(\mathcal{V}, \pi)$ and $(\widetilde{\mathcal{V}}, \widetilde{\pi})$ ($\widetilde{\mathcal{V}}^I$ is the dual of $\mathcal{V}^I$, so $\widetilde{\mathcal{V}}^I$ is also finite dimensional over $\mathbb{C}$). Therefore, we have

$$p_R(\mathcal{V}^I) = \mathcal{W}^{I \cap L}$$

and

$$p_{R'}(\widetilde{\mathcal{V}}^I) = \widecheck{\mathcal{W}}^{I \cap L}$$

Let $\ell \in \Lambda$ be such that

$$\mathcal{V}^I \cap \mathcal{V}(R) \subset \mathrm{Ker}(\pi_{\ell^{-1}(I \cap R)\ell})$$

and

$$\widetilde{\mathcal{V}}^I \cap \widetilde{\mathcal{V}}(R') \subset \mathrm{Ker}(\widetilde{\pi}_{\ell(I \cap R')\ell^{-1}})$$

(there always exists such an ℓ, see (D.3.10.1)). Then, thanks to (D.3.9), we have the direct sum decompositions

$$\mathcal{V}^I = \pi_{I \cap R}(\pi(\ell)(\mathcal{V}^I)) \oplus (\mathcal{V}^I \cap \mathcal{V}(R))$$

and

$$\widetilde{\mathcal{V}}^I = \widetilde{\pi}_{I \cap R'}(\widetilde{\pi}(\ell^{-1})(\widetilde{\mathcal{V}}^I)) \oplus (\widetilde{\mathcal{V}}^I \cap \widetilde{\mathcal{V}}(R')) \ .$$

To finish the proof of the theorem, it suffices to check that

$$< \widetilde{\mathcal{V}}^I \cap \widetilde{\mathcal{V}}(R'), \pi_{I \cap R}(\pi(\ell)(\mathcal{V}^I)) > = (0)$$

and that

$$< \widetilde{\pi}_{I \cap R'}(\widetilde{\pi}(\ell^{-1})(\widetilde{\mathcal{V}}^I)), \mathcal{V}^I \cap \mathcal{V}(R) > = (0) \ .$$

But we have

$$< \widetilde{\mathcal{V}}^{I \cap R} \cap \mathrm{Ker}(\widetilde{\pi}_{\ell(I \cap R')\ell^{-1}}), \ \pi_{I \cap R}(\pi(\ell)(\mathcal{V}^{I \cap R'})) > = (0)$$

and

$$< \widetilde{\pi}_{I \cap R'}(\widetilde{\pi}(\ell^{-1})(\widetilde{\mathcal{V}}^{I \cap R})), \ \mathcal{V}^{I \cap R'} \cap \mathrm{Ker}(\pi_{\ell^{-1}(I \cap R)\ell}) > = (0) \ .$$

Indeed, if $v \in \mathcal{V}^{I \cap R'}$ and $\widetilde{v} \in \widetilde{\mathcal{V}}^{I \cap R}$, we have

$$< \widetilde{\pi}_{I \cap R'}(\widetilde{\pi}(\ell^{-1})(\widetilde{v})), v > = < \widetilde{v}, \pi(\ell)(v) > = < \widetilde{v}, \pi_{I \cap R}(\pi(\ell)(v)) >$$

as $v = \pi_{I \cap R'}(v)$ and $\widetilde{v} = \widetilde{\pi}_{I \cap R}(\widetilde{v})$ and we have

$$\pi_{I \cap R} \circ \pi(\ell) = \pi(\ell) \circ \pi_{\ell^{-1}(I \cap R)\ell}$$

and

$$\widetilde{\pi}_{I \cap R'} \circ \widetilde{\pi}(\ell^{-1}) = \widetilde{\pi}(\ell^{-1}) \circ \widetilde{\pi}_{\ell(I \cap R')\ell^{-1}} \ .$$

Therefore, the proof of the theorem is completed. $\square$

THEOREM (D.3.13). — *Let us assume that there exists a decreasing sequence $(I_n)_{n \in \mathbf{Z}_{\geq 0}}$ of compact open subgroups of H which form a basis of neighborhoods of the identity element and a subset Λ of the center of L such that*

(i) *for each $n \in \mathbf{Z}_{\geq 0}$, I_{n+1} is a normal subgroup of I_n,*

(ii) *for each $n \in \mathbf{Z}_{\geq 0}$, I_n is in good position with respect to (R, L, R'),*

(iii) *for each $n \in \mathbf{Z}_{\geq 0}$ and each $\ell \in \Lambda$, we have $\ell(I_n \cap R)\ell^{-1} \subset I_n \cap R$ and $\ell(I_n \cap R')\ell^{-1} \supset I_n \cap R'$,*

(iv) *for each pair (J_1, J_2) of compact open subgroups of R and each pair (J_1', J_2') of compact open subgroups of R', there exists $\ell \in \Lambda$ such that*

$$\ell J_1 \ell^{-1} \subset J_2$$

and

$$\ell J_1' \ell^{-1} \supset J_2' \ .$$

Then, for each admissible representation $(\mathcal{V}, \pi)$ of H, the smooth representation $r_H^{L,Q}(\mathcal{V}, \pi) = (\mathcal{W}, \rho)$ is also admissible and its contragredient representation $(\widetilde{\mathcal{W}}, \widetilde{\rho})$ is canonically isomorphic to $r_H^{L,Q'}(\widetilde{\mathcal{V}}, \widetilde{\pi}) = (\check{\mathcal{W}}, \check{\rho})$, where $(\widetilde{\mathcal{V}}, \widetilde{\pi})$ is the contragredient representation of $(\mathcal{V}, \pi)$.

Moreover, if we denote by

$$\iota : (\check{\mathcal{W}}, \check{\rho}) \xrightarrow{\sim} (\widetilde{\mathcal{W}}, \widetilde{\rho})$$

the above canonical isomorphism and by

$$p_R : \mathcal{V} \longrightarrow\!\!\!\!\!\rightarrow \mathcal{V}/\mathcal{V}(R) = \mathcal{W}$$

and

$$p_{R'} : \widetilde{\mathcal{V}} \longrightarrow\!\!\!\!\!\rightarrow \widetilde{\mathcal{V}}/\widetilde{\mathcal{V}}(R') = \check{\mathcal{W}}$$

the canonical projections, we have

$$< \iota(p_{R'}(\widetilde{v})), \rho(\ell)(p_R(v)) > = \delta_Q^{-1/2}(\ell) < \widetilde{v}, \pi(\ell)(v) >$$

for each $n \in \mathbf{Z}_{\geq 0}$, each $v \in \mathcal{V}^{I_n}$, each $\widetilde{v} \in \widetilde{\mathcal{V}}^{I_n}$ and each $\ell \in \Lambda$ such that

$$\mathcal{V}^{I_n} \cap \mathcal{V}(R) \subset \mathrm{Ker}(\pi_{\ell^{-1}(I_n \cap R)\ell})$$

and

$$\widetilde{\mathcal{V}}^{I_n} \cap \widetilde{\mathcal{V}}(R') \subset \mathrm{Ker}(\widetilde{\pi}_{\ell(I_n \cap R')\ell^{-1}}) \ .$$

Proof: For each $n \in \mathbb{Z}_{\geq 0}$ and for $I = I_n$, the hypotheses of (D.3.8) and (D.3.12) are satisfied. Therefore, $\mathcal{W}^{I_n \cap L}$ is finite dimensional over $\mathbb{C}$ and we have a canonical isomorphism

$$\iota_n : \check{\mathcal{W}}^{I_n \cap L} \xrightarrow{\sim} (\mathcal{W}^{I_n \cap L})' = \widetilde{\mathcal{W}}^{I_n \cap L}$$

(see (D.1.10)).

As the decreasing sequence $(I_n \cap L)_{n \in \mathbb{Z}_{\geq 0}}$ of compact open subgroups of L form a basis of neighborhoods of the identity, $(\mathcal{W}, \rho)$ is admissible.

Let us check that the isomorphisms ι_n are compatible when n varies. First of all, as I_{n+1} is a normal subgroup of I_n, ι_{n+1} maps $\check{\mathcal{W}}^{I_n \cap L} \subset \check{\mathcal{W}}^{I_{n+1} \cap L}$ into $\widetilde{\mathcal{W}}^{I_n \cap L} \subset \widetilde{\mathcal{W}}^{I_{n+1} \cap L}$. Indeed, if $i \in I_n \cap L$, we have

$$\check{\rho}(i)(\check{\mathcal{W}}^{I_{n+1} \cap L}) = \check{\mathcal{W}}^{I_{n+1} \cap L}$$

and

$$\widetilde{\rho}(i)(\widetilde{\mathcal{W}}^{I_{n+1} \cap L}) = \widetilde{\mathcal{W}}^{I_{n+1} \cap L} ,$$

as $I_n \cap L$ normalizes $I_{n+1} \cap L$, and it suffices to prove that

$$\widetilde{\rho}(i^{-1}) \circ \iota_{n+1} \circ \check{\rho}(i) = \iota_{n+1} .$$

But let us fix $\ell \in \Lambda$ such that

$$\mathcal{V}^{I_{n+1}} \cap \mathcal{V}(R) \subset \mathrm{Ker}(\pi_{\ell^{-1}(I_{n+1} \cap R)\ell})$$

and

$$\widetilde{\mathcal{V}}^{I_{n+1}} \cap \widetilde{\mathcal{V}}(R') \subset \mathrm{Ker}(\widetilde{\pi}_{\ell(I_{n+1} \cap R')\ell^{-1}}) .$$

Then, for each $v \in \mathcal{V}^{I_{n+1}}$ and each $\widetilde{v} \in \widetilde{\mathcal{V}}^{I_{n+1}}$, we have

$$< \widetilde{\rho}(i^{-1}) \circ \iota_{n+1} \circ \check{\rho}(i)(p_{R'}(\widetilde{v})), \rho(\ell)(p_R(v)) >$$
$$= < \iota_{n+1}(p_{R'}(\widetilde{\pi}(i)(\widetilde{v}))), \rho(\ell)(p_R(\pi(i)(v))) >$$

with $\widetilde{\pi}(i)(v) \in \widetilde{\mathcal{V}}^{I_{n+1}}$ and $\pi(i)(v) \in \mathcal{V}^{I_{n+1}}$. Therefore we get

$$< \widetilde{\rho}(i^{-1}) \circ \iota_{n+1} \circ \check{\rho}(i)(p_{R'}(\widetilde{v})), \rho(\ell)(p_R(v)) >$$

$$= \delta_Q^{-1/2}(\ell) < \widetilde{\pi}(i)(\widetilde{v}), \pi(\ell)(\pi(i)(v)) >$$

$$= \delta_Q^{-1/2}(\ell) < \widetilde{v}, \pi(\ell)(v) >$$

$$= < \iota_{n+1}(p_{R'}(\widetilde{v})), \rho(\ell)(p_R(v)) > .$$

As we have

$$p_R(\mathcal{V}^{I_{n+1}}) = \mathcal{W}^{I_{n+1} \cap L}$$

and
$$p_{R'}(\widetilde{\mathcal{V}}^{I_{n+1}}) = \check{\mathcal{W}}^{I_{n+1} \cap L}$$

(see (D.3.8)), our assertion follows. Now let us fix $\ell \in \Lambda$ such that

$$\mathcal{V}^{I_n} \cap \mathcal{V}(R) \subset \mathcal{V}^{I_{n+1}} \cap \mathcal{V}(R) \subset \mathrm{Ker}(\pi_{\ell^{-1}(I_n \cap R)\ell}) \subset \mathrm{Ker}(\pi_{\ell^{-1}(I_{n+1} \cap R)\ell})$$

and

$$\widetilde{\mathcal{V}}^{I_n} \cap \widetilde{\mathcal{V}}(R') \subset \widetilde{\mathcal{V}}^{I_{n+1}} \cap \widetilde{\mathcal{V}}(R') \subset \mathrm{Ker}(\widetilde{\pi}_{\ell(I_n \cap R')\ell^{-1}}) \subset \mathrm{Ker}(\widetilde{\pi}_{\ell(I_{n+1} \cap R')\ell^{-1}})$$

(there always exists such an ℓ). Then, for each $v \in \mathcal{V}^{I_n} \subset \mathcal{V}^{I_{n+1}}$ and each $\widetilde{v} \in \widetilde{\mathcal{V}}^{I_n} \subset \widetilde{\mathcal{V}}^{I_{n+1}}$, we have

$$< \iota_{n+1}(p_{R'}(\widetilde{v})), \rho(\ell)(p_R(v)) > = \delta_Q^{-1/2}(\ell) < \widetilde{v}, \pi(\ell)(v) >$$

$$= < \iota_n(p_{R'}(\widetilde{v})), \rho(\ell)(p_R(v)) > .$$

As we have
$$\rho(\ell)(p_R(\mathcal{V}^{I_n})) = \mathcal{W}^{I_n \cap L},$$

$$p_{R'}(\widetilde{\mathcal{V}}^{I_n}) = \check{\mathcal{W}}^{I_n \cap L}$$

and
$$\iota_{n+1}(\check{\mathcal{W}}^{I_n \cap L}) \subset \mathcal{W}^{I_n \cap L} ,$$

(see (D.3.8)), it follows that the restriction of ι_{n+1} to $\check{\mathcal{W}}^{I_n \cap L} \subset \check{\mathcal{W}}^{I_{n+1} \cap L}$ is nothing else than ι_n.

The compatible system of the ι_n's defines an isomorphism

$$\iota : \check{\mathcal{W}} \longrightarrow \widetilde{\mathcal{W}} .$$

The last thing to check is that ι is a morphism of representations of L from $(\check{\mathcal{W}}, \check{\rho})$ into $(\widetilde{\mathcal{W}}, \widetilde{\rho})$. Let $\ell_0 \in L$. We want to prove that

$$\widetilde{\rho}(\ell_0^{-1}) \circ \iota \circ \check{\rho}(\ell_0)(\check{w}) = \iota(\check{w})$$

for each $\check{w} \in \check{\mathcal{W}}$ or, what is equivalent, that

$$\widetilde{\rho}(\ell_0^{-1}) \circ \iota \circ \check{\rho}(\ell_0)(\check{w}) = \iota(\check{w})$$

for each $\check{w} \in \check{\mathcal{W}}^{I_n \cap L}$ and each $n \in \mathbf{Z}_{\geq 0}$. Let us fix $n \in \mathbf{Z}_{\geq 0}$. Then there exists $m \in \mathbf{Z}_{\geq 0}$ such that

$$I_m \subset I_n \cap (\ell_0 I_n \ell_0^{-1}) .$$

Let us fix such an m and let $\ell \in \Lambda$ be such that

$$\mathcal{V}^{I_m} \cap \mathcal{V}(R) \subset \mathrm{Ker}(\pi_{\ell^{-1}(I_m \cap R)\ell})$$

and
$$\widetilde{\mathcal{V}}^{I_m} \cap \widetilde{\mathcal{V}}(R') \subset \mathrm{Ker}(\widetilde{\pi}_{\ell(I_m \cap R')\ell^{-1}}) \ .$$

For each $v \in \mathcal{V}^{I_n}$ and each $\widetilde{v} \in \widetilde{\mathcal{V}}^{I_n}$, we have $v \in \mathcal{V}^{I_m}$, $\pi(\ell_0)(v) \in \mathcal{V}^{I_m}$, $\widetilde{v} \in \widetilde{\mathcal{V}}^{I_m}$ and $\widetilde{\pi}(\ell_0)(\widetilde{v}) \in \widetilde{\mathcal{V}}^{I_m}$. Therefore, we have

$$< \widetilde{\rho}(\ell_0^{-1}) \circ \iota \circ \check{\rho}(\ell_0)(p_{R'}(\widetilde{v})), \rho(\ell)(p_R(v)) >$$

$$= < \iota(p_{R'}(\widetilde{\pi}(\ell_0)(\widetilde{v}))), \rho(\ell)(p_R(\pi(\ell_0)(v))) >$$

$$= \delta_Q^{-1/2}(\ell) < \widetilde{\pi}(\ell_0)(\widetilde{v}), \pi(\ell)(\pi(\ell_0)(v)) >$$

$$= \delta_Q^{-1/2}(\ell) < \widetilde{v}, \pi(\ell)(v) >$$

$$= < \iota(p_{R'}(\widetilde{v})), \rho(\ell)(p_R(v)) > \ .$$

But we have
$$\rho(\ell)(p_R(\mathcal{V}^{I_n})) = \rho(\ell)(\mathcal{W}^{I_n \cap L}) = \mathcal{W}^{I_n \cap L}$$
so that we get
$$\widetilde{\rho}(\ell_0^{-1}) \circ \iota \circ \check{\rho}(\ell_0)(p_{R'}(\widetilde{v})) = \iota(p_{R'}(\widetilde{v}))$$
for each $\widetilde{v} \in \widetilde{\mathcal{V}}^{I_n}$. As we also have

$$p_{R'}(\widetilde{\mathcal{V}}^{I_n}) = \check{\mathcal{W}}^{I_n \cap L} \ ,$$

this completes the proof of the theorem. $\qquad\qquad\qquad\qquad\square$

To finish this section, let us examine the transitivity properties of the induction and the restriction. In $\overline{H}$, let $\overline{R} \subset \overline{Q} \subset \overline{H}$ be closed subgroups such that

    $(\overline{A})$ $\overline{R}$ is normal in $\overline{Q}$,

    $(\overline{B})$ any compact subset of $\overline{R}$ is contained in a compact open subgroup of $\overline{R}$ (in particular, $\overline{R}$ is the union of its compact open subgroups),

    $(\overline{C})$ $\overline{H} = \overline{Q} \cdot \overline{H}^{\circ}$ (recall that $\overline{H}^{\circ} = R^{\circ}\backslash Q^{\circ}$),

    $(\overline{D})$ for any Haar measure $d\overline{r}$ on $\overline{R}$ and any $\overline{q} \in \overline{Q}$, we have

$$d(\overline{q} \, \overline{r} \, \overline{q}^{-1}) = \delta_{\overline{Q}}(\overline{q}) d\overline{r}$$

where $\delta_{\overline{Q}} : \overline{Q} \longrightarrow \mathbb{R}_{>0}$ is the modulus character of $\overline{Q}$ (thanks to $(\overline{B})$, $\overline{R}$ is unimodular).

We can apply all the constructions and all the results for $R \subset Q \subset H$ to $\overline{R} \subset \overline{Q} \subset \overline{H}$ ($\overline{H}$ is also unimodular). In particular, we have the functors

$$i_{\overline{H},\overline{Q}}^{\overline{H}} : \mathrm{Rep}_s(\overline{\overline{H}}) \longrightarrow \mathrm{Rep}_s(\overline{H})$$

and

$$r_{\overline{H}}^{\overline{\overline{H}},\overline{Q}} : \mathrm{Rep}_s(\overline{H}) \longrightarrow \mathrm{Rep}_s(\overline{\overline{H}})$$

where we have set

$$\overline{\overline{H}} = \overline{R}\backslash \overline{Q}$$

(again, $\overline{\overline{H}}$ is a unimodular, locally compact, totally discontinuous, separated topological group).

Let $R_1 \subset Q_1 \subset Q$ be the inverse images of $\overline{R} \subset \overline{Q} \subset \overline{H}$ by the canonical projection $Q \longrightarrow\!\!\!\!\!\rightarrow R\backslash Q = \overline{H}$. Then $R_1 \subset Q_1$ are closed subgroups of H. Let us assume that they satisfy the same hypotheses (A) to (D) as $R \subset Q$ (in fact, these hypotheses are automatically satisfied and R_1 is automatically unimodular; for (B), this has been explained to me by Waldspurger).

Again, we can apply all the constructions and all the results for $R \subset Q$ to $R_1 \subset Q_1$. In particular, we have the functors

$$i_{\overline{H},Q_1}^{H} : \mathrm{Res}_s(\overline{\overline{H}}) \longrightarrow \mathrm{Rep}_s(H)$$

and

$$r_H^{\overline{\overline{H}},Q_1} : \mathrm{Rep}_s(H) \longrightarrow \mathrm{Rep}_s(\overline{\overline{H}}) .$$

PROPOSITION (D.3.14). — *The functors* $i_{\overline{H},Q_1}^{H}$ *and* $i_{\overline{H},Q}^{H} \circ i_{\overline{\overline{H}},\overline{Q}}^{\overline{H}}$ *(resp.* $r_H^{\overline{\overline{H}},Q_1}$ *and* $r_{\overline{H}}^{\overline{\overline{H}},\overline{Q}} \,or_H^{\overline{H},Q})$ *from* $\mathrm{Rep}_s(\overline{\overline{H}})$ *to* $\mathrm{Rep}_s(H)$ *(resp. from* $\mathrm{Rep}_s(H)$ *to* $\mathrm{Rep}_s(\overline{\overline{H}}))$ *are canonically isomorphic.* □

(D.4) Cuspidal representations of H

Let Z be the center of H. In this section we will assume that

(i) H is countable at infinity,

(ii) there exists an open, normal subgroup H^1 of H such that H/H^1 is commutative, $Z \cap H^1$ is compact and ZH^1 is of finite index in H.

If $(\mathcal{V}, \pi) \in \mathrm{ob}\ \mathrm{Rep}_s(H)$ and if $(\widetilde{\mathcal{V}}, \widetilde{\pi})$ is the contragredient representation of $(\mathcal{V}, \pi)$, for each $v \in \mathcal{V}$ and each $\widetilde{v} \in \widetilde{\mathcal{V}}$, the locally constant function

$$\varphi_{\widetilde{v},v} : H \longrightarrow \mathbb{C},$$

$$h \longmapsto <\widetilde{v}, \pi(h)(v)> = <\widetilde{\pi}(h^{-1})(\widetilde{v}), v>$$

is called a **matrix coefficient** of $(\mathcal{V}, \pi)$.

If $(\mathcal{V}, \pi) \in \mathrm{ob}\ \mathrm{Rep}_s(H)$ admits a central character $\omega_\pi : Z \longrightarrow \mathbb{C}^\times$, for any matrix coefficient $\varphi_{\widetilde{v},v}$ of $(\mathcal{V}, \pi)$, we have

$$\varphi_{\widetilde{v},v}(zh) = \omega_\pi(z)\varphi_{\widetilde{v},v}(h) \quad (z \in Z,\ h \in H) .$$

If $(\mathcal{V}, \pi)$ is admissible and if $\varphi_{\widetilde{v},v}$ is a matrix coefficient of $(\mathcal{V}, \pi)$, $h \longmapsto \varphi_{\widetilde{v},v}(h^{-1})$ is a matrix coefficient of $(\widetilde{\mathcal{V}}, \widetilde{\pi})$.

A **quasicuspidal** (resp. **cuspidal**) representation of H is a smooth (resp. admissible) representation $(\mathcal{V}, \pi)$ of H such that each matrix coefficient $\varphi_{\widetilde{v},v}$ of $(\mathcal{V}, \pi)$ has compact support modulo Z (i.e. there exists a compact subset $\Omega_{\widetilde{v},v}$ of H such that $\text{Supp}(\varphi_{\widetilde{v},v}) \subset Z\Omega_{\widetilde{v},v} \subset H$). If $(\mathcal{V}, \pi)$ is cuspidal, the same is true for $(\widetilde{\mathcal{V}}, \widetilde{\pi})$ (see (D.2)).

LEMMA (D.4.1). — *Let $(\mathcal{V}, \pi) \in \text{ob } \text{Rep}_s(H)$. Then the following conditions are equivalent:*

(i) *$(\mathcal{V}, \pi)$ is quasicuspidal,*

(ii) *for any $v \in \widetilde{V}$ and any compact open subgroup I of H, the set*

$$K_{v,I} = \{h \in H \mid \pi(e_I)\pi(h)(v) \neq 0\}$$

is compact modulo Z.

Proof: For any $v \in \mathcal{V}$, any compact open subgroup I of H and any $\widetilde{v} \in \widetilde{\mathcal{V}}^I$, we have

$$\text{Supp}(\varphi_{\widetilde{v},v}) \subset K_{v,I} .$$

Indeed, we have

$$< \widetilde{v}, \pi(e_I)\pi(h)(v) > = < \widetilde{v}, \pi(h)(v) > \neq 0$$

for any $h \in \text{Supp}(\varphi_{\widetilde{v},v})$. Therefore, (ii) implies (i).

Conversely, let us assume that $(\mathcal{V}, \pi)$ is quasicuspidal and let I be a compact open subgroup of H. For any $v \in \mathcal{V}$, let $\mathcal{V}_{v,I}$ be the linear span of the vectors $\pi(e_I)\pi(h)(v)$ $(h \in H^1)$ in $\mathcal{V}^I \subset \mathcal{V}$. Then we have

$$\dim_{\mathbb{C}}(\mathcal{V}_{v,I}) < +\infty .$$

Indeed, if this is not the case, we can find a sequence $(h_n)_{n \in \mathbb{Z}_{>0}}$ of elements in H^1 such that the vectors $\pi(e_I)\pi(h_n)(v)$ $(n \in \mathbb{Z}_{>0})$ are linearly independent. Let $\mathcal{W} \subset \mathcal{V}^I$ be an arbitrary $\mathbb{C}$-vector subspace such that $\mathcal{V}^I$ is the direct sum of $\mathcal{W}$ and of the $\mathbb{C}$-linear span of the vectors $\pi(e_I)\pi(h_n)(v)$ $(n \in \mathbb{Z}_{>0})$. As we have a canonical splitting

$$\mathcal{V} = \mathcal{V}^I \oplus \text{Ker}(\pi(e_I)) ,$$

we can define $\widetilde{v} \in \mathcal{V}'$ by

$$< \widetilde{v}, \pi(e_I)\pi(h_n)(v) > = n$$

for any $n \in \mathbf{Z}_{>0}$ and by

$$\tilde{v} \mid (\mathcal{W} \oplus \mathrm{Ker}(\pi(e_I))) = 0 \ .$$

Obviously, we have

$$\tilde{v} \in (\mathcal{V}')^I \subset \tilde{\mathcal{V}} \subset \mathcal{V}'$$

and

$$\varphi_{\tilde{v},v}(h_n) = n$$

for each $n \in \mathbf{Z}_{>0}$. Therefore, $\mathrm{Supp}(\varphi_{\tilde{v}v}) \cap H^1$ cannot be compact as it must be (thanks to our second hypothesis on H, $Z\Omega_{\tilde{v},v} \cap H^1$ is compact) and our assertion is proved. Now we can choose functionals $\tilde{v}_1, \ldots, \tilde{v}_m$ in $\tilde{\mathcal{V}}^I = (\mathcal{V}^I)'$ which separate the vectors in the finite dimensional subspace $\mathcal{V}_{v,I}$ of $\mathcal{V}^I$. Then we have

$$K^1_{v,I} \overset{\mathrm{dfn}}{=\!=\!=} K_{v,I} \cap H^1 \subset \bigcup_{i=1}^{m} \mathrm{Supp}(\varphi_{\tilde{v}_i,v}) \cap H^1$$

and we get that $K^1_{v,I}$ is compact for any $v \in \mathcal{V}$. Finally, if $(h_1, \ldots, h_n)$ is a system of representatives of the classes in $ZH^1 \backslash H$, for any $v \in \mathcal{V}$, it is easy to see that

$$K_{v,I} \subset Z\left(\bigcup_{j=1}^{n} K^1_{\pi(h_j)(v),I} h_j \right)$$

and it follows that $K_{v,I}$ is compact modulo I.   □

PROPOSITION (D.4.2). — *If $(\mathcal{V}, \pi)$ is a quasicuspidal irreducible representation of H, then $(\mathcal{V}, \pi)$ is admissible and therefore cuspidal.*

Proof : Let us fix $v \in \mathcal{V} - \{0\}$. Then v generates $\mathcal{V}$. Therefore, for any compact open subgroup I of H, $\mathcal{V}^I = \mathrm{Im}(\pi(e_I))$ is the $\mathbf{C}$-linear span of the vectors $\pi(e_I)\pi(h)(v)$ ($h \in H$). In other words, if $(h_1, \ldots, h_n)$ is a system of representatives of the classes in $ZH^1 \backslash H$, $\mathcal{V}^I$ is the $\mathbf{C}$-linear span of its subspaces $\mathcal{V}_{\pi(h_j)(v),I}$ ($j = 1, \ldots, n$) with the notations of the proof of (D.4.1). Indeed, thanks to Schur's lemma (D.1.12), for any $w \in \mathcal{V}$, $\mathcal{V}_{w,I}$ is also the $\mathbf{C}$-linear span of the vectors $\pi(h)(w)$ for $h \in ZH^1$. This implies the finite dimensionality of $\mathcal{V}^I$.   □

Let $(\mathcal{V}, \pi)$ be an admissible representation of H. Then there is a natural left action σ of $H \times H$ on $\mathrm{End}_\mathbf{C}(\mathcal{V})$: if $h_1, h_2 \in H$ and $\nu \in \mathrm{End}_\mathbf{C}(\mathcal{V})$ it is defined by

$$\sigma(h_1, h_2)(\nu) = \pi(h_2) \circ \nu \circ \pi(h_1^{-1}) \ .$$

The smooth representation $(\mathrm{End}_\mathbf{C}(\mathcal{V})^\infty, \sigma^\infty)$ of $H \times H$ is admissible. Indeed, if I is a compact open subgroup of H,

$$\mathrm{End}_\mathbf{C}(\mathcal{V})^{I \times I} \subset \mathrm{End}_\mathbf{C}(\mathcal{V})^\infty$$

is canonically isomorphic to $\mathrm{End}_{\mathbb{C}}(\mathcal{V}^I)$ (if $\nu \in \mathrm{End}_{\mathbb{C}}(\mathcal{V})^{I \times I}$, ν admits the factorization

$$\mathcal{V} \xrightarrow{\ \pi(e_I)\ } \mathcal{V}^I \xrightarrow{\ \nu^I\ } \mathcal{V}^I \longleftrightarrow \mathcal{V}) .$$

We can also consider the smooth representation

$$(\widetilde{\mathcal{V}} \otimes_{\mathbb{C}} \mathcal{V}, \widetilde{\pi} \otimes_{\mathbb{C}} \pi) = (\widetilde{\mathcal{V}}, \widetilde{\pi}) \otimes_{\mathbb{C}} (\mathcal{V}, \pi)$$

of $H \times H$. Again it is admissible. Indeed, for any compact open subgroup I of H,

$$(\widetilde{\mathcal{V}} \otimes_{\mathbb{C}} \mathcal{V})^{I \times I} = \widetilde{\mathcal{V}}^I \otimes_{\mathbb{C}} \mathcal{V}^I$$

is canonically isomorphic to

$$(\mathcal{V}^I)' \otimes_{\mathbb{C}} \mathcal{V}^I .$$

We have a morphism

(D.4.3) $\qquad \alpha : (\widetilde{\mathcal{V}} \otimes_{\mathbb{C}} \mathcal{V}, \widetilde{\pi} \otimes_{\mathbb{C}} \pi) \longrightarrow (\mathrm{End}_{\mathbb{C}}(\mathcal{V})^{\infty}, \sigma^{\infty})$

in $\mathrm{Rep}_a(H \times H)$ given by

$$\alpha(\widetilde{v} \otimes v)(v_1) = <\widetilde{v}, v_1 > v$$

for any v, $v_1 \in \mathcal{V}$ and any $\widetilde{v} \in \widetilde{\mathcal{V}}$.

LEMMA (D.4.4). — *For each* $(\mathcal{V}, \pi) \in \mathrm{ob}\ \mathrm{Rep}_a(H)$, α *is an isomorphism.*

Proof : For any compact open subgroup I of H, $\alpha^{I \times I}$ is injective. Indeed, we can identify $\alpha^{I \times I}$ with the map

$$(\mathcal{V}^I)' \otimes_{\mathbb{C}} \mathcal{V}^I \longrightarrow \mathrm{End}_{\mathbb{C}}(\mathcal{V}^I),$$

$$v' \otimes v \longmapsto (v_1 \longmapsto < v', v_1 > v)$$

and $\mathcal{V}^I$ is finite dimensional over $\mathbb{C}$. $\qquad\qquad\square$

We also have a morphism

(D.4.5) $\qquad \beta : (\mathrm{End}_{\mathbb{C}}(\mathcal{V})^{\infty}, \sigma^{\infty}) \longrightarrow ((\mathbb{C} \otimes \mathcal{C}^{\infty}(H))^{\infty}, \rho^{\infty})$

of smooth representations of $H \times H$, where ρ is the action of $H \times H$ on $\mathbb{C} \otimes \mathcal{C}^{\infty}(H)$ defined by

$$(\rho(h_1, h_2)(f))(h) = f(h_1^{-1} h h_2)$$

for any h, h_1, $h_2 \in H$ and any $f \in \mathbb{C} \otimes C_c^\infty(H)$. This morphism β is defined by

$$\beta(\nu)(h) = \mathrm{tr}(\pi(h) \circ \nu) = \mathrm{tr}(\nu \circ \pi(h))$$

for any $\nu \in \mathrm{End}_{\mathbb{C}}(\mathcal{V})^\infty$ and any $h \in H$ (for any $\nu_1 \in \mathrm{End}_{\mathbb{C}}(\mathcal{V})^\infty$, there exists a compact open subgroup I of H such that $\nu_1 \in \mathrm{End}_{\mathbb{C}}(\mathcal{V})^{I \times I}$ and the trace

$$\mathrm{tr}(\nu_1) \overset{\mathrm{dfn}}{=\!=} \mathrm{tr}(\nu_1^I : \mathcal{V}^I \longrightarrow \mathcal{V}^I)$$

is well-defined).

It is easy to check that, for any $v \in \mathcal{V}$ and $\widetilde{v} \in \widetilde{\mathcal{V}}$, we have

(D.4.6) $$\beta \circ \alpha(\widetilde{v} \otimes v) = \varphi_{\widetilde{v},v} .$$

In particular, $\beta(\mathrm{End}_{\mathbb{C}}(\mathcal{V}^\infty)) \subset (\mathbb{C} \otimes C^\infty(H))^\infty$ is the $\mathbb{C}$-linear span of the matrix coefficients of $(\mathcal{V}, \pi)$.

PROPOSITION (D.4.7). — *Let us assume that Z is compact and let $(\mathcal{V}, \pi)$ be a cuspidal irreducible representation of H. Then, for any $f \in \mathbb{C} \otimes C_c^\infty(H)$, there exists a unique*

$$f_\pi \in \mathbb{C} \otimes C_c^\infty(H)$$

such that

$$\pi(f_\pi) = \pi(f) \quad (\text{in } \mathrm{End}_{\mathbb{C}}(\mathcal{V}))$$

and

$$\pi_1(f_\pi) = 0 \quad (\text{in } \mathrm{End}_{\mathbb{C}}(\mathcal{V}_1))$$

for any irreducible $(\mathcal{V}_1, \pi_1) \in \mathrm{ob}\,\mathrm{Rep}_s(H)$ which is not isomorphic to $(\mathcal{V}, \pi)$. Moreover, f_π^* is a $\mathbb{C}$-linear combination of matrix coefficients of $(\mathcal{V}, \pi)$. Here, we have set

$$g^*(h) = g(h^{-1}) \quad (h \in H)$$

for any $g \in \mathbb{C} \otimes C_c^\infty(H)$.

The uniqueness of f_π immediately follows from

LEMMA (D.4.8). — *Let $f \in \mathbb{C} \otimes C_c^\infty(H)$ not be identically zero. Then there exists at least one irreducible $(\mathcal{V}, \pi) \in \mathrm{ob}\,\mathrm{Rep}_s(H)$ such that $\pi(f) \neq 0$ (in $\mathrm{End}_{\mathbb{C}}(\mathcal{V})$).*

Proof: If $g \in \mathbb{C} \otimes C_c^\infty(H)$, we have

$$(g^* * g)(1) = \int_H |g(h)|^2 dh$$

and, if g is not identically zero, the same is true for $g^* * g$.

Let us denote by f_n the 2^n-th power of $f_0 = f^* * f$ for the convolution product. By induction on n, it is now easy to see that $f_n^* = f_n$, that

$$f_{n+1} = f_n^* * f_n$$

and that f_n is not identically zero for each non-negative integer. In other words, f_0 is not nilpotent in $\mathbb{C} \otimes \mathcal{C}_c^\infty(H)$.

Let us fix a compact open subgroup I of H such that $f \in \mathbb{C} \otimes \mathcal{C}_c^\infty(H//I)$. Then we also have f^*, $f_0 \in \mathbb{C} \otimes \mathcal{C}_c^\infty(H//I)$ and f_0 is not nilpotent in $\mathbb{C} \otimes \mathcal{C}_c^\infty(H//I)$. But the $\mathbb{C}$-algebra $\mathbb{C} \otimes \mathcal{C}_c^\infty(H//I)$ has countable dimension over $\mathbb{C}$ (H is countable at infinity), so there exists at least one maximal ideal $\mathfrak{m}$ of $\mathbb{C} \otimes \mathcal{C}_c^\infty(H//I)$ such that $f_0 \notin \mathfrak{m}$ (see [Jac], Ch. I, § 10, Thm. 2). If we fix such an $\mathfrak{m}$, $(\mathbb{C} \otimes \mathcal{C}_c^\infty(H//I))/\mathfrak{m}$ is an irreducible left $(\mathbb{C} \otimes \mathcal{C}_c^\infty(H//I))$-module on which the actions of f_0 and f are non-trivial. Now the lemma follows from (D.1.8). $\square$

Proof of (D.4.7) : The subspace $\mathbb{C} \otimes \mathcal{C}_c^\infty(H)$ of $(\mathbb{C} \otimes \mathcal{C}^\infty(H))^\infty$ is stable under ρ^∞. Let us denote by r the restriction of ρ^∞ to this subspace, so that $(\mathbb{C} \otimes \mathcal{C}_c^\infty(H), r)$ is a smooth representation of $H \times H$.

As Z is compact and $(\mathcal{V}, \pi)$ is quasicuspidal, we have

$$\beta(\mathrm{End}_{\mathbb{C}}(\mathcal{V})^\infty) \subset \mathbb{C} \otimes \mathcal{C}_c^\infty(H) \subset (\mathbb{C} \otimes \mathcal{C}^\infty(H))^\infty$$

(see (D.4.4) and (D.4.6)). Moreover, as $(\mathcal{V}, \pi)$ is admissible and irreducible, the two isomorphic representations $(\widetilde{\mathcal{V}} \otimes_{\mathbb{C}} \mathcal{V}, \widetilde{\pi} \otimes_{\mathbb{C}} \pi)$ and $(\mathrm{End}_{\mathbb{C}}(\mathcal{V})^\infty, \sigma^\infty)$ of $H \times H$ are irreducible, so that β is an embedding (β is not identically zero as there exist $v \in \mathcal{V}$ and $\widetilde{v} \in \widetilde{\mathcal{V}}$ such that $< \widetilde{v}, v > \neq 0$ and as $\varphi_{\widetilde{v}, v}(1) = < \widetilde{v}, v >$, see (D.4.4) and (D.4.6)).

Now let us consider the morphism

$$\beta' : (\mathbb{C} \otimes \mathcal{C}_c^\infty(H), r) \longrightarrow (\mathrm{End}_{\mathbb{C}}(\mathcal{V})^\infty, \sigma^\infty),$$

$$g \longmapsto \pi(g^*).$$

Then $\beta' \circ \beta$ is an endomorphism of the irreducible representation $(\mathrm{End}_{\mathbb{C}}(\mathcal{V})^\infty, \sigma^\infty)$ of $H \times H$. Therefore, $\beta' \circ \beta$ is a scalar operator (see (D.1.12)). Let λ be the corresponding scalar. We will show that $\lambda \neq 0$ and that we can take

$$f_\pi = \lambda^{-1}(\beta \circ \beta'(f^*))^* .$$

Let $(\mathcal{V}_1, \pi_1)$ be a smooth irreducible representation of H and let $v_1 \in \mathcal{V}_1 - \{0\}$. Let us denote by

$$\gamma_1 : (\mathbb{C} \otimes \mathcal{C}_c^\infty(H), r \mid \{1\} \times H) \longrightarrow (\mathcal{V}_1, \pi_1)$$

the morphism in $\mathrm{Rep}_s(H)$ defined by

$$\gamma_1(g) = \pi_1(g^*)(v) \ .$$

As the smooth representation $(\mathrm{End}_{\mathbb{C}}(\mathcal{V})^\infty, \rho^\infty \mid \{1\} \times H)$ of H is isomorphic to $\widetilde{\mathcal{V}} \otimes_{\mathbb{C}} (\mathcal{V}, \pi)$, it is isomorphic to a direct sum of copies of $(\mathcal{V}, \pi)$. The same is true for the subrepresentation

$$\gamma_1(\beta(\mathrm{End}_{\mathbb{C}}(\mathcal{V})^\infty, \rho^\infty \mid \{1\} \times H))$$

of $(\mathcal{V}_1, \pi_1)$. Therefore,

$$\gamma_1(\beta(\mathrm{End}_{\mathbb{C}}(\mathcal{V})^\infty)) = (0)$$

unless $(\mathcal{V}_1, \pi_1)$ is isomorphic to $(\mathcal{V}, \pi)$. Letting v_1 vary in $\mathcal{V}_1$, we get that

$$\pi_1((\beta \circ \beta'(g))^*) = 0$$

for any $g \in \mathbb{C} \otimes \mathcal{C}_c^\infty(H)$ unless $(\mathcal{V}_1, \pi_1)$ is isomorphic to $(\mathcal{V}, \pi)$. Furthermore, we have

$$\pi((\beta \circ \beta'(g))^*) = \beta' \circ \beta \circ \beta'(g) = \lambda \beta'(g) = \lambda \pi(g^*)$$

for any $g \in \mathbb{C} \otimes \mathcal{C}_c^\infty(H)$. It follows that λ cannot be zero. Indeed, there exists $g \in \mathbb{C} \otimes \mathcal{C}_c^\infty(H)$ such that $\beta'(g) = \pi(g^*) \neq 0$. As β is an embedding, $(\beta \circ \beta'(g))^* \neq 0$ and $\lambda = 0$ would contradict (D.4.8). Now it is obvious that $f_\pi = \lambda^{-1}(\beta \circ \beta'(f^*))^*$ has the required properties.   □

THEOREM (D.4.9). — *Let $(\mathcal{W}, \rho)$ be a cuspidal irreducible representation of H and let $(\mathcal{V}, \pi)$ be a smooth representation of H. Then there exists a canonical decomposition into a direct sum of two smooth representations of H,*

$$(\mathcal{V}, \pi) = (\mathcal{V}_\rho, \pi_\rho) \oplus (\mathcal{V}^\rho, \pi^\rho) \ ,$$

where $(\mathcal{V}_\rho, \pi_\rho)$ is quasicuspidal, where each irreducible subquotient of $(\mathcal{V}_\rho, \pi_\rho)$ is isomorphic to $(\mathcal{W}, \chi\rho)$ for some character $\chi : H \longrightarrow \mathbb{C}^\times$ trivial on $H^1 \subset H$ and where no irreducible subquotient of $(\mathcal{V}^\rho, \pi^\rho)$ is isomorphic to $(\mathcal{W}, \chi\rho)$ ($\chi : H \longrightarrow \mathbb{C}^\times$ a character trivial on $H^1 \subset H$).

Moreover, $(\mathcal{V}_\rho, \pi_\rho \mid H^1)$ is completely reducible.

We will give the proof of this theorem in two steps.

First step : proof of (D.4.9) *when Z is compact and $H^1 = H$.* For each compact open subgroup I of H, we can apply (D.4.7) to $(\mathcal{W}, \rho)$ and

$$f = e_I = 1_I/\mathrm{vol}(I, dh) \ .$$

We get a function

$$e_{I,\rho} \in \mathbb{C} \otimes \mathcal{C}_c^\infty(H) \ .$$

The uniqueness statement of (D.4.7) implies that

$$e_{I',\rho} * e_{I,\rho} = e_I * e_{I',\rho} = e_{I',\rho} * e_I = e_{I,\rho}$$

for any two compact open subgroups $I' \subset I$ of H and that

$$\delta_h * e_{I,\rho} * \delta_{h^{-1}} = e_{hIh^{-1},\rho}$$

for any $h \in H$ and any compact open subgroup I of H (recall that

$$(\delta_{h_1} * f * \delta_{h_2})(h') = f(h_1^{-1} h' h_2^{-1})$$

for any h_1, h_2, $h' \in H$ and any $f \in \mathbb{C} \otimes C_c^\infty(H)$).

Now it is clear that, for a fixed $v \in \mathcal{V}$, $\pi(e_{I,\rho})(v)$ is independent of I for a small enough compact open subgroup I of H. Let us denote by

$$\pi(e_\rho)(v) \in \mathcal{V}$$

the common value of $\pi(e_{I,\rho})(v)$ for small enough I's. Then

$$\pi(e_\rho) : (\mathcal{V}, \pi) \longrightarrow (\mathcal{V}, \pi)$$

is an idempotent in $\operatorname{End}_{\operatorname{Rep}_s(H)}(\mathcal{V}, \pi)$. Moreover, if $(\mathcal{V}', \pi')$ is another smooth representation of H and if $\nu : (\mathcal{V}, \pi) \longrightarrow (\mathcal{V}', \pi')$ is a morphism in $\operatorname{Rep}_s(H)$, we have

$$\nu \circ \pi(e_\rho) = \pi'(e_\rho) \circ \nu .$$

Let us set

$$(\mathcal{V}_\rho, \pi_\rho) = \operatorname{Im}(\pi(e_\rho))$$

and

$$(\mathcal{V}^\rho, \pi^\rho) = \operatorname{Ker}(\pi(e_\rho))$$

in $\operatorname{Rep}_s(H)$. Obviously, we have a direct sum decomposition

$$(\mathcal{V}, \pi) = (\mathcal{V}_\rho, \pi_\rho) \oplus (\mathcal{V}^\rho, \pi^\rho) .$$

We have seen in the proof of (D.4.7) that $(\operatorname{End}_{\mathbb{C}}(\mathcal{W})^\infty, \sigma^\infty \mid \{1\} \times H)$ is a direct sum of copies of $(\mathcal{W}, \rho)$ and that

$$e_{I,\rho} = \lambda^{-1}(\beta(\beta'(e_I^*)))^* \in \beta(\operatorname{End}_{\mathbb{C}}(\mathcal{W})^\infty)^*$$

for any compact open subgroup I of H. It follows that $(\mathcal{V}_\rho, \pi_\rho)$ is contained in the image of the morphism

$$(\operatorname{End}_{\mathbb{C}}(\mathcal{W})^\infty, \sigma^\infty \mid \{1\} \times H) \otimes_{\mathbb{C}} \mathcal{V} \longrightarrow (\mathcal{V}, \pi),$$

$$\nu \otimes v \longmapsto \pi(\beta(\nu)^*)(v)$$

in $\text{Rep}_s(H)$ and that $(\mathcal{V}_\rho, \pi_\rho)$ is therefore a direct sum of copies of $(\mathcal{W}, \rho)$.
If $(\mathcal{V}^\rho, \pi^\rho)$ contained subrepresentations

$$(\mathcal{V}', \pi') \subset (\mathcal{V}'', \pi'') \subset (\mathcal{V}, \pi)$$

such that $(\mathcal{V}'', \pi'')/(\mathcal{V}', \pi')$ is isomorphic to $(\mathcal{W}, \rho)$, we would have

$$\pi''(e_\rho) = \pi^\rho(e_\rho) \mid \mathcal{V}'' = 0$$

and

$$\rho(e_\rho) = \pi''(e_\rho) \text{ (modulo } \mathcal{V}') = 0 .$$

But it is clear that $\rho(e_\rho) \neq 0$ ($\mathcal{W}^I \neq (0)$ for a small enough compact open subgroup I of H) and we would have a contradiction. Therefore, no irreducible subquotient of $(\mathcal{V}^\rho, \pi^\rho)$ is isomorphic to $(\mathcal{W}, \rho)$. This completes the proof of the theorem when Z is compact and $H^1 = H$.   □

In the second step we will need the following lemma.

LEMMA (D.4.10). — *Let $(\mathcal{V}, \pi)$ be a smooth irreducible representation of H. Then $(\mathcal{V}, \pi \mid H^1)$ is the direct sum of finitely many smooth irreducible representations of H^1.*

Moreover, if $(\mathcal{V}', \pi')$ is another smooth irreducible representation of H such that at least one irreducible subquotient of $(\mathcal{V}', \pi' \mid H^1)$ is isomorphic to an irreducible subquotient of $(\mathcal{V}', \pi' \mid H^1)$, there exists a unique character $\chi : H \longrightarrow \mathbb{C}^\times$ which is trivial on H^1 such that $(\mathcal{V}', \pi')$ is isomorphic to $(\mathcal{V}, \chi\pi)$.

Proof : The subgroup ZH^1 of H is normal and of finite index. Let $h_1, \ldots, h_n$ be a system of representatives of the classes in H/ZH^1. Let us fix $v \in \mathcal{V} - \{0\}$. The non-degenerate $(\mathbb{C} \otimes \mathcal{C}_c^\infty(H))$-module $\mathcal{V}$ is generated by v (it is irreducible). Therefore, the non-degenerate $(\mathbb{C} \otimes \mathcal{C}^\infty(ZH^1))$-module is finitely generated (it is generated by $\pi(h_1)(v), \ldots, \pi(h_n)(v)$). Thanks to Zorn's lemma, the finitely generated non-degenerate $(\mathbb{C} \otimes \mathcal{C}_c^\infty(ZH^1))$-module $\mathcal{V}$ admits at least one irreducible quotient $\mathcal{V}_2 = \mathcal{V}/\mathcal{V}_1$. Let $(\mathcal{V}_2, \pi_2)$ be the corresponding quotient of $(\mathcal{V}, \pi|ZH^1)$ in $\text{Rep}_s(ZH^1)$. For each $i = 1, \ldots, n$, let σ_i be the automorphism of ZH^1 induced by the conjugation by h_i in H. Then $(\mathcal{V}_2, \pi_2 \circ \sigma_i)$ is an irreducible quotient of $(\mathcal{V}, (\pi|ZH^1) \circ \sigma_i)$ for $i = 1, \ldots, n$. But $\pi(h_i)$ induces an isomorphism from $(\mathcal{V}, \pi|ZH^1)$ onto $(\mathcal{V}, (\pi|ZH^1) \circ \sigma_i)$ and $(\mathcal{V}_2, \pi_2 \circ \sigma_i)$ is also an irreducible quotient of $(\mathcal{V}, \pi|ZH^1)$ for $i = 1, \ldots, n$. Let

$$\nu : (\mathcal{V}, \pi|ZH^1) \longrightarrow \bigoplus_{i=1}^n (\mathcal{V}_2, \pi_2 \circ \sigma_i)$$

be the sum of these quotient maps. It is not difficult to see that

$$\pi(H)(\operatorname{Ker}\nu) \subset \operatorname{Ker}\nu$$

($\pi(h_i)(\operatorname{Ker}\nu) \subset \operatorname{Ker}\nu$ for $i = 1, \ldots, n$). As $\nu \neq 0$ and $(\mathcal{V}, \pi)$ is irreducible, it follows that ν is an embedding. This implies that $(\mathcal{V}, \pi|ZH^1)$ is a direct sum of finitely many smooth irreducible representations of ZH^1. Now the restriction to H^1 of any smooth irreducible representation of ZH^1 is automatically irreducible, thanks to Schur's lemma (see (D.1.12)), and the first assertion of the lemma is proved.

Taking into account the first part of the lemma, the hypothesis on $(\mathcal{V}', \pi'|H^1)$ in the second part is equivalent to the non-vanishing of the finite dimensional $\mathbb{C}$-vector space

$$\mathcal{W} \overset{\mathrm{dfn}}{=\!=} \operatorname{Hom}_{\mathrm{Rep}_s(H^1)}((\mathcal{V}', \pi'|H^1), \ (\mathcal{V}, \pi|H^1)).$$

On $\mathcal{W}$ we have an action ρ of H/H^1 which is defined by

$$\rho(hH^1)(w) = \pi(h) \circ w \circ \pi'(h^{-1}) .$$

As H/H^1 is commutative, there exist $w \in \mathcal{W} - \{0\}$ and a character $\chi : H \longrightarrow \mathbb{C}^\times$ which is trivial on H^1 such that

$$\rho(hH^1)(w) = \chi(h)w ,$$

i.e.

$$w \circ \pi'(h) = (\chi\pi)(h) \circ w ,$$

for all $h \in H$. But this means that

$$w \in \operatorname{Hom}_{\mathrm{Rep}_s(H)}((\mathcal{V}', \pi'), (\mathcal{V}, \chi\pi))$$

and the second part of the lemma follows ($(\mathcal{V}, \chi\pi)$ and $(\mathcal{V}', \pi')$ are irreducible). $\qquad\square$

Second step: proof of (D.4.9) *for a general* H. Let $(\mathcal{W}_i, \rho_i)$ $(i = 1, \ldots, n)$ be the irreducible representations of H^1 occurring in $(\mathcal{W}, \rho|H^1)$ (see (D.4.10)). As H^1 is open and therefore closed in H, $(\mathcal{W}_i, \rho_i)$ is cuspidal for $i = 1, \ldots, n$. Applying n times the particular case of (D.4.9) that we have just proved to H^1, we get that

$$(\mathcal{V}, \pi|H^1) = (\mathcal{V}_\rho, (\pi|H^1)_\rho) \oplus (\mathcal{V}^\rho, (\pi|H^1)^\rho)$$

where $(\mathcal{V}_\rho, (\pi|H^1)_\rho)$ splits into a direct sum of irreducible representations of H^1 isomorphic to $(\mathcal{W}_i, \rho_i)$ for some $i = 1, \ldots, n$ and where no irreducible subquotient of $(\mathcal{V}^\rho, (\pi|H^1)^\rho)$ is isomorphic to $(\mathcal{W}_i, \rho_i)$ $(i = 1, \ldots, n)$.

For any $h \in H$, $\pi(h)$ maps any irreducible subquotient of $(\mathcal{V}, \pi|H^1)$ which is isomorphic to $(\mathcal{W}_i, \rho_i)$ for some $i = 1, \ldots, n$ onto a subquotient of $(\mathcal{V}, \pi|H^1)$ which is isomorphic to $(\mathcal{W}_i, \rho_i \circ \sigma_{h^{-1}})$ where $\sigma_{h^{-1}}$ is the automorphism of H^1 induced by the conjugation by h^{-1} in H. But $(\mathcal{W}_i, \rho_i \circ \sigma_{h^{-1}})$ is isomorphic to $(\mathcal{W}_j, \rho_j)$ for some $j = 1, \ldots, n$ (see the proof of (D.4.10)). Therefore, $\pi(h)$ maps $\mathcal{V}_\rho$ into $\mathcal{V}_\rho$ and $\mathcal{V}^\rho$ into $\mathcal{V}^\rho$. Let us denote by π_ρ and π^ρ the restrictions of π to $\mathcal{V}_\rho$ and $\mathcal{V}^\rho$ respectively. We have

$$(\mathcal{V}, \pi) = (\mathcal{V}_\rho, \pi_\rho) \oplus (\mathcal{V}^\rho, \pi^\rho)$$

and $\pi_\rho|H^1 = (\pi|H^1)_\rho$, $\pi^\rho|H^1 = (\pi|H^1)^\rho$. It follows immediately from the second part of (D.4.10) that this is the desired decomposition.   □

COROLLARY (D.4.11). — *Let $(\mathcal{W}, \rho)$ be a cuspidal irreducible representation of H and let $(\mathcal{V}, \pi)$ be a smooth representation of H. We assume that either $(\mathcal{V}, \pi)$ admits a central character or $(\mathcal{V}, \pi)$ is of finite length in $\mathrm{Rep}_s(H)$. Then, if $(\mathcal{W}, \rho)$ is a subquotient of $(\mathcal{V}, \pi)$, there exists a subrepresentation (resp. a quotient) of $(\mathcal{V}, \pi)$ which is isomorphic to $(\mathcal{W}, \rho)$.*

Proof : Thanks to (D.4.9), we can assume that $(\mathcal{V}, \pi) = (\mathcal{V}_\rho, \pi_\rho)$, so that $(\mathcal{V}, \pi|H^1)$ is completely reducible.

If $(\mathcal{V}, \pi)$ admits a central character, then $(\mathcal{V}, \pi|ZH^1)$ is also completely reducible. As ZH^1 is of finite index in H, $(\mathcal{V}, \pi)$ is also completely reducible. Indeed, let $(\mathcal{V}', \pi')$ be a subrepresentation of $(\mathcal{V}, \pi)$. We can find an idempotent ν in $\mathrm{End}_{\mathrm{Rep}_s(ZH^1)}(\mathcal{V}, \pi|ZH^1)$ such that $(\mathcal{V}', \pi'|ZH^1) = \mathrm{Im}(\nu)$. Let $h_1, \ldots, h_n$ be a system of representatives of the classes in H/ZH^1. Then the endomorphism

$$\bar{\nu} = \frac{1}{n} \sum_{i=1}^n \pi(h_i) \circ \nu \circ \pi(h_i^{-1})$$

is again an idempotent of $\mathrm{End}_{\mathrm{Rep}_s(ZH^1)}(\mathcal{V}, \pi|ZH^1)$ such that $\mathrm{Im}(\bar{\nu}) = (\mathcal{V}', \pi'|ZH')$. But it is easy to see that $\bar{\nu}$ commutes with $\pi(h)$ for any $h \in H$ and $(\mathcal{V}', \pi')$ is a direct summand of $(\mathcal{V}, \pi)$ as required. Therefore, the corollary is proved in that case.

If $(\mathcal{V}, \pi)$ is of finite length, we carry out the proof by induction on the length of $(\mathcal{V}, \pi)$. If $(\mathcal{V}, \pi)$ admits a central character, we can apply the previous argument. So we can assume that there exists $z \in Z$ such that

$$\pi(z) \neq \omega_\rho(z)\mathrm{id}_{\mathcal{V}}$$

where ω_ρ is the central character of $(\mathcal{W}, \rho)$ (see (D.1.12)). Let

$$\nu = \pi(z) - \omega_\rho(z)\mathrm{id}_{\mathcal{V}} \in \mathrm{End}_{\mathrm{Rep}_s(H)}(\mathcal{V}, \pi) .$$

Then ν is not identically zero and is not an isomorphism $((\mathcal{W}, \rho)$ is a subquotient of $(\mathcal{V}, \pi))$. Let us set

$$(\mathcal{V}', \pi') = \mathrm{Ker}(\nu),$$

$$(\mathcal{V}'', \pi'') = \mathrm{Coim}(\nu) = \mathrm{Im}(\nu)$$

and

$$(\mathcal{V}''', \pi''') = \mathrm{Coker}(\nu) .$$

These three smooth representations of H are of finite length and their lengths are strictly smaller than the length of $(\mathcal{V}, \pi)$. Moreover, $(\mathcal{W}, \rho)$ is a subquotient of $(\mathcal{V}', \pi')$ or $(\mathcal{V}'', \pi'')$ (resp. $(\mathcal{V}'', \pi'')$ or $(\mathcal{V}''', \pi''')$). Therefore, by induction hypothesis, $(\mathcal{W}, \rho)$ is a subrepresentation (resp. a quotient) of $(\mathcal{V}', \pi')$ or $(\mathcal{V}'', \pi'')$ (resp. $(\mathcal{V}'', \pi'')$ or $(\mathcal{V}''', \pi''')$) and this completes the proof of the corollary. $\qquad \square$

(D.5) Injective and projective objects in $\mathrm{Rep}_s(H)$; cohomology

We have seen that the abelian category $\mathrm{Rep}_s(H)$ is equivalent to the abelian category $\mathrm{Mod}_{nd}(\mathcal{C}_c^\infty(H))$ of non-degenerate left $(\mathbb{C} \otimes \mathcal{C}_c^\infty(H))$-modules (see (D.1.2)). Using this equivalence of categories, let us prove

LEMMA (D.5.1). — *The abelian category* $\mathrm{Rep}_s(H)$ *has enough injective and projective objects.*

Proof: Let $\mathcal{V} \in \mathrm{ob}\ \mathrm{Mod}_{nd}(\mathcal{C}_c^\infty(H))$. Then the $\mathbb{C}$-vector space

$$\mathrm{Hom}_\mathbb{Q}(\mathcal{C}_c^\infty(H), \mathcal{V})$$

has a natural structure of a left $(\mathbb{C} \otimes \mathcal{C}_c^\infty(H))$-module: for $f, g \in \mathbb{C} \otimes \mathcal{C}_c^\infty(H)$ and $w \in \mathrm{Hom}_\mathbb{Q}(\mathcal{C}_c^\infty(H), \mathcal{V})$, $f \cdot w \in \mathrm{Hom}_\mathbb{Q}(\mathcal{C}_c^\infty(H), \mathcal{V})$ maps g onto

$$(f \cdot w)(g) = w(fg) .$$

This module can be degenerate but we can consider its maximal non-degenerate submodule

$$\mathcal{W} = \mathcal{C}_c^\infty(H) \cdot \mathrm{Hom}_\mathbb{Q}(\mathcal{C}_c^\infty(H), \mathcal{V})$$

(we have

$$\mathcal{C}_c^\infty(H) * \mathcal{C}_c^\infty(H) = \mathcal{C}_c^\infty(H)) .$$

The canonical $(\mathbb{C} \otimes \mathcal{C}_c^\infty(H))$-linear map

$$\iota : \mathcal{V} \longrightarrow \mathrm{Hom}_\mathbb{Q}(\mathcal{C}_c^\infty(H), \mathcal{V}),$$

$$v \longmapsto (f \longmapsto f \cdot v)$$

is injective and factors through $\mathcal{W}$ (for any $v \in \mathcal{V}$, there exists a compact open subgroup I of H such that $v = e_I \cdot v$). Moreover, $\mathcal{W}$ is an injective object in $\mathrm{Mod}_{nd}(\mathcal{C}_c^\infty(H))$. Indeed, for any $\mathcal{V}_1 \in \mathrm{ob}\, \mathrm{Mod}_{nd}(\mathcal{C}_c^\infty(H))$, the morphism

$$\mathcal{C}_c^\infty(H) \otimes_{\mathcal{C}_c^\infty(H)} \mathcal{V}_1 \longrightarrow \mathcal{V}_1, \quad f \otimes v_1 \longmapsto f \cdot v_1,$$

is an isomorphism (it is obviously surjective and, if $\sum_\alpha f_\alpha \otimes v_{1,\alpha} \in \mathcal{C}_c^\infty(H) \otimes_{\mathcal{C}_c^\infty(H)} \mathcal{V}_1$, there exists a compact open subgroup I of H such that $e_I * f_\alpha = f_\alpha$ for all α with $f_\alpha \neq 0$ or $v_{1,\alpha} \neq 0$, so that

$$\sum_\alpha f_\alpha \otimes v_{1,\alpha} = e_I \otimes \sum_\alpha f_\alpha \cdot v_{1,\alpha})$$

and we have canonical isomorphisms

$$\mathrm{Hom}_{\mathbb{C} \otimes \mathcal{C}_c^\infty(H)}(\mathcal{V}_1, \mathcal{W}) \xrightarrow{\sim} \mathrm{Hom}_{\mathbb{C} \otimes \mathcal{C}_c^\infty(H)}(\mathcal{V}_1, \mathrm{Hom}_{\mathbb{Q}}(\mathcal{C}_c^\infty(H), \mathcal{V}))$$

$$\xrightarrow{\sim} \mathrm{Hom}_{\mathbb{C}}(\mathcal{C}_c^\infty(H) \otimes_{\mathcal{C}_c^\infty(H)} \mathcal{V}_1, \mathcal{V});$$

therefore, $\mathrm{Hom}_{\mathbb{C} \otimes \mathcal{C}_c^\infty(H)}(\mathcal{V}_1, \mathcal{W})$ is canonically isomorphic to $\mathrm{Hom}_{\mathbb{C}}(\mathcal{V}_1, \mathcal{V})$ and, for any monomorphism $\mathcal{V}_1 \hookrightarrow \mathcal{V}_2$ in $\mathrm{Mod}_{nd}(\mathcal{C}_c^\infty(H))$, the induced map

$$\mathrm{Hom}_{\mathbb{C} \otimes \mathcal{C}_c^\infty(H)}(\mathcal{V}_2, \mathcal{W}) \longrightarrow \mathrm{Hom}_{\mathbb{C} \otimes \mathcal{C}_c^\infty(H)}(\mathcal{V}_1, \mathcal{W})$$

is surjective.

For each compact open subgroup I of H, the left ideal

$$\mathbb{C} \otimes \mathcal{C}_c^\infty(H/I) = (\mathbb{C} \otimes \mathcal{C}_c^\infty(H)) * e_I \subset \mathbb{C} \otimes \mathcal{C}_c^\infty(H)$$

is a projective object in $\mathrm{Mod}_{nd}(\mathcal{C}_c^\infty(H))$. Indeed, for any $\mathcal{V}_1 \in \mathrm{ob}\, \mathrm{Mod}_{nd}(\mathcal{C}_c^\infty(H))$, the $\mathbb{C}$-linear map

$$\mathrm{Hom}_{\mathcal{C}_c^\infty(H)}(\mathcal{C}_c^\infty(H/I), \mathcal{V}_1) \longrightarrow e_I \cdot \mathcal{V}_1, \quad \nu_1 \longmapsto \nu_1(e_I),$$

is an isomorphism; therefore, for any epimorphism $\mathcal{V}_2 \twoheadrightarrow \mathcal{V}_1$ in $\mathrm{Mod}_{nd}(\mathcal{C}_c^\infty(H))$, the induced map

$$\mathrm{Hom}_{\mathcal{C}_c^\infty(H)}(\mathcal{C}_c^\infty(H/I), \mathcal{V}_2) \longrightarrow \mathrm{Hom}_{\mathcal{C}_c^\infty(H)}(\mathcal{C}_c^\infty(H/I), \mathcal{V}_1)$$

is surjective (the functor $\mathcal{V}_1 \longmapsto e_I \cdot \mathcal{V}_1 = \mathcal{V}_1^I$ is exact on $\mathrm{Mod}_{nd}(\mathcal{C}_c^\infty(H))$, see (D.1.5)). Now let $\mathcal{V} \in \mathrm{ob}\, \mathrm{Mod}_{nd}(\mathcal{C}_c^\infty(H))$ and, for each $v \in \mathcal{V}$, let $I_v \subset H_v$ be a compact open subgroup, so that $e_{I_v} \cdot v = v$. Then, the $(\mathbb{C} \otimes \mathcal{C}_c^\infty(H))$-linear map

$$\bigoplus_{v \in \mathcal{V}} \mathbb{C} \otimes \mathcal{C}_c^\infty(H/I_v) \longrightarrow \mathcal{V}, \quad \bigoplus_v f_v \longmapsto \sum_v f_v \cdot v,$$

is surjective and its source is a projective object in $\mathrm{Mod}_{nd}(\mathcal{C}_c^\infty(H))$. $\qquad \square$

Let $\mathrm{Vect}(\mathbb{C})$ be the category of $\mathbb{C}$-vector spaces. As $\mathrm{Rep}_s(H)$ is abelian and $\mathbb{C}$-linear, for each non-negative integer, we have the n-th **Yoneda extension bifunctor**

$$(\mathrm{D.5.2}) \qquad \mathrm{Ext}_H^n(-,-) : \mathrm{Rep}_s(H)^{opp} \times \mathrm{Rep}_s(H) \longrightarrow \mathrm{Vect}(\mathbb{C}) .$$

PROPOSITION (D.5.3). — (i) *For each* $(\mathcal{V}_1, \pi_1) \in \mathrm{ob}\ \mathrm{Rep}_s(H)$, *the functors*

$$\mathrm{Rep}_s(H) \longrightarrow \mathrm{Vect}(\mathbb{C}),\ (\mathcal{V}_2, \pi_2) \longmapsto \mathrm{Ext}_H^n((\mathcal{V}_1, \pi_1), (\mathcal{V}_2, \pi_2))$$

$(n \in \mathbb{N})$ *are the derived functors of the left exact functor*

$$\mathrm{Rep}_s(H) \longrightarrow \mathrm{Vect}(\mathbb{C}),\ (\mathcal{V}_2, \pi_2) \longmapsto \mathrm{Hom}_H((\mathcal{V}_1, \pi_1), (\mathcal{V}_2, \pi_2)) .$$

(ii) *For each* $(\mathcal{V}_2, \pi_2) \in \mathrm{ob}\ \mathrm{Rep}_s(H)$, *the contravariant functors*

$$\mathrm{Rep}_s(H) \longrightarrow \mathrm{Vect}(\mathbb{C}),\ (\mathcal{V}_1, \pi_1) \longmapsto \mathrm{Ext}_H^n((\mathcal{V}_1, \pi_1),\ (\mathcal{V}_2, \pi_2))$$

$(n \in \mathbb{N})$ *are the derived functors of the left exact contravariant functor*

$$\mathrm{Rep}_s(H) \longrightarrow \mathrm{Vect}(\mathbb{C}),\ (\mathcal{V}_1, \pi_1) \longmapsto \mathrm{Hom}_H((\mathcal{V}_1, \pi_1), (\mathcal{V}_2, \pi_2)) .$$

Here, $\mathrm{Hom}_H((\mathcal{V}_1, \pi_1), (\mathcal{V}_2, \pi_2))$ is the $\mathbb{C}$-vector space of morphisms from $(\mathcal{V}_1, \pi_1)$ to $(\mathcal{V}_2, \pi_2)$ in $\mathrm{Rep}_s(H)$, i.e.

$$\{\nu \in \mathrm{Hom}_{\mathbb{C}}(\mathcal{V}_1, \mathcal{V}_2) \mid \nu(\pi_1(h)(v_1)) = \pi_2(h)(\nu(v_1)),\ \forall\ h \in H,\ \forall\ v_1 \in \mathcal{V}_1\} .$$

In particular, the functors

$$\mathrm{Rep}_s(H) \longrightarrow \mathrm{Vect}(\mathbb{C}),\ (\mathcal{V}, \pi) \longmapsto \mathrm{Ext}_H^n((\mathbb{C}, 1), (\mathcal{V}, \pi))$$

$(n \in \mathbb{N})$ are the derived functors of the left exact functor

$$\mathrm{Rep}_s(H) \longrightarrow \mathrm{Vect}(\mathbb{C}),\ (\mathcal{V}, \pi) \longmapsto \mathcal{V}^H ,$$

and are also denoted by

$$(\mathrm{D.5.4}) \qquad \mathrm{H}^n(H, (\mathcal{V}, \pi)) = \mathrm{Ext}_H^n((\mathbb{C}, 1), (\mathcal{V}, \pi)) .$$

Proof of (D.5.3) : This is a direct consequence of the lemma (D.5.1). $\square$

PROPOSITION (D.5.5). — *Let $R \subset Q \subset H$, $\overline{H}$ and $H^\circ \subset H$ be as in (D.3).
Then the adjunction between the functors $r_H^{\overline{H},Q}$ and $i_{\overline{H},Q}^H$ (see (D.3.3)(i))
induces an isomorphism between the cohomological bifunctors*

$$((\mathcal{V}_1, \pi_1), (\overline{\mathcal{V}}_2, \overline{\pi}_2)) \longmapsto (\mathrm{Ext}_{\overline{H}}^n(r_H^{\overline{H},Q}(\mathcal{V}_1, \pi_1),\ (\overline{\mathcal{V}}_2, \overline{\pi}_2)))_{n \in \mathbb{N}}$$

and

$$((\mathcal{V}_1, \pi_1), (\overline{\mathcal{V}}_2, \overline{\pi}_2)) \longmapsto (\mathrm{Ext}_H^n((\mathcal{V}_1, \pi_1),\ i_{\overline{H},Q}^H(\overline{\mathcal{V}}_2, \overline{\pi}_2)))_{n \in \mathbb{N}}$$

on $\mathrm{Rep}_s(H)^{opp} \times \mathrm{Rep}_s(\overline{H})$.

Proof: As $r_H^{\overline{H},Q}$ and $i_{\overline{H},Q}^H$ are exact functors (see (D.3.3) (ii)) and as $r_H^{\overline{H},Q}$ is
left adjoint to $i_{\overline{H},Q}^H$, $r_H^{\overline{H},Q}$ maps projective objects of $\mathrm{Rep}_s(H)$ into projective
objects of $\mathrm{Rep}_s(\overline{H})$ and $i_{\overline{H},Q}^H$ maps injective objects of $\mathrm{Rep}_s(\overline{H})$ into injective
objects of $\mathrm{Rep}_s(H)$. Therefore, (D.5.5) is a direct consequence of (D.5.3).   □

COROLLARY (D.5.6). — *Let $R \subset Q \subset H$, $\overline{H}$ and H° be as in (D.3). Then
we have a canonical isomorphism between the cohomological functors*

$$(\overline{\mathcal{V}}, \overline{\pi}) \longmapsto (\mathrm{H}^n(\overline{H}, (\overline{\mathcal{V}}, \delta_Q^{1/2} \overline{\pi})))_{n \in \mathbb{N}}$$

and

$$(\overline{\mathcal{V}}, \overline{\pi}) \longmapsto (\mathrm{H}^n(H, i_{\overline{H},Q}^H(\overline{\mathcal{V}}, \overline{\pi})))_{n \in \mathbb{N}}$$

on $\mathrm{Rep}_s(\overline{H})$.

Proof : We have

$$r_H^{\overline{H},Q}(\mathbb{C}, 1) = (\mathbb{C}, \delta_Q^{-1/2})$$

and the corollary is a particular case of (D.5.5).   □

Let us finish this section with two useful technical results.

LEMMA (D.5.7). — *Let Z be the center of H and let $(\mathcal{V}_1, \pi_1)$ and $(\mathcal{V}_2, \pi_2)$
be two objects of $\mathrm{Rep}_s(H)$ which admit central characters ω_{π_1} and ω_{π_2}
respectively (see (D.1)). Then, if $\omega_{\pi_1} \neq \omega_{\pi_2}$, we have*

$$\mathrm{Ext}_H^n((\mathcal{V}_1, \pi_1), (\mathcal{V}_2, \pi_2)) = (0)$$

for all non-negative integers n.

*In particular, if $(\mathcal{V}, \pi) \in \mathrm{ob}\, \mathrm{Rep}_s(H)$ admits a central character ω_π and
if $\omega_\pi \neq 1$, we have*

$$\mathrm{H}^n(H, (\mathcal{V}, \pi)) = (0)$$

for all non-negative integers n.

Proof: Let $n \in \mathbf{N}$ and let $z \in Z$ be such that $\omega_{\pi_1}(z) \neq \omega_{\pi_2}(z)$. As $z \in Z$, we have

$$\pi_i(z) \in \operatorname{End}_{\operatorname{Rep}_s(H)}(\mathcal{V}_i, \pi_i) \quad (i = 1, 2) \,,$$

so that

$$\operatorname{Ext}_h^n(\pi_1(z^{-1}), \pi_2(z))$$

is an endomorphism of the $\mathbf{C}$-vector space

$$\operatorname{Ext}_h^n((\mathcal{V}_1, \pi_1), (\mathcal{V}_2, \pi_2)) \,.$$

But this last endomorphism must be at the same time the identity (if $(\mathcal{V}_2, \pi_2) \longrightarrow (\mathcal{V}_2^{\bullet}, \pi_2^{\bullet})$ is an injective resolution of $(\mathcal{V}_2, \pi_2)$, $(\pi_2(z), \pi_2^{\bullet}(z))$ is an endomorphism of this resolution and it is obvious that

$$\operatorname{Hom}_H(\pi_1(z^{-1}), \pi_2^{\bullet}(z)) = \operatorname{id}_{\operatorname{Hom}_H(\mathcal{V}_1, \mathcal{V}_2^{\bullet})})$$

and the multiple

$$\omega_{\pi_1}(z^{-1})\omega_{\pi_2}(z) \operatorname{id}$$

of the identity. Hence, we get

$$\operatorname{Ext}_H^n((\mathcal{V}_1, \pi_1), \ (\mathcal{V}_2, \pi_2)) = (0) \,.$$

$\square$

Let $I \subset H$ be a compact open subgroup and let $(\mathcal{W}, \rho)$ be a smooth representation of I. The **compactly induced representation of** $(\mathcal{W}, \rho)$ **from** I **to** H,

(D.5.8) $c\text{-}\operatorname{Ind}_I^H(\mathcal{W}, \rho) \in \operatorname{ob} \operatorname{Rep}_s(H) \,,$

is defined in the following way. Its space is the $\mathbf{C}$-vector space of locally constant functions with compact supports

$$v : H \longrightarrow \mathcal{W}$$

such that

$$v(ih) = \rho(i)(v(h))$$

for any $i \in I$ and any $h \in H$ and the left action of H on this space of functions is induced by the right translation of H on itself. It is easy to see that the obvious functor

$$c\text{-}\operatorname{Ind}_I^H : \operatorname{Rep}_s(I) \longrightarrow \operatorname{Rep}_s(H)$$

is left adjoint to the functor

$$\operatorname{Res}_H^I : \operatorname{Rep}_s(H) \longrightarrow \operatorname{Rep}_s(I), \ (\mathcal{V}, \pi) \longmapsto (\mathcal{V}, \pi|I) \,,$$

the adjunction morphisms being given by

$$(\mathcal{W}, \rho) \longrightarrow \operatorname{Res}_H^I(c\text{-}\operatorname{Ind}_I^H(\mathcal{W}, \rho))$$

$$w \longmapsto \left(h \longmapsto \left\{ \begin{array}{ll} \rho(h)(w) & \text{if } h \in I, \\ 0 & \text{otherwise} \end{array} \right. \right)$$

and by

$$c\text{-}\operatorname{Ind}_I^H(\operatorname{Res}_H^I(\mathcal{V}, \pi)) \longrightarrow (\mathcal{V}, \pi),$$

$$v \longmapsto v(1).$$

LEMMA (D.5.9). — *For each* $(\mathcal{W}, \rho) \in \operatorname{ob} \operatorname{Rep}_s(I)$, $c\text{-}\operatorname{Ind}_I^H(\mathcal{W}, \rho)$ *is a projective object of* $\operatorname{Rep}_s(H)$.

Proof : The functor Res_H^I is obviously exact. Therefore, as $c\text{-}\operatorname{Ind}_I^H$ is left adjoint to it, it suffices to prove that $(\mathcal{W}, \rho)$ is a projective object in $\operatorname{Rep}_s(I)$. But the category $\operatorname{Rep}_s(I)$ is semi-simple as I is compact (any epimorphism $(\mathcal{W}_1, \rho_1) \longrightarrow\!\!\!\!\rightarrow (\mathcal{W}_2, \rho_2)$ in $\operatorname{Rep}_s(I)$ has a section: take an arbitrary section s of the underlying epimorphism of $\mathbb{C}$–vector spaces and replace it by

$$w_2 \longmapsto \int_I \rho_1(i)(s(\rho_2(i^{-1})(w_2)))di$$

where di is the Haar measure on I which is normalized by $vol(I, di) = 1$) and the lemma follows. $\square$

(D.6) Unitarizable representations.

A smooth representation $(\mathcal{V}, \pi)$ of H is **Hermitian** (resp. **unitarizable**) if there exists an H-invariant, definite (resp. positive definite), Hermitian scalar product $(\ ,\ )$ on $\mathcal{V}$.

For any $\mathbb{C}$-vector space $\mathcal{W}$, let $\overline{\mathcal{W}}$ be the new $\mathbb{C}$-vector space with the same underlying additive group as $\mathcal{W}$ but with the new multiplication by scalars

$$(\lambda, w) \longmapsto \overline{\lambda}w$$

($\overline{\mathcal{W}}$ is canonically isomorphic to

$$\mathbb{C} \otimes_{(\lambda \longmapsto \overline{\lambda}), \mathbb{C}} \mathcal{W}) \ .$$

Then the **Hermitian contragredient** of $(\mathcal{V}, \pi) \in \operatorname{ob} \operatorname{Rep}_s(H)$ is the smooth representation

(D.6.1) $(\mathcal{V}^*, \pi^*) \stackrel{\text{dfn}}{=\!=\!=} (\widetilde{\overline{\mathcal{V}}}, \widetilde{\pi})$

$(\mathrm{Aut}_{\mathbb{C}}(\overline{\tilde{\mathcal{V}}}) = \mathrm{Aut}_{\mathbb{C}}(\tilde{\mathcal{V}}))$. The Hermitian contragredient of $(\mathcal{V}, \pi)$ is a contravariant functor, like the usual contragredient, and we have a functorial morphism in $\mathrm{Rep}_s(H)$

$$(\mathrm{D.6.2}) \qquad (\mathcal{V}, \pi) \longrightarrow (\mathcal{V}^{**}, \pi^{**}), \ v \longmapsto (v^* \longmapsto v^*(v)) \,,$$

which is an isomorphism if $(\mathcal{V}, \pi)$ is admissible.

LEMMA (D.6.3). — *Let $(\mathcal{V}, \pi)$ be an admissible representation of H.*

(i) *There is a canonical bijection between the set of H-invariant, definite, Hermitian scalar products $(\ ,\ )$ on $\mathcal{V}$ and the set of isomorphisms*

$$\iota : (\mathcal{V}, \pi) \xrightarrow{\ \sim\ } (\mathcal{V}^*, \pi^*)$$

in $\mathrm{Rep}_s(H)$ such that

$$\iota^* = \iota$$

(we identify $(\mathcal{V}^{**}, \pi^{**})$ to $(\mathcal{V}, \pi)$ via (D.6.2)).

(ii) *If $(\mathcal{V}, \pi)$ is irreducible, any two H-invariant, definite (resp. positive definite), Hermitian scalar products on $\mathcal{V}$ are proportional with a factor of proportionality in $\mathbb{R}^{\times}$ (resp. $\mathbb{R}_{>0}$).*

Proof: Part (ii) follows immediately from part (i) and (D.1.12).

To the H-invariant, definite, Hermitian scalar product $(\ ,\ )$ on $\mathcal{V}$, we associate the morphism

$$(\mathcal{V}, \pi) \xrightarrow{\ \iota\ } (\mathcal{V}^*, \pi^*), \ v_1 \longmapsto (v_2 \longmapsto (v_1, v_2))$$

(as usual, (v_1, v_2) is anti-linear in v_1 and linear in v_2). Obviously, it is a monomorphism. Now, for any compact open subgroup I of H, we have

$$\dim_{\mathbb{C}}((\mathcal{V}^*)^I) = \dim_{\mathbb{C}}((\tilde{\mathcal{V}})^I) = \dim_{\mathbb{C}}(\mathcal{V}^I)$$

(see (D.1.10)) and ι induces an isomorphism

$$\mathcal{V}^I \xrightarrow{\ \sim\ } (\mathcal{V}^*)^I \,.$$

Letting I vary, we obtain that ι is an isomorphism. The equality $\iota^* = \iota$ follows from the equality

$$\overline{(v_1, v_2)} = (v_2, v_1)$$

for all $v_1, v_2 \in \mathcal{V}$.

Conversely, to the isomorphism ι from $(\mathcal{V}, \pi)$ onto $(\mathcal{V}^*, \pi^*)$, we associate the H-invariant definite, sesquilinear scalar product

$$(v_1, v_2) = \iota(v_1)(v_2) \,.$$

This scalar product is Hermitian as long as $\iota^* = \iota$. $\square$

Examples of unitarizable smooth representations of H are given by the square-integrable irreducible representations. A **square-integrable representation** of H is an admissible representation $(\mathcal{V}, \pi)$ of H which admits a central character ω_π such that ω_π is unitary ($|\omega_\pi(z)| = 1$ for each z in the center Z of H) and such that

$$\int_{Z \backslash H} | < \widetilde{v}, \pi(h)(v) > |^2 \frac{dh}{dz} < +\infty$$

for any $v \in \mathcal{V}$ and any $\widetilde{v} \in \widetilde{\mathcal{V}}$. Here, dh and dz are arbitrary Haar measures on H and Z respectively and

$$h \longmapsto | < \widetilde{v}, \pi(h)(v) > |^2$$

is a locally constant function on H which is Z-invariant by translation as ω_π is unitary.

LEMMA (D.6.4). — *Any square-integrable irreducible representation of H is unitarizable.*

Proof: Let us fix an arbitrary non-zero vector $\widetilde{v}$ in $\widetilde{\mathcal{V}}$ and let us define

$$(v_1, v_2) = \int_{Z \backslash H} \overline{< \widetilde{v}, \pi(h)(v_1) >} < \widetilde{v}, \pi(h)(v_2) > \frac{dh}{dz}$$

for any $v_1, v_2 \in \mathcal{V}$. Here

$$h \longmapsto \overline{< \widetilde{v}, \pi(h)(v_1) >} < \widetilde{v}, \pi(h)(v_2) >$$

is a locally constant function on H which is Z-invariant by translation as ω_π is unitary and it is integrable thanks to the Cauchy–Schwarz inequality. Obviously, $(,)$ is an H-invariant, Hermitian scalar product on $\mathcal{V}$.

Moreover, for each $v \in \mathcal{V}$, we have

$$(v, v) \geq 0$$

with equality if and only if

$$< \widetilde{v}, \pi(h)(v) > = 0$$

for all $h \in H$, i.e. if and only if

$$v = 0$$

as $\{\pi(h)(v) \mid h \in H\}$ generates the $\mathbb{C}$-vector space $\mathcal{V}$ and as

$$< \widetilde{v}, \mathcal{V} > \neq \{0\}$$

by admissibility of $(\mathcal{V}, \pi)$ (see (D.2.3)). Hence, $(,)$ is positive definite. $\square$

COROLLARY (D.6.5). — *Any cuspidal irreducible representation of H with unitary central character is unitarizable.*

Proof : Such a representation is automatically square-integrable. □

Let us now examine the relations between unitarizability and complete reducibility.

LEMMA (D.6.6). — *Let $(\mathcal{V}, \pi)$ be a smooth representation of H, let $(\,,\,)$ be an H-invariant, positive definite, Hermitian scalar product on $\mathcal{V}$, let $\mathcal{W} \subset \mathcal{V}$ be a H-invariant $\mathbb{C}$-vector subspace and let $\mathcal{W}^{\perp}$ be its orthogonal subspace in $\mathcal{V}$ with respect to $(\,,\,)$. Then, if $(\mathcal{W}, \pi|\mathcal{W})$ is admissible, we have*

$$\mathcal{V} = \mathcal{W} \oplus \mathcal{W}^{\perp} .$$

Proof : The subspace $\mathcal{W}^{\perp}$ is also H-invariant. Let us set $\rho = \pi \mid \mathcal{W}$ and $\rho^{\perp} = \pi \mid \mathcal{W}^{\perp}$.

As $(\,,\,)$ is positive definite, we have

$$\mathcal{W} \cap \mathcal{W}^{\perp} = (0) .$$

As $(\mathcal{V}, \pi)$ is smooth, to prove that

$$\mathcal{V} = \mathcal{W} + \mathcal{W}^{\perp}$$

it suffices to prove that for any compact open subgroup I of H we have

$$\mathcal{V}^{I} = \mathcal{W}^{I} + (\mathcal{W}^{\perp})^{I} .$$

But, if we fix such an I, we have

$$\mathcal{W} = \mathcal{W}^{I} \oplus \operatorname{Ker} \rho(e_{I})$$

(see (D.1.3)) and

$$(v, w) = (\pi(e_{I})(v), w) = (v, \rho(e_{I})(w)) = 0$$

for any $w \in \operatorname{Ker} \rho(e_{I})$ and any $v \in \mathcal{V}^{I}$. Therefore, $(\mathcal{W}^{\perp})^{I}$ is the orthogonal subspace of $\mathcal{W}^{I}$ in $\mathcal{V}^{I}$ with respect to $(\,,\,)$. Moreover, by admissibility of $(\mathcal{W}, \rho)$, the $\mathbb{C}$-vector space $\mathcal{W}^{I}$ is finite dimensional and, if $(w_{1}, \ldots, w_{N})$ is an orthonormal basis of $\mathcal{W}^{I}$ with respect to $(\,,\,)$, for any $v \in \mathcal{V}^{I}$ the vector

$$v - \sum_{n=1}^{N} (w_{n}, v)w_{n} \in \mathcal{V}^{I}$$

is orthogonal to $\mathcal{W}^{I}$. Therefore, we have

$$\mathcal{V}^{I} = \mathcal{W}^{I} + (\mathcal{W}^{\perp})^{I}$$

as required. □

PROPOSITION (D.6.7). — *Let* $(\mathcal{V}, \pi)$ *be an admissible representation of* H *and let* $(\,,\,)$ *be an* H-*invariant, positive definite, Hermitian scalar product on* $\mathcal{V}$. *Then there exists at least one family* $(\mathcal{V}_s)_{s \in S}$ *of* H-*invariant* $\mathbb{C}$-*vector subspaces of* $\mathcal{V}$ *such that*

(i) *for any* $s' \neq s''$ *in* S, $\mathcal{V}_{s'}$ *is orthogonal to* $\mathcal{V}_{s''}$ *with respect to* $(\,,\,)$,

(ii) *we have*

$$\mathcal{V} = \sum_{s \in S} \mathcal{V}_s \,,$$

(iii) *for any* $s \in S$, *the* (*admissible*) *representation* $(\mathcal{V}_s, \pi \mid \mathcal{V}_s)$ *is irreducible.*

Moreover, any such family $(\mathcal{V}_s)_{s \in S}$ *satisfies the following extra property:*

(iv) *for any* (*admissible*) *irreducible representation* $(\mathcal{W}, \rho)$ *of* H, *the number of* $s \in S$ *such that* $(\mathcal{V}_s, \pi \mid \mathcal{V}_s)$ *is isomorphic to* $(\mathcal{W}, \rho)$ *is finite.*

Proof: Let $\mathcal{S}$ be the set of families

$$S := \{\mathcal{V}_s \mid s \in S\}$$

of H-invariant $\mathbb{C}$-vector subspaces of $\mathcal{V}$ satisfying the conditions (i) and (iii) of the proposition. This set $\mathcal{S}$ is partially ordered by the inclusion and, for any totally ordered subset $\mathcal{T}$ of $\mathcal{S}$

$$T = \bigcup_{S \in \mathcal{T}} S$$

belongs to $\mathcal{S}$. Therefore, by Zorn's lemma, we can find a maximal element S_{max} in $\mathcal{S}$. We want to prove that S_{max} satisfies condition (ii).

Let

$$\mathcal{V}_{max} = \sum_{s \in S_{max}} \mathcal{V}_s \subset \mathcal{V}$$

and let $\mathcal{V}_{max}^{\perp}$ be the orthogonal subspace of $\mathcal{V}_{max}$ in $\mathcal{V}$ with respect to $(\,,\,)$. The subspaces $\mathcal{V}_{max}$ and $\mathcal{V}_{max}^{\perp}$ are H-invariant and, by admissibility of $(\mathcal{V}, \pi)$, we have

$$\mathcal{V} = \mathcal{V}_{max} \oplus \mathcal{V}_{max}^{\perp}$$

(see (D.6.6)). Therefore, it suffices to prove that $\mathcal{V}_{max}^{\perp} = (0)$. Let us suppose the contrary. Then, by smoothness of $(\mathcal{V}_{max}^{\perp}, \pi \mid \mathcal{V}_{max}^{\perp})$, there exists a compact open subgroup I of H such that $(\mathcal{V}_{max}^{\perp})^I \neq (0)$ and, by admissibility of $(\mathcal{V}_{max}^{\perp}, \pi \mid \mathcal{V}_{max}^{\perp})$, the $\mathbb{C}$-vector space $(\mathcal{V}_{max}^{\perp})^I$ is finite dimensional. Among the H-invariant $\mathbb{C}$-vector subspaces $\mathcal{W}$ of $\mathcal{V}_{max}^{\perp}$ such that $\mathcal{W}^I \neq (0)$, let us choose one, $\mathcal{W}_0$, such that $\mathcal{W}_0^I$ is of minimal dimension. Let $\mathcal{V}_0$ be the intersection of all the H-invariant $\mathbb{C}$-vector subspaces of $\mathcal{V}_{max}^{\perp}$ containing

$\mathcal{W}_0^I$. We have $\mathcal{V}_0 \neq (0)$. Let $\mathcal{V}_1$ be an H-invariant $\mathbb{C}$-vector subspace of $\mathcal{V}_0$, we have

$$\mathcal{V}_0 = \mathcal{V}_1 \oplus (\mathcal{V}_0 \cap \mathcal{V}_1^{\perp}) \,,$$

where $\mathcal{V}_1^{\perp}$ is the orthogonal subspace of $\mathcal{V}_1$ in $\mathcal{V}$ with respect to $(\,,\,)$ (see (D.6.6)). In particular, we have

$$\mathcal{V}_0^I = \mathcal{V}_1^I \oplus (\mathcal{V}_0 \cap \mathcal{V}_1^{\perp})^I$$

and $\mathcal{V}_1^I = \mathcal{V}_0^I$ or $(\mathcal{V}_0 \cap \mathcal{V}_1^{\perp})^I = \mathcal{V}_0^I$ by minimality of the dimension of $\mathcal{V}_0^I = \mathcal{W}_0^I$. Therefore, $\mathcal{V}_1 = \mathcal{V}_0$ or $\mathcal{V}_0 \cap \mathcal{V}_1^{\perp} = \mathcal{V}_0$ by definition of $\mathcal{V}_0$. In other words, $\mathcal{V}_1 = \mathcal{V}_0$ or $\mathcal{V}_1 = (0)$ and we have proved that $(\mathcal{V}_0, \pi \mid \mathcal{V}_0)$ is irreducible. But this contradicts the maximality of S_{max} and the first part of the proposition is proved.

Let I be a compact open subgroup of H such that $\mathcal{W}^I \neq (0)$. The $\mathbb{C}$-linear map

$$\mathrm{Hom}_{\mathrm{Rep}_s(H)}((\mathcal{W}, \rho), (\mathcal{V}, \pi)) \longrightarrow \mathrm{Hom}_{\mathbb{C}}(\mathcal{W}^I, \mathcal{V}^I),$$

$$u \longmapsto u^I$$

is injective. Therefore, we have

$$\dim_{\mathbb{C}} \mathrm{Hom}_{\mathrm{Rep}_s(H)}((\mathcal{W}, \rho), (\mathcal{V}, \pi)) < +\infty$$

and the last assertion of the proposition is clear. □

REMARKS (D.6.8.1) If H admits a countable basis of neighborhoods of 1, for any admissible representation $(\mathcal{V}, \pi)$ of H, the $\mathbb{C}$-vector space $\mathcal{V}$ admits a countable basis ($\mathcal{V}$ is a countable union of finite dimensional $\mathbb{C}$-vector spaces) and for any family $(\mathcal{V}_s)_{s \in S}$ satisfying the properties of (D.6.7), S is countable.

(D.6.8.2) If $(\mathcal{W}, \rho)$ is an admissible irreducible representation of H, we have

$$\mathrm{End}_{\mathrm{Rep}_s(H)}(\mathcal{W}, \rho) = \mathbb{C}$$

(if H is countable at infinity, apply (D.1.12); in general, fix a compact open subgroup I of H such that $\mathcal{W}^I \neq (0)$, remark that we have an injective map

$$\mathrm{End}_{\mathrm{Rep}_s(H)}(\mathcal{W}, \rho) \hookrightarrow \mathrm{End}_{\mathbb{C} \otimes \mathcal{C}_c^{\infty}(H//I)}(\mathcal{W}^I),$$

$$u \longmapsto u^I,$$

and apply the classical Schur lemma to the irreducible (see (D.1.8)) left $(\mathbb{C} \otimes \mathcal{C}_c^{\infty}(H//I))$-module $\mathcal{W}^I$ of finite dimension over $\mathbb{C}$). Therefore, for any family $(\mathcal{V}_s)_{s \in S}$ satisfying the properties of (D.6.7), the number of $s \in S$ such that $(\mathcal{V}_s, \pi \mid \mathcal{V}_s)$ is isomorphic to $(\mathcal{W}, \rho)$ is equal to

$$\dim_{\mathbb{C}} \mathrm{Hom}_{\mathrm{Rep}_s(H)}((\mathcal{W}, \rho), (\mathcal{V}, \pi))$$

and is independent of $(\mathcal{V}_s)_{s \in S}$. □

More generally, let $(\mathcal{V}, \pi)$ be a smooth representation of H, endowed with an H-invariant, positive definite, Hermitian scalar product $(,)$. For any smooth irreducible representation $(\mathcal{W}, \rho)$ of H, we denote by $(\mathcal{V}_\rho, \pi_\rho)$ the image of the canonical morphism

$$\iota_\rho : (\mathcal{W} \otimes_{\mathbb{C}} \operatorname{Hom}_{\operatorname{Rep}_s(H)}((\mathcal{W}, \rho), (\mathcal{V}, \pi)), \rho \otimes 1) \longrightarrow (\mathcal{V}, \pi),$$

$$w \otimes u \longmapsto u(w).$$

LEMMA (D.6.9). — (i) *For any non-isomorphic admissible irreducible representations $(\mathcal{W}', \rho')$ and $(\mathcal{W}'', \rho'')$ of H, $\mathcal{V}_{\rho'}$ is orthogonal to $\mathcal{V}_{\rho''}$ with respect to $(,)$.*

(ii) *For any admissible irreducible representation $(\mathcal{W}, \rho)$ of H such that*

$$m(\rho) = \dim_{\mathbb{C}} \operatorname{Hom}_{\operatorname{Rep}_s(H)}((\mathcal{W}, \rho), (\mathcal{V}, \pi))$$

is finite, ι_ρ is a monomorphism and, therefore, $(\mathcal{V}_\rho, \pi_\rho)$ is non-canonically isomorphic to $(\mathcal{W}, \rho)^{m(\rho)}$.

Proof: Let $(\mathcal{V}', \pi')$ and $(\mathcal{V}'', \pi'')$ be arbitrary subrepresentations of $(\mathcal{V}, \pi)$ which are isomorphic to $(\mathcal{W}', \rho')$ and $(\mathcal{W}'', \rho'')$ respectively. Then $(\mathcal{V}'', \pi'')$ is admissible and unitarizable and is therefore isomorphic to $(\mathcal{V}''^*, \pi''^*)$ (see (D.6.3)). It follows that the morphism

$$(\mathcal{V}', \pi') \longrightarrow (\mathcal{V}''^*, \pi''^*),$$

$$v' \longmapsto (v'' \longmapsto (v', v''))$$

is identically zero ($(\mathcal{V}', \pi')$ and $(\mathcal{V}'', \pi'')$ are irreducible and non-isomorphic). In other words, $(\mathcal{V}', \pi')$ is orthogonal to $(\mathcal{V}'', \pi'')$. Part (i) follows.

To prove part (ii), we can assume that $m(\rho) \neq 0$. Then $(\mathcal{W}, \rho)$ is isomorphic to a subrepresentation of $(\mathcal{V}, \pi)$ and is therefore unitarizable. It follows that the representation

$$(\mathcal{W} \otimes_{\mathbb{C}} \operatorname{Hom}_{\operatorname{Rep}_s(H)}((\mathcal{W}, \rho), (\mathcal{V}, \pi)), \rho \otimes 1) ,$$

which is isomorphic to

$$(\mathcal{W}, \rho)^{m(\rho)} ,$$

is admissible and unitarizable. In particular, $\operatorname{Ker}(\iota_\rho)$ is a direct summand of this representation and we have

$$\operatorname{Hom}_{\operatorname{Rep}_s(H)}((\mathcal{W}, \rho), \operatorname{Ker}(\iota_\rho)) = (0)$$

if and only if $\operatorname{Ker}(\iota_\rho) = (0)$.

But, for any finite dimensional $\mathbb{C}$-vector space E, the $\mathbb{C}$-linear map

$$E \longrightarrow \mathrm{Hom}_{\mathrm{Rep}_s(H)}((\mathcal{W}, \rho), (\mathcal{W} \otimes_{\mathbb{C}} E, \rho \otimes 1)),$$

$$e \longmapsto (w \longmapsto w \otimes e),$$

is an isomorphism (see (D.6.8.2)). Therefore, by definition of ι_ρ, we have

$$\mathrm{Hom}_{\mathrm{Rep}_s(H)}((\mathcal{W}, \rho), \mathrm{Ker}(\iota_\rho)) = (0)$$

and part (ii) of the lemma is proved. $\square$

(D.7) Decomposition of representations into tensor products
Let $H_1, \ldots, H_s$ be a finite family of unimodular, locally compact, totally discontinuous, separated topological groups. If $(\mathcal{V}_j, \pi_j)$ is a smooth representation of H_j for $j = 1, \ldots, s$, we define their (exterior) **tensor product**

$$(\mathcal{V}, \pi) = (\mathcal{V}_1, \pi_1) \otimes \cdots \otimes (\mathcal{V}_s, \pi_s)$$

by

$$\mathcal{V} = \mathcal{V}_1 \otimes_{\mathbb{C}} \cdots \otimes_{\mathbb{C}} \mathcal{V}_s$$

and

$$\pi(h_1, \ldots, h_s)(v_1 \otimes \cdots \otimes v_s) = \pi_1(h_1)(v_1) \otimes \cdots \otimes \pi_s(h_s)(v_s) .$$

Then $(\mathcal{V}, \pi)$ is a smooth representation of

$$H = H_1 \times \cdots \times H_s .$$

LEMMA (D.7.1). — (i) *Let $(\mathcal{V}_j, \pi_j)$ be an admissible representation of H_j for $j = 1, \ldots, s$ and let $(\mathcal{V}, \pi)$ be their tensor product. Then $(\mathcal{V}, \pi)$ is admissible and its contragredient representation $(\widetilde{\mathcal{V}}, \widetilde{\pi})$ is canonically isomorphic to the tensor product of the contragredient representations $(\widetilde{\mathcal{V}}_j, \widetilde{\pi}_j)$ of $(\mathcal{V}_j, \pi_j)$ for $j = 1, \ldots, s$.*

(ii) *Let $(\mathcal{V}, \pi)$ be an admissible representation of H. Then $(\mathcal{V}, \pi)$ is irreducible if and only if there exist admissible irreducible representations $(\mathcal{V}_j, \pi_j)$ of H_j $(j = 1, \ldots, s)$ and an isomorphism*

$$(\mathcal{V}, \pi) \cong \bigotimes_{j=1}^{s} (\mathcal{V}_j, \pi_j)$$

in $\mathrm{Rep}_s(H)$. Moreover, if $(\mathcal{V}, \pi)$ is irreducible the $(\mathcal{V}_j, \pi_j)$ $(j = 1, \ldots, s)$ are, up to isomorphism, uniquely determined by $(\mathcal{V}, \pi)$.

Proof : By induction on s, it suffices to examine the case $s = 2$.

Let $I_1 \subset H_1$ and $I_2 \subset H_2$ be compact open subgroups and let $I = I_1 \times I_2 \subset H$. Then I is also a compact open subgroup and, if we choose the rational Haar measures dh_1, dh_2 and dh on H_1, H_2 and H respectively such that $dh = dh_1 \times dh_2$, we have a canonical isomorphism of $\mathbb{Q}$-algebras

$$\mathcal{C}_c^{\infty}(H//I) \cong \mathcal{C}_c^{\infty}(H_1//I_1) \otimes \mathcal{C}_c^{\infty}(H_2//I_2) \ .$$

Let $(\mathcal{V}_j, \pi_j)$ be a smooth representation of H_j $(j = 1, 2)$ and let

$$(\mathcal{V}, \pi) = (\mathcal{V}_1, \pi_1) \otimes (\mathcal{V}_2, \pi_2) \ .$$

Then the canonical map

$$\mathcal{V}_1^{I_1} \otimes_{\mathbb{C}} \mathcal{V}_2^{I_2} \longrightarrow \mathcal{V}^I$$

is an isomorphism $(\pi(e_I) = \pi_1(e_{I_1}) \otimes \pi_2(e_{I_2})$ and $\mathcal{V}^I = \pi(e_I)(\mathcal{V}))$. This isomorphism and the canonical map

$$(\widetilde{\mathcal{V}}_1, \widetilde{\pi}_1) \otimes (\widetilde{\mathcal{V}}_2, \widetilde{\pi}_2) \rightarrow (\widetilde{\mathcal{V}}, \widetilde{\pi})$$

induce the same map

$$(\mathcal{V}_1^{I_1})' \otimes_{\mathbb{C}} (\mathcal{V}_2^{I_2})' \simeq \widetilde{\mathcal{V}}_1^{I_1} \otimes_{\mathbb{C}} \widetilde{\mathcal{V}}_2^{I_2} \longrightarrow (\mathcal{V}^I)' \simeq \widetilde{\mathcal{V}}^I$$

(see (D.1.10)).

Part (i) of (D.7.1) is now obvious.

Let us assume that $(\mathcal{V}_j, \pi_j)$ is admissible for $j = 1, 2$ and that $(\mathcal{V}, \pi)$ is non-zero and reducible. Then there exist compact open subgroups $I_1 \subset H_1$ and $I_2 \subset H_2$ such that $\mathcal{V}^I$ is a non-zero, reducible, left $(\mathbb{C} \otimes \mathcal{C}_c^{\infty}(H//I))$-module, where $I = I_1 \times I_2$ (the I's form a basis of neighborhoods of 1 in H). It follows that the left $(\mathbb{C} \otimes \mathcal{C}_c^{\infty}(H_j//I_j))$-module $\mathcal{V}_j^{I_j}$ is non-zero for $j = 1$ and $j = 2$ and reducible for $j = 1$ or $j = 2$ (see [Bou] Alg. VIII. 7, Thm. 2). Therefore $(\mathcal{V}_j, \pi_j)$ is non-zero for $j = 1$ and $j = 2$ and reducible for $j = 1$ or $j = 2$ (see (D.1.8)). This proves the "if" part of (D.7.1) (ii).

Let $(\mathcal{V}, \pi)$ be an admissible irreducible representation of H. There exist compact open subgroups $I_1 \subset H_1$ and $I_2 \subset H_2$ such that $\mathcal{V}^I \neq (0)$, where $I = I_1 \times I_2$. Then $\mathcal{V}^I$ is an irreducible left $(\mathbb{C} \otimes \mathcal{C}_c^{\infty}(H//I))$-module (see (D.1.8)). Therefore, there exist irreducible left $(\mathbb{C} \otimes \mathcal{C}_c^{\infty}(H_j//I_j))$-modules V_j $(j = 1, 2)$ and an isomorphism

$$\mathcal{V}^I \cong V_1 \otimes_{\mathbb{C}} V_2$$

of left $(\mathbb{C} \otimes \mathcal{C}_c^{\infty}(H//I))$-modules (see [Bou] Alg. VIII.7, Prop. 8). Thanks to (D.1.8), there exists an admissible irreducible representation $(\mathcal{V}_j, \pi_j)$ of H_j

such that $\mathcal{V}_j^{I_j}$ is isomorphic to V_j as a left $(\mathbb{C} \otimes \mathcal{C}_c^\infty(H_j /\!/ I_j))$-module for $j = 1, 2$. The tensor product

$$(\mathcal{V}_1, \pi_1) \otimes (\mathcal{V}_2, \pi_2)$$

is an admissible irreducible representation of H (apply (D.7.1) (i) and the "if" part of (D.7.1) (ii) which are already proved) and

$$(\mathcal{V}_1 \otimes_{\mathbb{C}} \mathcal{V}_2)^I \cong V_1 \otimes_{\mathbb{C}} V_2$$

as left $(\mathbb{C} \otimes \mathcal{C}_c^\infty(H /\!/ I))$-module by construction. Hence, thanks again to (D.1.8), $(\mathcal{V}, \pi)$ is isomorphic to $(\mathcal{V}_1, \pi_1) \otimes (\mathcal{V}_2, \pi_2)$.

　　Finally, let $(\mathcal{V}_j, \pi_j)$ and $(\mathcal{W}_j, \rho_j)$ be admissible irreducible representations of H_j $(j = 1, 2)$ such that

$$(\mathcal{V}_1, \pi_1) \otimes (\mathcal{V}_2, \pi_2)$$

and

$$(\mathcal{W}_1, \rho_1) \otimes (\mathcal{W}_2, \rho_2)$$

are isomorphic in $\mathrm{Rep}_s(H)$. It follows from the unicity assertion of [Bou] Alg. VIII. 7, Prop. 8 that $\mathcal{V}_j^{I_j}$ and $\mathcal{W}_j^{I_j}$ are isomorphic non-zero irreducible left $(\mathbb{C} \otimes \mathcal{C}_c^\infty(H_j /\!/ I_j))$-modules $(j = 1, 2)$ as long as $I_1 \subset H_1$ and $I_2 \subset H_2$ are compact open subgroups such that $\mathcal{V}_1^{I_1} \neq (0)$ and $\mathcal{V}_2^{I_2} \neq (0)$. Applying (D.1.8) once more, we get that $(\mathcal{V}_j, \pi_j)$ is isomorphic to $(\mathcal{W}_j, \rho_j)$ $(j = 1, 2)$ and the proof of (D.7.1) (ii) is completed. □

(D.8) Comments and references

Sections (D.1) to (D.4) follow the lines of chapters I and II of [Be–Ze1]. Other references are [Car], [Cas 1] and [Ja].

　　The results of (D.5) come from the appendix of [Cas 2]. The lemma (D.1.5) is essentially due to P. Blanc (see [Bl]).

　　The standard reference for (D.7) is [Fla].

References

[Ar–Cl]   J. ARTHUR, L. CLOZEL. — *Simple algebras, base change and the advanced theory of the trace formula*, Ann. of Math. Studies, 120, Princetone University Press, 1989.

[Be–Ze 1]   I.N. BERNSTEIN, A.V. ZELEVINSKY. — Representations of the group $GL(n, F)$ where F is a local non-archimedian field, *Russian Math. Surveys* **31**, (1976), 1–68.

[Be–Ze 2]   I.N. BERNSTEIN, A.V. ZELEVINSKY. — Induced representations of reductive p-adic groups. I, *Ann. Scient. Ec. Norm. Sup.* **10**, (1977), 441–472.

[Bl]   P. BLANC. — Projectifs dans la catégorie des G-modules topologiques, *C.R. Acad. Sci. Paris* **289**, (1979), 161–163.

[Bo]   A. BOREL. — Admissible representations of a semi-simple group over a local field with vectors fixed under an Iwahori subgroup, *Invent. Math.* **35**, (1976). 233–259.

[Bo–Ha]   A. BOREL, G. HARDER. — Existence of discrete cocompact subgroups of reductive groups over local fields, *J. Reine Angew. Math.* **298**, (1978), 53–64.

[Bo–Wa]   A. BOREL, N. WALLACH. — *Continuous cohomology, discrete subgroups, and representation of reductive groups*, Ann. of Math. Studies 94, Princeton University Press, 1980.

[Bou]   N. BOURBAKI. — *Eléments de Mathématiques*, Hermann.

[Br–Ti]   F. BRUHAT, J. TITS. — Groupes réductifs sur un corps local. I, Données radicielles valuées, *Publ. Math. IHES* **41**, (1972), 5–251.

[Ca–Ei]   H. CARTAN AND S. EILENBERG. — *Homological algebra*, Princeton Math. Ser., 19, 1956.

[Ca]   R.W. CARTER. — *Finite groups of Lie type : conjugacy classes and complex characters*, John Wiley and sons, 1985.

[Car]   P. CARTIER. — Representations of p-adic groups : a survey, *Proc. Sym. Pure Math.* **33**, part 1, (1979), 111–155.

[Cas 1]   W. CASSELMAN. — Introduction to the theory of admissible representations of p-adic groups, preprint.

[Cas 2]   W. CASSELMAN. — A new non-unitarity argument for p-adic representations, *J. Fac. Sci. Univ. Tokyo* **28**, (1981), 907–928.

[Co]   L. COMTET. — *Analyse combinatoire*, PUF, 1970.

[De–Ka–Vi] P. DELIGNE, D. KAZHDAN, M.-F. VIGNÉRAS. — Représentations des algébres centrales simples p-adiques, in *Représentations des groupes réductifs sur un corps local*. Travaux en cours,
Hermann, (1984), 33–117.

[De–Mu]   P. DELIGNE, D. MUMFORD. — The irreducibility of the space
of curves of given genus, *Pub. Math. IHES* **36**, (1969), 75–110.

[Dr 1]   V.G. DRINFEL'D. — Elliptic modules, *Math. USSR Sbornik* **23**,
(1974), 561–592.

[Dr 2]   V.G. DRINFEL'D. — Elliptic modules II, *Math. USSR Sbornik*
31, (1977), 159–170.

[Dr 3]   V.G. DRINFEL'D. — The proof of Peterson's conjecture for
$GL(2)$ over a global field of characteristic p, *Funct. Anal. and
its Appl.* **22**, (1988), 28–43.

[Fla]   D. FLATH. — Decomposition of representations into tensor
products, *Proc. Sym. Pure Math.* **33**, part 1, (1979), 179–184.

[Fl]   Y. FLICKER. — Drinfeld moduli schemes and automorphic
forms, preprint, Ohio State University.

[Fl–Ka]   Y. FLICKER, D. KAZHDAN. — Geometric Ramanujan conjecture and Drinfeld reciprocity law, in *Number theory, trace formulas and discrete groups*, Academic Press, (1989), 201–218.

[Fo–Sc]   D. FOATA, M.-P. SCHÜTZENBERGER. — *Théorie géométrique
des polynômes eulériens*, Lecture Notes in Math., 138, Springer-
Verlag, 1970.

[Fon]   J.-M. FONTAINE. — *Groupes p-divisibles sur les corps locaux*,
Astérisque, 47–48, 1977.

[Gr]   M. J. GREENBERG. — Schemata over local rings : II, *Ann. of
Math.* **78**, (1963), 256–266.

[Gro]   A. GROTHENDIECK. — Le groupe de Brauer III : exemples et
compléments, in *Dix exposés sur la cohomologie des schémas*,
North Holland, (1968), 88–188.

[Ha]   R. HARTSHORNE. — *Residues and duality*, Lecture Notes in
Math., 20, Springer-Verlag, 1966.

[Haz]   M. HAZEWINKEL. — *Formal groups and applications*, Academic
Press, 1978.

[Ho]   R. HOWE. — The Fourier transforma dn germs of characters
(case of GL_n over a p-adic field), *Math. Ann.* **208**, (1974), 305–
322.

[Ho–Mo]   R.E. HOWE, C.C. MOORE. — Asymptotic properties of unitary
representations, *J. of Funct. Anal.* **32**, (1979), 72–96.

[Iw–Ma] N. IWAHORI, H. MATSUMOTO. — On some Bruhat decompositions and the structure of the Hecke rings of p-adic Chevalley groups, *Publ. Math. IHES* **25**, (1965), 237–280.

[Ja] H. JACQUET. — Représentations des groups linéaires p-adiques, *Theory of group representations and Fourier analysis*, CIME (1970), Edizioni Cremonese, (1971), 119–220.

[Jac] N. JACOBSON. — *Structures of rings*, Am. Math. Soc. Coll. Publ. 37, (1964).

[Ka] N.M. KATZ. — Slope filtrations of F-crystals, in *Journées de géometrie algébriques de Rennes*, Astérisque 63, (1979), 113–164.

[Kaz] D. KAZHDAN. — An introduction to Drinfeld's "shtuka", *Proc. Sym. Pure Math.* **33**, part 2, (1979), 347–356.

[Ko 1] R. KOTTWITZ. — Tamagawa numbers, *Ann. of Math.* **127**, (1988), 629–646.

[Ko 2] R. KOTTWITZ. — Sign changes in harmonic analysis on reductive groups, *Trans. Amer. Math. Soc.* **278**, (1983), 289–297.

[Ko 3] R. KOTTWITZ. — Base change for unit elements of Hecke algebras, *Compositio Math.* **60**, (1986), 237–250.

[La] J.-P. LABESSE. — Pseudo-coefficients très cuspidaux et K-théorie, *Math. Ann.* **291**, (1991), 607–616.

[Lan 1] R.P. LANGLANDS. — *Base change for $GL(2)$*, Annals of Math. Studies, 96, Princeton University Press, 1980.

[Lan 2] R.P. LANGLANDS. — *Les débuts d'une formule des traces stables*, Publ. Math. Univ. Paris VII, 13.

[Lau 1] G. LAUMON. — Letters to R. Kottwitz (12/2/1988, 12/7/1988) and to J. Arthur (2/24/1989, 3/7/1989).

[Lau 2] G. LAUMON. — Letter to J. Arthur (5/3/1989).

[MM] P.A. MACMAHON. — *Combinatorial analysis*, Chelsea Publishing Company, 1960.

[Ma] YU. I. MANIN. — The theory of commutative formal groups over fields of finite characteristic, *Russian Math. Surveys* **18**, (1963), 1–63.

[Mac] G.W. MACKEY. — *The theory of unitary group representations*, Chicago Lectures in Mathematics Series, 1976.

[Mar] G.A. MARGOULIS. — *Discrete subgroups of semi-simple Lie groups*, Ergebnisse der Mathematik und ihrer Grenzgebiete 3. Folge. Bd. 17, Springer-Verlag (1989).

[Na] M.A. NAIMARK. — *Normed algebras*, Wolters-Noordhoff Publishing, Groningen, the Netherlands (1970).

[Or] O. ORE. — On a special class of polynomials, *Trans. Amer. Math. Soc.* **35**, (1933), 559–584 (Errata, ibid., **36** (1934), 275).

[Ra] R. RAO. — Orbital integrals on reductive groups, *Ann. of Math.* **96**, (1972), 505–510.

[Re] J. REINER. — *Maximal orders*, Academic Press, 1975.

[Ro] J. ROGAWSKI. — Representations of $GL(n)$ and division algebras over a p-adic field, *Duke Math. J.* **50**, (1983), 161–196.

[Ru] W. RUDIN. — *Functional analysis*, McGraw-Hill Book Company (1974).

[Sa] H. SAITO. — *Automorphic forms and algebraic extensions of number fields*, Lectures in Mathematics, 8, Kyoto University, 1975.

[Sat] I. SATAKE. — Theory of spherical functions on reductive algebraic groups over p-adic field, *Publ. Math. IHES* **18**, (1963), 5–70.

[SGA3] M. DEMAZURE, A. GROTHENDIECK. — *Séminaire de géométrie algébrique du Bois-Marie, Schémas en groupes*, Lecture Notes in Math., 151, 152, 153, Springer-Verlag, 1970.

[Se 1] J.-P. SERRE. — *Corps locaux*, Hermann, 1968.

[Se 2] J.-P. SERRE. — Cohomologie des groupes discrets, in *Prospects in Mathematics*, Ann. of Math. Studies 70, Princeton University Press, (1971), 77–169.

[Se 3] J.-P. SERRE. — *Groupes algébriques et corps de classes*, Hermann, 1959.

[Sh] T. SHINTANI. — On liftings of holomorphic cusp forms, *Proc. Sym. Pure Math.* **33**, part 2, (1979), 97–110.

[Sp] T.A. SPRINGER. — *Linear algebraic groups*, Progress in mathematics 9, Birkhäuser (1980).

[Ti] J. TITS. — Reductive groups over local fields, *Proc. Sym. Pure Math.* **33**, part 1, (1979), 29–69.

[We 1] A. WEIL. — *Adeles and algebraic groups*, Progress in Math., 23, Birkhäuser, 1982.

[We 2] A. WEIL. — *Basic number theory, second edition*, Springer-Verlag, 1973.

[We 3] A. WEIL. — *L'intégration dans les groupes topologiques et ses applications*, Hermann, 1965.

[Wh] J.H.C. WHITEHEAD. — Combinatorial homotopy I, *Bull. Amer. Math. Soc.* 55, (1949), 213–245.

[Zi] T. ZINK. — *Cartier Theorie kommutativer formaler Gruppen*, Teubner-Texte zur Mathematik, 68, 1984.

Index

x-adic cohomology group : 31
Admissible representation : 160, 290
r-admissible (at the place o) : 55, 60
r-admissible (closed element) : 98
r-admissible (σ-closed element) : 101
Admit a central character : 289
Affine Weyl group : 181
Base change homomorphism : 83
q-binomial coefficient : 281
Brauer group : 249
Bruhat-Tits building : 133
Bruhat-Tits decomposition : 183
Cartan decomposition : 76
Central algebra (over a field) : 249
Central character : 289
Central division algebra (over a field) : 249
σ-centralizer : 93
Chamber (of the Bruhat-Tits building) : 133
Character : 291
Characteristic (of a Drinfeld module) : 5
Closed element : 87
σ-closed element : 96
Cohomology : 216, 322
Commutant : 250

Compactly induced representation : 326

σ-conjugate : 93

Constant term (along a parabolic subgroup) : 78

Contragredient representation : 288

Cuspidal (quasi-cuspidal) representation : 165, 312

Cyclic representation : 234

Cyclic vector : 234

Degree (of an isogeny) : 21

Dieudonné $\mathcal{O}_o$-module : 31

Dieudonné F_o-module : 31

Dieudonné module of a Drinfeld module : 35

ϖ_o-divisible scheme in $\mathcal{O}_o$-modules : 33

I-division point : 5

Drinfeld A-module : 4

Elliptic : 60

Elliptic at the place ∞ : 60

σ-elliptic : 102

Equivalent (central simple algebras) : 249

Euler-Poincaré characteristic : 217

Euler-Poincaré function : 132

Euler-Poincaré measure : 138

Facet (of the Bruhat-Tits building) : 133

Frobenius isogeny : 23

Good position (with respect to a parabolic subgroup) : 91

Good position (with respect to a triple) : 163, 297

Haar measure : 69

Hecke algebra : 13, 75, 129

Hecke eigenvalue (of a spherical representation) : 188

Hecke function : 101

Hecke operator : 15

Height (of an isogeny) : 20, 21

Hermitian contragredient : 327

Hermitian representation : 327

Induction functor : 160, 293

Inner twist : 68

Inner twisting : 68

Invariant at v of a central simple algebra (over a function field) : 254

K_P-invariant constant term (along a parabolic subgroup) : 91

$(\sigma\text{-}K_{P,r})$-invariant constant term (along a parabolic subgroup) : 97

Invariant of a central simple algebra (over a local field) : 254

Isogeny : 20

Isogeny (between Drinfeld modules) : 20

Iwahori subgroup : 129

Iwasawa decomposition : 77

Kottwitz function : 133

Lefschetz number (of a correspondance) : 52

Length function (on the affine Weyl group) : 181

Level-I structure : 6

Matrix coefficient : 311

Modular variety : 7

Multiplicity (of a slope) : 32

Non-degenerate module : 286

Norm : 93

σ-orbit : 96

Orbital integral : 57, 91

Parahoric subgroup (corresponding to a parabolic subgroup) : 131

Principal congruence subgroup of level n : 162

Principal series representation : 168

Rank (of a Drinfeld module) : 4

Rational Haar measure : 285

Representation : 284

Restriction functor (or modified Jacquet functor) : 160, 293

Satake isomorphism : 77

Satake transform : 77, 78

Simple algebra : 250

Slope (of a Dieudonné module) : 32

Smooth representation : 160, 284

Smooth vector : 284

Spectral decomposition : 238

Spectral measure : 237

Spherical representation : 187

Square-integrable representation : 220, 329

Stabilizer (of a vector in a representation) : 284

Steinberg representation : 193

Tamagawa measure : 70

Tamagawa number : 70

Tate module (of a Drinfeld module) : 28

Tensor product : 334

Transfer (of a conjugacy class) : 60

Transfer (of local or global Haar measures) : 69

Twisted orbital integral : 57, 96

(F, ∞, o)-type of rank d : 26

Unitarizable representation : 220, 327

Unitary continuous representation : 234

Unramified principal series representation : 178

Vertex (of the Bruhat-Tits building) : 133

Very cuspidal function : 133
Wall (of the Bruhat-Tits building) : 133
Weil (F, ∞, o)-pair of rank d : 23
Yoneda extension (vector space or bifunctor) : 216, 324